FORAGES

AN INTRODUCTION TO GRASSLAND AGRICULTURE

VOLUME I

7TH EDITION

FORAGES

AN INTRODUCTION TO GRASSLAND AGRICULTURE

VOLUME I

7TH EDITION

Edited by

Michael Collins
C. Jerry Nelson
Kenneth J. Moore
Robert F Barnes

WILEY Blackwell

This edition first published 2018

Edition History
Hughes, H. D., Maurice E. Heath and Darrel S. Metcalfe, 1st edition, The Science of Grassland Agriculture. Copyright by The Iowa State College Press, 1951
Hughes, H. D., Maurice E. Heath and Darrel S. Metcalfe, 2nd edition, The Science of Grassland Agriculture. Copyright by The Iowa State University Press, 1962
Heath, Maurice E., Darrel S. Metcalfe and Robert F Barnes 3rd edition, The Science of Grassland Agriculture. Copyright The Iowa State University Press, 1973
Heath, Maurice E., Robert F Barnes and Darrel S. Metcalfe. 4th edition, The Science of Grassland Agriculture. Copyright The Iowa State University Press, 1985
Barnes, Robert F, Darrell A Miller and C. Jerry Nelson, 5th edition, An Introduction to Grassland Agriculture. Copyright Iowa State University Press, 1995
Barnes, Robert F, C. Jerry Nelson, Michael Collins and Kenneth J. Moore, 6th edtion, Forages, An Introduction to Grassland Agriculture. Copyright Iowa State Press, 2003

Registered Office
John Wiley & Sons, Inc., 111 River Street, Hoboken, NJ 07030, USA

Editorial Office
111 River Street, Hoboken, NJ 07030, USA

For details of our global editorial offices, customer services, and more information about Wiley products visit us at www.wiley.com.

Wiley also publishes its books in a variety of electronic formats and by print-on-demand. Some content that appears in standard print versions of this book may not be available in other formats.

Library of Congress Cataloging-in-Publication Data

Names: Collins, Michael, 1951– editor. | Nelson, C. J. (Curtis J.), 1940– editor. | Moore, Kenneth J., editor. | Barnes, Robert F., 1933– editor.
Title: Forages / edited by Michael Collins, C. Jerry Nelson, Kenneth J. Moore, Robert F Barnes.
Description: Seventh edition. | Hoboken, NJ : Wiley, 2018– | Includes bibliographical references and index. | Description based on print version record and CIP data provided by publisher; resource not viewed.
Identifiers: LCCN 2017015856 (print) | LCCN 2017017201 (ebook) | ISBN 9781119300656 (pdf) | ISBN 9781119300663 (epub) | ISBN 9781119300649 (v.1 : cloth)
Subjects: LCSH: Forage plants. | Forage plants–United States.
Classification: LCC SB193 (ebook) | LCC SB193 .F64 2017 (print) | DDC 633.2–dc23
LC record available at https://lccn.loc.gov/2017015856

Cover Design: Wiley
Cover Images: Courtesy of Michael Collins

Set in 9/11pt AdobeGaramondPro by Aptara Inc., New Delhi, India

10 9 8 7 6 5 4 3 2 1

Millions of bison were present on the prairies of North America at the time of European settlement.

Contents

Preface

The concept of grassland agriculture is based on the premise that pastures and fields comprised of forage plants that are grazed or harvested mechanically are integral parts of a sustainable agricultural system. Their roles in livestock systems, particularly ruminants, are well documented in terms of nutritional value, improvements in fertility and quality of soils, and erosion control to protect soil and water resources. Today, forages are expected to provide even more ecosystem services associated with nutrient management, winter cover crops, biofuel production, wildlife benefits, well-being of pollinators, mitigating global climate change, and even providing aesthetic value to the landscape. The social values of agricultural landscapes, including pastures and haylands, have become a major public priority.

The public is not well aware of these values of grasslands or how best to provide these multiple functions, yet is very concerned with the quality and safety of the food products from grasslands while preserving the environment and its many components. The public is aware that farmers and ranchers, as land managers, are major stewards of these resources, and that government programs are often used to encourage agriculturalists to meet these expectations. Furthermore, only about 1% of the US population is currently involved in production agriculture, and there has been a gradual shift from general support of the traditional family farms to concerns about large "corporate" or "industrialized" farms that may lack the core values expected from the stewards of these grassland resources. Often the corporate farm is considered to be closely linked to private industry, with an accompanying negative perception of the profit incentive and management strategies.

Therefore our goal is to provide an up-to-date textbook that introduces students and professionals to the concepts and practical ways in which pastures and haylands can contribute to sustainable land management by providing quality food and fiber products while protecting the environment and providing the expected ecosystem services. The seventh edition of *Forages* continues to serve as a textbook for undergraduate students (Volume I), and a more advanced and comprehensive treatment serves as a major reference (Volume II). Volume I of the seventh edition is a thorough update of the sixth edition, to add new dimensions and purposes of managing grasslands and haylands. The outcome is the most comprehensive text available for students of undergraduate courses over a broad geographic area.

In the seventh edition we provide students with a good balance of scientific principles, to aid them in integrating the basic concepts, along with practical information on plant identification, growth characteristics, management, and utilization of these resources that can be used by agricultural practitioners. Grassland ecosystems are extremely complex, and this includes the plant–animal interface and the soil–climate–forage interface, all of which must be considered when making management decisions that maintain the environment and are socially acceptable. The explanatory coverage of the science behind these plant characteristics and responses will make the book applicable to many parts of the world, while more focused, region-specific management information relates mainly to North America and areas with similar forage species, climates, and soils.

Revisions from the sixth edition are of two main types. First, we have focused on a more consistent writing style and academic level to make the book even more suitable for undergraduates with both rural and urban backgrounds. We have included one editor as a co-author on each of the 19 chapters, to enhance consistency. And we have added color throughout the book both to make it more interesting and to greatly enhance readability and comprehension. Some subject matter has been condensed and placed in a more appropriate location for better continuity. In other chapters, emphasis on certain topics has been either increased or decreased so as to be current and consistent with the amount of technical knowledge available.

Second, we have addressed emerging topics that were not emphasized in the sixth edition. During the 14 years since that edition was published there have been several important changes in technology and its adoption, which are now addressed. One example is the prominent role of herbaceous annual and perennial species, many of which are warm-season grasses, as potential sources of lignocellulosic biomass energy. Many species that are receiving attention as bioenergy crops have dual uses as forages and/or conservation crops, but there are unique considerations for bioenergy use that need to be addressed. Other species such as *Miscanthus* have limited forage or conservation value, but have significant potential as dedicated energy crops. Yet their management for yield and quality is based on principles of plant growth and development that are similar to those for forages.

Rapidly increasing national interest in sustainability of agriculture, including environmental factors such as water quality, global climate change, eutrophication in the Gulf, support of wildlife, and other relevant topics required greater coverage. Forage species play important roles in these topics because of their erosion-limiting and soil-improving characteristics that can contribute to carbon sequestration in the soil, and to reducing run-off and nutrient loss to groundwater, streams, and rivers, while also providing other environmental and social benefits. Several of these involve inputs and analyses of economists and other social scientists to evaluate efficient energy use, increased fertilizer cost, increased emphasis on reducing loss of mobile nutrients to surface and ground water, and many emerging regulatory issues associated with human and food safety, animal rights, and manure management. There is also renewed interest in cover crops as components of row cropping systems.

We recognize the large role that private industry has assumed in forage improvement and seed production, so information on these topics has been distributed within the species-related chapters. Information relating to the importance of seed quality has been incorporated into the chapter on forage establishment. Essential information on forbs has been incorporated into associated chapters based on their comparative growth, quality, and management considerations. Agricultural engineers have brought large changes in mechanization of harvesting, processing, and storage technology, which are covered in the appropriate chapters.

The seventh edition has provided the opportunity to respond to the new technology and reflect the generational change that has taken place among forage scientists and teachers. We have added several new chapter authors who have become recognized leaders in the forage world. This will ensure that the book remains a contemporary and authoritative source that continues to be based on the cutting edge of scientific knowledge and its application in the field. We thank authors of a number of chapters in the fifth and sixth editions of *Forages: An Introduction to Grassland Agriculture* and *Forages: The Science of Grassland Agriculture* for their contributions.

Several prominent forage scientists reviewed and provided critical insights into the Compendium of Common Forages. In particular we would like to thank Drs. Don Ball (Auburn University), Gerald Evers (Texas A&M University), Al Frank (USDA Mandan, N.D.), John Jennings (University of Arkansas), and Chuck West (Texas Tech University) for their input on species adaptation, and Lynn Sollenberger (University of Florida) for his thorough and constructive review of the text.

On a sad note, we lost a critical member of our team when Dr. Robert F Barnes, known personally and professionally as "Bob", passed away in 2013. We have retained Bob as an editor since the four of us worked together in early discussions about how to revise the seventh edition, and many of his thoughts are incorporated in this volume. It was also appropriate due to the esteem in which he was held as a dedicated leader and scientist, and our genuine appreciation for the inspiration he provided for each of us personally. In addition, we wanted to honor his legacy for his long-term roles as a dedicated author and editor of *Forages*, beginning with the third edition, which was published in 1973. For these and other reasons, this volume is dedicated to his memory and expresses our gratitude for his mentoring and friendship.

As editors, we thank the authors who made Volume I of the seventh edition possible, many of them for the first time. Each has contributed new content and applications to their respective chapters. The change in authors also reflects the gradual turnover of scientific leaders in the discipline we know as grassland agriculture. Michael Collins carried administrative responsibilities. He or Jerry Nelson co-authored each chapter to maintain continuity and an appropriate level of coverage. Ken Moore contributed oversight to the content and had major responsibility for the appendix, glossary, and compendium. We hope that readers and students will find the subject to be interesting and helpful in their professional careers.

Michael Collins
C. Jerry Nelson
Kenneth J. Moore
Robert F Barnes

List of Contributors

DAVID J. BARKER is Professor of Horticulture and Crop Science at Ohio State University. He received his BS and MS from Massey University, New Zealand, and his PhD from the University of Nebraska. He teaches graduate and undergraduate forage classes. His research interests are pasture ecology and biodiversity, plant water relations, grazing management, and the interaction of these with soil fertility.

MICHAEL COLLINS is Emeritus Professor and Director of the Division of Plant Sciences at the University of Missouri, Columbia. He received his BS from Berea College, his MS from West Virginia University, and his PhD from the University of Kentucky. During a 36-year career in forage research and teaching at the University of Kentucky, Lexington and the University of Wisconsin, Madison, he taught an undergraduate/graduate forages course and advised graduate students. His research emphasized forage management and postharvest physiology of hay and silage, with an emphasis on forage quality.

MARVIN H. HALL is Professor of Forage Management at Penn State University, University Park. He received his BA from Bluffton College, his MS from the Ohio State University, and his PhD from the University of Minnesota. He conducts research, teaching, and extension in forage production and utilization of annual and perennial species. His main focus is on forage crop establishment, management, and utilization of forages.

EMILY A. HEATON is Associate Professor of Agronomy, Iowa State University. She received her BS and PhD from the University of Illinois at Urbana-Champaign. She researches warm-season perennial grasses grown in temperate climates, with a focus on plant–environment interactions and bioenergy applications. Her extension program addresses integration of perennials, particularly *Miscanthus* × *giganteus*, into subprime arable cropland.

NICHOLAS S. HILL is Professor of Crop Physiology and Management in the Department of Crop and Soil Sciences at the University of Georgia, Athens. He received his BS from the University of Wyoming, his MS from Montana State University, and his PhD from Kansas State University. His research focuses on plant fitness in mutualistic associations between endophytic fungi and cool-season pasture grasses. He teaches an introductory course in crop science.

ROBERT L. KALLENBACH is Professor and State Forage Extension Specialist in the Division of Plant Sciences and Assistant Dean of Agriculture and Natural Resources Extension in the College of Agriculture, Food, and Natural Resources at the University of Missouri, Columbia. He received his BS from Southwest Missouri State University, his MS from the University of Missouri, and his PhD from Texas Tech University. He has an extension and research appointment in forages, with a focus on winter feeding in forage–livestock systems.

WILLIAM O. LAMP is Associate Professor of Entomology at the University of Maryland, College Park, where he teaches and conducts research on integrated pest management. He received his BS from the University of Nebraska, his MS from Ohio State University, and his PhD from the University of Nebraska, and he gained postdoctoral experience at the University of Illinois. His research focuses on crop protection against insect pests through the integration of crop management practices and insect–plant interactions.

JENNIFER W. MacADAM is Professor at Utah State University where she teaches courses in forage production and pasture ecology, and plant structure and function. She received her BS from Missouri State University, and her MS and PhD from the University of Missouri, Columbia. She studies irrigated legume- and grass-based ruminant production systems and their environmental impacts.

ALI M. MISSAOUI is Assistant Professor of Forage and Biomass Breeding and Genetics at the University of Georgia. He received his BS from Oregon State University, his MS from Texas Tech University, and his PhD from the University of Georgia. He teaches an upper-level undergraduate/graduate course in forage management and utilization, and an advanced course in plant breeding. He conducts research on the genetic improvement of cool-season forage legumes and grasses adapted to the southeast USA, with a focus on stress tolerance and reduced anti-quality factors.

ROBERT B. MITCHELL is a Rangeland Scientist with the USDA Agricultural Research Service in Lincoln, Nebraska. He received his BS, MS, and PhD from the University of Nebraska. He has taught undergraduate and graduate courses in range improvement, ecology and conservation of natural resources, fire ecology, prescribed burning, and ecology of grazing lands systems. His research focuses on grassland management, invasive species management, and growth and development of perennial grasses and legumes.

KENNETH J. MOORE is Professor of Agronomy at Iowa State University. He received his BS from Arizona State University and his MS and PhD from Purdue University, He has taught graduate-level courses in field plot technique and design, pasture and grazing management, and forage quality and utilization. He advises graduate students and conducts research on pasture management and ecology.

C. JERRY NELSON is Emeritus Curators' Professor of Plant Sciences at the University of Missouri. He received his BS and MS from the University of Minnesota, and his PhD from the University of Wisconsin. His research has focused on growth of grasses and persistence of legumes, and he is now active in agriculture-related international programs, especially in Asia.

YOANA C. NEWMAN is Assistant Professor of Crop Science and Forage Extension Specialist in the Department of Plant and Earth Science at the University of Wisconsin, River Falls. She received her BA from the University of Rafael Urdaneta, Venezuela, and her MS and PhD from the University of Florida, Gainesville. She teaches an introduction to plant science course and forage crop production courses.

DAREN D. REDFEARN is Associate Professor and Forage and Crop Residue Specialist at the University of Nebraska–Lincoln. He received his BS from Texas Tech University and his MS and PhD from the University of Nebraska. His extension program areas include the enhanced use of crop residues and the incorporation of annual and perennial forage crops into existing beef production systems. His research program addresses the management, production, and utilization strategies for annual and perennial forage crops, the influence of crop residue management systems on establishment of annual forages, and the inclusion of these components in integrated crop–forage–livestock systems.

CRAIG A. ROBERTS is Professor and State Forage Extension Specialist in the Division of Plant Sciences at the University of Missouri. He received his BS from the University of North Alabama and his MS and PhD from the University of Arkansas. His research and extension appointment focuses on the identification of factors that affect forage quality and persistence, including the development of forage testing methodology and animal toxin management.

CRAIG C. SHEAFFER is Professor in the Department of Agronomy and Plant Genetics at the University of Minnesota. He received his PhD from the University of Maryland. His research focuses on perennial native and introduced legumes, grasses, and woody species and sustainable cropping systems. He directs the sustainable agriculture graduate program, and teaches undergraduate and graduate courses.

LYNN E. SOLLENBERGER is Distinguished Professor of Agronomy at the University of Florida, Gainesville. He received his BA from Messiah College, his MS from Pennsylvania State University, and his PhD from the University of Florida. He teaches undergraduate and graduate courses in grassland ecology, management, and utilization, and in grassland research techniques. His research focuses on ecosystem services of grasslands and evaluation of forages for use in grassland–livestock agroecosystems.

R. MARK SULC is Professor of Horticulture and Crop Science at Ohio State University, Columbus. He received his BS and MS from Iowa State University and his PhD from the University of Wisconsin–Madison. He is the Forage Extension Specialist with Ohio State University Extension. His research focuses on forage management, integrated pest management of forages, use of cover crops and annual forages to extend the grazing season, and integrated crop–livestock systems.

JEFFREY J. VOLENEC is Professor of Agronomy at Purdue University. He received his BS from the University of Wisconsin–Madison, and his MS and PhD from the University of Missouri. He teaches courses in forage management and crop physiology and ecology, and his research focuses on the physiological and biochemical mechanisms that control genetic differences in growth and stress tolerance of forages.

M. SCOTT WELLS is Assistant Professor and State Forage Extension Specialist in the Department of Agronomy and Plant Genetics at the University of Minnesota. He received his PhD from North Carolina State University. His research focuses on forage and crop systems.

CHARLES P. WEST is Thornton Distinguished Chair of Plant and Soil Science at Texas Tech University. He received his BS and MS from the University of Minnesota and his PhD from Iowa State University. He teaches undergraduate courses in forage management and environmental conservation, and a graduate course in pasture research techniques. He advises graduate students, and his research focuses on forage physiology and drought tolerance.

In Memoriam

Dr. Robert F Barnes, 1933–2013

To honor the contributions to this and earlier editions of *Forages*, and his dedicated national and international leadership in grassland agriculture, we dedicate Volume I of the seventh edition to Dr. Robert F Barnes. "Bob", as he was affectionately known, was born in Estherville, IA on February 6, 1933 and passed away on April 27, 2013 in Madison, WI. He graduated from Estherville Junior College in 1953, served in the Army as an intelligence specialist for 2 years, and graduated in Agronomy from Iowa State University in 1957. He earned his M.S. in Farm Crops from Rutgers University in 1959, and his PhD in Agronomy from Purdue University in 1963.

Bob began his stellar research and service career working with the US Department of Agriculture (USDA) at Purdue University, where he pioneered work on methodologies for measuring forage quality. Thereafter, he continued his research as Director of the USDA Regional Pasture Research Laboratory at Pennsylvania State University. He then served on the USDA National Program Staff for Forages and Range in Washington, DC until 1986, and as Associate Regional Administrator for the Southern Region of the USDA in New Orleans, LA. In 1986 he became Executive Vice President of the American Society of Agronomy, the Crop Science Society of America, and the Soil Science Society of America until he retired in 1999.

For more than four decades the sequential editions of *Forages* were synonymous with Bob's perspectives as he provided administrative leadership and worked diligently to ensure the high quality of each volume. He was instrumental in changing the fifth edition by splitting it into two volumes, one focused on undergraduate education and the other providing a comprehensive reference book for scientists and technical workers. That change was continued in the sixth and seventh editions.

In addition to the *Forages* books, Bob authored or co-authored more than 100 scientific publications and provided vision and leadership for progress in forage and grassland science in North America. Coupled with his vision and encouragement of others he had a major influence on young scientists and the growth of the forage industry. He led the development of the first broadly accepted definition of sustainability of agriculture, and facilitated the movement towards use of near-infrared radiation to measure forage quality. He helped to build partnerships in animal sciences, crop sciences, soil sciences, and social sciences in order to broaden the roles of forage scientists. As Executive Vice President he assisted the development of several conferences and detailed monographs on topics related to forage. Based on the series of successful International Grassland Congresses, he supported and provided leadership to initiate the series of successful International Crop Science Congresses, the first of which was held at Iowa State University in 1992.

However, our appreciation extends much further, because Bob was a quiet motivator through his genuine concern and quality of mentorship for each of us and for scores of other forage workers. Few individuals have had such a profound influence on our professional lives and our profession. Bob's leadership in grasslands and forages was effective through his associations with the USDA, the series of International Grassland Congresses, and his role as Executive Vice President of the American Society of Agronomy, the Crop Science Society of America, and the Soil Science Society of America. Throughout he maintained his interest in forages and championed the roles of forages and grasslands in each of the societies, and he was highly respected by government leaders and agricultural scientists in North America and around the world.

Peers recognized Bob's professional contributions with numerous accolades and awards, including the prestigious Fellow Award from both the American Society of Agronomy (1973) and the Crop Science Society of America (1973). He served as President of the Crop Science Society in 1985. The American Forage and Grassland Council honored him with the Merit Award (1973) and its distinctive Grasslander Award (2001). Bob served on the Boards of the Agronomic Science Foundation and International Grassland Congress for several years.

The Robert F Barnes Graduate Education Awards, begun in 2003, acknowledge his career and his vision to recognize graduate students who make outstanding paper presentations in the Forage and Grazinglands Division at the annual meetings of the Crop Science Society of America. These competitive awards recognize, encourage, and attract new educators, scientists, and practitioners to the division and its areas of study. They are funded largely by Dr. Barnes, and underscore his generosity and commitment to assisting others, especially young forage scientists, in their professional careers.

It is unfortunate that many younger scientists did not have the chance to experience the encouragement and friendly counsel we have received, or to see Bob's friendly smile and hear his characteristic chuckle. Our goal is to provide insight so that new students can be aware of this true gentleman and leader in forages. He was proud of the fact that his "middle name" was "F" and therefore did not need a period. His purpose was to live a life worthy of his calling as a Christian, husband, father, grandfather, professional colleague, and friend. He achieved these goals, in addition to his major scientific and professional contributions, by living humbly, rejecting passivity, accepting responsibility, listening carefully, and leading courageously.

Michael Collins
C. Jerry Nelson
Kenneth J. Moore

The Metric System

Base Units of the International System of Units

Measure	Base Unit	Symbol
Length	Meter	m
Mass	Kilogram	kg
Time	Second	s

Prefixes Used with Units of the Metric System

Multiples and Submultiples	Prefix	Value	Symbol
$1,000,000,000 = 10^9$	giga	one billion times	G
$1,000,000 = 10^6$	mega	one million times	M
$1,000 = 10^3$	kilo	one thousand times	k
$100 = 10^2$	hecto	one hundred times	h
$10 = 10^1$	deca	ten times	da
$0.\ 1 = 10^{-1}$	deci	one tenth of	d
$0.01 = 10^{-2}$	centi	one hundredth of	c
$0.001 = 10^{-3}$	milli	one thousandth of	m
$0.000,001 = 10^{-6}$	micro	one millionth of	μ
$0.000,000,001 = 10^{-9}$	nano	one billionth of	n

Metric to English Conversions

To convert metric to English multiply by	Metric unit	English unit	To convert English to metric multiply by
	Length		
0.621	kilometer (km)	mile (mi)	1.609
3.281	meter (m)	foot (ft)	0.305
0.394	centimeter (cm)	inch (in.)	2.54
0.0394	millimeter (mm)	inch (in.)	25.4
	Area		
0.386	kilometer^2 (km^2)	mile^2 (mi^2)	2.590
2.471	hectare (ha)	acre (A)	0.405

		Volume	
35.316	meter^3 (m^3)	foot^3 (ft^3)	0.028
1.057	liter (L)	quart (US liq.) (qt)	0.946
		Mass	
1.102	ton (1000 kg) (mt)	ton (US t)	0.907
2.205	kilogram (kg)	pound (lb)	0.454
0.00221	gram (g)	pound (lb)	454
		Yield	
0.446	ton/hectare (mt/ha)	ton/acre	2.24
0.446	megagram/hectare (Mg/ha)	ton/acre	2.24
0.892	kilogram/hectare (kg/ha)	pound/acre	1.12
		Temperature	
1.8C + 32	Celsius (C)	Fahrenheit (F)	(F – 32)/1.8

The Last of the Virgin Sod

We broke today on the homestead
The last of the virgin sod,
And a haunting feeling oppressed me
That we'd marred a work of God.

A fragrance rose from the furrow,
A fragrance both fresh and old;
It was fresh with the dew of morning,
Yet aged with time untold.

The creak of leather and clevis,
The rip of the coulter blade,
And we wreck what God with the labor
Of a million years had made.

I thought, while laying the last land,
Of the tropical sun and rains,
Of the jungles, glaciers, and oceans
Which had helped to make these plains.

Of monsters, horrid and fearful,
Which reigned in the land we plow,
And it seemed to me so presumptuous
Of man to claim it now.

So when, today on the homestead,
We finished the virgin sod,
Is it strange I almost regretted
To have marred that work of God?

—Rudolf Ruste

Flesh is Grass

Grass is the forgiveness of nature—her constant benediction. Fields trampled with battle, saturated with blood, torn with the ruts of cannon, grow green again with grass, and carnage is forgotten. Streets abandoned by traffic become grass-grown like rural lanes, and are obliterated. Forests decay, harvests perish, flowers vanish, but grass is immortal.

Unobtrusive and patient, it has immortal vigor and aggression. Banished from the thoroughfare and the field, it bides its time to return, and when vigilance is relaxed, or the dynasty has perished, it silently resumes the throne from which it has been expelled, but which it never abdicates.

It bears no blazonry or bloom to charm the senses with fragrance or splendor, but its homely hue is more enchanting than the lily or the rose. It yields no fruit in earth or air, and yet should its harvest fail for a single year, famine would depopulate the world.

The primary form of food is grass. Grass feeds the ox: the ox nourishes man: man dies and goes to grass again; and so the tide of life, with everlasting repetition, in continuous circles, moves endlessly on and upward, and in more senses than one, all flesh is grass.

(Excerpts from an address by John James Ingalls (1833–1900), Senator from Kansas from 1873 to 1891. Originally printed in the Kansas Magazine *in 1872.)*

PART I

CHARACTERISTICS OF FORAGE SPECIES

CHAPTER

1

Forages and Grasslands in a Changing World

C. Jerry Nelson, Kenneth J. Moore, and Michael Collins

Welcome to forages and grassland agriculture. The roles and importance of forages and grasslands for mankind have a long history and continue to change as societies evolve and new technologies are developed for the plant and animal sciences. The foundational **grasses, legumes,** and other **forbs** observed today are the result of natural evolution for adaptation and resilience, often with the presence of grazing animals, for over 10,000 years with the advent of sedentary agriculture by humans. These plant resources are fragile; when they are managed or mismanaged beyond their limits, they deteriorate and can be lost.

The focus of this book is to understand and appreciate the plant characteristics and fundamental principles that provide diversity among the major forage and grassland species and to describe their use and optimal management. The goals of this chapter are to provide background and future perspectives for grasslands in the USA and North America.

Grassland Terminology

With any subject, it is important to know and understand the terminology. As with other subjects, the terms and definitions (see Glossary) for grassland agriculture overlap and are intertwined. The main land and plant resources are **forage, pasture, range,** and **grassland.** Forage is defined by the International Forage and Grazing Terminology Committee (Allen et al., 2011) as "edible parts of plants, other than separated grain, that provide feed for animals, or can be harvested for feeding." It includes **browse** (buds, leaves, and twigs of woody species), **herbage** (leaves, stems, roots, and seeds of non-woody species), and **mast** (nuts and seeds of woody species). Thus *forage* is an inclusive term for plants and plant parts that are consumed in many forms by domestic livestock, game animals, and a wide range of other animals, including insects. Furthermore, production of forage involves several types of land use and is subdivided using more specific terms.

The term *pasture* is derived from the Latin *pastus* and is defined by the International Forage and Grazing Terminology Committee (Allen et al., 2011) as "an area in which grass or other plants are grown for the feeding of grazing animals." This broad context includes **pasturage** that more accurately means "the vegetation which animals graze." Thus pasture refers to the land area or **grazing management unit,** rather than to what is consumed. *Pastureland* refers to land, usually in humid areas, devoted to the production of both **indigenous** (i.e., native to the area) and introduced forage species that are harvested primarily by grazing. *Permanent pasture* refers to pastureland composed of perennial or self-seeding annual plants that are grazed annually, generally for 10 or more successive years. In contrast, **rangeland** refers to land, usually in arid or semi-arid areas, consisting of tall-grass and short-grass prairies, desert grasslands and shrublands that are managed extensively and grazed by domestic animals and wildlife.

Cropland forage is land devoted to the production of a cultivated crop (e.g., corn or winter wheat) that is harvested for silage or hay. **Cropland pasture** is cropland that is grazed for part of the year, such as grazing corn stalks after the grain is harvested or grazing leaves of winter wheat during winter and early spring before reproductive growth

Forages: An Introduction to Grassland Agriculture, Seventh Edition. Edited by Michael Collins, C. Jerry Nelson, Kenneth J. Moore and Robert F Barnes.

begins. In addition to grazing, cropland pastures are useful in row crop rotations as winter cover crops to reduce soil erosion. **Cover crops** such as red clover or winter rye in the north or ryegrass and crimson clover in the south are seeded in fall primarily to provide protective ground cover over winter. The crop can be grazed, harvested, or tilled into the soil in spring. In addition to erosion control and protection of water quality, cover crops have favorable effects on soil fertility, soil quality, water quality, weeds, pests, diseases, and biodiversity and wildlife in an agroecosystem.

Rangeland is land on which the indigenous vegetation consists predominantly of grasses, grass-like plants, forbs, or **shrubs** and is managed as a natural ecosystem. When non-native plants are seeded into rangeland, they are managed as part of the vegetation mix as if they were native species. **Range** is a more collective term that includes grazeable **forestland** or **forest range** that produces, at least periodically, an understory of natural herbaceous or shrubby vegetation that can be grazed. This use has raised interest in **agroforestry,** namely the use of cropland agriculture among trees until the tree canopy causes shade. **Silvopasture** describes an agroforestry practice that combines managed pastureland with tree production.

Cropland, forestland, pastureland, and rangeland are also the basis for land-use mapping units (Fig. 1.1). Terms for grazing lands and grazing animals have been prepared by the International Forage and Grazing Terminology Committee (Allen et al., 2011); many of these are included in the Glossary.

Grassland Agriculture

Grassland includes pastureland, rangeland, and cover crops used for grazing, and thus in general denotes all plant communities on which animals are fed, with the exception of crops sown annually (such as wheat, corn, cotton, or sugar beets) that may also be used as forage. More commonly, *grassland* is any plant community, including harvested forages, in which grasses and/or small-seeded legumes make up the dominant vegetation.

The term **grassland agriculture** describes a farming system that emphasizes the importance of grasses and legumes in livestock and land management, including manure management. Farmers who integrate row crops with hayfields and pastures on their farm and manage livestock production around their grassland resources are grassland farmers.

Success in grassland farming depends on maintaining a healthy soil–plant–animal biological system. Land, or more specifically soil, is basic to plant production and hence to all of life. Simply stated, plants absorb from the soil the mineral elements that are required by animals and humans. Plants also combine the natural resources of solar energy, carbon dioxide (CO_2), and water to form carbohydrates and other carbon compounds. Plants then blend nitrogen (N) with appropriate carbon chains to produce amino acids and proteins. However, no single food plant contains the nutrients in the same proportions as are required by animals or humans.

Herbivores (animals that can digest the fibrous tissue of plants) subsist primarily on plants and plant materials, converting grassland products to high-quality meat and milk foods that complement the nutritive value of plant products for humans (Fig. 1.2). Ruminants and other herbivores contribute to human well-being by producing meat and milk products that are rich sources of proteins, fats, vitamins, and minerals. In addition, ruminants provide non-food products of value, such as:

- Hides, wool, and horns for clothing, implements, and adornments
- Power for draft work or transportation
- Manure for fertilizer and fuels
- Benefits to humans, such as the pleasure derived from keeping animals as pets, observing wild animals in their

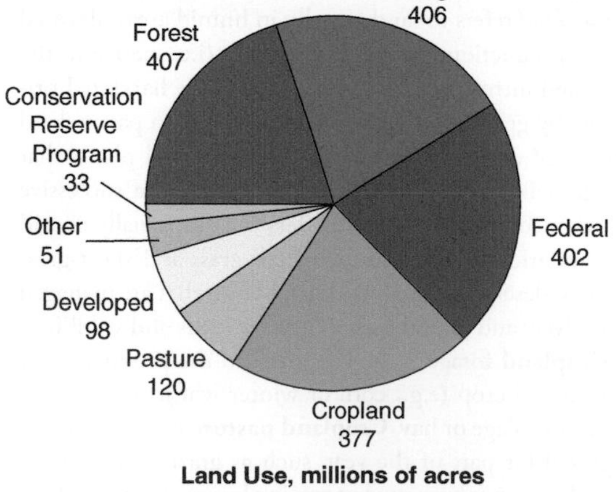

FIG. 1.1. Agricultural land use (million acres) in the contiguous USA. (Data from USDA Economic Information Bulletin No. 89, 2011.)

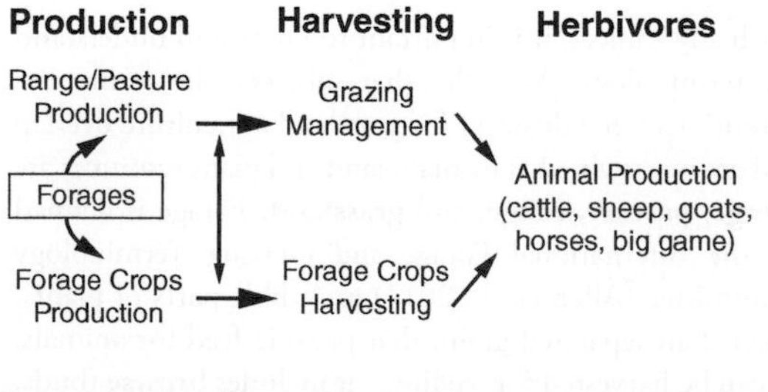

FIG. 1.2. The dual production-harvesting avenues that forage can follow during its harvest and conversion to useful products by herbivores. (Adapted from Vallentine, 2001.)

natural habitats, and using animals for competitive and sporting events, and hunting
- A biological means of harvesting desirable vegetation and removing unwanted vegetation.

Scientific Names of Forage Plants

Today grasslands and forage management are international in scope, and communication about forages occurs worldwide. The US Department of Agriculture (USDA) has active programs and strict regulations for evaluating introduced plants in regional testing sites that represent climatic areas. Seed of several grasses and legumes are produced in the USA and shipped to foreign markets. Similarly, the USA imports commercial seed of several forage species. These activities require accurate communication about forages based on universal terminology.

Most cultivated forages fit into two botanical families: the Poaceae (Gramineae), namely the grasses, and the **Fabaceae** (Leguminosae), namely the legumes. In addition, many other forbs, which are herbaceous (i.e., nonwoody) dicotyledonous plants (including legumes), and the leaves and buds of several trees and shrubs contribute to the nutritional requirements of ruminants. Each plant has its own scientific name. The binomial system of naming developed by the Swedish botanist Carl Linnaeus in the eighteenth century has been very effective, and is still the standard.

Each plant is known scientifically by its species name, which generally consists of two Latin words. The first word, the **genus,** always has a capitalized initial letter; the second word, the **species epithet,** is all lower case. The genus is similar to a surname and the species epithet to a first name. Thus, for example, *Medicago sativa* would be like Brown, John. The scientific name includes the authority, which is the abbreviated name of the person or persons who first classified the species. For example, *Medicago sativa* L. indicates that alfalfa was named by the Swedish botanist Linnaeus. If a plant is reclassified, the original authority is placed in parentheses, and the new authority follows it. For example, indiangrass, *Sorghastrum nutans* (L.) Nash, was first classified by Linnaeus and later reclassified by Nash.

Forage plants often have different common names in different regions of the USA and the world. For example, the name prairie beardgrass is occasionally used, but is not the approved common name for little bluestem *(Schizachyrium scoparium* [Michx.] Nash). Similarly, in the UK, alfalfa is called lucerne, and *Dactylis glomerata* L. is called cocksfoot instead of orchardgrass. The common and scientific names of many of the plants discussed in this book are listed in the Appendix.

The Early Role of Grasslands

Civilizations have had their origins on grasslands, and have vanished with its destruction. Grazing lands were vital to prehistoric nomadic peoples as hunting sites long before cattle and sheep were domesticated. Attempts to control the fate of humans by planting crops to provide for future needs, instead of remaining the victims of droughts or other calamities, must have taken place on grasslands, where the young calves, lambs, and kids that had been caught and tamed could find forage. After the nomads adopted a sedentary way of life and became food producers rather than food gatherers, the grasslands of these early peoples changed rapidly.

Early recognition of the value and fragilities of grass (i.e., forage resources) is noted in the Bible: "He makes *grass* grow for the cattle, and plants for man to cultivate—bringing forth food from earth" (Psalm 104:14). This is an early reference to the soil–plant–animal continuum to provide food, but it is unclear if the grass was managed. The close linkage between plants, animals, and humans is also referred to by Moses when he promised the Children of Israel that God "will provide *grass* in the fields for your cattle, and you will eat and be satisfied" (Deuteronomy 11:15). Psalm 103:15–16 compares grass to man and the transitory nature of human life: "As for man, his days are like *grass*, he flourishes like a flower of the field; the wind blows over it and it is gone, and its place remembers it no more." The shortage of grass was recognized as a symbol of desolation: "The waters … are dried up and the *grass* is withered; the vegetation is gone and nothing green is left" (Isaiah 15:6).

The theme of grass and grazing that runs throughout the Bible shows the early interrelationship of humans and nature, with the general view that nature was the major determinant of productivity. People knew little about how the grassland resource could be managed. As agriculture developed, populations of nomadic hunters and gatherers decreased, "apparently remaining only in areas where agriculture was unable to penetrate" (Harlan, 1975).

Early grassland management practices in Asia and Africa often consisted of communal grazing of livestock on large areas of native pastures. Such shared pastures were referred to as *commons*, a term generally used today for a public place for people. Commons were owned collectively, and one member could not exclude animals owned by another member. Consequently, overgrazing and conflict often resulted, leading to lower levels of animal production and subsequently to poorer human nutrition. At that time, cropland agriculture was becoming focused on planted monocultures with the expectation that farmers could overcome environmental and other constraints by management to increase yield.

The Evolution of Grassland Management

In Great Britain, the use of the scythe and the process of hay making date from 750 BC. The conversion of fresh green forage into dried hay, which could be stored with

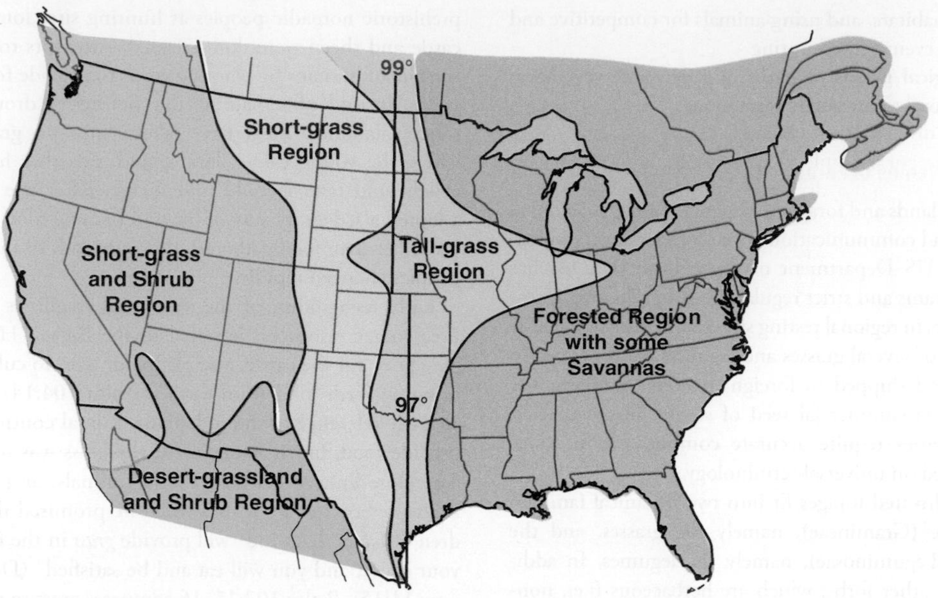

FIG. 1.3. Major grassland regions of the USA at present. Introduced species are dominant in northern areas along the West Coast and in areas of the east that were formerly wooded. Southern areas along the West Coast have primarily winter annuals in non-irrigated areas. (Adapted from Barnes, 1948.)

little change over time, was associated with a sedentary rather than nomadic agriculture. Winter survival of livestock depended on the success of the hay harvest. The growing of hay crops and the need for proper curing were described in detail by the Roman writer Columella in about AD 50.

The Anglo-Saxons began to enclose meadows in the Midlands of Great Britain around AD 800, probably using fences of stacked stones or hedges of thick or thorny plants. As early as 1165 the monks of Kelso were aware of the value of regular changes of pasture areas for the health of cattle and sheep. Around 1400, the monks of Couper were using crop rotations, alternating 2 years of wheat with 5 years of grass, a crop rotation practice that later came to be known as **ley** farming.

Red clover, which is a very important legume today, was cultivated in Italy as early as 1550, in western Europe somewhat later, in England by 1645, and in Massachusetts in 1747. Without knowing the causes, farmers recognized differences in value among the various forage species. We now know that red clover can fix nitrogen from the air, and that the addition of red clover or other legumes to the grass mixture increases both production and quality of the forage. The influence of red clover on civilization and European agriculture was probably greater than that of any other forage plant (Heath and Kaiser, 1985).

Native Grasslands of North America

Before colonization and the introduction of cattle, sheep, and goats by European settlers, heavy forest covered much of the eastern USA; about 40% of the total land area in the contiguous USA was grassland (Fig. 1.3). Around the year 1500 there were nearly 700 million acres (284 million ha) of grass-covered native prairie stretching from Ohio westward. In general, the most fertile, deep, rich, black soils developed under the vegetative growth of the prairies. The tall-grass prairies of the central and Great Lakes states were dominated by native grasses such as big bluestem, indiangrass, and switchgrass, which grew tall and dense. These prairie grass sods were so tough and thick that some farmers preferred not to plow them until the stand had been weakened for a few years by **overgrazing** and repeated mowing. Today, the Flint Hills area of eastern Kansas and the Osage Hills of Oklahoma are the only extensive areas of undisturbed native tall-grass prairie.

Further west, the shortgrass prairie originally extended from Mexico and Texas north into Canada, and east from the Rocky Mountains to mid-Kansas, Nebraska, and the Dakotas. Native short grasses such as wheatgrasses in the north and buffalograss and grama grasses in the south were in greatest abundance. Some tall grasses were intermixed in the region of transition and predominated toward the eastern margin, especially in sites with higher soil moisture levels.

The areas of grassland gradually changed several times over geological time, ranging in character from woodlands and forest during moist, cool eras to grasslands during more arid periods. For example, it is known that between 4000 and 8000 years ago a drying trend extended the arm of tall-grass prairie between the Ohio River and the Great Lakes all the way to the Appalachian Mountains. As less arid times returned, the forest encroached again and the prairie retreated westward, leaving behind soils of grassland origin and patches of relic prairie communities that still exist in New York and Pennsylvania.

The vast native grasslands of the USA were referred to as range and rangelands soon after the turn of the twentieth century. English settlers on the Atlantic Coast brought with them their term **meadow** for native grassland that was suitable for mowing. The French in Canada used the term prairie for similar grassland, and the Spanish in Florida used the word savanna. These various names for native grassland in North America have become a part of the American vocabulary, with each term having its own meaning.

Native Americans and Forages

A unique feature of the management of the vast grasslands of North America was the use of fire. References to burning of grasslands by Native Americans are found in the journals of many early explorers and settlers in the western USA. Early on, the range was shrubbier and had intermittent grasses. It is likely that Native Americans noted the effects of natural fires caused by lightning in summer when the soil and plants were dry, and then tried burning the overwintering residue earlier in the year. Today we know that the burning of grasslands in spring, as the plants are just beginning to grow, contributes to the abundance of productive grasses by removing old ungrazed forage, recycling minerals, and reducing the numbers of weeds and shrubs. The improved grassland led to increased numbers of American bison *(Bison bison* L.), also known as buffalo. Native Americans selectively burned large areas in order to entice the herds of buffalo and other wildlife to the improved areas, where they could be more easily hunted.

Fire Cleanses and Rejuvenates

The word *fire* conjures up thoughts of heat, tragedy, smoke, and dirty ashes. However, many native grasslands benefit from the burning of old stubble and residue in spring. A "good" fire is hot enough to burn the residue completely, but moves across the ground quickly so that the temperature of the soil and the meristematic regions of grasses at soil level remains low. Unless they are protected by thick bark, herbaceous plants, young shrubs, and young trees with meristems higher in the air are destroyed. The life cycles of insects and many pathogens are disrupted. Ashes on the soil surface absorb solar energy, which warms the soil and thus stimulates early growth of the surviving grasses. Plant minerals in the ashes are available to support new growth that is of very high quality. Researchers are still learning about the value of fire.

The relationship between soil, grass, fire, buffalo, and Native Americans developed over several thousand years, as a result of which these native grasslands provided great wealth. About 200 years ago the prairies supported 60 million buffalo, 40 million whitetailed deer, 40 million pronghorn antelope, 10 million elk, and hundreds of millions of prairie dogs, jackrabbits, and cottontail rabbits—all forage consumers. Grasslands were essential for buffalo, on which Native Americans depended for their existence. After the buffalo herds were destroyed by hunting on a massive scale in the 1800s, the prairie and Native Americans were subdued.

Forages in American Colonial Times

The first English settlers in the American Colonies found that their method of farming and producing food crops in the New World was minimally successful. The East Coast was covered with forests with little open grassland (Fig. 1.3). The few domestic animals that survived the long ocean voyage grazed the small pockets of native grasses, where they did well during the summer, but required shelters and supplemental harvested forage to survive the long, hard winters. As the number of livestock increased, the limited acreage of native pastureland and production of poor-quality hay made it difficult to carry animals through the winter. Gradually, the year-on-year grazing without rest weakened the tall-growing native species. This led to the introduction of short-growing grasses and clovers used previously in England. The introduced species had a longer growing season, were more productive, especially during the cool seasons of spring and fall, and were better adapted to close and repeated grazing.

By the early 1700s, the acreage of introduced grasses, somewhat open woodlands, and enclosed meadows was not keeping pace with the need for meat and milk. Croplands worn out by excessive tillage and then abandoned to weed fallow made poor pasture. Farmers continued to cut hay chiefly from natural meadows and marshy areas.

In England, between 1780 and 1820, many crude research trials on various grasses and legumes grown in small plots were conducted. Yield and nutritive value were determined, and the findings were published as a book that made its way to the USA, where these results were used for about 50 years.

In 1850, haying tools consisted of the scythe, a crude hay rake, and the pitchfork. Mechanization began with the sicklebar mower, followed by a harpoon-type fork for unloading hay from a wagon into the barn (1864), the hay loader for moving hay from a swath onto a wagon (1874), and the side-delivery rake to make a windrow from the swath (1893). However, hay making was still a difficult and time-consuming job. The baler with a pickup attachment (developed in the 1940s) made it easier to collect the dried hay and form a dense rectangular package that could be moved and stored efficiently. The big round baler (developed in the early 1970s) further reduced labor needs. Today, the use of dense, compact, rectangular bales weighing from 500 lb (225 kg) to over 1 ton (450 kg) are used for commercial sales and international trade. Later advances in electric fencing and watering systems facilitated the use of intensive grazing systems. Today, new technologies continue to shift and improve the nature of forage management and use.

Silage production, which is a process of fermenting plants in anaerobic conditions to preserve them with a high moisture content, was carried out in a crude form by the Egyptians and Greeks. They placed wet forage into a vertical structure (silo) made of stones or a covered pit, and packed it to reduce the oxygen content. Natural microorganisms used up the remaining oxygen to form organic acids that lowered the pH to prevent other organisms from rotting the material. Ensiling the wet forage sooner, before it was dry enough to store as hay, reduced the potential for weather damage and harvest losses in the field that decrease forage yield and quality. The modern era of ensiling crops began in the mid-1800s in Germany, perhaps due to the common practice of making sauerkraut to preserve cabbage. By 1900, silage making was being promoted in the USA, facilitated by improved storage structures, mechanization for harvest, and chopping the plant material into small pieces for improved ensiling.

The Merging of Grassland Cultures

The culture of grass in the USA evolved as a product of the Native American and European farming systems that produced the beginnings of American agriculture (Edwards, 1940). As the pioneer farmers began to push westward from New England in the late 1700s, they needed to clear heavy forest before crops and the introduced forage species could be grown. Interestingly, it was generally thought that land which supported only grass was inferior to that which supported tree growth. As settlers entered Ohio and western areas where there was a choice between forest and prairie, forest-covered soils were favored. Forest also provided security for the pioneer farmers; it sheltered the game that was a major source of meat, and it supplied timber for cabins, stock shelters, fuel, and fences.

Fencing materials were important to the settlers, and fences on the prairies were not practicable until the invention of barbed wire in 1867 and its rapid introduction into US agriculture. The pioneers hesitated to migrate onto the large prairies of seemingly endless tall grass. The vastness was overwhelming and left the impression that it could not be subdued. Low-cost fencing and the steel moldboard plow opened up new opportunities. Meanwhile, settlers of Spanish origin were entering the south-west from Mexico.

George Stewart's chapter in Senate Document No. 199 provides a classic historical documentation of range resources in the USA (Stewart, 1936, p. 2):

> The western range is largely open and unfenced, with control of stock by herding; when fenced, relatively large units are enclosed. It supports with few exceptions only native grasses and other forage plants, is never fertilized or cultivated, and can in the main be restored and maintained only through control of grazing. It consists almost exclusively of land which, because of relatively meager precipitation and other adverse climatic conditions, or rough topography or lack of water for irrigation, cannot successfully be used for any other form of agriculture. In contrast, the improved pastures of the East and Middlewest receive an abundant precipitation, are ordinarily fenced, utilize introduced forage species, ... cultivation for other crops, and are often fertilized to increase productivity, and are renewed following deterioration.

The impression of early explorers was that the growth of grasses on these vast prairie areas would endlessly support countless herds and flocks. However, two factors eventually upset the resilience of this grassland resource: first, the Spanish heritage of rearing cattle in large herds, and second, the increased demand for meat after the discovery of gold in California in 1848. Livestock on the prairies had previously been raised largely for hides, tallow, and wool, but that changed with the rapid migration of people to the West. After 1870 the number of large herds of cattle increased rapidly from central Texas northward and westward (Edwards, 1940). The influence of the colonial Spanish on the use of grasslands of the southwestern USA is best summarized by Stewart (1936, p. 122):

> The tremendous growth in range cattle, however, carried with it a weakness that in the end proved fatal. It was based on a husbandry transplanted from Mexico, which brought to English-speaking people for the first time in history the practice of rearing cattle in great droves without fences, corrals, or feed. ... Cattle instead of grass came to be regarded as the raw resource, and the neglected forage began to give way before the heavy and unmanaged use to which it was subjected.

The steel moldboard plow and the tractor spread the development of agriculture, particularly in the Midwest and tall-grass prairie region. However, this led to widespread conversion of grassland, with its rich soils, to cropland even in dry areas, and eventually to the Dust Bowl

The Resiliency of Nature

When cropland in the eastern USA is abandoned or pastureland is not managed, it is encroached, first with annual weeds and eventually with deciduous trees (Fig. 1.4). The ecosystem wants to regain balance with nature (i.e., to develop a mixture of plant species that coexist and are in long-term equilibrium with the environment). When land use by humans is far from this natural equilibrium, resources must be expended in terms of reseeding, fertilizing, controlling weeds and pests, and cutting or grazing management to keep the system as close as possible to the desired condition. The further that condition is from the natural equilibrium, the greater the management cost; "fighting nature" too far from the natural equilibrium becomes non-economic. Can you determine whether an abandoned cornfield in eastern Nebraska would revert to a more natural state as quickly as would one in central Pennsylvania? Would tree encroachment be a greater problem for a pasture site in western Missouri or in eastern Kentucky?

Fig. 1.4. Beef cattle grazing a managed pasture of smooth bromegrass and orchardgrass during autumn in Wisconsin. Note the deciduous hardwood trees in the background that provide shelter for the livestock and shade in summer, but the trees and shrubs would invade and dominate the ecosystem if the pasture was not managed correctly. (Photo courtesy of Michael Collins.)

of the 1930s. Vavra et al. (1993) conclude that "the westward expansion of the US was characterized by exploitation of natural resources." Interestingly, this exploitation was supported by the American public through federal legislation and policies. For example, the initial Homestead Act of 1862, followed by others in 1873 and 1877, encouraged westward expansion by allocating publicly owned grassland at very low cost for settlers for growing crops.

The influx of a farming population into the rangeland areas resulted in even greater demand for animal and meat production from the remaining rangeland. The public rangeland, which was in many respects a commons, was grazed by animals owned by nearby farmers and ranchers. In contrast with the culture and management of the Native Americans, there was little incentive to conserve the available forage and land for later use. This lack of property rights led to what Hardin (1968) called the "tragedy of the commons." The boon in unlimited livestock grazing lasted only 20 years, and by 1880 continued overstocking had reduced the carrying capacity of most of the western range. Dry summers coupled with severe winters resulted in the loss of 30–80% of the cattle in the Northern Plains during the winters of 1886 and 1887.

Soon thereafter the cattle industry in the Great Plains changed from an open-range, exclusive-use enterprise to a ranch-based industry that coexisted with cropland. The Taylor Grazing Act of 1934 began to control grazing on government-owned grasslands, and ended the series of Homestead Acts.

Although the interactions between livestock grazing and other elements of the Great Plains ecosystem are complex, research indicates that grazing of domestic livestock at conservative levels on sensitive western rangelands can be sustainable (Vavra et al., 1993), and that in most cases it is similar to grazing by buffalo. Sustainability must be the primary goal of any forage management plan, particularly for the Great Plains, where 63% of the area consists of grazing lands. At the same time, factors such as urbanization, climatic changes, reduction in the use of fire, increases in the number of woody species, introduction of alien plant species, and other human activities have had major influences on this natural resource.

Forage, Range, and Grasslands Today

Approximately 75 years ago there was an average of about 10 acres (4 ha) of cropland per person in the USA. By 30 years ago that figure had been reduced to about 5 acres (2 ha), and today it averages less than 2 acres (0.8 ha) per person. Meanwhile, average farm size increased, and the number of farms decreased from a peak of 6.8 million in 1935 to 2.1 million by 2012, and remains at approximately the same level today. About 8 million people were living on farms and ranches in 1978, but this figure had decreased to about 5 million by 1987, and to 3.2 million in 2012, as farm size gradually increased. As a result, the farm and ranch population today has declined to less than 1.2% of the total US population. However, forage and grazinglands remain an important part of US agriculture.

Currently, pasture and non-forested rangeland represent about 29% of the 2.3 billion acres (1.04 billion ha) of land area of the USA. The proportion of land used in agriculture has decreased from about 63% in 1950 to about 50% at the time of writing. Decreases occurred mainly in cropland, pastures, and range, especially grazed forestland. At the same time, non-agricultural uses increased to 49% of the total land base, primarily due to increases in national parks and national wilderness and wildlife areas, mainly in Alaska. In 2015, hay was produced on an estimated 56.2 million acres (22.8 million ha), of which about 32% was alfalfa and alfalfa mixtures. The average yield of alfalfa and alfalfa mixtures was 3.5 tons/acre (7.8 Mg/ha) whereas that for other hay was 2.1 tons/acre (4.7 Mg/ha). Corn and sorghum silage grown on over 7 million acres (2.8 million ha) added to the total value from forages.

More than 40% of the rangeland, mainly in the far west, is owned by the US government, and much of it is grazed under contract with local ranchers. The amount needed, appropriate use, and management of these public lands continue to be debated.

Today, much of the woodland and forested area of the eastern USA has been cleared and is used for crop production. Except for a few pockets of native unplowed prairie, many of which are being preserved, the eastern half of the USA (east of 99° longitude) uses mainly introduced forage species for pasture and hay production (Fig. 1.5).

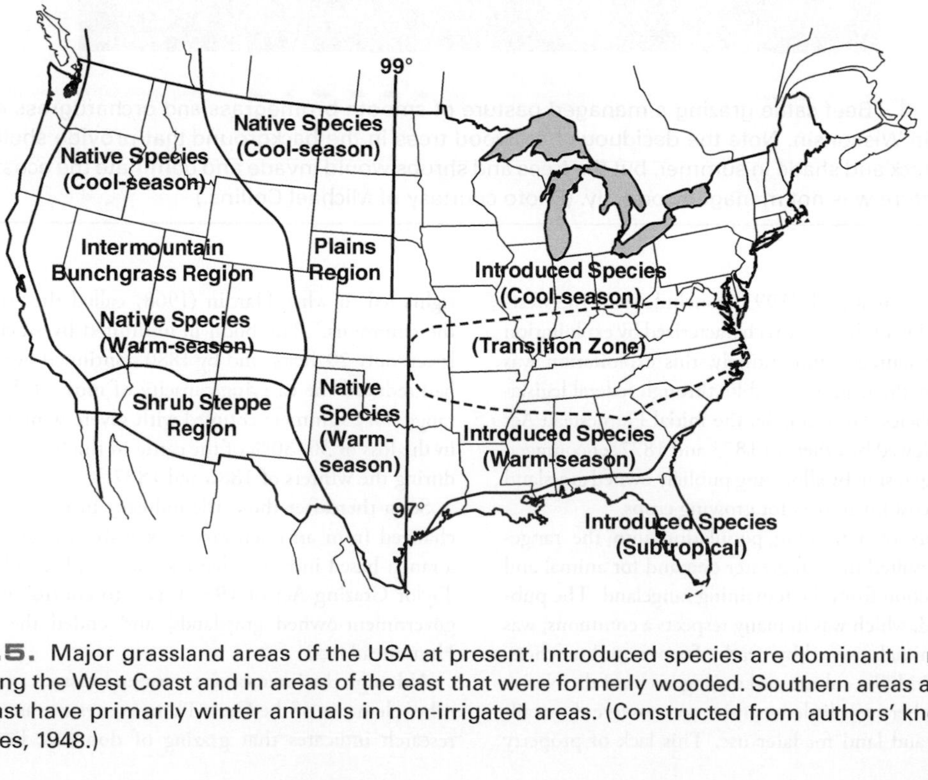

Fig. 1.5. Major grassland areas of the USA at present. Introduced species are dominant in northern areas along the West Coast and in areas of the east that were formerly wooded. Southern areas along the West Coast have primarily winter annuals in non-irrigated areas. (Constructed from authors' knowledge and Barnes, 1948.)

FIG. 1.6. Dairy cows utilizing a managed pasture of cool-season grasses and legumes in Wisconsin. The large barn includes the milking area and provides shelter during winter. The silos near the barn are for storing silage, mainly corn or grass silage, for winter feeding when pastures are not growing. Pasturing dairy cattle is a way to reduce the costs of harvesting, packaging, storing, and feeding forage. It also reduces problems with waste management because the manure is distributed on the pasture. (Photo courtesy of Michael Collins.)

Permanent pastures in the north central and northeastern regions are dominated by white clover and kentucky bluegrass, both of which are introduced species. Improved pastures (Fig. 1.6) include grasses such as timothy, orchardgrass, smooth bromegrass, perennial ryegrass, and reed canarygrass, and legumes such as red clover, white clover, birdsfoot trefoil, and alfalfa, all of which have been introduced. Each cool-season species has its optimum area of adaptation within the general region, depending on environmental stresses and the management system used.

Pastures in the southern part of this region are dominated by bermudagrass, dallisgrass, bahiagrass, and johnsongrass, all of which are introduced warm-season grasses (Fig. 1.5). Few perennial legumes show good adaptation to the South because of the long, warm summers, but white clover is widespread. Winter annuals such as crimson clover and several other clovers have also been introduced. In the subtropical areas of Florida and along the Gulf Coast there are many introduced species. In addition, in the southeast there are large areas of silvopasture where pine forests have an understory of grasses and legumes. These forages are usually burned on a regular basis to maintain productivity.

Tall fescue dominates in the region where the distributions of warm- and cool-season species overlap, creating what is known as the transition zone. Although tall fescue is physiologically a cool-season species, it evolved in North Africa and can survive the warm, dry summers and moderate winters of the area. In addition, an endophytic fungus lives in the sheaths of tall fescue leaves, enabling the plant to resist insects and enhancing its drought resistance and rooting capabilities. Chemicals produced as a result of the plant–fungus association also reduce the palatability to livestock, so plants are less likely to be severely defoliated. In southern areas, most of the tall fescue grown has the endophytic fungus. In the northern areas, where stresses are lower, endophyte-free cultivars are often used because they give better animal performance. In the transition zone there is interest in using various warm-season grasses, both native and introduced, to enhance summer production.

The plains grasslands west of 99° longitude, where rainfall is lower, are still dominated by a range of native short grasses, depending on their adaptation to cold. Species such as wheatgrasses and wildryes are common in the north. With good management and modest rainfall, the southern

Table 1.1. Numbers of ruminant livestock in the USA

Animal	1978	1987	1994	2002	2012
Cattle (million)	116.4	102.1	101.1	95.5	90.0
Beef cows (million)	38.7	33.8	34.6	33.4	29.0
Dairy cows (million)	10.9	10.5	9.5	9.1	9.2
Sheep (million)	12.4	10.7	9.8	6.3	5.4

Source: US Department of Agriculture, 2016.

areas have mainly native grasses such as blue grama and buffalograss.

The southwestern shrub-steppe region consists of grama grasses, some annuals, and shrubs. If there is poor grazing management the shrubs can become dominant.

Along the West Coast, especially in the south where climates are more Mediterranean, winter annuals predominate in non-irrigated areas. Along the northern coastline, the grasslands are mainly introduced cool-season species.

The intermountain bunchgrass region is located between the Cascade and Rocky Mountains. In northern areas where precipitation mainly falls as snow, the growing season is cool and short, and grasslands mainly consist of bluebunch wheatgrass and idaho fescue, both of which are native species, and numerous forbs. Downy bromegrass, a winter annual species, is an invader. Some of the northern areas have been cultivated from time to time. In the south, several native grasses predominate, but shrubs are primary invaders, especially in the absence of fire.

Forages in the National Economy

Grassland agriculture is highly dependent on a reliable source of forage as the primary feed base for ruminant livestock. Resource inputs used when raising cattle and sheep are widely scattered, both in location and in ownership, because livestock raising is a land-based, forage-utilizing enterprise. Cattle numbers in the USA peaked in 1975, with 135 million head, and then declined gradually (Table 1.1), mainly due to a reduction in the number of beef cattle as annual per-capita consumption of beef has decreased from 87 lb (40 kg) in 1978 to 57 lb (26 kg) in 2012. This occurred primarily due to higher costs and growing health concerns. Interestingly, the number of dairy cows has remained relatively steady, but production per cow has increased by about 13% over the past 10 years. Sheep numbers peaked in 1935 and then declined significantly, with a reduction of over 56% between 1978 and 2012. These numbers may increase somewhat in the future depending on the ethnic market that is developing. Overall, the total number of ruminant livestock has gradually declined, decreasing by about 25% between 1978 and 2012. During that same time period, annual per-capita consumption of pork remained relatively steady at about 46 lb (21 kg), whereas that of poultry increased from 54 lb (25 kg) to 98 lb (45 kg).

Data on the use of feedstuffs for livestock and poultry in the USA allow the proportion of the diet provided by forage for various classes of livestock to be calculated (Table 1.2). Ruminant animals depend on forage as the basic feed source, but very little forage is fed to poultry, and minimal amounts are fed to swine. However, for beef cattle, before being fed out on high-concentrate rations, about 96% of the feed is in the form of forages. In general, dairy cattle use a lower proportion of forages in the diet than do beef cattle, mainly because dairy cattle need large amounts of concentrate to produce milk for several years, compared with a period of only a few months when beef cattle are finished on a high-concentrate diet.

Table 1.2. Use of concentrates and forages in livestock and poultry rations in the USA

	Proportion of ration (%)		Proportion of total concentrate (%)	Proportion of total feed (%)
	Concentrate	Forages		
All livestock and poultry	36	63	100.0	100.0
All dairy cattle	39	61	16.6	16.0
All beef cattle	17	83	25.8	56.9
Beef cattle on feed	72	28	20.7	10.7
Other beef cattle	4	96	5.1	46.2
Sheep and goats	9	91	0.4	1.9
Hens and pullets	100	0	12.4	4.6
Turkeys	100	0	3.3	1.3
Broilers	100	0	9.3	3.5
Swine	85	15	30.0	13.2
Horses and mules	28	72	2.2	2.9

Source: Adapted from Council for Agricultural Science and Technology, 1980.

Table 1.3. Estimated value of forages consumed by ruminant livestock

Animals	Receipt as feed costs[a] (%)	Feed units as forage[b] (%)	Cash receipts as forage value[c]	1996 Cash receipts[d] (US$ million)	Forage value[e] (US$ million)
Beef cattle	70	83	0.581	36,094	20,971
Sheep and wool	70	91	0.637	680	433
Dairy cattle (milk)	50	61	0.305	20,997	6404
Total forage value					27,808

[a]From Hodgson, 1974.
[b]From Council for Agricultural Science and Technology, 1980.
[c]Obtained by multiplying column 2 (receipt as feed costs) by column 3 (feed units as forage).
[d]Reported in US Department of Agriculture, 1999.
[e]Obtained by multiplying column 4 (cash receipts as forage value) by column 5 (1996 cash receipts).

Hodgson (1974) calculated the value of forage based on proportional feed costs for each class of livestock (Table 1.3). Using estimated feed costs for livestock production, rather than feed value, underestimates the true value of forages. Using feed costs based on cash receipts for 1998 (US Department of Agriculture, 1999), the total value of forages of US$ 27.8 billion (Table 1.3) exceeded the cash value of other crops. About 75% of the total value of forages was for beef cattle and 23% was for dairy cattle. The value of hay alone in 2014 was US$ 19.1 billion (US$ 10.6 billion for alfalfa and US$ 8.5 billion for other hay), which was exceeded only by corn and soybeans, which were valued at US$ 53.0 billion and US$ 39.5 billion, respectively (US Department of Agriculture, 2016). Unfortunately, no data are available for annual values of pasture and rangeland.

Sustainable Pastures and Hayfields

The number of farms and the number of people farming are gradually declining, while at the same time the challenge of providing adequate supplies of safe and wholesome food and other **ecosystem services** for an ever-expanding population is increasing. Large-scale farming, often termed "corporate" or "industrial" farming, has raised genuine suspicion and mistrust about a food system that appears to be being driven by large companies and its stockholders, not by local farmers. The need for sustainability of agriculture is a high priority involving the agricultural community, the public, and focused government programs to ensure that resources are conserved in order to be available for and meet the needs of subsequent generations. Most farmers and ranchers are aware of the basic issues, and the need to adopt practices that contribute to sustainability. Yet the public, which is geographically and historically removed from agriculture, is concerned and often reacts in ways that are based on emotion or hearsay, and with little scientific background. New technologies and the growing role of industry in marketing seed, chemical fertilizers, pesticides, and genetically modified crops have heightened public awareness and the need for regulations.

Public concerns have caused regulations to be put in place by the US Department of Agriculture (USDA), the Environmental Protection Agency (EPA), and the Food and Drug Administration (FDA) to ensure that the food, water, and air are tested and determined to be safe. There was a strong need for clarity about sustainable agriculture so that rational and clear communication could be used. Sustainable agriculture was legally defined in U.S. Code Title 7, Section 3103 (US Government Publishing Office, 2006, p. 1406) as "an integrated system of plant and animal production practices having a site-specific application that will over the long term:

- Satisfy human food and fiber needs
- Enhance environmental quality and the natural resource base upon which the agricultural economy depends
- Make the most efficient use of non-renewable resources and on-farm resources and integrate, where appropriate, natural biological cycles and controls
- Sustain the economic viability of farm operations
- Enhance the quality of life for farmers and society as a whole."

In summary, the basic goals of sustainable agriculture are economic profitability, environmental preservation, and social responsibility, each of which has intrinsic or inherent value (Fig. 1.7). Although the terms are valid, it is unclear what approach should be used to measure and then determine whether a farm is sustainable. As well as serving as the major source of feed nutrients for wild and domestic animals, grassland agriculture provides many complementary benefits for the environment and helps to gain social acceptance of farming practices. In addition to food, it is well known that forages and grasslands contribute to human well-being by:

- Providing raw materials for clothing and other textiles
- Soil erosion control

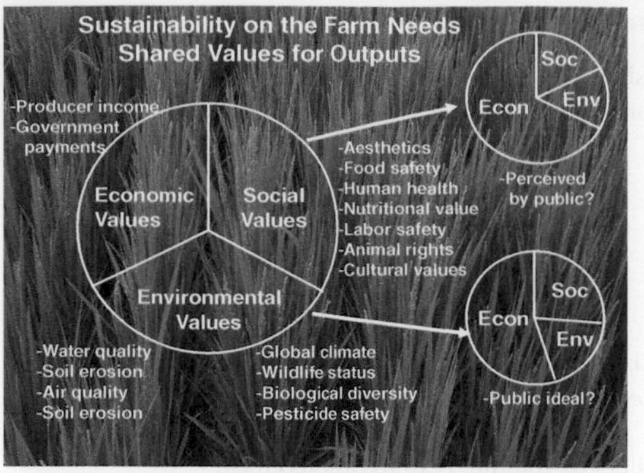

FIG. 1.7. Sustainable agriculture involves economic return to the farmer and environmental preservation that are achieved in ways which are socially acceptable. The left portion gives each component equal emphasis, but that probably does not reflect reality. The public believes that farmers are less concerned about environmental and social values than they are about economic values. However, the ideal is site specific and unknown. (Adapted from Nelson, 2007.)

- Improvement of soil structure and fertility of arable lands
- Water conservation and protection
- Providing habitat for biologically diverse plants and wildlife
- Protecting the environment from pollutants such as sediment, windblown soil, municipal and farm wastes, and some toxic substances
- Providing a source of outdoor recreation and pleasure
- Providing biomass for conversion to energy as a renewable resource
- Providing feedstocks for manufacturing products.

Economic return is relatively easy to define in monetary units that are used for farm management. Everyone wants clean water, but each person will differ with regard to what they perceive to be its "value." Everyone wants farmers to have a clean and safe working environment and to treat animals humanely, but how are those factors described and valued? Environmental factors are difficult to quantify and value. Social values differ markedly among people, and vary depending on culture, age, and economic status. It is easy to envision how they differ among rural and urban residents. The conclusions drawn by activists and others about sustainability are frequently based on valid concerns (e.g., animal rights), but often are not defined or supported by scientific research. This leads to mistrust and conflict.

In the USA, farm and ranch sizes are increasing due to mechanization and the known increase in efficiency associated with size. This often develops into a "business style" in which the owner or manager makes the decision while the employees do the work. The farms are usually focused on production of one or only a few commodities on large fields of monocultures with little diversity. This contrasts markedly with the smaller "family farm", where there is diversity of crops and animals on each farm, use of pastures and hayfields in rotations, and labor and decision making undertaken by the farmer and family members. Many researchers in social and agricultural sciences are investigating the values of the various environmental and social factors involved in order to compare the various scenarios with regard to sustainability.

Sustainability of agriculture tends to be addressed differently by individuals and their governments. Countries with high incomes, such as Germany, Switzerland, and other European countries, restrict the amalgamation of land and subsidize the high proportion of small landowners (farmers) in order to maintain a good income from diversity of crops and livestock on a small land area. Conversely, Canada, Australia, and Brazil, which have very large farms, have policies more similar to those of the USA. Other countries, such as some African and Asian countries, have low personal incomes, and agriculture production is prioritized in the short term to feed the people even though the conditions are not sustainable. Gradually as incomes increase there is an increased public emphasis on environmental factors. As incomes rise further, other sustainability concerns are added sequentially, such as food safety, followed by taste and nutritional value, and eventually animal rights and biodiversity, including wildlife. Thus the concept of sustainability changes as a function of both measurable outcomes and individual perspectives.

As the world population continues to grow from today to more than 9 billion by 2050, there is a need for an increase in food production of about 70%. Many areas of the world are arid, and for other reasons related to soil or climate are not suitable for crop production, but can support ruminant livestock for meat, milk, and animal fiber production. A sound national grassland philosophy must be goal oriented and supported by government policies. As any nation becomes self-sufficient in diet, first in plant-based products and then in animal-based products, the public's demands for protection of the environment, for landscapes with enhanced aesthetic value, and for easy access for leisure and recreation will increase. Meeting these needs for sustainability requires producers to develop management practices that emphasize and balance the multiple uses of the grassland resource.

Adjusting to Climate Change

It is clear that climate change is related to the total increase in greenhouse gases such as carbon dioxide (CO_2) (77% of the total), methane (CH_4) (14% of the total), and nitrous oxide (N_2O) (8% of the total). About 14% of the total emission worldwide is from all aspects of agriculture and 17% is from deforestation. Forages and pastures make some contribution to the 14% from agriculture, but much less per acre than crops. Even so, they should be part of the solution.

The atmospheric CO_2 concentration increased rapidly from about 290 parts per million (ppm) in 1900 to nearly 400 ppm in 2015, largely as a result of the burning of fossil fuels (57%) and the reduction of forests (17%) that normally use CO_2 and store carbon as wood. The atmospheric concentration of CH_4 from ruminant animals and wetlands also increased rapidly from about 800 parts per billion (ppb) in 1900 to more than 2000 ppb in 2015, and the atmospheric concentration of N_2O, lost mainly from chemical fertilizers, increased from about 1000 ppb in 1900 to around 1300 ppb in 2015. Although CO_2 receives the most attention, a molecule of CH_4 is about 25 times stronger and a molecule of N_2O is about 300 times stronger as a greenhouse gas than a molecule of CO_2. Methane is probably the major factor for forages and pastures, as N_2O levels are decreased when legumes are used, and CH_4 levels can be reduced if ruminants have higher-quality diets.

Greenhouse gases are expected to cause an increase in air temperature, mainly night temperature, of 3°F (2°C) near the equator and up to 9°F (5°C) nearer the poles. Projections are that the growing season will be longer and that annual precipitation will be similar to current levels, but more precipitation will occur as severe storms with longer dry periods between storms. It is expected that there will be more droughts, more floods, and more inundation of shorelines with the melting of polar ice. Thus plants and animals will be exposed to more heat and drought stress, and will face more challenges with regard to insects (and pollinators) and diseases. The timing of field operations and effective storage and preservation of quality products will be more critical.

Management of pastures, livestock, and hayland production will need to adjust to the changes in both the short and long term. Seeding can be done using minimal tillage and less use of fossil fuels. Cultivars with improved drought and flood tolerance will help during drought and storms. Using appropriate mixtures of grasses and legumes will reduce erosion and will cause less N_2O release, as legumes fix nitrogen. Ruminants should have high-quality diets that reduce methane production in the rumen. The grazing season will need to be extended to reduce dependence on harvested and stored forage, but requirements for shade and clean water for livestock in the pastures will increase. With incentives, some managers will focus on increasing **soil organic matter** content as a way to sequester CO_2 to reduce atmospheric concentration, especially with long-term rotations and permanent grasslands.

In the longer term, water will be more restricted and this will eventually reduce irrigation of cropland and forages such as alfalfa. Large aquifers such as the Ogallala that extend underground from South Dakota to Texas are gradually decreasing due to crop irrigation, and that land area may revert back to grasslands to restore soil carbon, and provide feed for ruminant livestock and biofuel for energy. More beef and milk production will be based on high-quality pastures to reduce methane production from livestock and fossil fuel use for harvest and preservation. Marginal land sites for crops will be shifted to forages to prevent erosion, improve water quality, and provide wildlife habitat. There will be economic incentives to enhance the organic matter content of the soil, and use of minimal tillage to save energy and reduce losses of greenhouse gases from the soil.

Grasslands and Energy Issues

Grasslands can contribute to energy needs. Switchgrass has demonstrated potential for producing large amounts of plant **biomass**, a raw agricultural commodity of above-ground growth that can be processed into a solid or liquid fuel and other organic feedstocks to offset the use of fossil fuels. Biomass crops are a source of renewable energy, as they recycle CO_2 from the environment to offset that produced by burning the fuel. Most of these biofuel crops are efficient, as they are perennial and provide ground cover to reduce soil loss. Many industries are restricted with regard to the amounts of gaseous emissions that they can release, and know their "carbon footprint", which is the net amount of CO_2 emitted. Industries that produce high levels of CO_2 can offset these emissions by contracting with landowners who manage crops and grassland to sequester the CO_2 to gain the benefits of additional carbon incorporated into soil organic matter. Due to their perennial nature, forage-based systems can be managed for livestock production and increased carbon sequestration in the soil (Council for Agricultural Science and Technology, 2000).

The Need for Knowledge-Based Management

Regardless of the situations that develop, there will be a continued need for forages and grasslands, since in many ways this form of land use is the best alternative. However, the management of the soil–land–animal resources will continue to be economically productive while conserving the environment, in both cases in a socially acceptable way. New technologies such as drones, genetically modified plants and animals, and improved pesticides will be available. However, the fundamental decisions will require a strong basic understanding of the principles of soils, plants, ruminant livestock, and economics, and that is the main purpose of this book.

As time passes, the value of forages and grassland resources to the world and to national and individual well-being will be defined and redirected based on new technologies and alterations in human needs and expectations as changes occur in the physical and social climate. The future of forages and grasslands will be knowledge based and will require continued changes in management and attitudes. Our goal for this book is to provide a technical foundation for sound and rational decision making both now and in the future.

Summary

Since the beginning of sedentary agriculture, forages and grazinglands have contributed to the food supply for domesticated animals and wildlife that are used for human diets or serve as draft animals. In addition to supplying meat, milk, and power, these animals contribute manures and nitrogen to improve the soil resource for annual grain crops in rotations. The early Northern European immigrants to the eastern areas used fenced pastures and harvest strategies for feed supplies over winter. The western areas, influenced by the Spanish, used large herds of animals owned by several ranchers who moved the cattle and sheep over large land areas.

Forage and pasture contribute about 61% of the diet for dairy cattle, 83% for beef cattle, and 91% for sheep and goats, being supplemented by grains to increase production. Forages improve soil properties, maintain water quality, and protect the environment and wildlife. Future emphasis will be on providing environmental services together with use as biofuels. Populations in less developed countries will continue to demand more animal protein in their diets. However, since crop plants will compete for the best soils, the animal products will be produced on lower-productivity soils and with lower inputs. New technologies and forage species are needed to manage forages and livestock to effectively support the demand for animal products and conservation of resources.

Questions

1. Discuss some early forage practices in Europe, including Great Britain, that were transferred to America by the colonists.
2. Discuss the role of native grasslands in the development of the western USA, and its implications for modern grassland agriculture.
3. Discuss three primary factors that have contributed to the prominence of introduced forage species in the eastern area of the USA.
4. What constitutes sustainability of a pasture? Would the answer be the same for a nearby hayfield?
5. Explain the importance of the soil–plant–animal biological system and its significance to sustainability.
6. Explain how global change will affect forage and grasslands in your state.
7. What is the role of forages as an overall source of feed for livestock?
8. List and discuss at least five trends related to forage and pasture production that have occurred, that are occurring, or that will occur.
9. Why are forages undervalued in the overall food system?
10. Why is the Latin name important when communicating about forage plants?

References

Allen, VG, C Batello, EJ Berretta, J Hodgson, M Kothmann, X Li, J McIvor, J Milne, C Morris, A Peeters, and M Sanderson. 2011. An international terminology for grazing lands and grazing animals. Grass Forage Sci. 66: 2–28.

Barnes, CP. 1948. Environment of natural grassland, pp. 45–49. In A Stefferud (ed.) Grass: The Yearbook of Agriculture, 1948. US Government Printing Office, Washington, DC.

Council for Agricultural Science and Technology. 1980. Forages: Resources for the Future. Rep. 108. Council for Agricultural Science and Technology, Ames, IA.

Council for Agricultural Science and Technology. 2000. Storing Carbon in Agricultural Soils to Help Mitigate Global Warming. Issue Paper 14. Council for Agricultural Science and Technology, Ames, IA.

Edwards, EE. 1940. American agriculture—the first 300 years. In Farmers in a Changing World: The Yearbook of Agriculture, pp. 171–276. US Government Printing Office, Washington, DC.

Hardin, G. 1968. The tragedy of the commons. Science 162:1243–1248.

Harlan, JR. 1975. Crops and Man. American Society of Agronomy, Madison, WI.

Heath, ME, and CJ Kaiser. 1985. Forages in a changing world. In ME Heath, RF Barnes, and DS Metcalfe (eds.), Forages: The Science of Grassland Agriculture, 4th ed., pp. 3–11. Iowa State University Press, Ames, IA.

Hodgson, HJ. 1974. Importance of forages to livestock production. In HB Sprague (ed.), Grasslands of the United States, pp. 43–56. Iowa State University Press, Ames, IA.

Nelson, CJ. 2007. Sustainability of agriculture: issues, observations and outlook, pp. 1–24. In MS Kang (ed.), Agricultural and Environmental Sustainability: Considerations for the Future. Hayworth Press, New York.

Stewart, G. 1936. History of range use, pp. 119–133. In The Western Range. Senate Document 199, 74th Congress.

US Department of Agriculture. 1999. Agricultural Statistics. US Government Printing Office, Washington, DC.

US Department of Agriculture. 2016. Agricultural Statistics. US Government Printing Office, Washington, DC.

US Government Publishing Office. 2006. Title 7, Ch. 6, United States Code-2006 Edition. US Government Printing Office, Washington, DC.

Vallentine, JE. 2001. Grazing Management. Academic Press, San Diego, CA.

Vavra, M, WA Laycock, and RD Pieper. 1993. Ecological Implications of Livestock Herbivory in the West. Society of Range Management, Denver, CO.

CHAPTER

2

Structure and Morphology of Grasses

Kenneth J. Moore and C. Jerry Nelson

Grasses are grouped into 650 to 785 genera containing around 10,000 species; about 170 genera and 1400 species grow in the USA (Watson and Dallwitz, 1992). The grass family has the most species with the C_4 photosynthetic pathway, which is characteristic of warm-season grasses. The C_3 pathway is characteristic of cool-season grasses. As a result of having both photosynthesis pathways, the grass family has a wider range of adaptation to temperature and rainfall than any other plant family, with members being found in the humid tropics, subpolar regions, arid areas, and at high elevations. All of the cereal crops and about 75% of the species used as forages are grasses. Grasses are the dominant vegetation on much of the world's **rangeland.**

Grasses are either **annuals** or **perennials**, almost all are **herbaceous**, and they vary widely in structure and growth habit. Those in the short-grass prairie, such as blue grama and buffalograss, rarely grow taller than 8 to 10 in., whereas bamboo may exceed 60 ft. Sugarcane, corn, and forage sorghum also represent the larger grasses. Warm-season perennial grasses such as big bluestem and indiangrass may exceed 6 ft even in temperate zones in the northern USA and Canada. Tropical forage grasses, such as napiergrass or elephantgrass in Florida, may be 9 ft tall.

Grasses generally are the backbone or core of successful forage management systems, but each species is unique in terms of growth habit and environmental adaptation. These differences are associated with their general **morphology** (growth habit and structure) and **physiology** (growth and metabolic processes). The objectives of this chapter are to introduce the structure of grasses, describe their growth processes, and show how the structure and growth habit of plants affect their adaptation to specific environments and management practices.

Seedling Development

Seedling development is a critical factor in forages because an adequate population or stand of plants needs to be established in order to be productive. Despite the wide range of adaptation among grasses, the processes of seed germination and seedling development are similar for all grasses. The seed dispersal unit of grasses is a **caryopsis**. It is enclosed by a pericarp rather than a true seed coat. Since the pericarp develops from the ovary wall it is considered to be a single-seeded fruit, but is usually referred to as a seed despite its botanical description. The **embryo** of a grass seed consists of an **embryo axis** and a single **cotyledon** called the **scutellum** (Fig. 2.1). The scutellum secretes enzymes to digest the starchy **endosperm**, which provides energy for the germinating and emerging seedling. The embryo axis is connected to the scutellum at the **scutellar node**, which by definition is the first node. The embryo axis develops into the seedling. The endosperm in the caryopsis of most forage grasses is relatively small, so there is little stored energy. Shallow planting is critically important as it allows grass seedlings to emerge and develop quickly before the reserves are used up and growth must depend on photosynthesis. At this stage the plant becomes an **autotroph** and is able to obtain or synthesize the nutrients that it needs for further growth.

Germination and Emergence

Following imbibition of water, the first sign of germination is the emergence of the **radicle** (primary root) that grows through the **coleorhiza**, a protective sheath for the root tip (Fig. 2.2). Early in seedling development up to five **seminal roots** may emerge from the shoot area just above the scutellar node. The primary and seminal root systems

Forages: An Introduction to Grassland Agriculture, Seventh Edition. Edited by Michael Collins, C. Jerry Nelson, Kenneth J. Moore and Robert F Barnes.

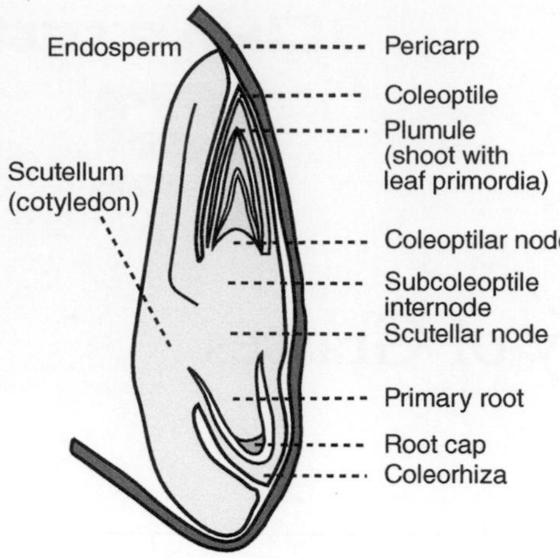

FIG. 2.1. A longitudinal section through a grass embryo before germination, showing various embryonic structures. The embryo consists of the scutellum and the embryo axis, which includes the shoot apex, embryonic leaves, and root tissues. The embryo axis is the growing part of the embryo. The endosperm, to the left of the scutellum, is several times larger in volume than the embryo axis. The line in the scutellum represents the vascular strand. (Adapted from Chapman, 1996. Reproduced with permission of CAB International, Wallingford, UK.)

are very important early on for water uptake and nutrient absorption. After the radicle emerges, the epicotyl elongates to emerge from the seed and grow toward the soil surface. The pointed **coleoptile**, which is a tubular sheath attached to the second node, protects the tender **plumule** as it elongates through the soil (Moser, 2000). Coleoptile growth ceases when it reaches light at the soil surface and the tip spreads open to allow the true leaves to emerge.

Development of the Root System

The primary and seminal root systems of seedlings are short-lived, so a longer-term **adventitious root** system must develop for seedlings to become fully established and grow rapidly. The adventitious roots form at the coleoptilar node and at nodes immediately above it (Fig. 2.2).

Hoshikawa (1969) found that most cool-season grasses have **festucoid** and most warm-season grasses have **panicoid** seedling development. The coleoptile of festucoid seedlings elongates from near the seed to reach the soil surface. However, in panicoid seedlings the subcoleoptilar internode, which is between the scutellar and **coleoptilar nodes** (Fig. 2.1) and is sometimes called a **mesocotyl**, also elongates to push the coleoptilar node and the short coleoptile upward to the soil surface (Fig. 2.3). Thus the coleoptile still protects the emerging shoot, but it reaches the surface largely due to elongation of the subcoleoptile internode. This method of emergence places the rooting sites, located at the coleoptilar node and higher nodes, of warm-season grasses very close to the soil surface.

Cool-season grasses generally are easier to establish under sporadic rainfall conditions than warm-season grasses. Adventitious roots rarely grow into dry soil. For example, blue grama, a panicoid-type grass adapted to the Great Plains, emerges quickly but needs additional rains to keep the soil surface moist for 2 to 4 days between 2 and 8 weeks after germination so that adventitious roots can develop (Wilson and Briske, 1979). If the surface remains dry, seedlings die when they are 6 to 10 weeks old when the temporary root system is no longer functional (Hyder et al., 1971). In contrast, crested wheatgrass, which has festucoid seedling development, establishes under these conditions because the coleoptilar node remains at the seeding depth, further below the soil surface, where the improved water status facilitates adventitious root development (Table 2.1). Once initiated on either festucoid or panicoid species, the adventitious roots grow downward rapidly to provide water and minerals, and the seedling is considered to be established (Fig. 2.2).

Regardless of the emergence pattern, the **shoot apex** or **terminal meristem**, located at the tip of the stem in the seed (Fig. 2.1) and the seedling (Fig. 2.3), remains below the soil surface. At this location the apex, which is protected by the soil from strong changes in air temperature, functions mainly to initiate new nodes, leaves, and **axillary buds** for tiller, rhizome, or stolon production.

Plant Structures and Development

Despite having some commonality as young seedlings, growing plants develop modifications that allow each species to be adapted to specific environmental conditions and management practices. For example, grasses that **tiller** actively, producing large amounts of leaf area near the soil level, can tolerate closer and more frequent grazing than species that produce fewer tillers and have leaf area displayed further above the soil surface. Differences in **morphology** also aid plant identification.

Leaves

Leaves are borne on the stem, one at each node, but are projected alternately in two rows on opposite sides of the stem (Fig. 2.4). Each leaf consists of a **sheath, blade, ligule** (which includes the **collar**), and in some cases **auricles** (Fig. 2.5A). The sheath surrounds the stem above the node where it is attached. Its margins usually are overlapping (open), although in some species they are united (closed) into a cylinder for a part or all of the distance between the node and the blade (Fig. 2.5B). The sheath is green and

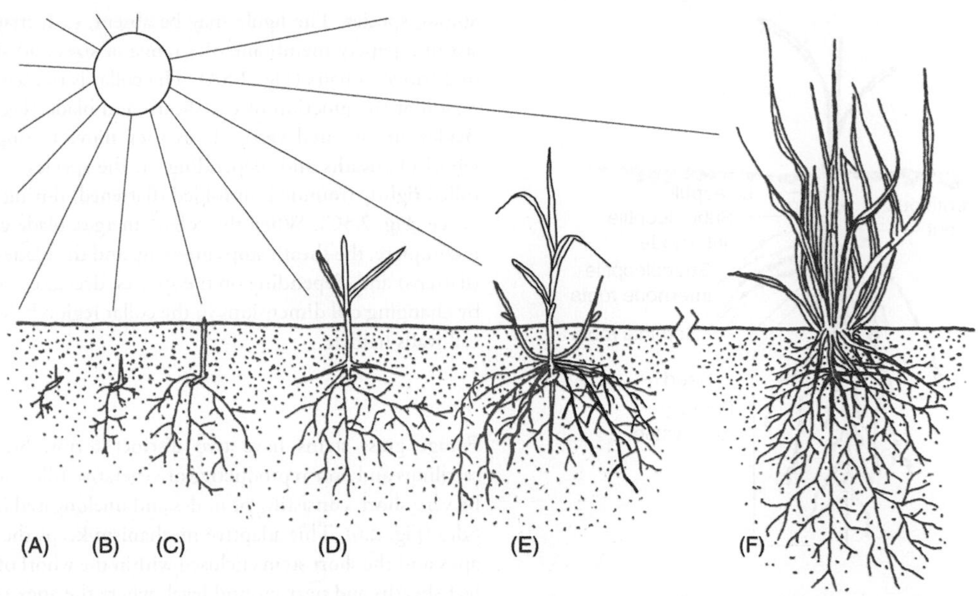

Fig. 2.2. The establishment process for a festucoid grass. (A) After the seed has imbibed water, the radicle (seedling root) and epicotyl (seedling shoot) emerge, indicating successful germination. (B) Mainly coleoptile elongation pushes the shoot tip toward the soil surface; the seminal root system appears. (C) Elongation of the coleoptile ceases as the tip reaches light at the soil surface; the shoot (leaf blade) extends past the coleoptile and the seminal root system continues to expand. (D) The seedling shoot develops more leaves and adventitious roots develop, initially at the coleoptilar node. (E) The main shoot continues to add leaves, new tillers develop from basal nodes, many adventitious roots develop from basal nodes, and the seminal root system begins to deteriorate; the grass plant is established at this point. (F) Continued development results in a fully established grass plant with numerous tillers, a dominant adventitious root system, and the potential to produce rhizomes or stolons.

photosynthetic, but it functions mainly to physically support the blade, transport materials between the blade and the stem, and store food reserves.

Factories Making Factories

When a factory becomes inefficient and needs to be replaced, rarely is it the job of the old factory to build the new one, but that is the case with grasses. Leaves are the factories—the photosynthetic organs that capture the sunlight and carbon dioxide (raw materials) to make the sugars that support growth of the leaf and other parts of the plant (factory output). At the same time, leaves are harvested as pasture or hay.

Managers decide when the leaf (factory) has functioned for long enough and when it needs to be reduced or destroyed (the animal is allowed to graze the whole leaf, part of the leaf, or it is harvested). Managers also decide whether the entire factory is to be harvested or grazed, or whether only the upper stories should be removed.

The leaf area remaining after harvest (the lower stories) and stored energy (in the basement) help to rebuild new leaf area (the efficient new factory) from new tillers or the shoot apices of existing vegetative tillers that are near the soil surface and have not been removed. The result is an efficient new photosynthetic factory.

Net increase in growth of the plant (output from the factory) occurs if the leaves are allowed to produce photosynthate (sugars) for sufficient time to rebuild stored energy and rebuild leaf area.

Can you describe why frequent and close defoliation reduces the vigor of the grass plant?

The leaf blades are parallel veined and typically flat and narrow. Some species have auricles (small ear-like appendages) projecting from the blade near the collar area at the junction of the sheath and the blade (Fig. 2.5A). The ligule, an appendage located where the sheath and blade join, aids plant identification as its presence and form vary

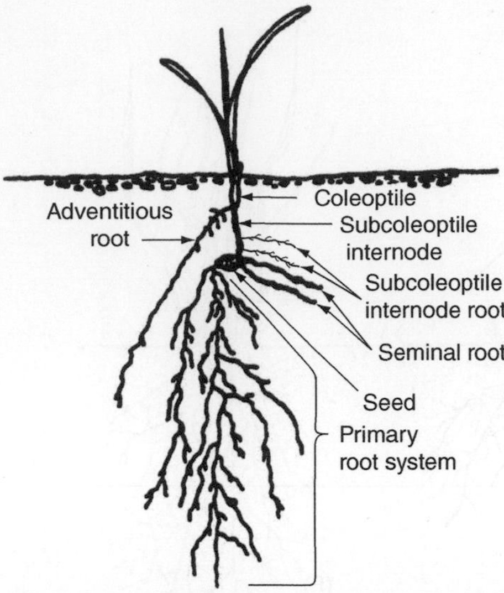

FIG. 2.3. Schematic diagram of seedling of a panicoid grass plant, showing the position and length of the coleoptile, subcoleoptile internode, and primary, seminal, and adventitious root systems. The primary and seminal roots arising from the seed are short-lived, leaving the plant totally dependent on the adventitious roots that arise from the nodes of the shoot axis near the soil level. Festucoid grasses develop similarly (see Fig. 2.2), but they have a long coleoptile and a short or nonexistent subcoleoptilar internode. (Adapted from Newman and Moser 1988. Reproduced with permission of Cop Science Society of America.)

among species. The ligule may be absent, or it may consist of a papery membrane of a range of sizes and shapes, or a fringe of hairs (Fig. 2.5A). The collar is the hardened region at the junction of the sheath and blade (Fig. 2.4). Blades are oriented vertically as they move through the whorl of sheaths and, depending on the species, may be rolled tightly (rounded) or folded (flattened) during emergence (Fig. 2.5C). When the collar emerges, blade growth is complete, the sheath stops growing, and the blade opens (flattens) and, depending on the species, decreases in angle by changing cell dimensions in the collar region before the cell walls harden.

Stems

Forage grass stems have two distinct forms. Stems of seedlings and non-reproductive or vegetative tillers tend to be very short, consisting of nodes and unelongated internodes (Fig. 2.6). This adaptive mechanism keeps the shoot apex and the short stem enclosed within the whorl of older leaf sheaths and near ground level, where the apex escapes removal by grazing or cutting. When there is a flowering stimulus, usually caused by daylength or temperature conditions, the shoot apex differentiates into the reproductive structure **(inflorescence)**. At nearly the same time, the internodes below it elongate as a result of the activity of **intercalary meristems** located at the base of the internodes. This separates the nodes and pushes the inflorescence upward through the whorl of sheaths, eventually exposing it at the top of the plant (Fig. 2.4). It is usually only the upper internodes that elongate, while the lower ones remain unelongated at the base of the plant.

Elongated stems **(culms)** of a flowering grass plant are divided into distinct nodes and internodes (Fig. 2.4). The

Table 2.1. Morphological features of seedlings 22 days after planting at three depths in a greenhouse

Species and feature	Planting depth (in.) 0.8	1.6	2.4
Blue grama			
Emergence (%)	100	58	0
Coleoptile length (in.)[a]	0.2	0.3	–
Subcoleoptile internode length (in.)[a]	0.7	1.2	–
Depth to adventitious roots (in.)	0.1–0.2	0.3–0.4	
Crested wheatgrass			
Emergence (%)	91	88	20
Coleoptile length (in.)[a]	0.9	1.3	1.7
Subcoleoptile internode length (in.)[a]	0	0	0
Depth to adventitious roots (in.)	0.8	1.0–1.6	1.9–2.4

Source: Adapted from Hyder et al., 1971.

[a]The combined length of the coleoptile and the subcoleoptile internode represents the maximum depth of protected emergence.

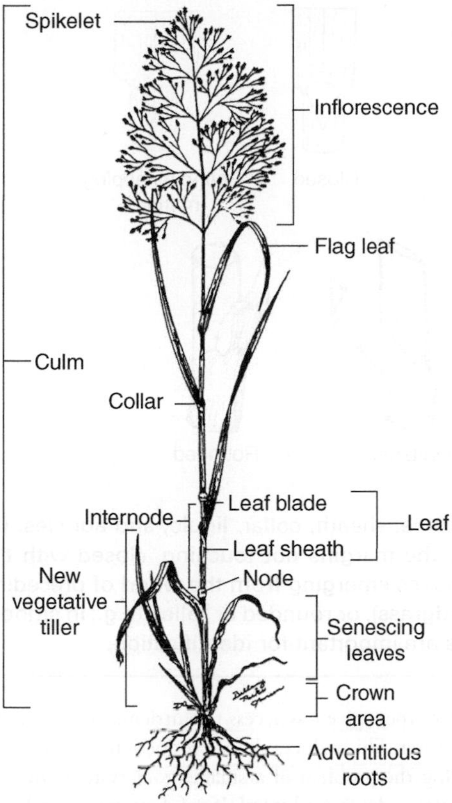

FIG. 2.4. Schematic diagram of a generalized grass plant showing a reproductive tiller (culm), a vegetative tiller (internodes not elongated, arising from the left side of the crown area), the leaf parts, inflorescence, crown, and the upper part of the adventitious root system. (Drawn by Bellamy Parks Jansen; adapted from Stubbendieck et al., 1997.)

internodes of forage grasses are usually hollow, but may be pithy and solid as, for example, in big bluestem and forage sorghum. The node or joint is always solid. Leaves have vascular connections between the sheath and stem at the nodes. During vegetative regrowth in summer, the internodes of new tillers in reed canarygrass, smooth bromegrass, and some other grasses often elongate slightly, elevating the shoot apex to 4–8 in. above soil level, even though the apex remains vegetative and continues to initiate new leaves. Close grazing in summer may remove the apex of these species, but both have **rhizomes** (Fig. 2.7A), which aid regrowth and persistence.

The vertical stems of many perennial grasses have thickened lower internodes that form a **crown** where storage carbohydrates and proteins accumulate. Axillary buds on lower nodes in this crown area develop into new tillers, rhizomes, or stolons. The combination of an energy source (storage in the lower stem) and active **meristems** (axillary buds) near ground level allows plants to persist through the winter or dormant seasons, and allows regrowth to commence under suitable environmental conditions. In most grasses, axillary buds at the crown develop into tillers, although in some grasses (e.g., big bluestem and reed canarygrass) axillary buds located in leaf axils higher on the culm give rise to aerial tillers.

Brace roots and other **adventitious roots** initiated just above soil level arise from the base of the intercalary meristem, a zone of cell division and elongation just above the node that is responsible for stem internode elongation. Cells of the stem intercalary meristem remain meristematic, capable of division and elongation, until the stem has matured. This allows lodged stems to turn upward again because cell growth of the intercalary meristem is more rapid on the lower side.

In addition to tillers and reproductive stems or culms, grasses such as quackgrass, johnsongrass, kentucky bluegrass, switchgrass, reed canarygrass, and many others have rhizomes (Fig. 2.7A). Rhizomes are stems that grow horizontally below ground. They originate from an axillary bud on the crown, generally below ground level, are often sharply pointed, and grow laterally underground. Rhizomes have an axillary bud located at each node behind a scale-like leaf **(cataphyll)**. Rhizomes often contain a large reserve of stored foods that support growth of axillary buds into new tillers, and are usually the overwintering part of perennial rhizomatous grasses. Rhizomes allow thin stands to develop into a dense sod, but this character also makes plants such as quackgrass and johnsongrass weed-like and difficult to control. Some grasses, such as tall fescue, have only short rhizomes, which give the canopy a loose, bunch-like appearance. Grasses without rhizomes (e.g., orchardgrass and little bluestem) have little lateral spread and form tight bunches.

Stolons are creeping, lateral stems located above ground (Fig. 2.7B). They originate from axillary buds near soil level and, like rhizomes, have identifiable nodes and internodes. Stolons may form nodal roots and produce an axillary bud at each node from which new shoots or stolon branches can arise. As with rhizomes, grasses propagate vegetatively by means of stolons and form a sod. Bermudagrass and buffalograss are examples of stoloniferous species.

Roots

Established grasses have an adventitious, fibrous root system. Each new tiller develops adventitious roots from lower nodes shortly after it has emerged above ground and developed leaf area. Thus each tiller begins like a new seedling because it depends on another root system (the mother tiller) temporarily until it has developed its own adventitious root system. After developing leaf area and its

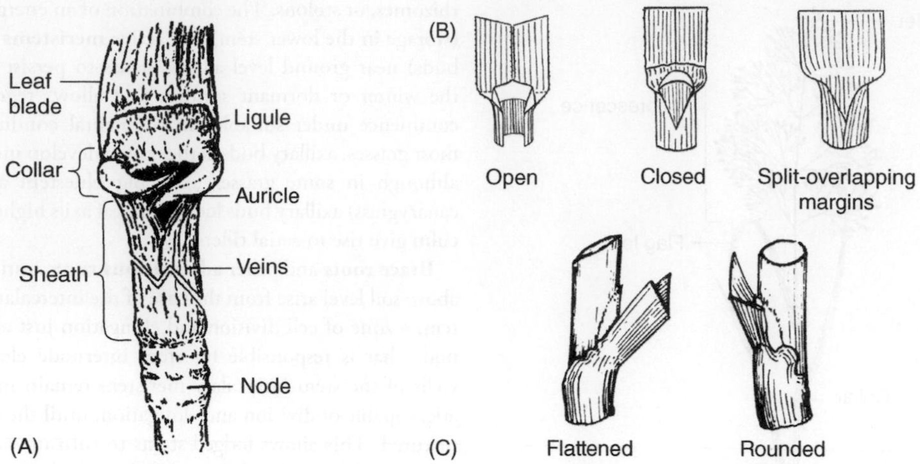

FIG. 2.5. (A) Grass leaf morphology showing the leaf blade, sheath, collar, ligule, and auricles. (B) Sheaths of specific grasses are characteristically open with the margins not touching, closed with the margins fused, or split with overlapping margins. (C) New leaves emerging from the whorl of preceding leaves are characteristically flattened or folded (e.g., in orchardgrass), or rounded or rolled (e.g., in smooth bromegrass and tall fescue). These leaf morphology features are important for identification.

own root system, the daughter tiller is largely independent, although still attached to the mother tiller.

The root system of grasses is heavily branched, especially in the upper soil horizons, making it well adapted for using intermittent rainfall, holding soil particles together to aid soil conservation, and taking up top-dressed fertilizers. Depth of rooting is favored in soils that have low physical strength and good aeration. Deeper rooting enables the plants to extract water from a larger soil volume, which aids drought resistance. Species also differ with regard to the depth and distribution of roots in the same soils. For example, smooth bromegrass roots deeper than orchardgrass, and orchardgrass roots deeper than kentucky bluegrass and timothy; not surprisingly, these species rank in the same order for drought resistance.

Stoloniferous plants such as bermudagrass and bahiagrass form adventitious roots at nodes of the stolons (Fig. 2.7B). These roots anchor the stolon and take up water and nutrients for that part of the plant. Likewise, roots that form at rhizome nodes help the plant to obtain water and nutrients (Fig. 2.7A).

Warm-season grasses, especially the tall-growing prairie species such as big bluestem and switchgrass, generally have fewer roots, but they are larger in diameter and grow deeper into the soil than the roots of cool-season grasses (Weaver, 1926). The added depth is important in relation to the adaptation of these grasses to drier regions because the soil volume occupied by the root system is high, giving it access to more soil water.

Close and frequent defoliation reduces both shoot and root development (Fig. 2.8) because there is less leaf area to produce the carbohydrate necessary for root production. Shallower roots give less access to nutrients and especially to soil water. This reduces shoot production, further compounding the problem of a smaller root system, and the grass enters a downward spiral. If defoliation is not relaxed, the plants weaken and may eventually die. This principle affects both cool- and warm-season grasses (Dawson et al., 2000). For example, reducing the frequency or delaying cutting of sand bluestem, a warm-season species, led to higher levels of carbohydrate in the roots, a larger root system, and higher forage yield (Table 2.2).

Inflorescence

The grass inflorescence (Fig. 2.4) consists of a group or cluster of **spikelets**, which are the basic reproductive unit. Spikelet characteristics and the organization of the inflorescence offer convenient traits for identifying grasses (Fig. 2.9A, B, C).

The **spike** inflorescence characteristic of wheat, western wheatgrass, and perennial ryegrass has a strong central **rachis.** Spikelets are **sessile** (i.e., they are attached directly to the rachis) (Fig. 2.9A).

The **raceme** differs from the spike in that spikelets are connected to the rachis by short stalks called pedicels (Fig. 2.9B). The simple raceme is the least common inflorescence type in grasses, yet there are numerous modifications. For example, the raceme of big bluestem has a sessile fertile spikelet positioned with a sterile spikelet that is on a pedicel. Crabgrass has a cluster of (digitate) racemes.

The panicle, the most common grass inflorescence, is branched and has pedicelled spikelets (Fig. 2.9C). Panicles may be open (highly branched) and diffuse, as in smooth bromegrass, kentucky bluegrass, and switchgrass.

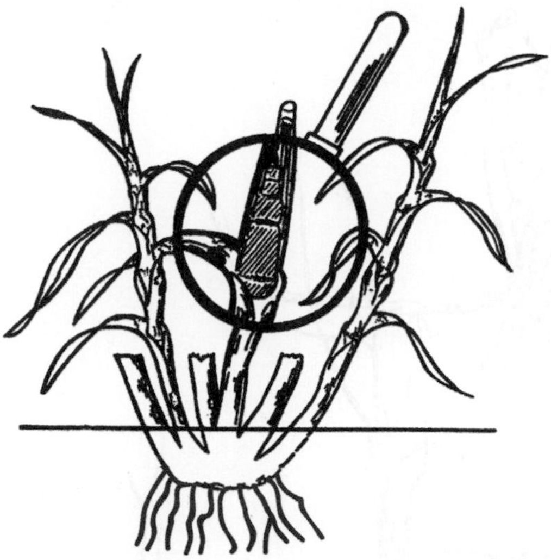

FIG. 2.6. The shoot apex (growing point) of a vegetative tiller remains near the soil level, where it is protected while producing leaves and unelongated internodes. Under suitable environmental conditions, the tiller becomes reproductive, and then the shoot apex differentiates into the inflorescence, and the stem internodes elongate to elevate the inflorescence to the top of the canopy. The magnified area shows the partially elongated internodes of the reproductive tiller and the developing inflorescence. Bulges on the tillers can be detected by palpation to estimate the height of the shoot apex region or inflorescence. (Adapted from Teel, 1957, with permission.)

Alternatively, panicles with short branches and pedicels can be compact, almost resembling spikes, as in timothy and pearlmillet.

The spikelet consists of the two subtending **glumes** or **bracts** and everything contained within. Depending on the species, spikelets (Fig. 2.9D, E) can have one to many **florets**, each consisting of a **lemma, palea**, and the floral parts (Fig. 2.9F). After fertilization, the **ovary** develops into a caryopsis. The two glumes are attached at the base of the spikelet, one on each side of the rachilla, the central axis of the spikelet. In species that have more than one floret per spikelet, a portion of the rachilla often remains attached to the floret as it dries and breaks off (dehisces). Glumes enclose the florets, providing photosynthate and protection to the seed.

Florets

Grasses may have one to several florets per spikelet (Fig. 2.9D, E). The larger (outer) bract of the floret is the lemma, which may have an awn attached; the smaller (inner) bract is the palea, which is usually partially enclosed by the lemma (Fig. 2.9G). The flowers are usually small and perfect—that is, they have a functional **pistil** and **stamens** (Fig. 2.9F)—although separate **staminate** and **pistillate** florets are characteristic of some species. **Monecious** species such as corn have separate staminate and pistillate florets on the same plant, while **diecious** species such as buffalograss have staminate and pistillate florets on different plants.

Grass flowers usually have three stamens. The single pistil has an ovary containing one **ovule** (Fig. 2.9F) and a **style** that commonly terminates in a feathery two-parted **stigma.** In many cross-pollinated grasses, two or sometimes three **lodicules** are located between the ovary and the lemma and palea at the base of the floret. These sac-like structures swell by absorbing water to help to force open the lemma and palea to facilitate wind **pollination.**

Most forage grasses are cross-pollinated by wind, but some are self-pollinated within the floret. Species such as kentucky bluegrass and buffelgrass reproduce by **apomixis,** an asexual form of seed set in the floret without union of male and female gametes. Apomixis produces a seed that is genetically identical to the mother plant, a characteristic that is very important in the development of new varieties of these species. Grasses that are self-pollinated or apomictic generally have less genetic diversity among plants within a cultivar than do cross-pollinated species.

Flowering

Many temperate grasses flower only once a year because they require a specific sequence of a short photoperiod (autumn), a cold period (winter), and long days (spring) for the shoot apices to differentiate and develop into the inflorescence. This sequence occurs naturally in temperate environments. Older tillers, those produced in spring and early summer, are stimulated by cold during winter to flower the following year. Tillers initiated late in the year are often not far enough advanced to be induced, and will remain vegetative through the next year. Thus these grasses flower only once per year. Grasses in more tropical regions and some grasses in temperate regions (e.g., timothy) may have a long-day requirement but have a weak or no cold temperature **(vernalization)** requirement, and can flower more than once during the growing season.

How Far Does Pollen Go?

Most grasses are cross-pollinated by wind, so pollen dispersal and capture are important. In order to obtain high-quality seed, producers want to avoid pollen from plants of the same species that are not in the seed field. (Would you want the neighbor's scrub bull running with your purebred cow herd?) So the seed producer "isolates" seed fields to avoid unwanted pollen. Pollen is shed by most grasses for a short period each day (often for only 4–5 hours) for about a week. Wind-blown pollen is buoyant but slightly heavier than air;

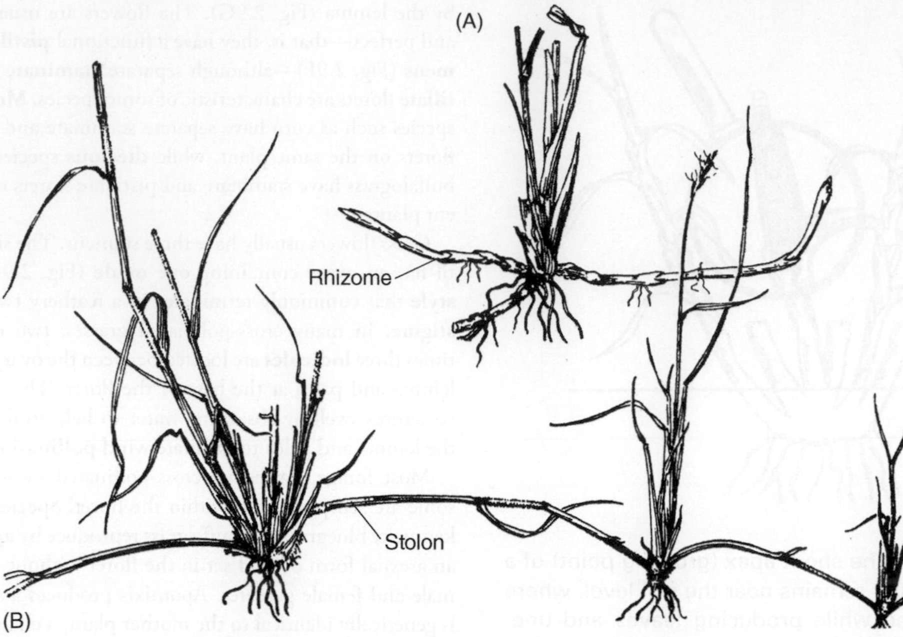

Fig. 2.7. Grasses may have underground horizontal stems called rhizomes (A) and/or above-ground horizontal stems called stolons (B) that help the plant to spread laterally for the purposes of vegetative propagation and sod formation. The internodes and nodes of these stems are often covered by incomplete leaves (in the case of stolons) or scale-like leaves (in the case of rhizomes). Axillary buds, which are formed at nodes behind these unique leaves, can lead to further branching or to the development of new plants. Note that both types of plants also form vertical tillers. (Adapted from Dayton, 1948.)

it is deposited downwind in the first 100–175 ft. Less than 10% of the pollen is dispersed further than 425 ft.

To avoid contamination, the Association of Official Seed Certifying Agencies recommends that Foundation Seed fields of cross-pollinated grasses be isolated 900 ft from the nearest pollen source for that particular grass. The distance is less for Registered and Certified seed fields, and for apomictic grasses.

How would pollen from another source affect the genetic quality of the seed? Why are the required distances less for apomictic grasses? How do receiving plants capture the pollen?

If flowering is stimulated, the shoot apex stops producing new leaves and differentiates into the initial cells to develop the inflorescence. A tiller that forms an inflorescence dies after the seed is produced, so a new tiller must be ready to replace it. This leads to two important considerations. First, the grass plant is perennial because it produces new tillers each year, yet each individual tiller lives for only 12–18 months because it will receive the flowering stimulus, flower, and die. Second, in spring there is a mixture of older reproductive tillers and younger vegetative tillers (often 40–60% of the total tillers) that do not flower and that will contribute to growth and regrowth through the season.

The root system for a reproductive tiller also dies, contributing organic matter and open channels in the soil to give the characteristic soil structure of grasslands. Grass plants are comprised of tillers that arise from the crown area and tend to grow vertically, or rhizomes or stolons that also arise from the crown but grow horizontally. The grass sward continues to grow after reproduction and perennates year after year through the initiation and growth of new tillers.

Fruit or Caryopsis

The seed unit of forage grasses is the **caryopsis** (Fig. 2.9G). The caryopsis is a specialized fruit because the seed coat consists of layers from the ripened ovary wall that are fused onto the outer wall of the single ovule. In most forage grasses the caryopsis remains enclosed by the lemma and palea (Fig. 2.9G), even during seed harvest and processing. The lemma and palea and the seed coat (ovary wall) protect the caryopsis against mechanical damage, moisture loss in storage, and attack by biological pests. The caryopsis

Fig. 2.8. Root development of grasses with (A) no defoliation, (B) moderate defoliation, and (C) close, continuous defoliation. Root development depends on photosynthate produced by the leaf area, and leaf area depends on water and nutrients (especially nitrogen) that are absorbed from the soil. (Adapted from Walton, Production and Management of Cultivated Forages, 1st Ed., © 1983. Reprinted by permission of Pearson Education, Inc., New York.)

consists largely of endosperm, which is the starchy tissue used to maintain and protect the embryo during storage and to provide energy for the seedling during germination and emergence.

Cloning—Nothing New for Grasses

Perennial grasses live year after year—or do they? We tend to think that perennial grasses live for many years, growing in the summer and going dormant in the winter, rather like a hibernating bear. However, the same living cells and tissue in grasses do not live year after year.

For a grass to perennate, it must produce new buds on the base of the plant. These buds form new tillers, rhizomes, or stolons that produce new leaves and stems, new roots, and more new buds, so essentially a perennial grass vegetatively develops new plants every year. Winter survival of crowns, rhizomes, and other critical plant tissue, followed by vegetative propagation, is what allows a grass to persist indefinitely. Although some tissue may live for a couple years, the grass plant keeps reproducing vegetatively—that is, cloning itself—which makes it perennial.

Can you figure out how other perennial plants, such as alfalfa or a shrub, live year after year? (Hint: Consider which parts of these plants are actually living tissue.)

Many grasses, including big bluestem and indiangrass, are difficult to seed using mechanical equipment because the lemma and palea are **pubescent** (hairy) and have appendages such as **awns**, rachis sections, and pedicels that make the harvested seed light and fluffy. Mechanical removal of the hair and appendages facilitates seeding, but the caryopsis is small and may be affected if the protective lemma and palea are damaged. To produce a vigorous

Table 2.2. Root development and weight of total non-structural carbohydrate (TNC, food reserve) in roots of sand bluestem plants defoliated at different times and frequencies (roots were sampled in October)

Month(s) of defoliation	Root weight	Root area	Root length	Weight of TNC in roots
	Percentage of October value			
June, July, Aug.	67 a*	58 a	57 a	66 a
June, Aug.	67 a	57 a	57 a	65 a
June only	86 c	81 c	84 c	87 c
July only	75 b	67 b	68 b	75 b
Aug. only	80 bc	72 b	70 b	75 b
Oct. only	100 d	100 d	100 d	100 d

Source: Adapted from Engel et al., 1998.
*Means within a column followed by the same letter are not significantly different ($P > 0.05$).

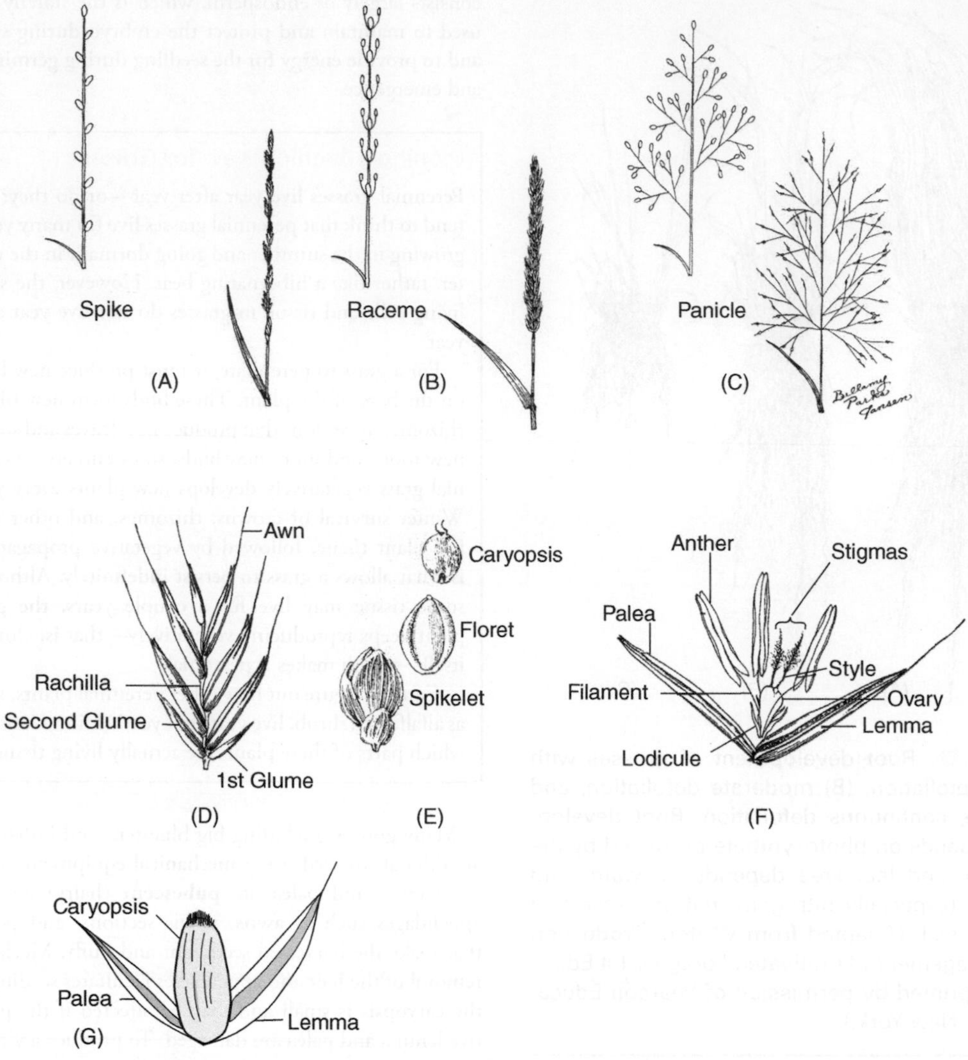

FIG. 2.9. Reproductive structures of grasses. Diagrammatic and actual inflorescence and the flag leaf (A, B, C) that is attached at the node below the inflorescence. (D) Spikelet with six florets arranged sequentially on the rachilla or central stalk. (E) Spikelet with only one floret, with glumes removed to show the floret, and with the lemma and palea removed to show a caryopsis. (F) Grass floret at anthesis showing the parts that are normally enclosed by the lemma and palea. (G) Mature floret (seed unit) showing the lemma, palea, and the caryopsis that develops from the ovary. (A, B, and C drawn by Bellamy Parks Jansen, adapted from Stubbendieck et al., 1997; D, E, and F adapted from Dayton, 1948.)

seedling, the caryopsis should be protected during germination and emergence.

Role of Meristems in Tiller Growth

Grass swards are composed of many plants, each of which is made up of interconnected tillers. Dry matter productivity of grasses depends on the product of tiller density (number per unit area) and the average weight per tiller. When vegetative stands of grasses such as tall fescue are thin (i.e., few tillers per unit area), production can be increased by nitrogen (N) fertilization (Fig. 2.11). When soil N is low, the first increment of N applied mainly stimulates tillering and helps to cover the soil more completely and capture more light. With higher increments of N, the growth response shifts to increased leaf length and tiller weight. Understanding the meristems responsible for tillering and growth per tiller, and how they can be managed, is of major importance in forage production.

FIG. 2.10. A cloud of pollen released from crested wheatgrass.

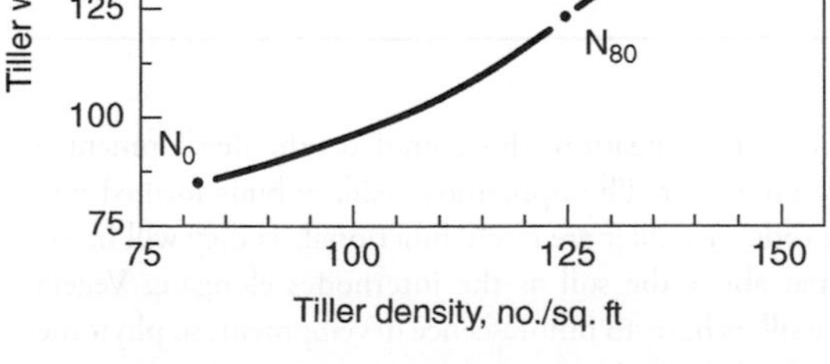

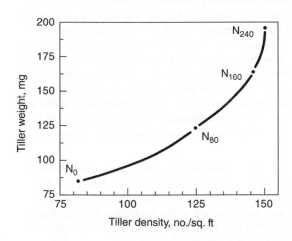

FIG. 2.11. Tiller weight of tall fescue in the field in response to annual N fertilization rates of 0–240 lb/A. Note that there is an increase in yield (product of tiller weight and tiller density) with increase in N application, but tiller density is increased most by lower rates until the maximum density is approached, and then tiller weight is increased most by higher rates of N. (Adapted from Nelson, 2000. Reproduced with permission of CAB International, Wallingford, UK.)

The basic repeating unit of growth of a grass tiller is the **phytomer.** Each phytomer consists of a node, an internode, a leaf sheath, a leaf blade, and an axillary bud (Moore and Moser, 1995) (Fig. 2.12). The node and axillary bud are at the lower part of the phytomer, and the sheath and blade are associated with the upper node. The internode is between the lower and upper nodes. Vegetative tillers have an active shoot apex that initiates the phytomer components, and there is essentially no internode elongation. Once induced to become reproductive, new phytomers cease to be produced, the shoot apex differentiates into an inflorescence, and the intercalary meristems elongate the stem (Fig. 2.12). When a reproductive tiller is fully expanded, it has no more meristematic areas, and has thus reached its capacity for growth. As the seed mature, the reproductive tiller dies.

The shoot apex initiates new leaves, lays down the cells to develop into axillary buds, and develops the cells for nodes. As such, it forms the phytomer in sequence and regulates the rate at which the grass plant grows or develops. The vegetative shoot apex has two to six short leaf **primordia** that are sequential in age. Examining the terminal region longitudinally (Fig. 2.13) allows one to envision the chronology of events for phytomer development. The apex shown in Fig. 2.13 has a fully developed leaf attached at node n and has just initiated the **flag leaf** primordium at node n+4. The tiller is transitioning from vegetative to

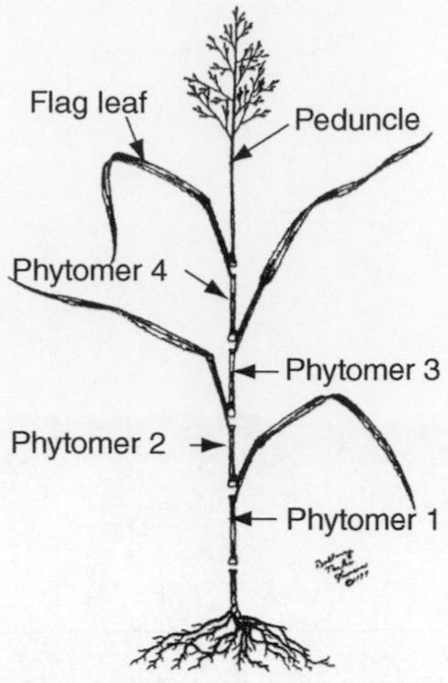

FIG. 2.12. Illustration of a grass plant showing a culm or reproductive tiller and its arrangement of phytomers. Each phytomer consists of a leaf blade, a leaf sheath, the node where the sheath is attached, the internode below, and the axillary bud. Axillary buds on upper nodes generally remain dormant. Tillers would form from axillary buds on lower phytomers produced much earlier during vegetative growth. Phytomer 1 was produced first. After the components of phytomer 4 were initiated, the shoot apex (terminal meristem) differentiated into the inflorescence. (Drawn by Bellamy Parks Jansen; adapted from Moore and Moser, 1995.)

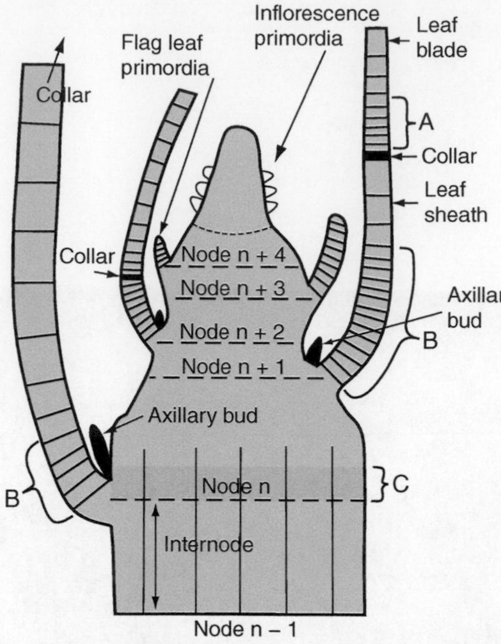

FIG. 2.13. Longitudinal view of the terminal meristem area showing leaf initiation at node n+4, leaf development, and growth zones in the tiller. (A) Blade intercalary meristem (where cells divide and elongate). (B) Sheath intercalary meristem. (C) Stem intercalary meristem. Axillary buds form in the axil of each leaf, but develop only if the plant is vigorous, the stand is thin, or the top growth is cut to open the canopy. Each axillary bud has a shoot apex that is vegetative. Note that the shoot apex is elongated as it is differentiating to form the central axis and branch primordia for the inflorescence. (Adapted from Nelson, 2000. Reproduced with permission of CAB International, Wallingford, UK.)

reproductive growth, so the apex is elongated and is developing primordia for the branches and spikelets of the inflorescence, and internodes of the lower stem are beginning to elongate. During spring, previous to this stage, a phytomer was initiated at sequential nodes every few days, so the leaf at node n may be several weeks older than the flag leaf.

The youngest, newest phytomer at a given node (e.g., node n+4) is initiated as the leaf primordia at the previous node (node n+3) begin to elongate to form a blade and a sheath that are separated by an immature collar. As cells divide and elongate above the collar, they push the blade through the tube of older sheaths. Blade growth continues until much of it is exposed, and then sheath growth commences (node n+2) and pushes the blade above the whorl. As sheath growth becomes dominant, blade growth stops but the blade continues to be pushed out of the whorl by sheath elongation, which stops when the collar reaches the light.

The axillary bud is initiated in the axil (the point where the sheath is attached to the node) of each leaf at about the time when the tip of the blade is exposed. Except for internode elongation, this completes the development of the phytomer. The uppermost axillary buds formed on a reproductive tiller are rarely functional, as they will be elevated above the soil as the internodes elongate. Vegetative tillers have no inflorescence development, so phytomer production continues.

Grassland managers should always be aware of the growth stages of the grass plants, and especially the location of the shoot apices in the canopy. Grazing early in spring, before the apex differentiates into the reproductive structure and before stem elongation occurs, gives high-quality leafy forage and high rates of animal performance. Cutting for hay is usually delayed until the inflorescence has been elevated and new tillers have been initiated. Delaying cutting until flowering increases the yield markedly because

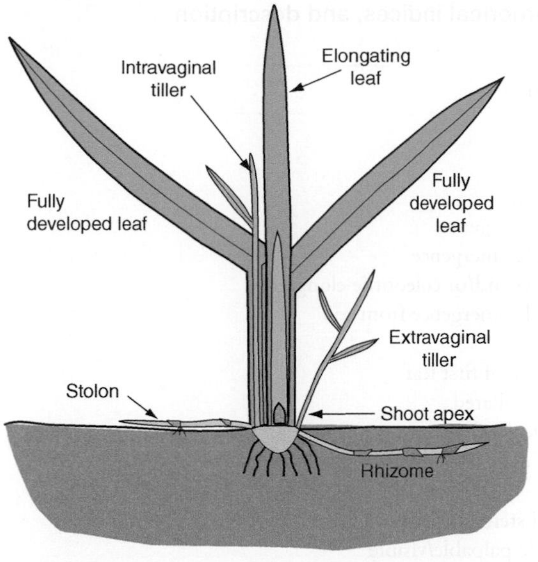

FIG. 2.14. Schematic diagram of grass plant showing the origin and positions of tillers, stolons, and rhizomes. Each develops from an axillary bud located at the base of a stem, but the growth orientation and structures differ. Note the position of the shoot apex near soil level and the next youngest growing leaf that is encircled by the elongating leaf. The tillers will form adventitious roots soon after they have three leaves.

of the added stem and inflorescence growth, but it reduces forage quality (see Chapter 14).

Grasses can produce **intravaginal** and **extravaginal tillers** (Fig. 2.14). An intravaginal tiller grows vertically on the inside of the leaf sheath to emerge at the collar area of the subtending leaf. An extravaginal tiller grows laterally through the older leaf sheaths to emerge a short distance from the main stem. Rhizomes and stolons originate from similar lateral buds and are extravaginal (Fig. 2.14). It is not known how an axillary bud directs its growth to form a tiller, rhizome, or stolon, but species differ and there are environmental factors involved. For example, kentucky bluegrass produces tillers during summer but rhizomes during fall.

Intravaginal tiller production, characteristic of bunchgrasses such as orchardgrass, big bluestem, and wheatgrasses, causes the individual plants to appear as distinct clumps or bunches with open spaces between them. Conversely, sod-forming grasses such as kentucky bluegrass, smooth bromegrass, and reed canarygrass often have extravaginal tillers and rhizomes. Bermudagrass produces intravaginal tillers, rhizomes, and stolons. Such grasses gradually spread to fill in open areas and form dense sods.

Some tall grasses with short rhizomes and extravaginal tillers, such as tall fescue and big bluestem, form loose bunches. Grasses that are relatively short have rhizomes (e.g., kentucky bluegrass) or stolons (e.g., bermudagrass), and have abundant unelongated tillers that tolerate frequent and close cutting or grazing better than do bunchgrasses. In contrast, bunchgrasses allow other seed to germinate in open spaces, may be more compatible with associated legumes or other plants, have highest production when harvested or grazed infrequently, and provide easier accessibility and better habitat for wildlife.

Developmental Stages

Knowing the stage of morphological development of forage grasses is important for characterizing maturity, predicting quality, and communicating about recommended management practices. The latter is critical for timing and documenting pesticide treatments according to label guidelines. Forage maturity of grass swards is often described in qualitative terms, such as height or stage of reproductive development, by observing the most advanced tillers. In this manner, the stage can be classified as follows:

1. *Vegetative:* Unelongated tillers are producing only leaves.
2. *Elongation:* Internodes are starting to elongate and elevate shoot apices.
3. ***Boot stage:*** Most inflorescences are located in the sheath of the flag leaf.
4. *Heading:* Inflorescences have emerged and are expanding.
5. ***Anthesis:*** Flowers are shedding pollen.
6. *Mature seed:* Inflorescences are fully developed and seed are ripe.

Although these terms help us to visualize grass maturity, this approach does not consider the total population of tillers that comprise the pasture or hayfield. Quantitative approaches have been developed for grasses that are relatively easy to use, are descriptive, and make predicting forage quality more reliable. In one common method (Moore et al., 1991), the life cycle of individual tillers is divided into five primary growth stages: germination, vegetative, elongation, reproductive, and seed development and ripening (Table 2.3). Each stage is given a mnemonic code and a numerical index.

The germination stage is applicable only to emerging seedlings. A detailed system for establishment of grasses (Moser et al., 1993) that includes stages of root and shoot development is more helpful for assessing establishment progress, timing of herbicide applications, or evaluating other establishment practices.

Classification of established grasses begins with the vegetative stage, which is based on the number of emerged leaf blades (Table 2.3). Each stage can be indexed between

Table 2.3. Developmental stages, their numerical indices, and description for perennial grasses

Stage	Index	Description
Germination		
G_0	0.0	Dry seed
G_1	0.1	Imbibition
G_2	0.3	Radicle emergence
G_3	0.5	Coleoptile emergence
G_4	0.7	Mesocotyl and/or coleoptile elongation
G_5	0.9	Coleoptile emergence from soil
Vegetative–leaf development		
VE or V_0	1.0	Emergence of first leaf
V_1	$(1/N) + 0.9$[a]	First leaf collared
V_2	$(2/N) + 0.9$	Second leaf collared
V_n	$(n/N) + 0.9$	*N*th leaf collared
Elongation–stem elongation		
E_0	2.0	Onset of stem elongation
E_1	$(1/N) + 1.9$	First node palpable/visible
E_2	$(2/N) + 1.9$	Second node palpable/visible
E_n	$(n/N) + 1.9$	*N*th node palpable/visible
Reproductive–floral development		
R_0	3.0	Boot stage
R_1	3.1	Inflorescence emergence/first spikelet visible
R_2	3.3	Spikelets fully emerged/peduncle not emerged
R_3	3.5	Inflorescence emerged/peduncle fully elongated
R_4	3.7	Anther emergence/anthesis
R_5	3.9	Post-anthesis/fertilization
Seed development and ripening		
S_0	4.0	Caryopsis visible
S_1	4.1	Milk
S_2	4.3	Soft dough
S_3	4.5	Hard dough
S_4	4.7	Endosperm hard/physiological maturity
S_5	4.9	Endosperm dry/seed ripe

Source: Moore et al., 1991.

[a]Where n is the event number (number of leaves or nodes) and N is the number of events within the primary stage (total number of leaves or nodes developed). General formula is $P + (n/N) - 0.1$, where P is the primary stage number (1 or 2 for vegetative and elongation, respectively) and n is the event number. When $N > 9$, the formula $P + 0.9(n/N)$ should be used.

1.0 and 1.9, based on the actual number of leaves present divided by N, the number of leaves that will eventually appear on the tiller prior to elongation. In practice, N is equal to the number of visible leaves in the vegetative stage or the number of palpable nodes in the elongation stage. Just prior to internode elongation, all leaves are visible, $n/N = 1$, and the index is 1.9. Similarly, elongation stages are based on the number of nodes that are palpable or visible. Again, if N is known, an index from 2.0 to 2.9 can be constructed based on past experience with fully developed stems of the same species or by calculating the index after inflorescences begin to emerge at stage R_1. Reproductive and seed development stages are based on specific morphological events.

Collecting a representative sample of the standing grass is essential for accurate staging. Random groups of about 50 tillers should be sampled from a small area from several different areas in the field. The stage can be calculated based on the number or weight of tillers. In both cases, each collected tiller is assigned a stage number according to the criteria listed in Table 2.3. The numerical index for each stage is then multiplied by the number of tillers assigned to that stage. The resulting numbers are added, and are then divided by the total number of tillers sampled. Hence the mean stage by count (MSC) is simply the mean of the stages of individual tillers in the sample—that is:

$$MSC = \sum (S \cdot N)/C$$

where S = stage index, N = the number of tillers in stage S, and C = the total number of tillers in the sample. The mean stage can be based on tiller weight (MSW) in a similar manner, except that total dry weight (D) of tillers in each stage is used in place of N, and weight of the total sample (W) is used in place of C. This universal system quantitatively characterizes most bunch- and sod-forming grass swards.

Both count- and weight-based methods give a good estimate of the growth stage for decision making, but the count method is easier to use in the field. Sometimes new tillers (with a low stage index) may emerge into older canopies (with a high stage index), causing the stage to "decrease." When using the count method, each tiller adds equally. When using the weight method, a new, small tiller adds relatively little to the total compared with a large, more mature tiller. For predicting or communicating about quality of the forage, determining the stage according to stem weight is preferable.

Summary

Seedlings and older grass plants have a fibrous root system with leaves that arise from the shoot apex. Most grasses require vernalization to change the stem apex to form a reproductive stem and flowers in spring. Summer growth of cool-season species consists mainly of leaves that are of high forage quality, but the canopy is shorter, and with no stems the yield is considerably lower. Excess spring growth can be harvested for hay or silage followed by grazing during summer and fall. Some species, such as tall fescue, produce long leaves in fall that resist cold and can be grazed through early winter to reduce the need for stored feed. The shoot apex produces axillary buds that can grow into upright tillers to increase yield. Some grasses form stolons or rhizomes to spread the plants laterally.

Shorter-growing grasses, such as kentucky bluegrass (rhizomes) and bermudagrass (stolons and rhizomes), spread laterally, retain leaf area near the soil surface, and are better adapted to close and frequent defoliation. In contrast, grasses with upright leaves, such as orchardgrass and big bluestem, are adapted to less frequent defoliation and longer rest periods. The morphology of grass plants also affects their ability to grow in mixture with legumes. White clover has stolons and a low growth habit that fits well in mixture with kentucky bluegrass. Conversely, alfalfa has upright growth and needs to have rest periods between grazings or harvests, so it fits well with orchardgrass. Knowledge of the structure and morphology of grasses provides valuable insights into the most appropriate management.

Questions

1. Describe the difference between festucoid and panicoid seedlings and how that difference affects the depth of root initiation.
2. Draw and label the parts of a grass leaf.
3. Compare and contrast the functions of an intravaginal tiller, an extravaginal tiller, a stolon, and a rhizome.
4. Explain the role of axillary buds in persistence of grasses.
5. Draw the shoot apex area of a grass tiller and identify the intercalary meristems.
6. Describe the process of stem elongation of a grass tiller from the vegetative stage until the inflorescence has emerged.
7. Draw and describe the various types of grass inflorescences.
8. Draw and label the parts of a grass spikelet that has four florets and a spikelet that has only one floret.
9. Create a hypothetical set of data and calculate a mean growth stage by count for a forage grass.
10. What is the difference between a sod-forming grass and a bunch-forming grass? Why does one form a sod while the other does not?
11. Describe a collar, seminal root, caryopsis, lodicule, unelongated tiller, perfect flower, diecious species, scutellum, culm, and glume.
12. Describe what features help to make a grass plant grazing tolerant.
13. What is the crown of a grass plant and what is its function?
14. A sample of 50 grass tillers evaluated for stage of development had 9 tillers in V_2, 15 in V_3, 17 in E_0, and 9 in E_1 according to the descriptors in Table 2.3. Three nodes were palpable when R_0 occurred. What was the MSC for the sample? How would you describe this stage of development to a forage producer?

References

Chapman, GP. 1996. The Biology of Grasses. CAB International, Wallingford, UK.

Dawson, LA, SJ Grayston, and E Paterson. 2000. Effects of grazing on the roots and rhizosphere of grasses. In G Lamaire, J Hodgson, A. de Moraes, C. Nabinger, and PC de F. Carvalho (eds.), Grassland Ecophysiology and Grazing Ecology, pp. 61–84. CAB International, Wallingford, UK.

Dayton, WA. 1948. The family tree of Gramineae. In USDA Yearbook of Agriculture, pp. 637–639. US Government Printing Office, Washington, DC.

Engel, RK, JT Nichols, JL Dodd, and JE Brummer. 1998. Root and shoot responses of sand bluestem to defoliation. J. Range Manage. 51:42–46.

Hoshikawa, K. 1969. Underground organs of the seedlings and the systematics of the Gramineae. Bot. Gaz. 130:192–203.

Hyder, DN, AC Everson, and RE Bement. 1971. Seedling morphology and seeding failures with blue grama. J. Range Manage. 24:287–292.

Moore, KJ, and LE Moser. 1995. Quantifying developmental morphology of perennial grasses. Crop Sci. 35:37–43.

Moore, KJ, LE Moser, KP Vogel, SS Waller, BE Johnson, and JF Pedersen. 1991. Describing and quantifying growth stages of perennial forage grasses. Agron. J. 83:1073–1077.

Moser, LE. 2000. Morphology of germinating and emerging warm-season grass seedlings. In KJ Moore and BE Anderson (eds.), Native Warm-Season Grasses: Research Trends and Issues, pp. 35–47. CSSA Special Publication 30. Crop Science Society of America, Madison, WI.

Moser, LE, KJ Moore, MS Miller, SS Waller, KP Vogel, JR Hendrickson, and LA Maddux. 1993. A quantitative system for describing the developmental morphology of grass seedling populations. In Proceedings of the 17th International Grassland Congress, Palmerston North, New Zealand, pp. 317–318.

Nelson, CJ. 2000. Shoot morphological plasticity of grasses: leaf growth vs. tillering. In G Lamaire, J Hodgson, A. de Moraes, C. Nabinger, and PC de F. Carvalho (eds.), Grassland Ecophysiology and Grazing Ecology, pp. 101–126. CAB International, Wallingford, UK.

Newman, PR and LE Moser. 1988. Seedling root development and morphology of cool-season and warm-season forage grasses. Crop Sci. 28:148–151.

Stubbendieck, J, SL Hatch, and CH Butterfield. 1997. North American Range Plants, 5th ed. University of Nebraska Press, Lincoln, NE.

Teel, MR. 1957. Brome grass can die out too. Hoard's Dairyman, p. 470, May 10, 1957.

Walton, PD. 1983. Production and Management of Cultivated Forages. Reston Publishing Company, Inc., Reston, VA.

Watson, L and MJ Dallwitz. 1992. The Grass Genera of the World. CAB International, Wallingford, UK.

Weaver, JE. 1926. Root Development of Field Crops. McGraw-Hill, New York.

Wilson, AM, and DD Briske. 1979. Seminal and adventitious root growth of blue grama seedlings on the Central Plains. J. Range Manage. 32:209–213.

CHAPTER 3

Structure and Morphology of Legumes and Other Forbs

Robert B. Mitchell and C. Jerry Nelson

Forbs can be defined generally as **herbaceous** broadleaf plants. In grassland science, they are one of five plant functional groups (grasses, grass-like plants, forbs, shrubs, and succulents) and are represented by numerous species, including legumes. In humid areas, **legumes** are often discussed separately from forbs. Many legumes have superior forage quality, most fix atmospheric N_2, and, as highly desired forbs, they are often seeded into and managed as important components of pastures and hayfields. Similarly, the brassicas are often discussed separately because they are seeded for specific uses. Many forbs exist in pastures and rangelands, where they contribute to yield, species diversity, nutrients for livestock, and food and habitat for wildlife.

The Fabaceae, or legume family, consists of about 600 genera and 12,000 species that include summer annuals (e.g., korean lespedeza, soybean), winter annuals (e.g., arrowleaf clover, hairy vetch), biennials (e.g., sweetclover) and many perennials (e.g., alfalfa, red clover) of economic importance. Seeded legumes are a unique subgroup of forbs that have a narrower range of adaptation than do grasses, and usually require greater management inputs, such as fertilizer, pesticides, and harvest frequency, to maintain production and persist.

The Asteraceae is the largest plant family of forbs, consisting of over 1500 genera with more than 19,000 species, but other families also contain important forbs. These forbs, which commonly grow in pastures, hayland, and rangeland, include summer annuals (e.g., sunflower, artichoke), winter annuals (e.g., chickweed), a few biennials and many perennials (e.g., aster, dandelion, chicory). This wide range of forb species, many of which are reseeding annuals (e.g., redroot pigweed, common ragweed), fill niches in the **canopy** and contribute to the forage resource. These opportunists (invaders) in the grassland ecosystem are usually not a major concern unless they have undesirable characteristics, such as low forage quality, or are toxic.

The Brassicaceae, or mustard family, consists of about 350 genera and more than 4000 species of forbs, including some annuals (e.g., turnip, kale) that are seeded annually and used for grazing.

Our goal is to introduce the growth and development of seeded legumes and other forbs, consisting of invaders and seeded species that contribute to production and ecosystem functions of pastures, hayfields, and rangeland. Each species has unique growth characteristics that must be taken into consideration in order to manage them effectively so as to utilize their unique values in the overall livestock production system.

The Legumes

The term *legume* indicates the type of fruit (a pod) that is characteristic of plants of this family (Fig. 3.1). A legume is a monocarpellary (one-chamber) fruit containing a single seed or a single row of seed that dehisces along both sutures

Forages: An Introduction to Grassland Agriculture, Seventh Edition. Edited by Michael Collins, C. Jerry Nelson, Kenneth J. Moore and Robert F Barnes.

Fig. 3.1. Different types of legume seed pods: (A) siratro; (B) soybean; (C) birdsfoot trefoil; (D) hairy indigo; (E) bigflower vetch (Photos courtesy of Albert Kretschmer and Michael Collins).

or ribs. **Dehiscence** is a good mechanism for natural seed dispersal, but is a problem in commercial seed production because mature pods open or shatter easily and seed are lost. Legumes may be **annual, biennial, or perennial,** and may be herbaceous or woody, depending on the lifespan and growth form of individual plants.

Most forage legumes grow symbiotically with N-fixing bacteria which are present in nodules that form on the roots. Thus legumes are valuable components in forage mixtures with grasses and in crop rotations with grain crops to decrease dependence on fertilizer N. With regard to animal nutrition, legumes are usually higher in **crude protein** and have higher **intake** and **digestibility** in ruminants than do grasses.

The **morphology** (i.e., plant form or shape) of forage legumes differs greatly from that of grasses (see Chapter 2). Distinct morphological differences exist among legume species that allow identification and help to explain their adaptation to specific environmental or management conditions.

Leaves

Leaves of legumes are arranged alternately on the stem, and are usually connected to the stem by a **petiole.** If a single leaf blade is attached directly to the petiole, the leaf is described as *unifoliolate*. In compound leaves, three (trifoliolate) or more leaf blades are individually connected to the petiole by a short stalk called a **petiolule** (Fig. 3.2). Legume leaves often have large stipules, which are leaf-like appendages attached directly to the stem near the junction with the petiole (Fig. 3.2F).

Compound leaves can be pinnate (Fig. 3.2 A, B), where the petiolule of the central leaf blade is longer than those of the lateral leaflets. Alfalfa and sweetclover have pinnately compound trifoliolate leaves. On palmately compound leaves (Fig. 3.2 D, E), all of the leaf blades have equally short petiolules.

Stems

Stems of legume species vary greatly in length, diameter, amount of branching, and woodiness. The **shoot apex** or terminal bud of the stem is always located at the tip, which is near the top of the canopy in upright legumes such as alfalfa, sweetclover, and korean lespedeza. The shoot apex of these species is usually removed by harvest or grazing, so plants must regrow from **axillary buds** at lower nodes on the stem, especially those buds at nodes near the soil surface.

Fig. 3.2. Different types of legume leaves: (A) white clover, (B) *Lespedeza bicolor*; (C) birdsfoot trefoil; (D) kura clover; (E) red clover; some "lucky" clover leaves have four leaflets (inset); (F) korean lespedeza on left, and common lespedeza on right, (G) crownvetch, (H) white clover. Alfalfa varieties with multi-foliolate leaves are available (See the box "Is Multi-Foliolate Better?") (Photos courtesy of Michael Collins).

Is Multi-Foliolate Better?

Alfalfa and the clovers typically have three leaf blades on each petiole, but several years ago scientists noted that some alfalfas had four, five, or more blades on the same petiole. Red clover can have as many as eight blades arising from the same petiole. It was thought that these plant types would have a higher leaf-to-stem ratio and improved forage quality. The small increases in leaf-to-stem ratio in multifoliolate alfalfa plants did produce small increases in dry matter digestibility by ruminants. However, maturity stage at harvest still had a much larger effect on quality.

Today, some alfalfa cultivars are marketed for their "multi-leaf" character, but it is known that factors other than leafiness are important for forage quality. For example, trifoliolate cultivars with higher levels of crude protein and digestibility are now available. Genetic improvement of the digestibility of alfalfa stems is feasible and desirable, as stem digestibility is lower than leaf digestibility, and decreases faster with maturity.

To date, the multifoliolate character of red clover has not been exploited. Does the multi-leaf character also exist in white clover?

In red and alsike clovers, the long petioles display the leaf blades at the top of the canopy, well above the shoot apex, which remains low in the canopy (Fig. 3.3). At vegetative growth stages, animals graze mostly leaf blades and petioles and may not remove the shoot apex, allowing it to continue to produce new growth. This is one reason why red clover is popular in pasture mixes with perennial grasses. Species such as white clover have prostrate stolons with the shoot apices near ground level (Fig. 3.4). Normal cutting or grazing removes only the leaf blades and petioles, and has little effect on the shoot apex.

An axillary bud is located in the axil of each leaf, where it is protected between the stipules (Fig. 3.2). These buds, each of which has an intact shoot apex that is similar to that on the main stem, can break dormancy. Light penetrating the canopy stimulates growth of these buds to contribute vegetative growth (branching) or to form flower buds. If the shoot apex is removed or damaged by cutting or grazing, apical dominance is reduced and axillary buds at the **crown** can develop to provide regrowth.

Roots

Most herbaceous legumes have prominent taproots (Fig. 3.5G), from which fine secondary roots or large branches may arise. For example, most alfalfa and sweetclover plants have only a few secondary roots, with the primary root surviving as a taproot until the entire plant dies. Roots of red clover, a short-lived perennial, are more branched than those of alfalfa (Fig. 3.3), but plants normally survive for only about 2 years because the primary root becomes diseased, causing the plant to die.

White clover also has a small taproot as a seedling, but it becomes diseased and dies within 2 years. Prior to that time this stoloniferous species develops **adventitious roots** from

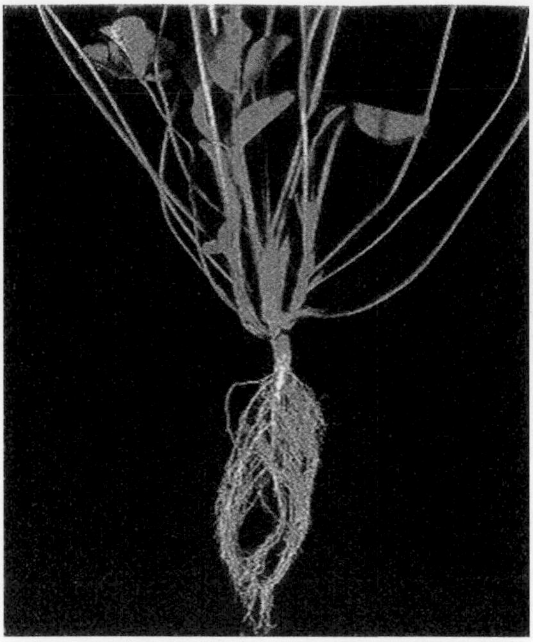

FIG. 3.3. The primary stem of red clover does not elongate, but it produces additional shoots from axillary buds that elongate. Note the long petioles that display the leaf blades at the top of the canopy, where they may be removed by grazing with little effect on the shoot apices at the tip on the short shoots. (Photo courtesy of University of Kentucky.)

nodes of stolons to form a fibrous-rooted plant (Fig. 3.4). Each year stolons elongate and produce new adventitious roots at nodes, while older stolon segments and roots die (Westbrooks and Tesar, 1955). Thus young white clover plants can be quite drought resistant because the taproot can penetrate up to 40 in. or more. Later, the plant is less drought resistant because adventitious roots proliferate mainly in the upper soil horizons.

Roots of biennial and perennial legumes serve as major storage organs for reserve substances, primarily carbohydrate and N-containing compounds (see Chapter 4). These substances are accumulated from photosynthesis and other metabolism when the leaf area is high, and are then used later to support regrowth of buds after herbage and leaf area have been removed by cutting or grazing (see Chapter 5).

Inflorescences

Legume flowers are usually arranged in a **raceme** (as in pea), in a spike-like (compact) raceme (as in alfalfa and sweetclover), in a very compact raceme or **head** (as in the true clovers), or in an **umbel** (as in birdsfoot trefoil and crownvetch). The number of flowers per **inflorescence** varies greatly among legume species. The short raceme of alfalfa may contain up to 20 flowers, each producing 8–10 seed per pod, whereas the longer raceme of sweetclover may contain over 60 flowers, usually with only one seed per pod (Fig. 3.1). Clovers typically have 60–120 flowers in a head (Fig. 3.6), each producing a pod with one seed. Birdsfoot trefoil usually has 4–7 flowers per umbel, with each developing into an elongated brown pod that extends

FIG. 3.4. White clover spreads by growth of stolons with adventitious roots developing at the nodes. The leaf blades are supported by long petioles; the flower heads are supported by long peduncles that arise from the nodes. Axillary buds located behind scale leaves at the nodes can develop into new stolon branches. (Photo courtesy of Chuck West, Texas Tech University. Drawing adapted from Isely, 1951.)

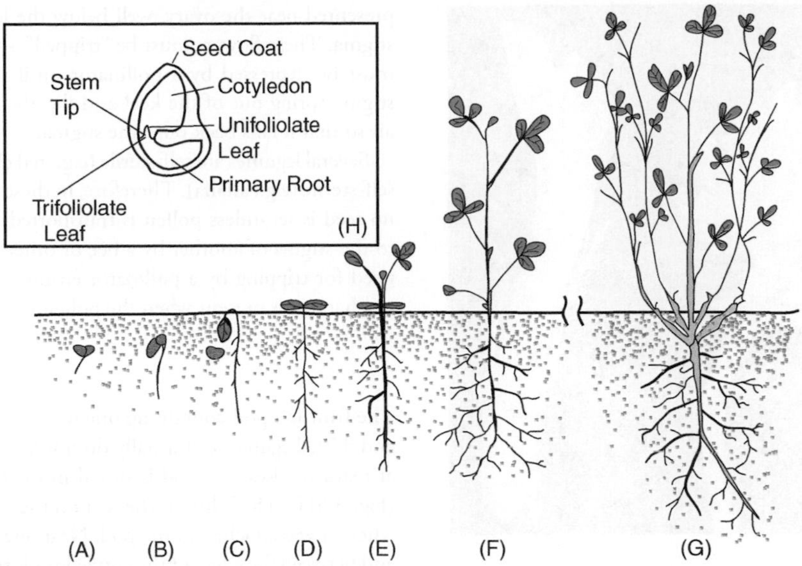

FIG. 3.5. Seedling development of a legume, such as alfalfa, with epigeal emergence: (A) The seed imbibes water and the primary root emerges. (B) The hypocotyl becomes active, and forms an arch to penetrate the soil. (C) Elongation of the hypocotyl stops when the arch reaches the light. (D) The arch straightens and the cotyledons open for photosynthesis, exposing the epicotyl that was protected as it was moved through the soil. (E) The primary root continues to elongate and enlarge, developing some secondary (branch) roots. A unifoliolate leaf develops, followed by the first trifoliolate leaf. The shoot apex is located between the stipules of the last developed leaf. (F) The cotyledons fall off. Axillary buds at the cotyledonary nodes swell to develop into new shoots. The stem continues to elongate, producing a leaf at each node. (G) Contractile growth has occurred, forming a crown, and taproot morphology is developing. The crown is forming, with branches clearly evident from buds at the cotyledonary node and in axils of the unifoliolate and first trifoliolate leaves. The crown will continue to enlarge because each new branch has unelongated internodes near or below soil level that have incomplete leaves and axillary buds at the nodes; these provide sites for regrowth following cutting. (H) Alfalfa seed dissected longitudinally to show one cotyledon, primary root, shoot apex, and embryonic leaves.

outward from the stalk in the shape of a bird's foot, and contains 5–15 seed (Fig. 3.7).

Flowers

Flowers of common legume species have irregular floral petals that are characteristically papilionaceous (i.e., arranged in a manner that resembles a butterfly) (Fig. 3.8). Each flower has a corolla consisting of five petals—a standard, two wings, and a keel that consists of two petals that are more or less united (Fig. 3.8). The calyx, consisting of five green sepals at the base of the flower, surrounds and protects the corona in the bud stage. The keel, so named for its boat-like shape, encloses the **ovary, stigma, style,** and **stamens.** There are usually 10 stamens, one being free whereas the filaments of the other nine are joined at the base to form an envelope that encloses the ovary and the long, slender style.

The corolla tube, formed by partial joining of the five petals near their bases, varies in length among species. For example, the corolla tube of red clover, which is sometimes 0.4 in. or more in length, is relatively long in relation to the small size of these flowers (Fig. 3.6). In alsike clover, white clover, sweetclover, and alfalfa, the tube is much shorter. The length of the corolla tube affects cross-pollination, since some bees and other pollinators cannot reach the nectar secreted by glands at the base of the tube.

Some legumes, including soybean and pea, are self-pollinating and almost completely self-compatible. They are pollinated within the keel when the anthers open and pollen rolls out and contacts the stigma. However, many forage legumes have short stamens such that pollen is

Fig. 3.6. Flowers of red clover are borne on heads that develop from the shoot apices of branches. The long corolla tube (the light-colored cylinder below the petals) has pubescent sepals at the base and is about 0.4 in. long. (Photo courtesy of Norman L. Taylor.)

presented near the ovary, well below the longer style and stigma. These flowers must be "tripped"—that is, the keel must be depressed by a pollinator until the anthers and stigma spring out of the keel and flip the pollen into the air so that it falls back onto the stigma.

Several legumes are self-sterile (e.g., red clover) or mostly self-sterile (e.g., alfalfa). Therefore, in these species, little or no seed is set unless pollen is transported from one plant to the stigma of another by a bee or other pollinator. The need for tripping by a pollinator ensures that a transport mechanism is present when the pollen is released.

Fruit and Seed

The fruit is a pod containing one to several seed (Figs 3.1 and 3.7). Legume seed usually do not have an endosperm at maturity. Reserve food is stored in the two **cotyledons** (Fig. 3.5H). The hilum is the scar on the seed at the point where it was attached to the pod. Near one end of the seed and between the edges of the cotyledons is the **embryo axis** (Fig. 3.5H), which consists of the **epicotyl** and primary root. Each seed is enclosed by a testa (seed coat).

Many forage legumes have **hard seed**, in which case the seed is alive but the seed coat is impervious and restricts water imbibition, thus limiting germination. For example, over a period of several years in Colorado it was found that about 22% of the commercial seed of alfalfa was classified as hard. Over time in the soil, these seed coats become softened by microorganisms, wetting and drying, and freezing

Fig. 3.7. Developmental stages of an inflorescence of birdsfoot trefoil, including the umbel with several florets and seedpods.

Fig. 3.8. A typical legume flower, partridge pea, with a large standard petal at the top, two wing petals on the sides, and two fused keel petals. Five sepals and 10 stamens (one free and nine partially fused) surround much of the pistil, which consists of the ovary, style, and partially exposed stigma. (Photo courtesy of Rob Mitchell.)

and thawing, which allows the seed to absorb water and germinate. Thus volunteer stands of legumes occur, even when the area has not been seeded for several years.

The germination of hard seed, which must be listed on the seed tag, can be increased by storage, by blending with seed lots that have a lower frequency of hard seed, or by **scarification** (Bass et al., 1988). Scarification is a mechanical process that involves scratching the seed coat to allow imbibition (uptake) of water, but rigorous treatment can injure the cotyledons or embryo axis, leading to abnormal seedlings and loss of vigor.

Conversely, annual legumes need to re-establish each year in permanent pastures or hayfields. Hard seed can be an asset for annuals because the seed crop produced in a given year may germinate over several years. However, this does not work well with alfalfa, where **autotoxicity** from established alfalfa plants interferes with germination, root growth, and development of young alfalfa seedlings (see Chapter 11).

Seedling Development

Most forage legumes have epigeal emergence, during which the cotyledons are pushed above soil level (*epi* = above) by the elongating **hypocotyl** (Fig. 3.5). In contrast, the cotyledons of legumes that exhibit hypogeal emergence remain below ground (*hypo* = below) at planting depth, and emerge due to growth of the epicotyl (the stem and leaves). For both types of emergence, the first sign of germination is the emergence of the primary root through the seed coat. If emergence is epigeal, the hypocotyl (that part of the embryo axis between the primary root and the cotyledonary node) lengthens and forms an arch that penetrates the soil (Fig. 3.5C).

Light causes the emerged arch to straighten, the cotyledons to open, and **photosynthesis** in the cotyledons to begin. Before photosynthesis begins, seedlings are dependent on reserves stored in the cotyledons for growth and respiration. The first true leaf can be unifoliolate (as in alfalfa and clovers) or trifoliolate (as in birdsfoot trefoil and crownvetch). In annual lespedezas, the first two true leaves are unifoliolate. Subsequent leaves of forage legumes are consistent with the characteristic of the species.

A few forage legumes, such as hairy vetch, exhibit hypogeal emergence during which the hypocotyl is not active. Instead, the epicotyl forms an arch that is pushed through the soil by internode elongation until it reaches the light. Hypogeal emergence is advantageous if a very young seedling is cut or frosted back to soil level. Legumes with hypogeal emergence have one or more nodes below ground with axillary buds; thus they have both a meristem and a residual energy source (cotyledons) to regrow. Epigeal seedlings in the same situation do not have axillary buds or an energy source below ground (Fig. 3.5).

The advantages of epigeal emergence include protection of the epicotyl by the cotyledons that enclose the shoot apex and young leaves during emergence and, in particular, the added role of the cotyledons in early photosynthesis (Shibles and MacDonald, 1962). Cotyledons of cultivated forage legumes usually expand their area by up to 10-fold during and shortly after emergence, and remain photosynthetically active for 2 to 3 weeks. After becoming established, axillary buds at the cotyledonary node and one or two nodes just above the cotyledonary node develop shoots to begin formation of the crown (Fig. 3.5G) (Teuber and Brick, 1988).

Contractile Growth

About 6–8 weeks after emergence, many perennial legumes that exhibit epigeal emergence begin contractile growth. In this unique process the first node, where the cotyledons were attached and where the crown forms, is gradually pulled back below the soil surface (Fig. 3.5 F, G). The phenomenon is believed to be caused by a lateral expansion of cells of the hypocotyl and upper primary root, causing these cells and structures to shorten vertically, which lowers the crown. The first node and the developing crown are pulled about 0.25 in. below the soil surface for white clover, about 0.5 in. below the soil surface for red clover and birdsfoot trefoil, and about 0.8 in. below the soil surface for alfalfa. Contractile growth of sweetclover may exceed 1.5 in. The **winter hardiness** of these legumes tends to be directly related to their crown depth. Deeper crowns may allow the soil to provide more cold protection of the axillary buds that are needed for spring growth (see Chapter 5).

As each new shoot of alfalfa or red clover develops from an axillary bud on the crown, several nodes and short internodes are produced below or near the soil surface (Fig. 3.3 and Fig. 3.5G). Each node has an axillary bud to provide a site near or below soil level for the next regrowth when the stem is grazed or cut (Nelson and Smith, 1968). In contrast, summer regrowth of birdsfoot trefoil arises from axillary buds well above the crown. However, each year trefoil plants develop somewhat dormant axillary buds on the crown, at or below ground level, that survive the winter and provide spring growth.

Developmental Stages

It is important to determine the stage of development of forage crops in order to characterize maturity, predict quality, and communicate about management practices. Forage maturity may be qualitatively characterized according to height or stage of floral development by observing the most advanced shoots. For example, stage of development of a legume field is commonly noted as vegetative, early bud, late bud, early bloom, mid-bloom, and so on. Although these terms help us to visualize and communicate about the maturity of the forage, this approach does not accurately characterize the total population of shoots that comprise the hayfield or pasture.

For some legume species, quantitative approaches have been developed that are relatively easy to use, are descriptive, and make predicting forage quality more reliable. Methods of calculating the mean morphological stage of alfalfa development based on the number or weight of stems in different stages have been described by Fick and Mueller (1989). Ten development stages represent four broad categories: vegetative, flower bud development, flowering, and seed production (Table 3.1).

Table 3.1. Developmental stages, their numerical indices, and descriptions for alfalfa

Index	Stage	Description
Vegetative stages		
0	Early vegetative	Stem length ≤ 6 in., no visible buds, flowers, or seed pods
1	Mid-vegetative	Stem length 6–12 in., no visible buds, flowers, or seed pods
2	Late vegetative	Stem length ≥ 12 in., no visible buds, flowers, or seed pods
Flower bud development		
3	Early bud	1–2 nodes with visible buds, no flowers or seed pods
4	Late bud	≥ 3 nodes with visible buds, no flowers or seed pods
Flowering		
5	Early flower	1 node with one open flower, no seed pods
6	Late flower	≥ 2 nodes with open flowers, no seed pods
Seed production		
7	Early seed pod	1–3 nodes with green seed pods
8	Late seed pod	≥ 4 nodes with green seed pods
9	Ripe seed pod	Nodes with mostly brown mature seed pods

Source: Fick and Mueller, 1989.

Table 3.2. Characteristics of brassica forages adapted to North America

Type	Seeding to harvest (days)	Harvestable parts	Aerial growth	Regrowth potential	Cultivar(s)
Turnip (and hybrids)	60–90	Tops and root	Leaves	Yes	Purple Top, Tyfon
Forage rape	60–90	Tops	Leaves and stems	Yes	Rangi, Winfred
Kale: marrow stem	90–150	Tops	Leaves and stems	No	Maris Kestral, Gruner, Ring
Kale: stemless	60–90	Tops	Leaves	Yes	Premier
Swedes (rutabagas)	90–150	Tops and root	Leaves	No	Calder, Doon, Major

Obtaining accurate stage information depends on the collection of a representative sample of the standing crop. Generally all stems (shoots) more than 1.5 in. long are considered to be growing, and should be sampled from at least four separate areas (1 ft^2 each) or from a specified distance along four random rows within a field. Each shoot in the sample is assigned a stage number using the criteria listed in Table 3.1. The numerical index for each stage (0–9) is then multiplied by the number of shoots assigned to that stage. The resulting numbers are added together, and are then divided by the total number of shoots in the sample. The mean developmental stage by count (MSC) is simply the weighted mean of the stages of individual shoots in the sample, and is calculated as follows:

$$MSC = \sum(S \bullet n)/C$$

where S is the stage index, n is the number of shoots in stage S, and C is the total number of shoots in the sample. The mean stage can be based on shoot weight (MSW) by substituting total dry weight of shoots in each stage (D) for n, and total dry weight of the sample (W) in place of C.

Both methods give a good estimate of the growth stage for hay-making decisions, but the counting procedure is easier to use in the field. Sometimes new alfalfa shoots (with low stage numbers) may emerge into older canopies (with high stage numbers), causing the calculated stage to "decrease." To predict or communicate about the quality of the forage, it is better to determine the stage according to stem weight (Fick and Mueller, 1989).

Other Forbs

This section addresses the morphology, structure, and growth habit of non-legume forbs. Information on the morphology of forbs associated with range and grassland is based primarily on Stubbendieck et al. (2011), and that of seeded annual forbs, primarily brassicas, is based on Smith and Collins (2003).

As a group, non-legume forbs that are seeded comprise a relatively small number of forage species, and they are grown on far fewer acres in North America than are either grasses or legumes. In most pastures and rangeland, and in some hayfields, annual and perennial non-legume forbs are natural invaders (i.e., they are not seeded) that occupy small open areas among the dominant plants. In these cases the forbs contribute to the forage supply, increase the diversity of the vegetation, and support pollinators. Some forbs, including some legumes, have undesirable characteristics associated with invasiveness, low forage quality, and toxicities.

The primary importance of seeded, annual non-legume forbs such as turnip in North America arises from their use as specialty crops (Table 3.2). These plantings fill a time or quality niche to complement a forage production system. In Europe and other areas, however, forbs such as artichoke or brassicas are major components within cultivated forage systems. In North America, annual brassicas have been the focus of increased interest over the past 40 years (Jung et al., 1986) as a specialty pasture crop or as a cover crop in row crop systems that provides winter ground cover and reduces soil erosion. These crops have potential in many areas of the USA and Canada.

The morphology and growth patterns of most forbs are similar to those of legumes, but there is more species-to-species variation and they lack the ability to fix atmospheric N. An understanding of the morphological basis of the growth and persistence of these seeded and invading forbs is important for determining their roles in a forage or grazing system. A knowledge of these morphological characteristics is also critical for determining how to reduce the abundance of undesirable forbs in the mixture.

Leaves

In addition to the alternate leaf arrangement that is found in legumes, some other forbs have opposite leaf arrangement (i.e., two leaves attached at the same node), as in stiff sunflower, or a whorled leaf arrangement, as in whorled milkweed. Some forbs have a combination of leaf arrangements. For example, western ragweed has alternate leaf arrangement on the upper portion of the plant and opposite leaf arrangement on the lower portion. Leaf attachment on some forbs is clasping (i.e., the blade partially surrounds the stem), as in shell-leaf penstemon, decurrent (i.e., the blades extend down the stem and fuse together), as in musk thistle, attached to the stem by a petiole, as in plains sunflower, or sessile (i.e., lacking a petiole), as in hoary vervain.

Low-growing forbs such as plantain and dandelion have very short stems with numerous leaves from nodes appearing on the very short internodes. In dense stands of legumes or grasses, the leaves of dandelion are long and extend upward to intercept radiation. However, in stands of legumes and grasses with low density the leaves are shorter and extend laterally to cover more ground area and intercept more radiation. These prostrate leaves have a long lifespan, usually escape mowing and grazing, and remain horizontal during winter to increase survival. The spreading leaves near the soil surface shade regrowth sites (axillary buds) in the crown areas of nearby grasses and legumes. This reduces tillering of grasses and lateral spreading of legume crowns of these preferred species.

Brassicas such as turnip have upright leaves suspended by long petioles that arise from the base of the plant (Fig. 3.9). Early in plant development the new leaves are initiated by the shoot apex and grow rapidly, with virtually no elongation of the stems and slow growth of the root. Their tolerance of frost allows them to grow during cooler periods of the fall or early winter. In contrast to most grass and legume forages, the forage quality of brassica leaves does not decline significantly with advancing maturity, and some produce large leaves that can be grazed or stored for winter feed.

Fig. 3.9. A mature turnip plant showing the large root-like structure (tuber) that provides additional high-quality forage for grazing. Top growth is usually grazed off preferentially before the tubers are pushed out and consumed. (Photo courtesy of Michael Collins.)

Stems

Although highly variable in terms of stem characteristics, two general types of forbs in grassland communities are commonly recognized. *Caulescent* forbs, such as the sunflowers, typically have an erect or ascending growth habit with long internodes, and produce a prominent stem that extends well above the soil surface. In contrast, *acaulescent* forbs, such as plantain and dandelion, have very short internodes and appear stemless until flowering. This results in differences in response to management, with tall caulescent forbs usually being found on range sites and in hayfields, whereas acaulescent forbs occur in pastures because they often have a shoot apex near soil level and prostrate leaves that allow them to survive frequent and close grazing. Similarly, the seeded brassicas such as turnip also appear stemless while young, but the upper internodes elongate later, just prior to flowering.

Roots and Root-Like Structures

Similar to legumes, most biennial and perennial forbs in grassland systems have prominent taproots, with fine secondary roots or large branches arising from the primary root. Annual forbs in grasslands typically have a well-defined, slender taproot with minimal lateral branching. This difference provides a way to determine whether a mature forb is an annual or a perennial. As with annual lespedeza, the root systems of many annual forbs (especially summer annuals) do not have to serve as a storage organ for carbohydrates for winter survival. Like perennial legumes, the roots of biennial and perennial forbs must serve as storage organs for carbohydrates and N-containing compounds in order to survive winter and initiate growth in the spring.

Some forbs, such as turnip, have prominent belowground structures that appear to be roots but which are actually a modified underground stem called a **tuber**. The true root of turnip, with its lateral roots, extends downward from the base of the large tuber. **Corms** are also modified underground stems in which the bulbs are large modified buds. Leaves of annual brassica species that grow large "roots" have very short stems and long petioles that display the leaves in an upright manner (Fig. 3.9). The leaves provide photosynthate for growth of the large and prominent root-like structures that serve as excellent storage organs for the plant, filled primarily with carbohydrates to support future growth of perennial species. In forage management the tubers are grazed or can be collected and stored for later feeding.

Inflorescences

Inflorescence type is highly variable in grassland forbs, but similar to legumes. Most forbs have heads (e.g., the sunflowers) or racemes (e.g., the ragweeds). Some forbs have **cymes** (convex or flat-topped inflorescences), as in many

FIG. 3.10. Leadplant is a deciduous shrub native to the North American prairies, and is highly palatable and nutritious for livestock and wildlife. Legumes such as leadplant have a spicate raceme, with the flowers first emerging at the base of the inflorescence. Unlike typical legumes, leadplant flowers have a single petal, an unusual characteristic in legumes, as indicated by the name of the genus *Amorpha* (meaning "formless" or "deformed"). (Photo courtesy of Rob Mitchell.)

of the milkweeds, a spike-like inflorescence, as in kochia, umbels, as in water hemlock, or an umbel of **cyathiums,** as in snow-on-the-mountain. As with legumes, some forbs are determinate, with the inflorescence being produced from a terminal bud, whereas others are indeterminate, with the inflorescences being produced from axillary buds along the stem.

Flowers

The type and number of flowers per inflorescence vary greatly from species to species. The sunflowers have a compound flower with both ray and disk flowers. The ray flowers radiate around the disk flowers, which are embedded in the receptacle. Members of the sunflower genus, *Helianthus,* typically have 10–30 ray flowers per head. Other forbs have complete flowers borne on axillary branches. The importance of legume and forb flowers for the support of much needed pollinators such as bees is becoming increasingly evident. Typical recommendations for improving pollinator habitat include planting legumes and forbs, which are good sources of pollen and nectar (Fig. 3.10).

Fruit and Seed

The different fruit types of forbs vary dramatically. The fruit type for all members of the Asteraceae family is an **achene,** often with a pappus (a feathery attachment). Other common fruit types include **utricles**, as in kochia, **follicles**, as in the milkweeds, and a **capsule**, as in leafy spurge. Many annual forbs produce an abundance of seed that remain viable in the soil seed bank for years. For example, prostrate pigweed, an annual, has been estimated to produce 6 million seed per plant, and redroot pigweed seed have remained viable in the soil seed bank for more than 40 years. These seed properties help forb species to occupy niches in grazing lands and provide excellent food sources for wildlife. However, they also contribute to the problems associated with their weed-like nature. The seed oil content of many forbs (e.g., canola) can exceed 40%, resulting in forb seed, primarily brassicas (e.g., canola, pennycress), being evaluated for biodiesel and jet fuel production.

Forb Lifespan

Forbs occur as annuals (summer and winter), biennials, and perennials. Summer annuals (e.g., annual lespedeza) emerge in the spring or summer and complete their life cycle by fall. Winter annuals emerge in the fall or winter and complete their life cycle and seed production by summer. Seed production is the only mechanism for survival of invasive annuals, so a high proportion of the plant's energy is invested in reproduction. Seeded annuals such as brassicas usually do not produce seed if they are grazed properly.

Annuals

Summer annual forbs such as kochia are good seed producers, and generally are the first plants to invade a site following disturbances such as overgrazing. Kochia provides good forage in the vegetative stages, but can accumulate high concentrations of nitrate, especially during periods of water stress. If kochia is grown for forage, the plants must produce seed before harvest in order to recolonize the area in subsequent years. In its region of use, kochia flowers from July to September, and the seed matures in September and October, so grazing pressure must be managed during spring and early summer to maintain shoot apices for seed production. If kochia poses a competition problem—for example, when establishing perennial grasses—an effective form of control is to prevent seed production by mowing to remove elevated shoot apices.

Most seeded brassicas are annuals that are planted in spring (e.g., swedes, kale) or mid-summer (e.g., rape, turnip) for grazing in November or December (Fig. 3.11). Leaves are not very cold tolerant, and will die when frosted.

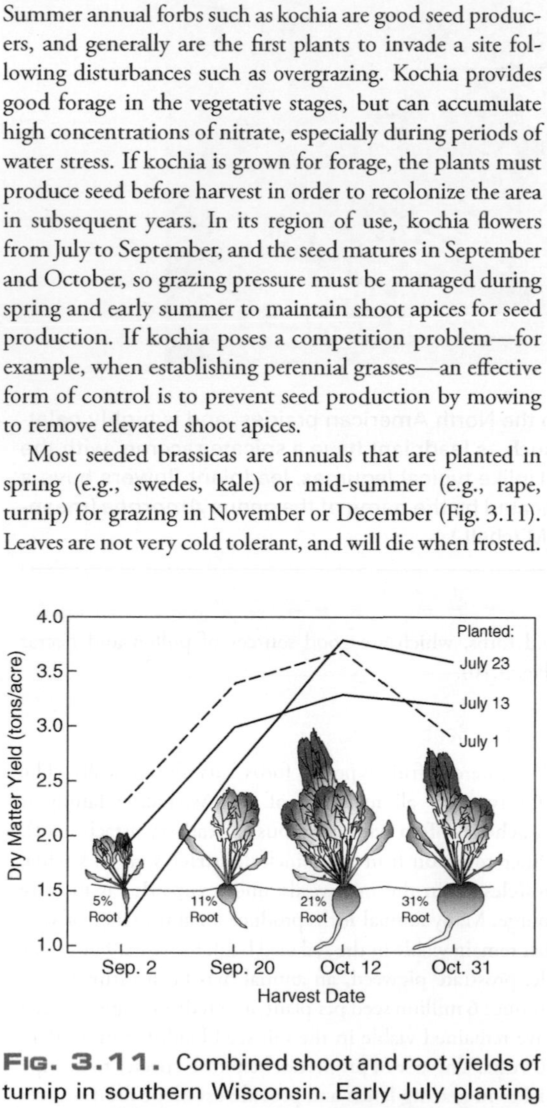

FIG. 3.11. Combined shoot and root yields of turnip in southern Wisconsin. Early July planting dates provided more yield in September, but late July planting dates resulted in higher yields being available at the end of October. The root component made up about one-third of the total dry matter at the end of October.

Animals graze the tops before they begin to graze the roots. Most brassicas do not overwinter successfully, so should be grazed or have the roots harvested in winter.

Biennials

Biennial forbs complete their life cycle in 2 years. Biennials grow vegetatively during the first year and accumulate carbohydrates in below-ground storage organs. They require **vernalization** to induce flowering. The low temperatures during winter following the first growing season vernalize the plants and prepare them for flowering. During the second year the plants flower, produce seed, and die. Most types of sweetclover are biennials, but remain in grasslands due to production of hard seed. In long-term grasslands, biennial forbs, such as the legume sweetclover, usually perennate from the soil seed bank. Thus each year the population contains both first-year and second-year plants.

Sugarbeet, which is a true biennial, can be sown in the late summer or fall, will accumulate carbohydrates in the root, and will be vernalized during winter. Then, if the roots are not grazed or harvested for sugar, the plants left in the field will produce seed the next fall.

Perennials

Perennial forbs persist for several years, but they die back to near the soil surface annually and regrow the following spring from axillary buds on perennating tissues. Although the above-ground portion of forbs is by definition herbaceous, the below-ground portion is variable but is often woody. For example, the below-ground portion of forbs can be a corm with a taproot (as in dotted gayfeather), rhizomes or a spreading **caudex** (as in Missouri goldenrod), rhizomes and stolons (as in heath aster), a deep taproot with rhizomes (as in black samson), a taproot with a woody caudex (as in shell-leaf penstemon), or deep woody rhizomes with numerous buds (as in leafy spurge). Each of these perennating tissues is instrumental in determining the competitive ability of the plant, and helps to determine the niche that each species will fill. For example, the combination of the corm, taproot, and lateral roots found in dotted gayfeather provides a set of unique structures that allows this species to occupy and thrive on more **xeric** upland sites.

Forage Quality of Forbs

Many producers consider some forbs undesirable because their competition reduces grass and legume production. In addition, some forbs decrease livestock gain, reduce animal reproductive success, or cause livestock death losses, all due to the presence of organic toxins or excessive levels of minerals such as selenium. These negative associations are sometimes erroneous. For example, western ragweed is considered to have little forage value for grazing livestock, and often increases in abundance when rangelands are grazed. In fact, research in mixed and tallgrass prairies indicates that western ragweed has little effect on grass production. Increases in its abundance result from reductions in grass competitive ability caused by other stresses, such as overgrazing (Vermeire and Gillen, 2000).

Stress factors such as overgrazing also have an impact on animal losses due to poisonous plants. Holechek (2002) reviewed 36 grazing experiments in North America to evaluate the relationship between grazing pressure and livestock losses due to poisonous plants, many of which are forbs. Annual death losses of livestock due to poisonous plants averaged 2.0% and 4.8% under moderate and heavy grazing pressures, respectively. Holechek concluded that most livestock losses due to poisonous plants could be prevented with good management. Although these problems do occur, it is important to understand which species are problematic and when the problem can be reduced with appropriate grazing management.

Forb Ecology and Livestock Value

In grasslands, forbs have the ecological role of an opportunist or niche filler. Forb seed are present in the soil seed bank, and their relative abundance in the plant community is determined by factors such as water, space, and disturbance. After establishment, forbs tend to persist in grazing systems because they produce large quantities of seed, many produce secondary compounds that reduce palatability and intake, and some may be poisonous to grazing animals. Animal utilization of these forbs depends on which species are present and the relative abundance of associated legume and grass species. Some forbs have excellent forage quality. For example, redroot pigweed, common lambsquarters, common ragweed, dandelion, white cockle, and young Jerusalem artichoke are comparable in forage quality to seeded legumes and cool-season grasses (Buxton and Fales, 1994). As with seeded forages, the forage quality of forbs is highest when the plants are in the immature stages. As forbs mature, they decrease in **palatability** and forage value, and species with thorns reduce accessibility to animals. The decrease is often faster than for seeded forages, which suggests that forb utilization at young stages is important. As previously mentioned (see Forage Quality of Forbs), some forbs can have negative effects on forage production and quality.

Seeded forbs such as brassicas also fill niches mainly by providing grazing at times when other forage species are not productive. Overall performance is best when these annuals are stockpiled for use in late fall or early winter. Controlled grazing systems, limiting animals to relatively small areas through the use of strip grazing, are preferred. This reduces wastage due to trampling and soiling, and improves potential regrowth.

Late fall use must be managed carefully to prevent excessive loss due to freezing of the top growth. Freezing injury occurs after a few days of exposure to night temperatures near 10°F. Animals consume the frozen forage, but decomposition proceeds rapidly after it thaws. Roots of turnips and swedes persist after the top growth has deteriorated, thereby extending the grazing season (Koch and Karakaya, 1998). Cattle tend to uproot turnips when a portion of the root is above the soil surface. Sheep usually do not uproot the plants, but consume only the above-ground portion of the roots.

Location and Role of Meristems

The location and functions of the shoot apex, or growing point, are major factors in the management of legumes and forbs. As with grasses, the shoot apex of legumes and other forbs initiates new leaves, contains cells that develop into axillary buds, develops the cells for nodes, and regulates the division and enlargement of the internode cells that elongate the stems to increase plant height. Elongated growth spreads the leaf area over a larger vertical distance to improve light interception and photosynthesis (see Chapter 4). Thus the shoot apex regulates both the growth rate and the morphology of the legume or forb plant.

The **determinate** or **indeterminate** flowering of legumes and forbs affects their management. For example, red clover is determinate because the shoot apices develop into an inflorescence (head) that stops stem growth. Also many of the axillary buds develop into stems and terminate in a head. In contrast, the shoot apex of a white clover stolon rarely flowers but continues to grow laterally, demonstrating indeterminate growth, even when many flower stalks are produced by axillary buds. Birdsfoot trefoil is an indeterminate species in that the shoot apices of the main stems and axillary branches continue to grow vegetatively while the axillary buds on the lateral branches flower profusely. The ability of indeterminate species to flower during periods of defoliation by animals enables these plants to persist better under grazing.

When alfalfa or an upright perennial forb is cut, the shoot apices in the upper part of the canopy are removed and the plant must regrow from axillary buds at the base of the stems. As with grasses, the stem must be able to replace itself and perhaps add another stem to expand the crown. In contrast to grasses, however, the root system of many perennial legumes and other grassland forbs can live for several years and support several cycles of new stem production.

Annual legumes and forbs must produce seed, but they also depend on the seed being deposited in an environment and canopy that are amenable to germination and seedling development the following year. Fortunately, most bunchgrasses allow other seed to germinate in open spaces, which improves the diversity of the pasture, range, or hayfield to improve yield, quality, and habitat for wildlife (see Chapter 19). Annual forbs such as the brassicas are managed to avoid seed production; therefore they need to be reseeded.

Perennation of Birdsfoot Trefoil

Birdsfoot trefoil is a desirable perennial legume that does not cause livestock bloat, and decumbent types are well adapted for grazing. Individual plants of this species are relatively short-lived, having a lifespan of only about 3 years in the northern USA and even less at more southern locations. Loss of plants to root and crown diseases means that plants must produce seed at some regular interval and the canopy has to be open for a period of time in spring to allow seedlings to establish.

A few years ago, scientists working for the Agricultural Research Service of the United States Department of Agriculture found birdsfoot trefoil in Morocco that produced rhizomes. This trait was bred into germplasms adapted to the USA. As with other birdsfoot trefoils, the plants did not store much carbohydrate in the roots during summer, but they produced abundant rhizomes in the fall. Testing is needed to determine how rhizome formation and new plant development are affected by competition from other species, fertilizer nutrients, and summer and fall management.

The rhizomatous types produce seed, but rhizomes should reduce dependence on seed production in warmer climates where flowering is less profuse and natural bee activity is often low. Rhizomes may help to maintain stands in more arid environments where natural seed germination and establishment in dry soil are problematical.

Management Implications for Mixtures

Legumes and other forbs are often grown in mixtures with grasses, but differ markedly from the grasses in morphology and do not go through the various growth stages at the same rate. For example, it is difficult to interseed cool-season legumes such as red clover into established warm-season grasses such as switchgrass, big bluestem, and indiangrass due to the highly competitive environment, and they typically provide little benefit to livestock (Keyser et al., 2016). In addition, the plants in each group differ in palatability. Species for mixtures should be matched for maturity and palatability. Other forbs will volunteer in areas where they can, and those with a growth habit that matches the maturity of the legume and grass component can persist.

Management practices can be modified to maintain desirable but less competitive species. For example, red clover is commonly grown with timothy in many northern regions for hay production, whereas alfalfa is commonly grown with orchardgrass. Orchardgrass reaches each development stage earlier than timothy, and alfalfa reaches each developmental stage earlier than red clover. During summer regrowth, however, timothy commonly produces new reproductive stems. This habit makes it less well suited for grazing compared with orchardgrass, tall fescue, or kentucky bluegrass, all of which remain vegetative during summer and fall, with the shoot apex protected near soil level. This non-flowering growth habit of grasses is more compatible with white clover or birdsfoot trefoil.

Illinois Bundleflower Finds Its Place

Illinois bundleflower is a deep-rooted, warm-season perennial legume native to the plains and prairies of the USA. The plants are both drought tolerant and winter hardy. They grow upright to reach a height of 2–4 ft., and have pinnately compound leaves with up to 200 small blades per leaf, seed heads with numerous flowers, and high seed yield. Research conducted at the University of Missouri showed that Illinois bundleflower is high yielding among native legumes, and is similar to introduced legumes in forage quality. It is also palatable to all classes of livestock. Even so, its primary use is to produce food for wildlife. The large seed is an excellent source of energy for birds such as the northern bobwhite quail (*Colinus virginianus* L.).

Illinois bundleflower grows best with bunch-type perennial warm-season grasses such as little bluestem. The combination of little bluestem and Illinois bundleflower provides excellent habitat for bobwhite quail. During winter the stiff, vertical structure of little bluestem provides thermal and predator protection, while the seed from Illinois bundleflower provides energy. During the growing season, little bluestem provides protection from predators, and the flowers of Illinois bundleflower attract insects. This is ideal brood-rearing habitat for bobwhite quail. The open spaces between bunches allow easy movement in well-protected areas, and the insects provide the high levels of protein necessary for chick growth. The combination of a native grass, a native legume, and a native bird is a natural fit for the plains and prairies.

Animals that are grazing summer regrowth of an orchardgrass–alfalfa mixture remove only leaf blades of orchardgrass but remove the upper shoot apices from alfalfa, forcing the legume to repeatedly regrow from the crown (Fig. 3.5). If summer regrowth of white clover and tall fescue is grazed, animals remove mainly leaf blades from tall fescue and leaf blades and petioles from white clover (Fig. 3.4), leaving the stem and shoot apices of both species near soil level, so they can grow and continue to initiate new leaves. When management is detrimental to a particular plant species, other forb species or annual grasses may increase in proportion. Although these encroaching species can contribute, forage yield or quality may be lower compared with the planted species (Buxton and Fales, 1994).

Palatability differences among species can affect the proportions in a mixture. Taste or physical factors such as thorns or sharp-edged leaves deter grazing of specific plants. Plants that are less grazed have more photosynthetic area remaining and may increase in abundance compared with more palatable competitors. Animal taste is not understood, but differences in grazing preferences may be due to chemicals such as tannin or alkaloids that alter acceptance and intake. Preference for or avoidance of certain plants is a behavior learned by young animals by observing what the mother eats. Thus the knowledge can be passed on from generation to generation (Vallentine, 2000).

Understanding how each forage plant grows helps to determine how it can be best managed to meet the needs of the producer. The balance among species can be managed by seeding practice, grazing or cutting management, fertilization practices, and herbicides. For example, fertilizing with P and K favors the legume component over other forbs and grasses, whereas fertilizing with N favors the grass and other forbs over legumes. Undesirable forbs that are low in palatability or poisonous may need to be removed with herbicides or a timely fire. Several management principles are similarly related to plant development, and are discussed throughout this book.

Summary

Annual and perennial legumes are characterized by a taproot with several stems growing from the crown near soil level. Upright-growing legumes (e.g., alfalfa, red clover) regrow from axillary buds at the crown, and this is repeated with each defoliation. Longer time intervals between defoliations are critical for plant persistence. Conversely, prostrate-growing legumes regrow from axillary buds along stolons (e.g., white clover) or prostrate stems (e.g., birdsfoot trefoil) near ground level, allowing them to tolerate frequent grazing. Persistence of upright legumes depends on the longevity of the taproot. Prostrate legumes persist by overwintering of stolons or rhizomes. Many annual legumes and birdsfoot trefoil persist by natural seed production.

Forbs, especially legumes, are broad-leaved plants that can be valuable contributors to forage needs. Non-legume annual forbs such as turnip are planted as fall cover crops to lengthen the grazing season, but in most pastures and hayfields forbs are opportunists that volunteer to fit niches and contribute forage. Legumes produce high-quality forage and fix atmospheric N that is available to associated grasses and subsequent crops, but they are more sensitive to management practices than are most grasses. Therefore management of grass–legume mixtures depends on matching the growth habits of the species and giving priority to the legume components.

Questions

1. What does the term *legume* mean?
2. How do legumes in pastures and hayfields contribute to the economy of the USA?
3. List the major morphological characteristics that can be used to distinguish between forage legumes.
4. What functions do cotyledons serve in young legume seedlings?
5. Why are fields of legumes for seed production generally isolated some distance from other fields of the same species?
6. Distinguish between a stolon and a rhizome.
7. Explain the role of axillary buds in persistence of legumes and other forbs.
8. Why is it important to be able to identify the growth stages of legumes?
9. Briefly explain why it may be more difficult to manage a grass–legume–forb mixture than a monoculture of a perennial legume.
10. Discuss the importance of forbs to native rangeland and to humid pastureland.
11. Why is it important to manage some forb species in rangeland and pasture?

References

Bass, LN, CR Gunn, OB Hesterman, and EE Roos. 1988. Seed physiology, seedling performance, and seed sprouting. In AA Hanson, DK Barnes, and RR Hill, Jr. (eds.), Alfalfa and Alfalfa Improvement, pp. 961–983. Agronomy Monograph 29. American Society of Agronomy, Madison, WI.

Buxton, DR, and SL Fales. 1994. Plant environment and quality. In GC Fahey et al. (eds.), Forage Quality, Evaluation, and Utilization, pp. 155–199. American Society of Agronomy, Madison, WI.

Fick, GW, and SC Mueller. 1989. Alfalfa: Quality, Maturity, and Mean Stage of Development. Information Bulletin 217. Department of Agronomy, College of Agriculture and Life Sciences, Cornell University, Ithaca, NY.

Holechek, JL. 2002. Do most livestock losses to poisonous plants result from "poor" range management? J. Range Manage. 55:270–276.

Isely, D. 1951. The Leguminosae of the north-central United States: I. Loteae and Trifolieae. Iowa State Coll. J. Sci. 25:439–482.

Jung, GA, RA Byers, MT Panciera, and JA Shaffer. 1986. Forage dry matter accumulation and quality of turnip, swede, rape, Chinese cabbage hybrids and kale in the eastern USA. Agron. J. 78:245–253.

Keyser, PD, ED Holcomb, CM Lituma, GE Bates, JC Waller, CN Boyer, and JT Mulliniks. 2016. Forage attributes and animal performance from native grass interseeded with red clover. Agron. J. 108:373–383.

Koch, DW, and A Karakaya. 1998. Extending the Grazing Season with Turnips and Other Brassicas. Cooperative Extension Service Bulletin B-1051. University of Wyoming, Laramie, WY.

Nelson, CJ, and D Smith. 1968. Growth of birdsfoot trefoil and alfalfa. II. Morphological development and dry matter distribution. Crop Sci. 8:21–25.

Shibles, RM, and HA MacDonald. 1962. Photosynthetic area and rate in relation to seedling vigor of birdsfoot trefoil (*Lotus corniculatus* L.). Crop Sci. 2:299–302.

Smith, DH, and M Collins. 2003. Forbs. In RF Barnes, CJ Nelson, M Collins, and KJ Moore (eds.), Forages. Volume 1: An Introduction to Grassland Agriculture, 6th ed., pp. 215–236. Iowa State Press, Ames, IA.

Stubbendieck, J, SL Hatch, and NM Bryan. 2011. North American Wildland Plants, 2nd ed. University of Nebraska Press, Lincoln, NE.

Teuber, LR, and MA Brick. 1988. Morphology and anatomy. In AA Hanson, DK Barnes, and RR Hill, Jr. (eds.), Alfalfa and Alfalfa Improvement, pp. 125–162. Agronomy Monograph 29. American Society of Agronomy, Madison, WI.

Vallentine, JF. 2000. Grazing Management, 2nd ed. Academic Press, San Diego, CA.

Vermeire, LT, and RL Gillen. 2000. Western ragweed effects on herbaceous standing crop in Great Plains grasslands. J. Range Manage. 53:335–341.

Westbrooks, FE, and MB Tesar. 1955. Taproot survival of ladino clover. Agron. J. 47:403–410.

CHAPTER

4

Physiology of Forage Plants

Jennifer W. MacAdam and C. Jerry Nelson

Forage physiology describes the integrated function of forage plants, from biochemical pathways operating at the cellular level to plant growth and development in the field. The most fundamental biochemical pathways in plants are **photosynthesis** and **respiration**.

Photosynthesis captures solar energy and stores it in simple sugars, referred to as **photosynthate**. Photosynthesis is the first step in the food chain that supports all animals, including humans. Photosynthate can either be used immediately in metabolism, to provide the building blocks of plant tissues, or it can be stored for future use. Respiration releases the energy stored in photosynthate, allowing it to be used for cellular metabolism. This energy is needed to build proteins, lipids, and complex carbohydrates, to fuel the growth of roots and shoots, to absorb mineral nutrients from the soil, and to support the enzymatic reactions that maintain mature tissue.

The goal of this chapter is to demonstrate how underlying plant processes interact with management and the environment to determine the productivity, nutritive value, and persistence of forages and grasslands.

Forages and the Productivity of Agricultural Land

The majority of land available for agriculture is too steep, wet, dry, rocky, or vulnerable to wind or water erosion to be used for row crop production. Grazing of forage plants by ruminants can be the most beneficial use of this land, because perennial forage grasses and legumes provide permanent ground cover that harvests solar energy, slows wind, aids water infiltration, adds significant organic matter and nitrogen (N) to improve soil health, and can—through grazing—support the production of high-quality protein for human consumption. During grazing, most of the nutrients contained in the forages consumed by livestock are recycled to the soil as urine and dung. Thus perennial forages can preserve or improve the least favorable agricultural land while supporting food production by converting solar energy into feed for animals.

The Photosynthetic Process

Photosynthesis occurs in **chloroplasts**, which are located in the cells of all green tissues of plants, but especially in the leaves. Leaves have the optimal structure for maximizing the interception of light and the absorption of carbon dioxide (CO_2) that is converted into photosynthate. Chloroplasts are filled with an aqueous matrix, called the **stroma**, which contains a system of **thylakoid membranes** that are folded into stacks called **grana** (Fig. 4.1).

The thylakoid membranes are flattened tubes in which clusters of chlorophyll and carotenoid pigments are embedded next to enzyme complexes. In the initial photochemical reactions of photosynthesis, light increases the energy of the pigment clusters, causing the release of electrons that are captured by the enzyme complexes. Some of the chemical energy generated as these electrons are passed along an electron transport chain on the thylakoid membranes is sequestered inside the thylakoid membranes, and then released in a controlled process that is used to form adenosine triphosphate (**ATP**). The final step in the photochemical reactions of photosynthesis is the capture of the chlorophyll electrons and their energy in the reduction of nicotinamide adenine dinucleotide phosphate (**NADPH**).

Both ATP and NADPH are formed in the stroma of the chloroplast, where they are subsequently used to reduce CO_2 to sugars. Nitrate (NO_3^-) and sulfate (SO_4^{2-}) are also reduced in chloroplasts using ATP. However, N constitutes

Forages: An Introduction to Grassland Agriculture, Seventh Edition. Edited by Michael Collins, C. Jerry Nelson, Kenneth J. Moore and Robert F Barnes.

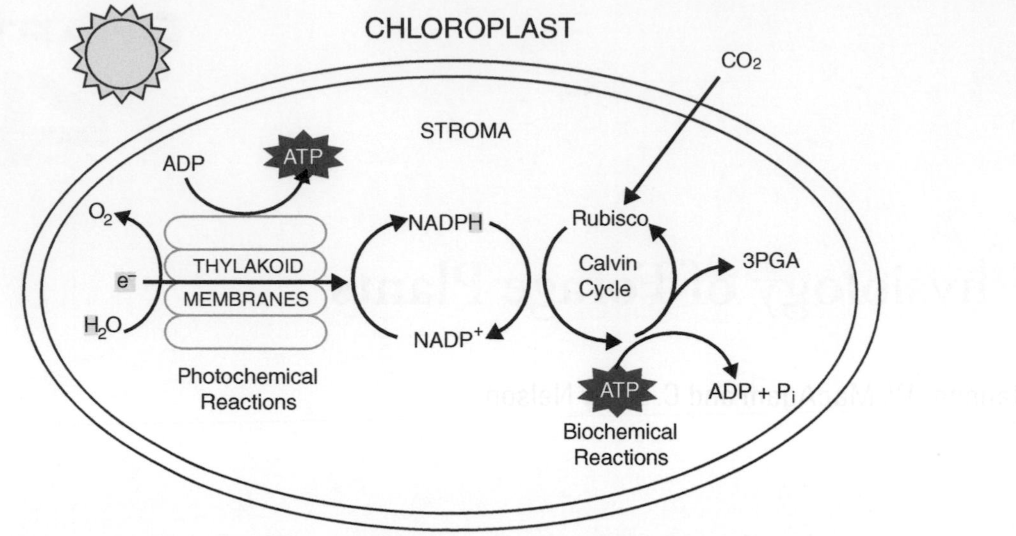

$$6\,CO_2 + 6\,H_2O \xrightarrow{\text{Sunlight}} C_6H_{12}O_6 + 6O_2$$

FIG. 4.1. In chloroplasts, the photochemical reactions of photosynthesis capture solar energy, while the biochemical reactions of photosynthesis use the energy for carbohydrate synthesis. (Adapted from MacAdam, 2009.)

just 2–3% of plant dry weight and sulfur (S) only 0.2%, whereas carbon (C) represents about 45% of the dry weight of plants.

Carbohydrates are formed in chloroplasts by adding CO_2 to the 5-C sugar ribulose-1,5-bisphosphate (RuBP) through the activity of the enzyme ribulose bisphosphate carboxylase/oxygenase (**rubisco**) (Fig. 4.1). Rubisco is an abundant enzyme, constituting about 40% of the soluble protein in the leaves of **cool-season** or **C_3 plants**. Functioning as a carboxylase, rubisco adds CO_2 to RuBP to form a highly unstable 6-C compound that immediately splits to form two molecules of the 3-C compound 3-phosphoglycerate (**3PGA**) in the Calvin cycle (Fig. 4.1). Because the first measured product of photosynthesis was a 3-C compound, plants that use only this basic photosynthetic pathway are referred to as C_3 plants.

During daylight hours, when the rate of photosynthesis is high, most of the new 3PGA is used to form glucose molecules that are linked together to form **starch**, which accumulates in the stroma. At night, the starch is broken down again to glucose and then to phosphorylated 3-C sugars (triose phosphates) that are exchanged for inorganic P (P_i) from the **cytosol** (Fig. 4.2). In the cytosol, two molecules of triose phosphate are combined to form glucose or fructose, both 6-C sugars. One fructose molecule and one glucose molecule are linked to form sucrose for transport through the **phloem** (Fig. 4.2).

Sucrose can be transported throughout the plant via the phloem to sites that require energy for respiration, growth, or storage. Cattle prefer tall fescue hay cut in the late afternoon to hay cut early in the morning (Fisher et al., 1999), in part because the gradual accumulation of sucrose and other **non-structural carbohydrates** in leaves during the day results in higher nutritional value and palatability.

Photorespiration

Rubisco also catalyzes the addition of molecular oxygen (O_2) to RuBP (Fig. 4.3). For every two molecules of O_2 that are added to a molecule of RuBP, two molecules of a 2-C compound, 2-phosphoglycolate, are formed, along with two molecules of 3PGA. This reaction is the "oxygenase" function of rubisco, because O_2 rather than CO_2 is a substrate of the reaction. Metabolism of the 2-phosphoglycolate creates one more 3PGA molecule, but also results in the loss of one CO_2 molecule from the plant.

During drought, reducing the loss of water from the leaves becomes a higher priority than absorbing CO_2 for photosynthesis, and when plants close their stomata to prevent dehydration, the CO_2 levels inside the leaves become depleted. This favors the reaction of rubisco with O_2 and the subsequent loss of CO_2, which is termed **photorespiration** and is most likely to occur in hot, dry weather.

O_2 is present at much higher concentrations in the atmosphere (21%) than is CO_2 (0.04%), but rubisco has a

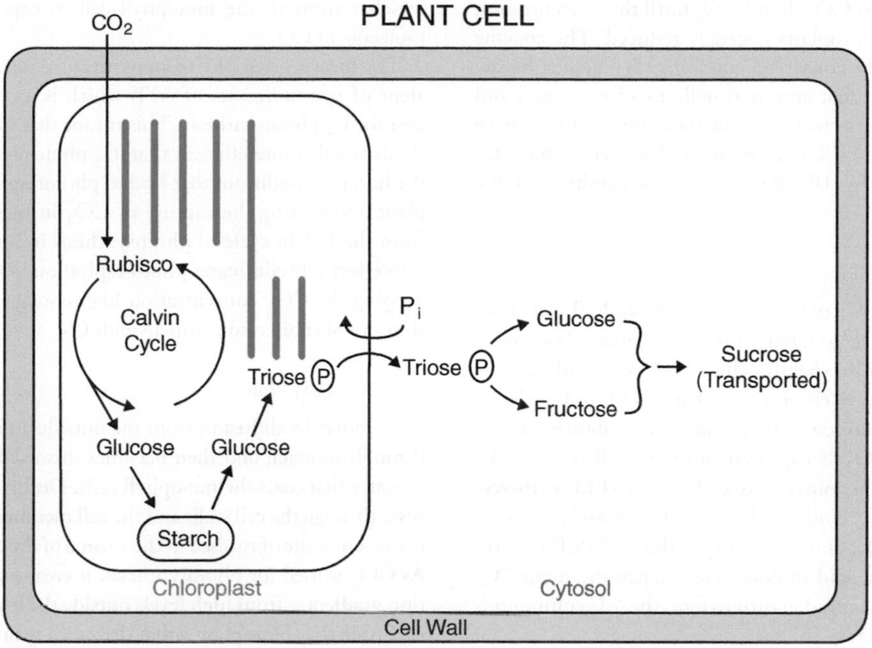

FIG. 4.2. Carbohydrates formed by photosynthesis are used to make sucrose, which is the form in which photosynthate is translocated via the phloem from sources to sinks. (Adapted from MacAdam, 2009.)

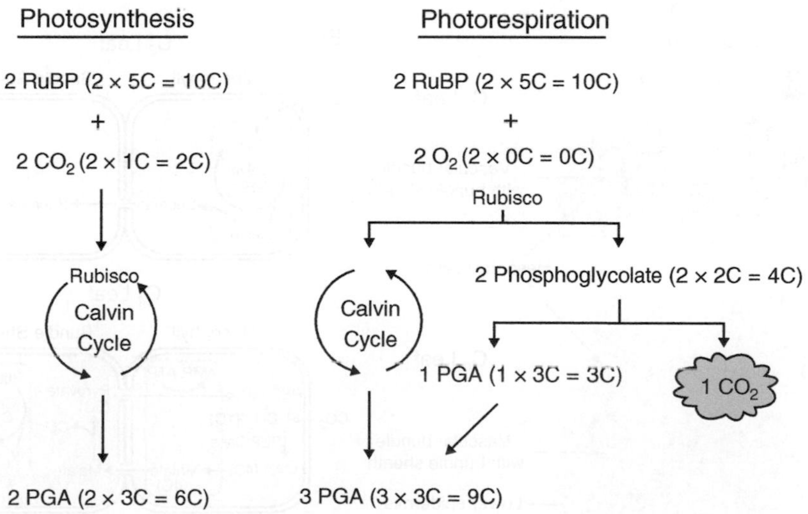

FIG. 4.3. Photorespiration occurs when the internal leaf concentration of CO_2 becomes so low that rubisco adds O_2 instead of CO_2 to RuBP. The result is the loss of one CO_2 molecule for every two O_2 molecules added to RuBP. (Adapted from MacAdam, 2009.)

higher affinity for CO_2 than for O_2 until the concentration of CO_2 near chloroplasts is greatly reduced. The enzyme rubisco is highly conserved, meaning that it has hardly changed since it first appeared millions of years ago, and attempts to reduce the oxygenase function of rubisco have been unsuccessful. Photorespiration effectively reduces C_3 photosynthesis by 10–50% or more depending on the temperature.

C_4 Photosynthesis

In the leaves of C_3 plants, only the **mesophyll** cells have well-developed chloroplasts, but in **C_4 plants** both mesophyll and **bundle sheath** cells, which surround the vascular bundles, have chloroplasts (Fig. 4.4A). In C_4 plants, nearly all the rubisco is found in bundle sheath chloroplasts, while CO_2 is captured in mesophyll cells by the enzyme phosphoenolpyruvate carboxylase **(PEP carboxylase)** (Fig. 4.4B). Unlike rubisco, PEP carboxylase has no reaction with O_2, and adds CO_2 to the 3-C PEP to form the 4-C organic acid oxaloacetate—hence the name "C_4 plants"—which is used in turn to form the 4-C compounds malate or aspartate.

The 4-C compounds diffuse through **plasmodesmata** from the mesophyll into the bundle sheath cells, where CO_2 is released and is then absorbed by the chloroplasts. In the bundle sheath cells, CO_2 is released and reacts with rubisco in the Calvin cycle, just as in C_3 plants (Fig. 4.4B). In C_4 bundle sheath cells the release of CO_2 from the 4-C acid results in formation of the 3-C molecule pyruvate, which returns to the mesophyll cell to capture another molecule of CO_2.

The formation of PEP from pyruvate requires the equivalent of two molecules of ATP, which is an extra energy cost for C_4 photosynthesis. This means that C_4 photosynthesis is only more efficient than C_3 photosynthesis under the hot, dry conditions that lead to photorespiration in C_3 plants. Separating the capture of CO_2 in mesophyll cells from the Calvin cycle of photosynthesis in bundle sheath cells effectively eliminates photorespiration in C_4 plants by keeping the CO_2 concentration high enough at the reaction site of rubisco to compete with O_2.

The Role of Stomata

CO_2 moves by diffusion from the outside air into the leaf through stomata, and then becomes dissolved in the film of water that coats the mesophyll cells. Dissolved CO_2 diffuses through the cell walls and the cell membranes to reach the reaction site of rubisco in the stroma of the chloroplasts. As CO_2 is used for photosynthesis, it creates a concentration gradient—from high levels outside the leaf to low levels inside the chloroplast—that drives CO_2 uptake.

Stomata are pores in the epidermis of plants (Fig. 4.4) that open when the concentration of CO_2 is depleted through photosynthesis, which means that stomata normally open during the day and close at night. The concentration of water vapor inside the leaf tissue is constantly replenished by water from the roots, so when the stomata open to allow CO_2 to diffuse in, water vapor

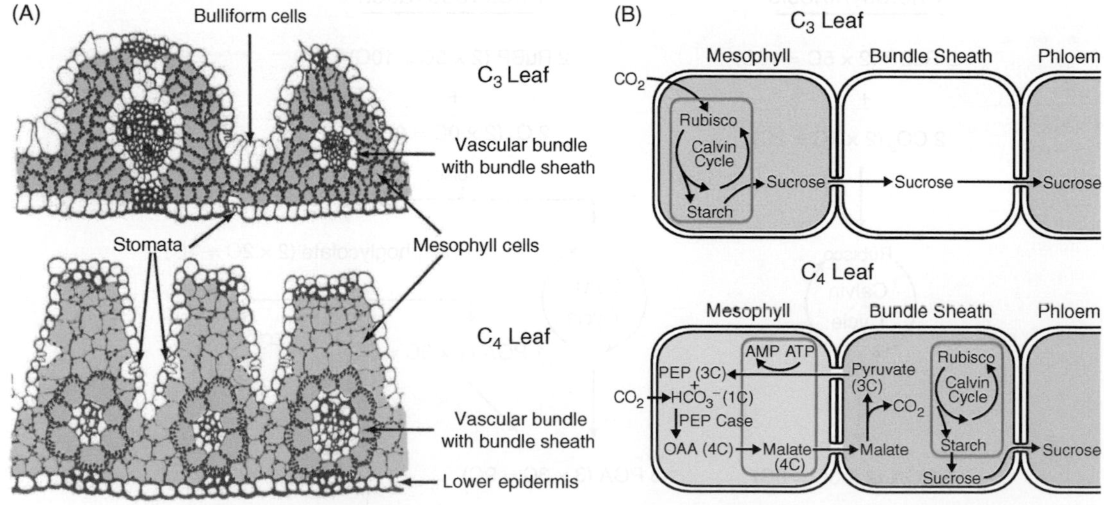

Fig. 4.4. The Calvin cycle of photosynthesis takes place in the mesophyll of C_3 leaves and in the bundle sheath cells of C_4 leaves (green cells, A and B), while the mesophyll of C_4 leaves captures CO_2 as 4-C compounds (blue cells, A and B) that are transported to bundle sheath cells. (Figure 4.4A: Drawings of tall fescue and cordgrass hybrid leaves adapted from Burr and Turner, 1933. Figure 4.4B: Adapted from MacAdam, 2009.)

diffuses out. This evaporation of water from the leaves is termed **transpiration**. However, when drought occurs, the stomata close during the day to reduce the transpiration of water and prevent wilting.

There are often high levels of solar radiation during drought that continue to drive photosynthesis. When the stomata of C_3 plants are partially or completely closed, ongoing photosynthesis uses the available CO_2, thus reducing the CO_2 concentration inside the stroma and causing photorespiration to occur. In contrast, the more efficient CO_2 uptake and transfer system of C_4 plants concentrates CO_2 in the bundle sheath cells even when the stomata are partially closed. The internal CO_2 concentration becomes lower in C_4 plants than in C_3 plants when the stomata are closed, creating a steeper CO_2 concentration gradient from the outside air to the stroma, and therefore increasing the rate of CO_2 uptake when C_4 plants open their stomata. This means that the photosynthetic efficiency of C_4 plants is greater than that of C_3 plants under hot, dry conditions.

Water use efficiency (WUE) is the amount of biomass produced per volume or weight of water used. Because CO_2 uptake and use are more efficient in C_4 plants, C_4 photosynthesis occurs with less water loss, so the WUE of C_4 plants is greater than that of C_3 plants. The C_4 grasses of the tall-grass prairie, such as switchgrass and big bluestem, root deeper than C_3 grasses. This adaptation allows these C_4 plants to avoid drought by extracting water from deeper in the soil profile during hot, dry periods, further improving their productivity during periods of hot, dry weather.

The CO_2 concentration in the atmosphere is expected to increase from the current level of 0.04% to 0.06% by the middle of this century (US Environmental Protection Agency, 2016) unless effective mitigation measures are implemented. As the CO_2 level rises, the concentration gradient from the outside to the inside of leaves will also increase, so the rate of diffusion of CO_2 into chloroplasts will increase. In locations where the climate does not become hotter and drier, the photosynthetic efficiency and productivity of C_3 plants could increase by 30% or more through reduction of photorespiration. The efficiency of C_4 plants will probably increase little if at all as the atmospheric CO_2 concentration rises, causing them to lose some of their ecological advantage.

Cacti and Life after Dark

Cacti are an important component of the vegetation of deserts and very arid rangelands, but how do they manage to grow and survive with a lower water supply than other plants? Cacti, pineapple, and several other arid plant species have developed a variation on capturing CO_2 for C_3 photosynthesis called crassulacean acid metabolism (CAM). The stomata of CAM plants open at night rather than during the day to minimize transpiration losses, and CO_2 is added to PEP by PEP carboxylase to form 4-C organic acids as in C_4 plants (Fig. 4.4B). These organic acids are stored in the vacuole at night. Then, during the day, when sunlight is available to drive photosynthesis, the stomata remain closed to conserve water. The 4-C acids, in the same way as in C_4 plants, release their CO_2 in chloroplasts, where it is captured by rubisco and used for C_3 photosynthesis. Pyruvate, the 3-C residual, is used to regenerate PEP that is ready to capture more CO_2 when the stomata open the following night. As in C_4 plants, there is an extra energy cost for converting pyruvate to PEP. The result is that these plants grow slowly but are extremely water efficient.

Radiation Effects

Light intensity influences the rate of the photochemical reactions of photosynthesis that occur on the thylakoid membranes to produce ATP and NADPH. Full sun is the maximum intensity of solar radiation outdoors near midday, which is approximately 2000 μmol photons m^{-2} s^{-1}. At low light intensities, as the interception of solar radiation increases, photosynthesis increases in a nearly linear manner, but C_3 photosynthesis begins to saturate at lower light intensities than in C_4 plants (Fig. 4.5A).

In C_3 plants, the rates of NADPH and ATP synthesis continue to increase as the light intensity increases, but the rate of CO_2 diffusion into the leaves for the biochemical reactions becomes limiting, so rubisco begins to react with O_2, causing increased photorespiration. In contrast, the photosynthesis of C_4 plants continues to increase with increasing radiation intensity, mainly because PEP carboxylase keeps CO_2 levels low in the mesophyll cells, creating a steeper CO_2 gradient from the outside to the inside of leaves, and thus enhancing the rate of CO_2 diffusion into the leaf.

The efficiency of CO_2 fixation at low radiation intensity is similar for C_3 and C_4 plants, but in full sun the photosynthetic rate of C_4 species may be nearly double that of C_3 species. Using cutting and grazing management to optimize radiation interception by C_3 and C_4 plants is important for production of pastures and hay crops.

Temperature Effects

Whereas the direct effect of light on photosynthesis is on the photochemical reactions that lead to the formation of ATP and NADPH, the effect of temperature on photosynthesis is a function of the biochemical reactions that occur in the stroma. The rate of change with a 10°C increase in temperature is called the Q_{10} temperature coefficient. Rates of biological enzyme reactions, including photosynthesis and respiration, typically double with a 10°C (18°F) increase in temperature. Most C_3 grasses and legumes can fix CO_2 at temperatures near freezing, and the net

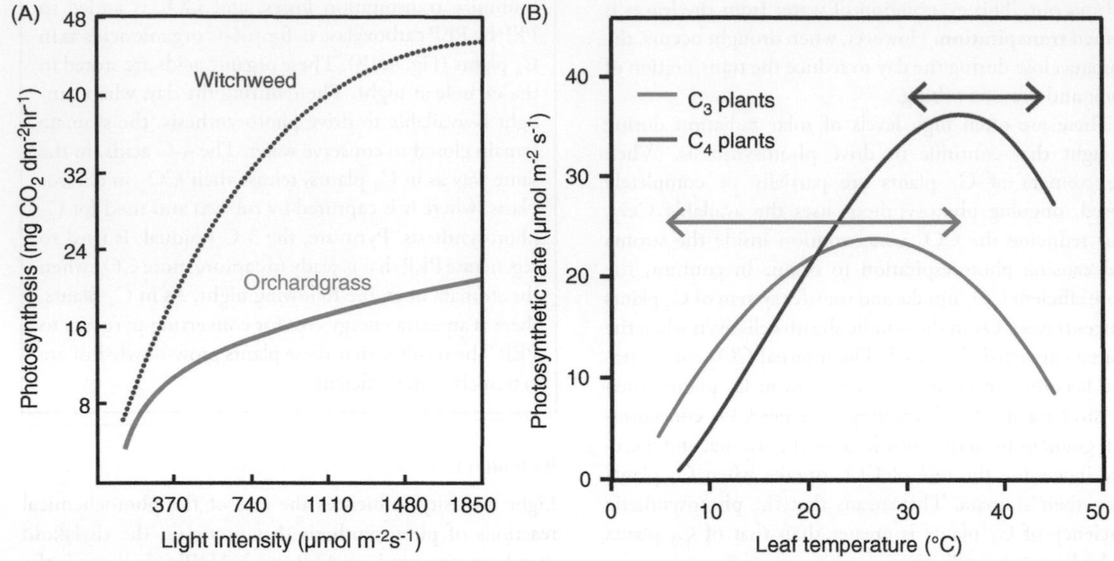

FIG. 4.5. A. In low light, the rate of photosynthesis increases linearly with increasing light intensity, but the greater efficiency of delivery of CO_2 to rubisco in C_4 plants (e.g., witchweed) compared with C_3 plants (e.g., orchardgrass) results in higher rates of C_4 photosynthesis in full sun. B. The photosynthesis of C_3 plants is more efficient at low temperatures than is that of C_4 plants, but at temperatures above 30°C, C_3 photosynthesis is reduced by photorespiration, which is effectively eliminated in C_4 plants. (Fig. 4.5A adapted from Singh et al., 1974. Fig. 4.5B adapted from Yamori et al., 2014, reproduced with permission of Springer.)

photosynthetic rate (photosynthesis minus respiration and photorespiration) of C_3 plants reaches a maximum at temperatures in the range 20–25°C (68–77°F) (Fig. 4.5B). Photosynthesis is reduced at temperatures above 30°C (86°F) because the solubility of CO_2 in the cytosol decreases more with increasing temperature than does the solubility of O_2. The result is that as the ratio of CO_2 to O_2 in the stroma decreases, photorespiration is favored, and the net photosynthesis rate in C_3 plants decreases rapidly. In contrast, the photosynthesis rate of C_4 plants is low at 10°C (50°F) because C_4 plants have relatively low concentrations of rubisco and other C_3 photosynthetic enzymes. Levels of these Calvin cycle enzymes are sufficient at higher temperatures (under which conditions their activity is higher) since the CO_2 is captured by the C_4 pathway (Fig. 4.5B). C_4 photosynthesis increases to a maximum at 35–40°C (95–104°F), and then decreases as proteins become destabilized by excess heat. The higher temperature optimum for C_4 plants is mainly due to control of photorespiration.

Leaf Anatomy and Forage Quality

Plant cells that are metabolically active, such as mesophyll cells, are thin-walled, contain abundant sugars and proteins, and are rapidly degraded by rumen microbes. Leaves of grasses with C_3 photosynthesis, such as tall fescue, have several layers of chloroplast-lined mesophyll cells that radiate from widely spaced veins (Fig. 4.4A). In contrast, leaves of C_4 grasses such as bermudagrass have closely spaced veins often surrounded by only one layer of mesophyll cells, because mesophyll cells function only to capture CO_2 and transport it to bundle sheath cells around the veins in C_4 plants.

The major veins of all grasses contain **xylem** elements with reinforced walls for water transport, clusters of phloem sieve tubes, and companion cells. There are also bundles of thick-walled fiber cells located at the epidermis above and below veins, and these sometimes extend from the upper to the lower epidermis to provide strength and rigidity. One or more minor veins are interspersed with each major vein; these may only contain phloem, to facilitate the transport of photosynthate from the leaves to the other organs of the plant. Minor veins are positioned closer together in C_4 grass leaves because the efficiency of photosynthesis is high, requiring fewer mesophyll cells per unit of photosynthate.

The thick walls of fiber cells are composed largely of cellulose, which is a carbohydrate, so fiber cells are relatively low in protein. Cellulose strengthens the fiber cell walls, and the deposition of lignin in fiber cell walls makes them rigid, so veins function as both the circulatory system and the skeleton of leaves. Lignin is also deposited in response to diseases.

Lignin strongly resists microbial degradation, resulting in slow digestion in the rumen, where the structural carbohydrates from cell walls are digested by microbes. The proportion of lignified xylem elements and fiber cells, which constitute the fiber fraction of the forages used as animal feed, is higher in C_4 than in C_3 grasses because the proportion of veins to mesophyll cells is higher in C_4 plants. The rate of fiber digestion in the rumen decreases as the concentration of lignin increases, and the fiber from C_4 grasses has a higher lignin concentration than the fiber from C_3 grasses (Akin, 1989).

The PEP carboxylase that is present in C_4 leaves is a smaller molecule than rubisco, and in C_4 leaves rubisco is sequestered in the bundle sheath cells where it is used very efficiently. Therefore less rubisco, which is a major protein, is needed in C_4 leaves than in C_3 leaves to achieve the same or higher rates of photosynthesis. This, combined with the greater number of veins and the associated increased proportion of fiber, results in a lower protein concentration in C_4 grasses than in C_3 grasses.

Due to their lower protein content, C_4 grasses can produce more biomass per unit of fertilizer N, making their use of nitrogen more efficient, but the lower protein concentration reduces the nutritive value of C_4 grasses for ruminants. Moreover, rumen degradation of C_3 grasses is generally faster and more complete than that of C_4 grasses.

Shaping Up a Grass Leaf

Leaf blades of grasses consist of different cell types that determine leaf size, shape, and morphology. Grass blades are vertical during emergence. After emergence, the angle of the blade and sheath increases. Parallel veins are apparent as ridges along grass leaves, and consist of bundles of fiber cells grouped above and below major veins to form the vascular bundle. This forms a girder-like structure to control the arching of the leaf blade and its final orientation to the sun. The final angle affects the number of leaves that can productively intercept light, and therefore determines the critical **leaf area index (LAI)**.

When drought stressed, many grass leaves reduce their surface area by rolling the blade. When water stressed, the large thin-walled bulliform cells located between veins (Fig. 4.4) lose turgor and shrink in diameter, causing the leaf blade to roll inward forming a long, thin cylinder with the upper surface on the inside. Reducing the leaf area that is exposed to light and heat reduces water loss through the stomata. If the plant is re-watered before the leaf is damaged, the bulliform cells will restore turgidity and the leaf will unroll.

The long slender leaves of grasses, supported by parallel veins that are continuous from base to tip, have a higher fiber content than the small round or oval leaflets that comprise legume leaves, which are supported by a midrib and a network of smaller veins. Since legumes all have C_3 photosynthesis and can synthesize nitrogen through associations with rhizobia bacteria, they also typically contain higher levels of protein than grasses. Although legume stems can develop high levels of fiber, forage legumes generally have significantly greater protein content, digestibility, and intake than forage grasses.

Translocation of Carbohydrates

Plant organs that provide resources to other tissues in the same plant are termed "sources", and those that receive the resources are termed "sinks." Sucrose from a source cell, such as a mesophyll cell, is loaded into the phloem for long-distance transport, termed translocation (Fig. 4.2 and 4.4B). Sucrose from the phloem is unloaded at a "sink", such as a developing seed or a root, where it is used for respiration, growth, or storage. When plants are defoliated by cutting or grazing, the process is reversed and storage organs such as roots and residual stubble become sources and growing shoots become sinks.

While they are still growing, the young leaves of shoots are sinks that import resources from older leaves. As their surface area increases, leaves begin to produce photosynthate in excess of their own needs, and they then export this excess to even younger leaves at the shoot apex, or to nearby tillers or branches developing from axillary buds. As each leaf matures, its exported photosynthate is used for root growth or storage while newer leaves support growth at the shoot apex which initiates new leaves. Grass leaves attached at nodes located at the base of the stem support the growth of axillary buds as they form new tillers, rhizomes, or stolons, as well as the growth of new roots.

New growth at the top of the canopy shades the lower, older leaves, eventually causing them to senesce (i.e., gradually die). **Senescence** is an active process during which nonstructural carbohydrates and proteins are broken down to form sucrose and amino acids, which are then translocated to storage or to younger growing tissues. Mobile mineral nutrients such as N, P, K, and Mg are also translocated from older to younger leaves to support the growth of the youngest leaves at the top of the canopy, where they can provide the greatest benefit to the plant.

The strength of a sink for photosynthate is a function of sink size, growth rate, and distance from the source. Grasses develop fibrous root systems with minimal storage capacity, and thus are more dependent on current photosynthate. When a grass stand is harvested or grazed, leaf area is decreased and root growth stops almost completely and does not recover until the canopy has partially regrown (see Chapter 2). Without a taproot, grasses are dependent

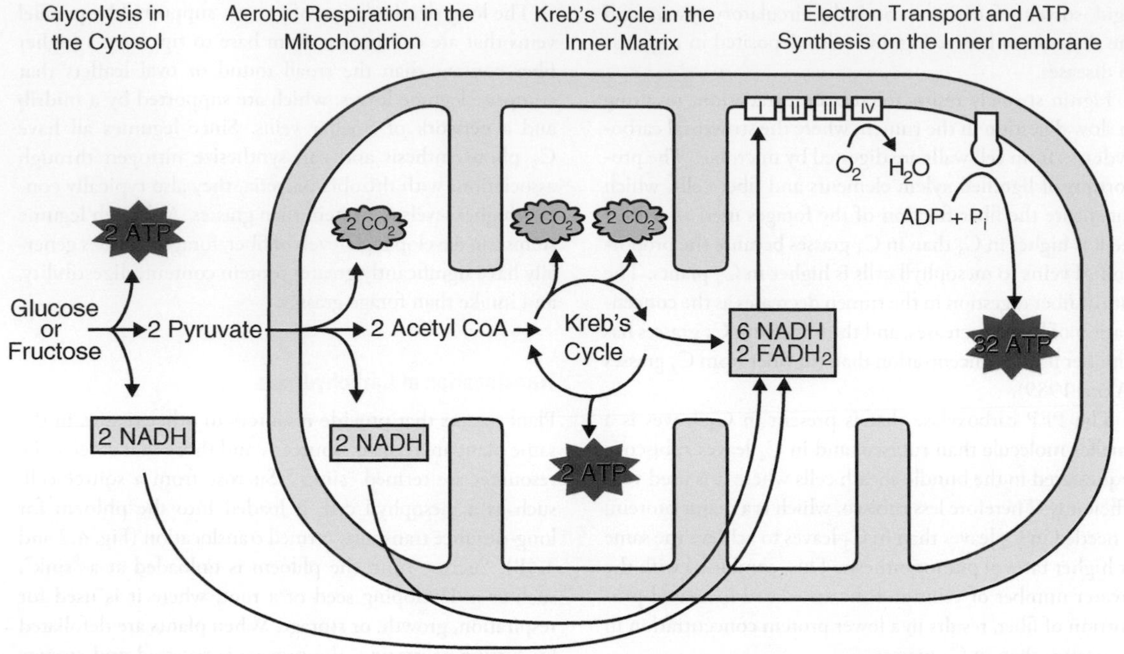

$$C_6H_{12}O_6 + 6O_2 \longrightarrow 6CO_2 + 6H_2O$$

Fig. 4.6. Respiration of glucose or fructose begins with glycolysis outside the mitochondria forming two molecules of pyruvate and two molecules of ATP. In mitochondria, each pyruvate is used to form a molecule of acetylCoA that enters the Kreb's cycle, with the loss of all carbon as CO_2. In electron transport, NADH and $FADH_2$ are used to form ATP, with a total theoretical yield of 32 molecules of ATP from each glucose or fructose molecule. (Adapted from MacAdam, 2009.)

on the leaf area remaining after harvest to supply photosynthate for initial leaf and root regrowth. This makes stubble height an important consideration, especially in the management of pastures.

Leaf growth is affected by even mild water stress, which appears to be a protective mechanism that allows the plant to slow leaf expansion and thus additional transpiration while maintaining photosynthesis (see Chapter 5). A reduction in the rate of growth of leaves and stems during stress allows more photosynthate to be partitioned to the roots, supporting increased growth in root length during drought, and effectively changing the balance between potential water uptake and water loss. Conversely, when adequate water and nutrients are available, shading reduces photosynthesis but not leaf expansion, so less photosynthate is allocated to the roots for growth.

Aerobic Respiration

Although aerobic respiration in plants is sometimes referred to as "dark respiration" to distinguish it from photorespiration, aerobic (O_2-requiring) respiration occurs continuously in the **mitochondria** of all living cells (Fig. 4.6). The summary equation for aerobic respiration, in which CO_2 is respired and the energy stored in the chemical bonds of glucose molecules is used to form nicotinamide adenine dinucleotide (**NADH**) and ATP, is essentially the reverse of the summary equation for photosynthesis (Fig. 4.1). The structure of NADH only differs from the structure of NADPH formed in photosynthesis by a single phosphate group (P_i). The molecules that enable photosynthesis and cellular metabolism to occur, namely ATP and NADPH, cannot be transported from cell to cell, so sucrose is translocated instead, to provide a substrate for respiration that in turn provides distant cells with ATP and NADH.

Non-structural carbohydrates such as starch, fructan, and sucrose are reduced to their subunits, consisting of glucose and fructose molecules, before they are respired. The initial phase of respiration is glycolysis, which occurs in the cytosol. Glycolysis can supply a small amount of energy in the absence of oxygen, and becomes the major source of energy in oxygen-deprived tissue, a state termed **hypoxia**, such as occurs in the roots of waterlogged plants. Glycolysis releases just two molecules of ATP per glucose, and yields two molecules of the 3-C compound pyruvate. If oxygen is unavailable, pyruvate is metabolized to generate the raw material (NAD^+) needed for glycolysis to continue, resulting in the production of lactic acid or ethanol, both of which contain unused energy. Glycolysis is much less efficient than aerobic respiration, and the byproducts can accumulate to levels that are toxic to plants. This inefficient respiration is one of the reasons why flooding and ice encasement can injure plants (see Chapter 5).

Oxygen is needed in order for pyruvate to be used in the next phases of aerobic respiration, which take place in the mitochondria. This is why aeration is needed in solution culture (hydroponics), and why plants that are tolerant of flooding can form a continuous network of intercellular air channels (aerenchyma) connecting the leaves, stem, and roots to carry oxygen, which allows them to avoid depending solely on glycolysis.

Mitochondria, like chloroplasts, have an outer and an inner membrane that enclose an inner aqueous enzyme-rich matrix analogous to the stroma (Fig. 4.6). Instead of an inner thylakoid membrane system, the inner membrane of mitochondria has an enlarged surface area that folds and protrudes into the matrix, where it functions in electron transport and ATP synthesis, in a similar way to the thylakoid membranes.

Pyruvate is taken up from the cytosol by mitochondria, and it passes through both the outer and inner membrane and into the matrix. As it is processed, the three carbon atoms of pyruvate are oxidized in succession and released as CO_2 molecules, at the same time using the released energy to form NADH and a similar molecule, $FADH_2$. These molecules have "reducing power" and are used to form ATP.

Electron transport, which is the last phase of aerobic respiration, takes place on the inner membrane and in the matrix of the mitochondria. NADH and $FADH_2$ donate their high-energy electrons to one of a series of protein complexes (I–IV) that are successively reduced and oxidized to form ATP. The process is similar to the synthesis of ATP in photosynthesis, except that NADH donates electrons at the start of electron transport, rather than being created at the end. As the last step the electrons from NADH or $FADH_2$ are donated to oxygen, reducing it to water (H_2O). Approximately 32 ATP molecules can in theory be formed for every glucose molecule that is completely respired in the cell.

Aerobic respiration is central to the biochemistry of plants, not only because it generates the energy needed for the metabolism that allows cells to function, but because the intermediate compounds formed during the process can leave the mitochondria to be used in other biochemical pathways. For instance, α-ketoglutarate is the "carbon skeleton" to which an amino ($-NH_2$) group is added to form glutamate, which is one of the 20 amino acids and one that is particularly important in the synthesis of the transported forms of fixed N in legumes.

Respiration uses 30–80% or more of new photosynthate each day, depending on the temperature and other environmental factors. Plant yield is essentially the difference between photosynthesis and respiration, so improving the efficiency of respiration could improve yield. The energy generated by aerobic respiration can conceptually be separated into that used for the synthesis of new tissues, termed **growth respiration**, and that used to repair and ensure the proper functioning of mature, non-growing tissues, termed **maintenance respiration**. Respiration of mature, stored seed or of a fully developed leaf where no growth is taking place is purely maintenance respiration, whereas respiration in a young growing root tip, leaf base, or developing seed is largely growth respiration.

Growth Respiration

The rate of growth respiration is directly linked to growth rate. A portion of the sugar transported to regions of cell division (meristems) and cell expansion is respired in order to provide the energy needed to assemble other carbohydrates into cell walls. In addition to building cell walls, respiratory intermediates are used to synthesize lipids, nucleic acids, and proteins. Growth respiration usually adds weight to the plant and occurs where cells are dividing or elongating, or in cells that are synthesizing secondary cell walls and lignin as they mature.

The respiratory cost of the synthesis of new tissue depends on tissue composition. About 1.21 g of glucose is needed, both as a building material and as the substrate for growth respiration, to synthesize 1 g of cellulose (cell wall) or 1 g of starch (Penning de Vries et al., 1983). Protein synthesis is more costly due to the need to take up and reduce NO_3^-, requiring about 2.48 g of glucose to synthesize 1 g of protein or nucleic acid. About 3 g of glucose are needed to synthesize 1 g of lipid, and 2.12 g of glucose to synthesize 1 g of lignin.

Organic acids, which are used in photosynthesis, respiration, and stomatal guard cell regulation, are more oxidized than glucose, which means that their synthesis is more energetically favorable. Therefore only about 0.91 g of glucose is required to synthesize 1 g of organic acids. Root uptake of mineral ions such as K^+, Mg^{2+}, calcium (Ca^{2+}), and phosphate ($H_2PO_4^-$) only costs about 0.05 g of glucose per gram of mineral. By weight, good-quality dried forages are composed of 50–80% carbohydrates, 10–25% proteins, 2–5% lipids, and 6–12% minerals (Schwab et al., 2006).

In a study of young barley leaves, respiration was measured as oxygen (O_2) uptake in regions of cell division and elongation (Fig. 4.7; Thompson et al., 1998). While respiration per cell (open blue squares) was constant from the youngest tissue at the base of the leaf to differentiating tissue toward the leaf tip, respiration per unit of protein (closed blue squares) was greatest in the youngest, fastest-growing tissue at the leaf base, where the cells were smallest.

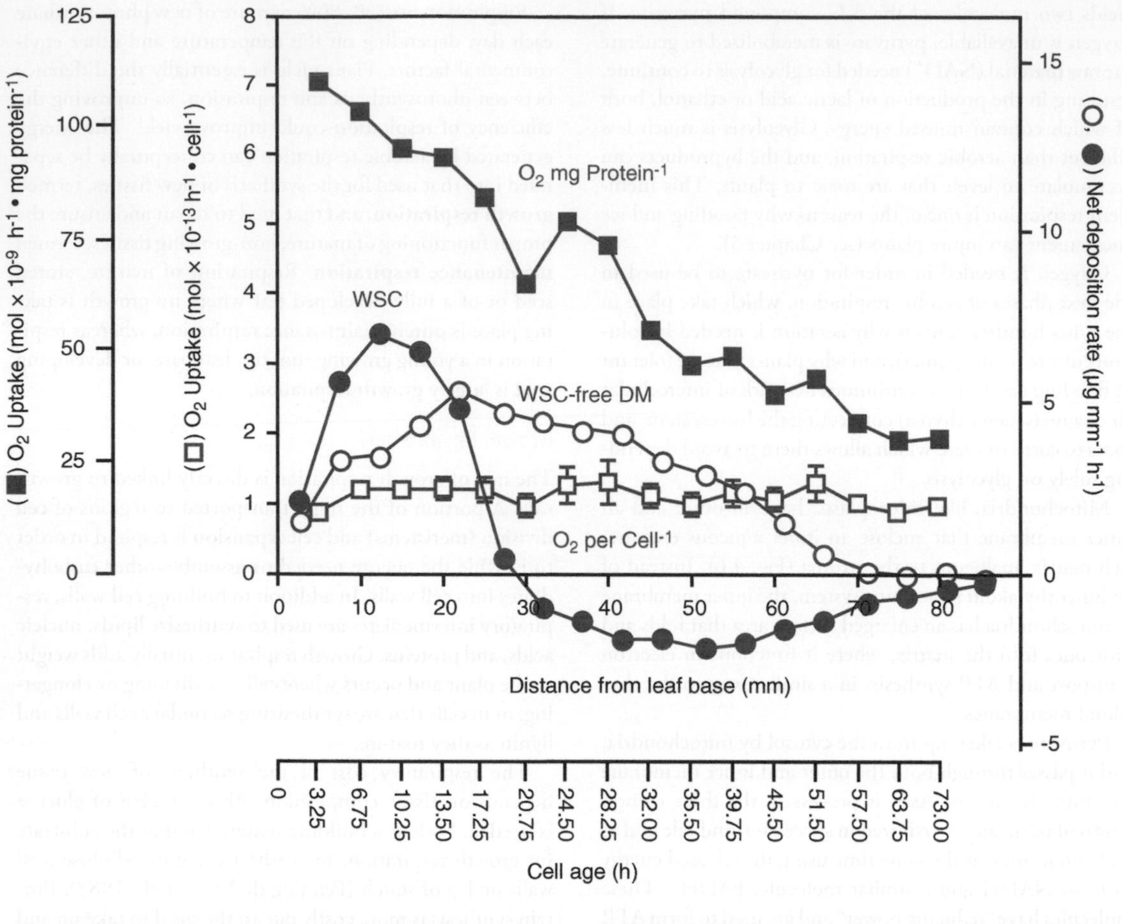

FIG. 4.7. The length of growth regions (cell division, elongation, and differentiation) at the base of grass leaves is similar in barley and tall fescue leaves. Growth respiration is highest in the region of cell division (2–10 mm), at the base of the leaf, while the deposition of water-soluble carbohydrates (WSC) peaks in the elongation zone (2–25 mm), where rapid cell wall synthesis is needed to support cell growth. A net loss of WSC in differentiating leaf tissue (25–80 mm) is used for the accumulation of structural dry matter (DM). (Adapted from Thompson et al., 1998, and Allard and Nelson, 1991. Reproduced with permission of the American Society of Physiologists and CCC Republication.)

As the cells elongated and then differentiated, the protein concentration in successive leaf segments decreased, as did the respiration rate.

The growth of elongating tall fescue leaves occurs over a similar distance, so data for the deposition of water-soluble carbohydrates (WSC; closed red circles) and the accumulation of other macromolecules in leaf tissue (WSC-free dry matter; open red circles) have been overlaid on the same graph (Fig. 4.7; Allard and Nelson, 1991). Combining these spatially similar grass leaf data allows us to see that the highest rate of deposition of photosynthate (WSC) occurs in the elongation zone, where cell wall growth must keep up with cell expansion. In more mature tissue, WSC in newly elongated cells is used for the synthesis of secondary cell walls in differentiating xylem and fiber cells.

It has been determined that leaf blades of smooth bromegrass that are comprised of 16% protein, 6.5% mineral, 2.5% lipid, 4% organic acids, and 71% cell wall (96% carbohydrate and 4% lignin) require about 1.39 g of glucose for the synthesis of 1 g of leaf tissue dry weight. When N for the growth of new leaves is recycled from senescing lower (older) leaves (grass tillers usually only support three mature leaves at one time), the amino ($-NH_2$) group of protein is already formed, so biosynthesis of new protein costs only 1.62 g of glucose per gram. This reduces total synthesis costs to 1.26 g of glucose per gram of new leaf. Leaves of legumes such as alfalfa have a higher protein content and are therefore more costly to synthesize than grass leaves. Similarly, due to the higher protein concentration in leaves, more glucose is required to form leaves than stems, and more glucose is needed to form stems than roots, which are low in protein.

Most forages are herbaceous perennials that persist from year to year but maintain only their below-ground shoot and root biomass over the winter in cold-temperate climates. This allows them to adapt new growth to the prevailing nutrient and climatic conditions during each growing season. Growth temperature influences the composition of the tissue formed, which in turn affects growth respiration. For example, at temperatures above the optimum for leaf development, cell wall growth is slowed and a higher proportion of the cell wall is lignin, leading to an increase in the proportion of leaf protein relative to cell wall. In this case, the respiratory efficiency of growth is reduced because both lignin and protein are more expensive to synthesize than cellulose.

Maintenance Respiration

Once formed, all living tissue needs to be maintained and repaired. Maintenance respiration includes that needed for synthesis of proteins and lipids to replace those that have broken down with time or to alter the rate of metabolic activity, and maintenance of ion concentrations, such as K^+, which is needed to stabilize cellular proteins. The rate of maintenance respiration is strongly affected by temperature, being very low at 0°C (32°F), but increasing as the temperature rises (Fig. 4.8). Of the 30–80% of photosynthate that is used for aerobic respiration each day, most is used for maintenance respiration of the large amount of mature tissue.

Some plant species that are well adapted to northern regions do not survive in the South because of excessive maintenance respiration. Lower night-time temperatures in the North, in dry climates or at high altitudes, decrease maintenance respiration costs. These same conditions may cause sufficient chilling injury to the membranes of perennial C_4 plants to restrict their adaptation in these locations (see Chapter 5).

Rates of maintenance respiration are closely related to the age of plant tissue and the metabolic activity of an organ. Many enzymes degrade rapidly, some within a few hours, especially at high temperature, requiring maintenance respiration for continued resynthesis. Maintenance respiration begins as a small component of total respiration in young seedlings, and then becomes a higher proportion of overall respiration as more tissue stops growing and matures. Senescence can require the breakdown (catabolism) of macromolecules of older tissue and the transport of their subunits or of nutrient ions to newer growth or storage. This temporarily increases maintenance

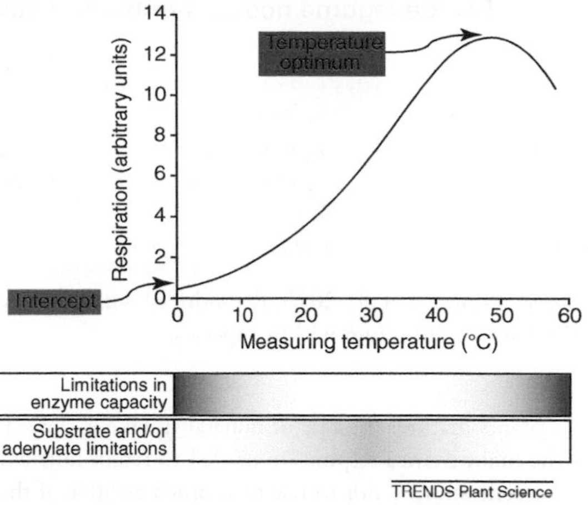

FIG. 4.8. Maintenance respiration of mature tissues peaks at approximately 45°C. At very low (violet) and very high (red) temperatures, respiration is limited by the rate of enzyme activity. From about 10°C to 45°C respiration is limited by the supply of carbohydrates (substrate) or ADP (adenylate). (Adapted from Atkin and Tjoelker, 2003. Reproduced with permission of Elsevier.)

respiration, but respiration declines again as the protein content of the senesced tissue decreases.

While growth respiration is required for plant productivity, it would be beneficial to reduce maintenance respiration. Wilson (1982) successfully selected for low aerobic respiration rate in mature leaf blades of perennial ryegrass, which led to decreased maintenance respiration and an increase in forage yield. The carbohydrate conserved was available for growth, and was expressed as an increase in production of new tillers.

Inorganic Nutrient Uptake

Mineral nutrient ions in the soil solution, such as K^+, Mg^{2+}, Ca^{2+}, $H_2PO_4^-$, SO_4^{2-}, and NO_3^-, come into contact with roots passively via water uptake and move to the plasma (cellular) membrane of root cells without restriction. The internal anatomy of plant roots forces these mineral nutrients to move across the plasma membrane and into cells, often requiring the energy of ATP for active uptake, before they are transported into the xylem and pulled into shoots in the transpiration stream.

The requirement for active uptake allows the plant to selectively accumulate some nutrients (K^+, $H_2PO_4^-$, and NO_3^-) to much higher concentrations than those at which they occur in the soil solution, and to effectively exclude less desirable elements, such as sodium (Na^+). This selectivity is critical, but the use of ATP is a respiratory cost.

Table 4.1. Forage legume nodule symbionts showing specificity for nodulation of a host

Genus	Species	Biovar[a]	Host genus	Host common name
Mesorhizobium	*haukuii*		*Astragalus*	Milkvetch
Rhizobium	*leguminosarum*	*trifolii*	*Trifolium*	Clovers
Rhizobium	*leguminosarum*	*viciae*	*Vicia*	Vetches, peas
Mesorhizobium	*loti*		*Lotus*	Trefoil
Ensifer	*meliloti*		*Medicago, Melilotus*	Alfalfa, sweetclovers

Source: Mousavi et al., 2015. Reproduced with permission of Elsevier.
[a]Biovars are groupings within a species.

As plants mature, the rate of mineral uptake is slowed, but the maintenance respiration needed to retain nutrient ion concentrations tends to rise to as much as 50% of the total aerobic respiration of roots. The rates are higher for forages with taproots that store carbohydrates and proteins.

Nitrogen Uptake from the Soil

In warm (> 10°C or 50°F) soils with sufficient O_2 available for soil microbes, ammonium (NH_4^+) is rapidly converted to NO_3^-. Plants readily take up NO_3^- and store the excess in the **vacuoles** of cells. However, before the NO_3 can be used in the plant, it must be reduced to NH_4^+, which requires NADPH from photosynthesis or NADH from respiration. In some plants, absorbed NO_3^- will be reduced to NH_4^+ and used to form nitrogen compounds in the roots. In others, especially after fertilization, when NO_3^- is abundant in the soil solution, the excess NO_3^- moves to the shoot in the xylem stream and is accumulated in the stem and leaves, where it can eventually be reduced. Excessive amounts of transiently stored NO_3^- in plants are potentially harmful to animals because NO_3^- is reduced to nitrite (NO_2^-) by microbes in the rumen. Nitrite that is absorbed into the blood interferes with the ability of hemoglobin to transport O_2 in the blood (see Chapter 16).

Nitrogen Assimilation

In legumes, the nitrogen required for plant growth can be taken up from the soil as ammonium (NH_4^+) or nitrate (NO_3^-). When these ions are too deficient to support growth, atmospheric dinitrogen (N_2) is fixed into ammonia (NH_3) in a symbiotic relationship with soil bacteria from the genera *Ensifer, Rhizobium, Bradyrhizobium, Mesorhizobium*, or *Sinorhizobium* (all of which are referred to as **rhizobia**).

Symbiotic Nitrogen Fixation

A **symbiotic** (mutually beneficial) relationship exists between many legume species and rhizobia, in which the rhizobia receive carbohydrates and other nutrients from the plant that are then used to reduce N_2 from the air to usable forms of nitrogen for the plant, such as amides and transported amino acids. If they are not enclosed in a root nodule, rhizobia do not fix N, because low-O_2 conditions are required for the N-fixing bacterial enzyme nitrogenase to function, and the pores of a well-drained soil contain atmospheric O_2.

Rhizobia are attracted to the roots of legumes by chemical signals specific to the legume species, and infection occurs only when there is a correct match between a rhizobium species and a legume species (Table 4.1). In response to a chemical interchange between a legume root and compatible rhizobia, cells in the cortex of the root multiply to create a nodule primordium in anticipation of infection.

Rhizobia enter the young root through root hairs, which are outgrowths of the cell walls of epidermal cells (Fig. 4.9). Rhizobia attach to the root hair cell wall, and the root hair may curl to enclose them as the rhizobia degrade the cell wall of the root hair. An infection thread, which is an ingrowth of the root hair plasma membrane, encloses the dividing bacteria and grows through the root tissue into the

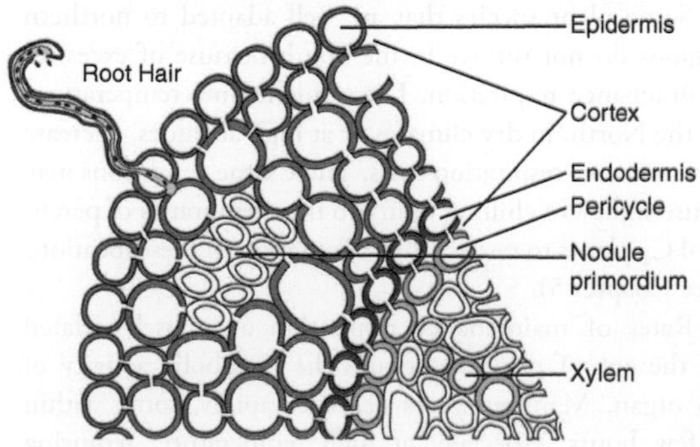

FIG. 4.9. Legume root cross-section illustrating the invasion of a root hair by soil-living rhizobia. An ingrowth of the plasma membrane, the infection thread, encloses rhizobia during growth through successive cell layers to the developing nodule primordium. (Adapted from MacAdam, 2009.)

cortex until it reaches the newly divided root cells that comprise the developing nodule. The bacteria, still enclosed in remnants of the plasma membrane, multiply further in the nodule cells. In this low-O_2 environment they alter their form to become N-fixing **bacteroids**. Nodule primordia are initiated in the cortex of the root, adjacent to lobes of the xylem, and branches of xylem and phloem develop along with the nodule for translocation of water and nutrients.

Nitrogenase, the nitrogen-fixing enzyme, is synthesized by bacteroids but is irreversibly inactivated by O_2. However, it requires considerable amounts of ATP, which has to be generated in nodule cells by aerobic respiration. To supply sufficient O_2 for respiration while at the same time protecting nitrogenase from O_2, **leghemoglobin,** an O_2 carrier, is synthesized jointly by bacteroids and the plant. It captures O_2 entering nodule cells before the O_2 reaches the nitrogenase complex, and then releases it as needed for aerobic respiration to drive nitrogen fixation. Like hemoglobin in blood, leghemoglobin with captured O_2 imparts a pink pigmentation to healthy nodules.

The N_2 gas used in fixation diffuses from the air through the soil and into the nodule to reach the active site of the nitrogenase enzyme. Each N_2 molecule is reduced stepwise using 16 ATP molecules from aerobic respiration. In the first step, N≡N is reduced to HN=NH. In the second step, HN=NH is reduced to $H_2N–NH_2$, and in the third step $H_2N–NH_2$ is reduced to two molecules of ammonia (NH_3). In legumes of temperate origin, the ammonium (NH_4^+) ion is added to glutamate to form glutamine and/or asparagine for transport, and in legumes of tropical origin, glutamine is further converted to ureides for transport. If legumes are fertilized with inorganic nitrogen, such as NO_3^-, they will take it up via their roots and utilize it as readily as do grasses, and nitrogen fixation in the nodules will decrease.

The inoculation process takes up to 4 weeks to result in nodulation and N fixation, so young seedlings are dependent on seed reserves and soil nitrogen. Low to moderate amounts of fertilizer nitrogen do not interfere with nodule formation, and as soil nitrogen becomes depleted, active N_2 fixation will occur. However, nitrogen fertilization will also benefit weeds that can compete with legume seedlings. It is rarely economical to fertilize established forage legumes or legume–grass mixtures with nitrogen.

Is Biological Nitrogen Fixation Better or Cheaper?

When nitrogen (N) fertilizer is produced commercially, ammonia (NH_3) is formed by reacting N gas (N_2) with methane (CH_4) under conditions of high temperature and high pressure. The NH_3 that is formed is readily converted into other N fertilizers, such as calcium ammonium nitrate. When incorporated into the soil, the ammonium (NH_4^+) released is converted to nitrate (NO_3^-) ions by soil microbes, so NO_3^- is the form commonly encountered by plant roots. Producing N fertilizer from N_2 and CH_4 costs about 3 g of glucose per gram of N (International Fertilizer Industry Association, 1998). The cost to the plant of uptake and assimilation of NO_3^- into usable organic N ($–NH_2$; the amino group) is about 8 g of glucose per gram of N (Gutschick, 1981), giving a total energy cost of 11 g of glucose per gram of N used. The cost for legume fixation of N_2 to $–NH_2$ is about 12 g of glucose per gram of N (Gutschick, 1981), mainly from higher rates of root respiration in nodulated legumes. However, 5.5 g of CO_2 equivalent greenhouse gas emissions occur for every gram of N applied as fertilizer, due to manufacture, transportation, machinery use during application, and direct loss of N_2O from the fertilized soil (International Fertilizer Industry Association, 1998), even when carbon sequestration from elevated crop production has been deducted (Kim and Dale, 2008).

Legumes support symbiotic N_2 fixation for their own benefit, providing from 30% to 95% of total plant N requirements; the rest of the N used by legumes comes from the soil. Often grasses growing next to legumes are greener, which suggests better N availability than for grasses growing further away. There is some evidence for direct N transfer between intermingled grass and legume roots, but nodule turnover may be most important in species such as birdsfoot trefoil. That species has determinate nodules that are sloughed after a harvest or a killing frost in the autumn, and must be reinitiated as part of regrowth. Other species, such as alfalfa and white clover, have indeterminate nodules that show a reduction in activity when plants are cut or grazed, and recover as the plants regrow.

Nodules are not considered to be nitrogen storage organs, and nodule mass is too small compared with root organic matter to account for a significant portion of N mineralization following harvest. Nitrogen transfer can occur indirectly from leaf drop or death of the legume plant followed by microbial mineralization of organic matter. However, the most significant and effective redistribution of fixed N occurs as a result of animals grazing legumes and depositing urine and dung within the same pasture.

Organic Food Reserves

When photosynthesis exceeds growth and maintenance respiration, legumes and grasses store carbohydrates in readily available forms in various plant parts. The principal storage organ may be the root (as in alfalfa), the stolons (as in white clover), the rhizomes (as in smooth bromegrass), or the leaf and stem bases (as in orchardgrass and other bunchgrasses). Plants also accumulate storage proteins, and the accumulation of these carbohydrate and protein ("organic") reserves is coordinated.

Compounds that are synthesized as storage forms of carbohydrates and proteins are used to support respiration and growth when leaf area is insufficient to support growth in early spring or after cutting or grazing. At these times, the reserves are reconverted to sucrose and amino acids and are translocated to meristematic sites to support growth. They are also used to develop heat and cold resistance, to support respiration and metabolism during periods of **dormancy**, and to provide carbohydrates and N required for flower and seed formation. These organic reserves are a readily accessible buffer to support critical metabolic needs such as the initial development of plant photosynthetic capacity.

Starch and Fructan Accumulators

Starch, a polymer of glucose, is the primary non-structural carbohydrate stored in roots, rhizomes, and stolons of legumes and C_4 grasses such as big bluestem, switchgrass, and bermudagrass. Legumes and all grasses also store starch in seed, and transiently accumulate starch in chloroplasts during photosynthesis. Amyloplasts, which are similar to chloroplasts but do not develop thylakoid membranes or the enzymes needed for photosynthesis, are found in storage tissue of these species and can accumulate starch as insoluble granules.

In contrast with C_4 grasses, C_3 grasses such as orchardgrass, perennial ryegrass, and tall fescue accumulate **fructan**, a polymer of fructose, in their vegetative tissues. Fructans (also referred to as "fructosans") are chains of fructose molecules that vary in length and have a single glucose molecule at one end. Fructans are water-soluble and accumulate in the cell vacuole. Enzymes for fructan metabolism function at lower temperatures than those for starch metabolism, and fructan storage may have contributed to the adaptation of C_3 grasses to cool-temperate climates.

Seasonal Cycles

The initial growth of shoots, tillers, and fibrous roots in perennial forages is rapid in the spring, and occurs at the expense of storage (Fig. 4.10a). With sufficient leaf growth, excess photosynthate accumulates in storage organs until the first cutting or grazing. Within each cycle of harvest and regrowth, the leaf area for photosynthesis is initially low, so stored carbohydrates are drawn from storage organs such as the roots of perennial legumes, to support this new shoot growth (Fig. 4.10b). In autumn, in preparation for overwintering in temperate climates, growth slows as well-adapted species and overwintering storage organs such as basal stubble of grasses and the rhizomes, stolons, and taproots of legumes become strong sinks for photosynthate storage (Fig. 4.10c).

Seasonal trends of carbohydrate storage in roots help to explain why birdsfoot trefoil can be cut or grazed frequently, but not closely. The carbohydrates that are stored over winter in birdsfoot trefoil roots are used to support spring growth, but photosynthate is not used to restore the roots' carbohydrates after flowering in birdsfoot trefoil as it is in alfalfa and red clover (Fig. 4.10). Instead, birdsfoot trefoil continues to show active growth, even during seed fill and storage. Unlike alfalfa or red clover cut at bloom stage, there is no pool of stored carbohydrates in the taproot of birdsfoot trefoil. Therefore each time birdsfoot trefoil is cut or grazed, a tall stubble with leaves and axillary buds is needed to support shoot regrowth.

As in grasses, new stems that form from axillary buds on the stubble of birdsfoot trefoil will have functioning leaves to furnish the carbohydrates needed for continued regrowth. The similarity of desirable stubble heights and flexible cutting frequencies make birdsfoot trefoil an excellent companion for many C_3 bunchgrasses in mixtures. Storage remains at a low level in birdsfoot trefoil until growth slows down in autumn, when photosynthate is partitioned from the shoots to storage in the roots (Fig. 4.10).

For persistence of birdsfoot trefoil, it is absolutely critical that grazing or cutting ceases from early autumn until complete dormancy, usually marked by the first killing frost. This is the only period when birdsfoot trefoil stores carbohydrates and proteins for spring regrowth.

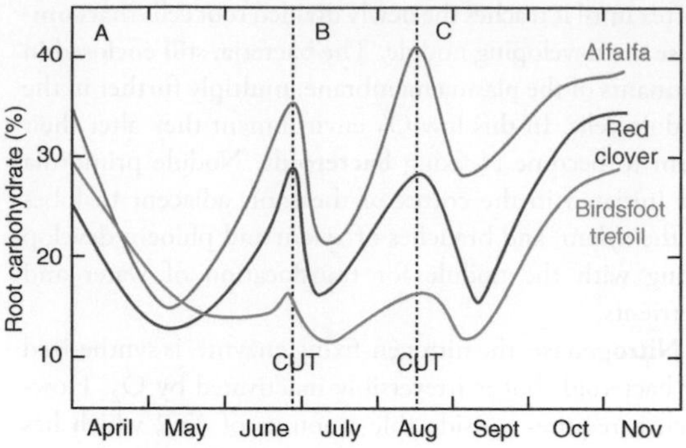

FIG. 4.10. Changes in the content of root storage carbohydrates in field-grown alfalfa, red clover, and birdsfoot trefoil. A. Root carbohydrate levels are high in early spring for all three species, and then decrease as the roots serve as a source to support initial shoot growth. In late spring, root carbohydrate reserves are restored in alfalfa and red clover before the first harvest. B. In birdsfoot trefoil, root carbohydrates are not restored regardless of management during the summer. C. In autumn, as all forage legumes become dormant, root carbohydrates are restored from photosynthesis by the leaves that develop in late summer, if no further harvests are taken. (Adapted from Smith, 1962, with permission of *Crop Science.*)

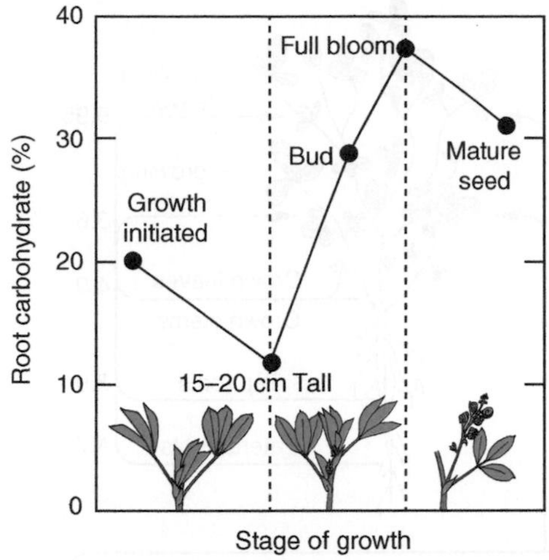

FIG. 4.11. Alfalfa root carbohydrate storage decreases from the initiation of spring growth or summer regrowth until sufficient leaf area has developed to produce excess photosynthate. Root storage increases with continuing shoot development and increased photosynthetic capacity through the bud (shown) and bloom stages. Seed fill and the initiation of new shoots from the crown combined with the maintenance of older leaves deplete storage carbohydrates because the requirements of seed storage are added to maintenance respiration and senescence. (Wisconsin field data from Graber et al., 1927. Alfalfa stage drawings from Fick and Mueller, 1989.)

In alfalfa roots, depletion of organic reserves occurs during the early vegetative phase of each growth cycle, until the shoots are 6–8 in. tall, when there is enough leaf area to produce photosynthate to fully support respiration and continuing shoot growth (Fig. 4.11). When alfalfa shoots exceed 6–8 in., photosynthate is produced in excess of the needs for ongoing respiration and growth in shoots, and is translocated to the crown and roots for storage (Fig. 4.11).

Accumulation in the roots continues as the shoot grows and flower buds form, reaching the highest level of storage near full bloom if the plants have not already been harvested (Fig. 4.11). In seed fields, some carbohydrate is used from alfalfa roots between full bloom and mature seed stages because the leaves are aging, seed are developing rapidly, and new shoots are being initiated from axillary buds on the crown, which means that total respiration of uncut plants will be high (Fig. 4.11).

High temperature may increase growth rate in well-adapted plant species, or simply increase maintenance respiration more than photosynthesis and hasten maturity. In either case, less storage is likely to occur in mid-summer. Conversely, low temperatures, particularly at night when photosynthesis has stopped, reduce shoot growth and respiration more than they reduce photosynthesis, so carbohydrate storage is enhanced by low temperature, low N supply, and moderate drought stress.

Nitrogen Reserves

In a classic study of storage carbohydrates, Graber et al. (1927) noted significant changes in the N compounds stored in alfalfa taproots during regrowth in spring or following harvest. Due to their low concentration, N reserves were thought to be less important than carbohydrate reserves. However, it is now understood that young regrowth is very high in protein, especially in the form of enzymes, and therefore it is high in N compared with other tissues. N_2 fixation in the nodules of legumes declines dramatically after harvest, to the point where the rate of N_2 fixation cannot meet the N needs of new shoots. During this period, even if inorganic nitrogen (i.e., NO_3^- or NH_4^+) is available in the soil, amino acids from plant-N storage sources are critical for the initiation of new shoot tissues.

In a study of perennial ryegrass, 27% of the N in new leaves was found to be remobilized from N stored in stubble and roots in the 20 days following harvest (Fig. 4.12); however, remobilized N was the *sole source* of N to support regrowth during the first 4 days following harvest. Even when uptake of inorganic N from the rooting medium occurred, it was slow until the leaves were able to supply the roots with new photosynthate (Volenec et al., 1996).

The rate of leaf elongation in grasses such as tall fescue is generally not carbohydrate limited, but is often limited by N availability. In grasses, N stimulates cell division, which is fundamental to leaf growth. Detailed analyses of tall fescue leaves demonstrated that N fertilization increased the rate of cell division as well as the number of dividing cells. However, N had little effect on the rate of cell elongation and final cell size (MacAdam et al., 1989), indicating that N regulates the number of cells, while the elongation of new cells uses large amounts of carbohydrate for respiration and cell wall synthesis (Fig. 4.7). Clearly, both N reserves and carbohydrate reserves are important for regrowth.

Greater N storage in the form of protein in roots increases the stress tolerance of forage legumes. Proteins accumulate in taproots as plants harden for winter (see Chapter 5), with the greatest accumulation occurring in alfalfa and the least in red clover. This stored N is used for growth in spring and after cutting. Volenec et al. (1996) found that, in alfalfa, 39% of N in the first 24 days of regrowth came from N storage in roots and crowns, while the balance was supplied by uptake (Fig. 4.12). In alfalfa, red clover, and yellow sweetclover, greater protein accumulation in autumn was associated with increased winterhardiness. Potassium content in legumes increases with protein content, so the value of protein accumulation for winterhardiness may explain the increase in alfalfa

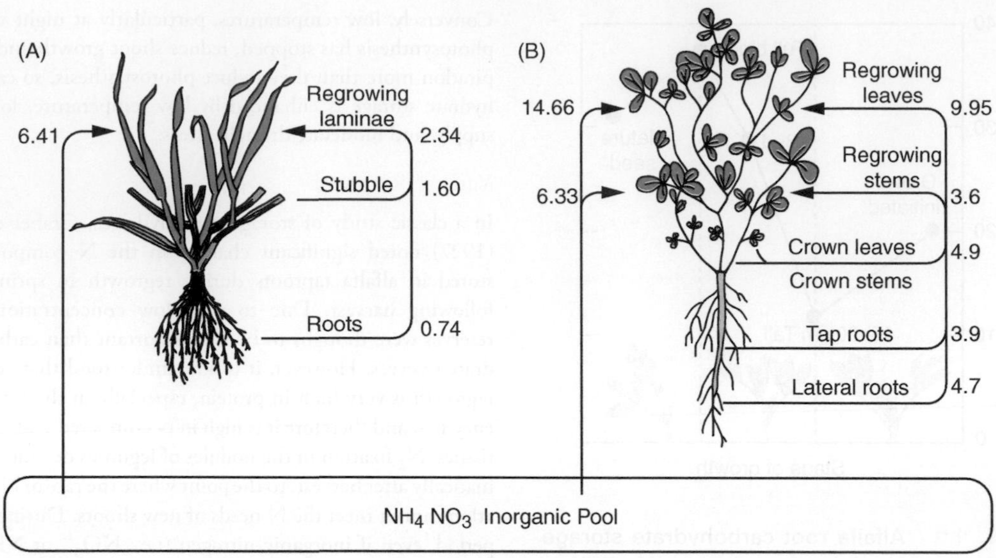

FIG. 4.12. Organic nitrogen, expressed as milligrams of N per plant, was reallocated from storage in roots and stubble during the first 20 and 24 days of perennial ryegrass and alfalfa regrowth, respectively. Organic N withdrawn from storage sources (right arrows) equals deposition in new leaves and stems (N sinks). Perennial ryegrass (A) relied solely on organic N reserves for the first 6 days of regrowth, and alfalfa (B) mainly relied on reallocation for the first 10 days of regrowth. The remaining 73% (perennial ryegrass) or 61% (alfalfa) of N used during this regrowth period was provided as inorganic N from the rooting medium. (Redrawn from Volenec et al., 1996, with permission of John Wiley & Sons.)

winterhardiness with increasing rates of K^+ fertilization in this species.

Proteins and NO_3^- are the forms in which N is most often accumulated in the storage organs of herbaceous plants. Rubisco is not just the enzyme that adds CO_2 to RuBP to "fix" carbon in photosynthesis—it is effectively the most important N storage protein in leaves. Rubisco and other leaf proteins gradually degrade as the leaves mature, releasing N in the form of amino acids for transport to the meristems for use in the synthesis of new leaf tissues needed to intercept sunlight at the top of the canopy.

Managing the Canopy

To maximize net photosynthesis and therefore productivity, the canopy of leaves in a pasture or hayfield should be managed so as to maximize light interception over the course of the growing season. Defoliation by cutting or grazing reduces leaf area, which must be regenerated in order to intercept radiation. After defoliation the period of rapid increase in forage shoot dry matter continues until the leaves in the canopy intercept about 95% of the sunlight.

The leaf area index (LAI) is the ratio of the leaf area to the land area that it covers, and the critical leaf area index is the LAI at 95% light interception. For plants with leaves that are oriented horizontally, such as white and red clover, the critical LAI is 3–5; for alfalfa, with tall stems and smaller leaflets, it is 5–6; and for grasses with vertically oriented leaf blades, such as orchardgrass and perennial ryegrass, it is 7–10.

After 95% light interception has been achieved, shading will cause the lower leaves to senesce, so while new leaves continue to be added at the top of the canopy, older leaves are shaded and will die, so there is little net increase in usable forage mass. To maximize the productivity of a stand, unless there are other management goals to consider, dry matter should be harvested as soon as the canopy has achieved 95% light interception. The leaf area or reserves of carbohydrates and proteins remaining after harvest will determine the length of the lag phase before rapid regrowth occurs and the time to the next harvest.

Canopy structure

Leaf angle markedly affects light penetration into a grass or legume canopy (Fig. 4.13). The leaves of grasses are narrow and emerge more or less vertically at the top of the canopy. This results in a favorable leaf arrangement for light penetration and a high critical LAI, and places the leaves with the greatest requirement for photosynthate at the top where light interception is unimpeded.

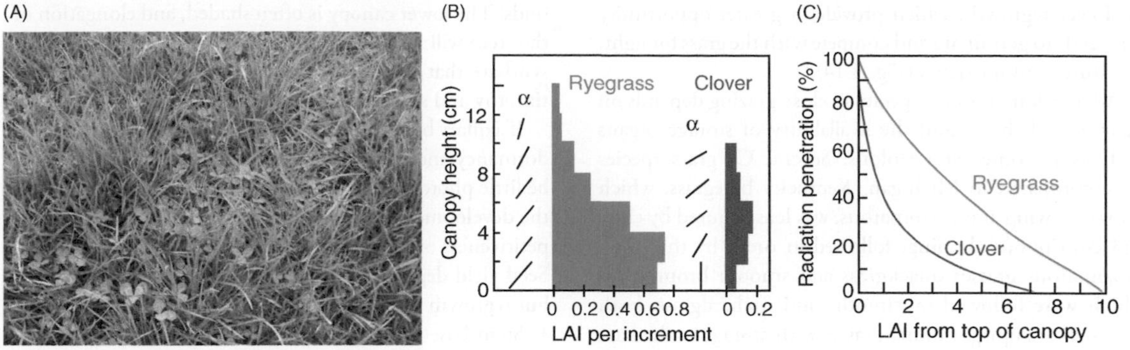

Fig. 4.13. In a perennial ryegrass–white clover canopy (A), the grass leaf area is concentrated at the bottom of the canopy (B), and leaf angles (α) become more upright from the bottom to the top of the leaf canopy to aid light penetration. White clover leaf area is concentrated near the middle of the canopy, and leaf angles become more horizontal from the bottom to the top. Light is distributed effectively throughout the dense grass canopy (C), only becoming reduced to a penetration of 25% at an LAI of 6, whereas light penetration in white clover drops to 25% at an LAI of only 2. (Adapted from Loomis and Williams, 1969, with permission of ACSESS.)

In contrast, the leaflets of white clover are folded early in development, and are then moved by petiole extension to the top of the canopy, where the blade unfolds to be displayed almost horizontally. In this flat-leafed arrangement, most radiation is intercepted by the young leaf blades in the upper part of the canopy, with less light getting through to the lower leaves, resulting in a low critical LAI. This effectively shades weeds, but it also causes older leaves with shorter petioles to senesce and die.

Forage dry matter will not be used efficiently for animal production unless it is grazed or cut as soon as the critical LAI is reached. White clover has prostrate stolons and is therefore tolerant of close grazing, but is not as shade-tolerant as legumes such as red clover, which has an upright stem and a higher critical LAI.

There is relatively little carbohydrate storage in the fibrous root system of grasses. In bunchgrasses, the highest concentration of carbohydrates is in the leaf sheaths and stem bases (the **pseudostem**) in the vegetative stages, and in the lower stem in the reproductive stages. Therefore it is critical to gauge stubble height in order to retain sufficient storage carbohydrates and basal leaf area for the support of regrowth.

If the leaf blades but not the leaf sheaths of a grass tiller are removed by moderate grazing, the effect on regrowth rate and on root biomass is minimal because the basal storage sites were not removed (Fig. 4.14). However, if severe grazing or close cutting remove a significant portion of the sheath area or pseudostem of grass tillers, a significant proportion of the stored carbohydrates and proteins will be removed, leaving less reserves to support root maintenance and shoot regrowth. A reduction in storage reserves results

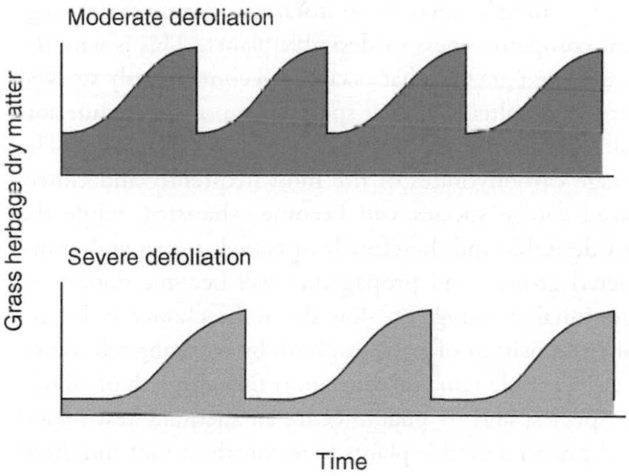

Fig. 4.14. Under well-managed defoliation of grasses (top panel), sufficient stubble remains to supply stored carbohydrates for initial regrowth, so the lag phase for dry matter accumulation following grazing or cutting is minimal. With severe defoliation (bottom panel), when leaf and stem base storage tissues have been removed, grass regrowth is slow to begin, extending the lag phase and reducing the seasonal productivity of the stand. In both cases, defoliation has occurred each time the critical LAI (95% light interception) was reached. (Adapted from Walton, 1983, with permission of Pearson Education.)

in slower regrowth, which provides a greater opportunity for weeds to germinate and compete with the grass for light, moisture, and nutrients (Fig. 4.14).

Adaptation of forage species to close grazing depends on their growth habit and the availability of storage organs such as rhizomes and stolons. Several C_3 grass species were compared in Michigan. Kentucky bluegrass, which is low-growing and rhizomatous, was least injured by close and continuous clipping, followed in order by the taller, rhizomatous grasses quackgrass and smooth bromegrass, which were followed by timothy and orchardgrass, both of which are upright bunchgrasses with storage in the base of the stem (Harrison and Hodgson, 1939). In North Carolina, with regard to C_4 species, dallisgrass, a tall bunch-type warm-season grass, was injured more by close cutting than were carpetgrass and bermudagrass, which are both stoloniferous and have canopies with many leaves near the soil surface (Lovvorn, 1945).

Continuous vs. Rotational Stocking

Cutting or grazing of forage plants that have a reduced capacity to support shoot and root regrowth, either from the remaining leaves or from storage, undermines the long-term competitiveness of desirable plants. This is why the unrestricted grazing that occurs in a continuously stocked pasture will alter the plant species composition, while animal production will probably be reduced gradually. The storage carbohydrates of the most frequently and closely grazed forage species will become exhausted, while the least desirable and therefore least grazed species with unrestricted growth and propagation will become dominant. Rotational stocking can slow down the change in botanical composition of pastures, both by reducing selectivity, so that stubble remains to support the regrowth of all forage species, and by guaranteeing an adequate rest period for the most desirable plants to restore their root and shoot food reserves.

Location of Meristems

Managing a canopy involves more than managing light interception for photosynthesis, and accumulation of reserve carbohydrates and N compounds in storage organs. Perennial forage plants need active meristems to provide new growth or regrowth after cutting or grazing. Both legumes and grasses have a shoot apex at the top of each stem, but the stems of most grasses are not elongated until the reproductive growth stage, when the seedhead (inflorescence) appears (see Chapter 2).

The most critical management stage for some grasses occurs during reproductive growth, when internode elongation elevates the inflorescence to a height at which it can be removed by cutting or grazing. When a shoot apex becomes an inflorescence, it will no longer produce leaves, so regrowth must come from tillers initiated from axillary buds. The lower canopy is often shaded, and elongation of the stem will have caused a transient redirection of photosynthate that suppresses tillering in some species, such as timothy and smooth bromegrass.

If axillary buds at the base of grass stems have not broken dormancy and developed into tillers by harvest, there will be little photosynthetic tissue in the stubble and a delay in the development of new tillers. The low LAI reduces competitiveness or creates an excessive lag phase for regrowth. Seed yield depends on the number of reproductive tillers, but regrowth depends on vegetative tillers.

Stem base carbohydrate levels in timothy are minimal when the grass is cut at the boot or stem elongation (SE) stage (Fig. 4.15). There are few axillary tillers present at the stem elongation stage, since they form close to anthesis (inflorescence emergence), after stem growth is complete. Thus cutting timothy at the boot stage can delay regrowth by 2 weeks or more. The reproductive (mother) tiller is no longer producing new leaves, and the few vegetative tillers

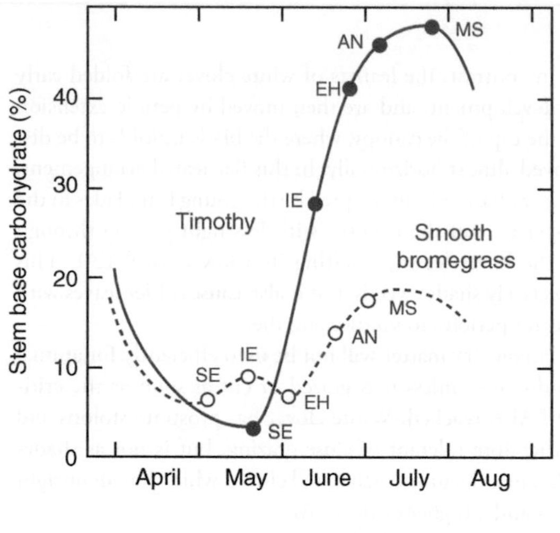

Fig. 4.15. Total non-structural carbohydrates in the stem bases of two C_3 grasses, timothy and smooth bromegrass, at successive stages of development in the field in Wisconsin. SE, beginning of stem elongation; IE, inflorescence emergence (or boot stage); EH, early heading; AN, early anthesis; MS, mature seed. Although the pattern of carbohydrate storage is similar, timothy reaches each growth stage later in the season than smooth bromegrass. As with root carbohydrate storage for legumes (Fig. 4.10), these grasses are most vulnerable to mismanagement when their carbohydrate storage is lowest, and will regrow and persist best if harvesting is managed with an understanding of the carbohydrate storage pattern. (Adapted from Smith et al., 1986, with permission of Kendall Hunt Publishing Company.)

that have broken dormancy have minimal energy reserves, and can be shaded by basal leaves or associated species.

Cutting or grazing *before* the initiation of stem elongation removes only leaf blades, leaving the shoot apex intact, which means that the stem can continue to elongate to produce an inflorescence. Delaying cutting until *after* anthesis will allow time for the axillary tillers to begin growth, and the regrowth rate will be improved.

Like alfalfa, timothy and smooth bromegrass are both good forages for hay, but they often fail to persist in mixtures with alfalfa when managed for high alfalfa hay quality. The reproductive stems of these grasses are often partially elongated when alfalfa is ready to harvest at the early flower stage, but the growth of new grass tillers is suppressed. In smooth bromegrass, significant storage of carbohydrate in the stem base occurs after inflorescence emergence. Although the crude protein concentration of forages declines with maturation, to improve the persistence of timothy or smooth bromegrass in mixtures with alfalfa, harvest should be delayed until anthesis of these grasses.

Orchardgrass recovers rapidly when cut at almost any growth stage, even when it is cut at the boot stage of stem elongation as part of an alfalfa mixture. The flowering shoots with active intercalary meristems for stem elongation appear to exert less apical dominance than those of timothy or smooth bromegrass, so new basal tillers are produced throughout the spring period. Therefore shoots at different stages of development are present at any given time. When harvested, new leaves develop rapidly on the axillary tillers, so photosynthesis of the canopy is only temporarily interrupted, making orchardgrass a potentially more persistent grass companion for alfalfa.

Summary

Plants need energy, meristems, and water for growth. Energy drives the assembly of carbohydrates, nitrogen (from the soil or the atmosphere), and minerals to support respiration and synthesis of new tissue in meristems. Water transports substances, cools the plant, and expands the cells. Depending on the species, plants grow within a range of temperatures, from 0°C (32°F) to above 38°C (100°F). Physiology influences growth rates, flowering, seed production, and resistance to or tolerance of abiotic and biotic stresses. Forage quality depends on rates of leaf and stem growth, including cell wall synthesis and the accumulation of chemical compounds that affect animal performance.

A knowledge of physiological processes forms the basis for understanding how genetics and management affect the rates and efficiency of critical growth processes for climate adaptation. For example, due to their enzymes, cool-season grass and legume species have lower photosynthesis rates than warm-season grasses at temperatures above about 27°C (80°F), but higher rates at 4°C (40°F). Perennial forage legumes store starch and proteins in their roots to support regrowth after harvest, whereas perennial cool-season grasses store carbohydrate and proteins in their stem bases and depend more on leaf area for photosynthesis during regrowth. Stomata in the leaves help the plants to control water stress.

Questions

1. What are the factors that influence the photosynthetic rate of C_4 grasses and allow it to continue to increase as the temperature rises beyond the optimal range for C_3 grasses?
2. What are the characteristics of forage nutritive value that cause cool-season grasses to be of higher quality than warm-season grasses?
3. What causes the total non-structural carbohydrate content of leaves to change between late afternoon and the following morning?
4. Will warm-season or cool-season grasses be more negatively affected by increasing biosphere CO_2 levels? Which will adapt better in a location where maximum summer temperatures and drought also increase?
5. Describe three situations where uninformed management of perennial forage crop carbohydrate reserves can undermine persistence. What are the recommended practices?
6. What is the principal storage organ for organic reserves in (a) alfalfa, (b) white clover, and (c) smooth bromegrass?
7. How does plant storage and plant use of carbohydrates in the roots of alfalfa differ from that in the roots of birdsfoot trefoil during the growing season? How should the cutting management of these two forages differ so as to optimize persistence based on midsummer patterns of carbohydrate storage?
8. What is the difference between grazing to the proper height and overgrazing of a grass such as tall fescue in terms of plant carbohydrate storage, and what is the effect of each on the initial regrowth rate?
9. What is the critical LAI of white clover and of tall fescue? Why are these values different? How would this difference affect the potential accumulation of usable biomass in a monoculture pasture of each species?
10. Where is the meristem for a stem or branch located, relative to the rest of the branch? Provide an example in which the initial rate of regrowth is affected by the location of the reproductive stage of growth at cutting.

References

Akin, DE. 1989. Histological and physical factors affecting digestibility of forages. Agron. J. 81:17–25.

Allard, G, and CJ Nelson. 1991. Photosynthate partitioning in basal zones of tall fescue leaf blades. Plant Physiol. 95:663–668.

Atkin, OK, and MG Tjoelker. 2003. Thermal acclimation and the dynamic response of plant respiration to temperature. Trends Plant Sci. 8:343–351.

Burr S, and DM Turner. 1933. British Economic Grasses. Edward Arnold, London.

Fick, GW, and SC Mueller. 1989. Alfalfa: Quality, Maturity, and Mean Stage of Development. Information Bulletin 217. Department of Agronomy, College of Agriculture and Life Sciences, Cornell University, Ithaca, NY.

Fisher, DS, HF Mayland, and JC Burns. 1999. Variation in ruminants' preference for tall fescue hays cut either at sundown or at sunup. J. Anim. Sci. 77:762–768.

Graber, LF, NT Nelson, WA Luekel, and WB Albert. 1927. Organic Food Reserves in Relation to the Growth of Alfalfa and Other Perennial Herbaceous Plants. Wisconsin Agricultural Experiment Station Research Bulletin 80. University of Wisconsin, Madison, WI.

Gutschick, VP. 1981. Evolved strategies in nitrogen acquisition by plants. Am. Nat. 118:607–637.

Harrison, CM, and CW Hodgson. 1939. Response of certain perennial grasses to cutting treatments. J. Am. Soc. Agron. 31:418–430.

International Fertilizer Industry Association. 1998. The Fertilizer Industry, World Food Supplies and the Environment. International Fertilizer Industry Association and United Nations Environmental Programme, Paris.

Kim, S, and B Dale. 2008. Effects of nitrogen fertilizer application on greenhouse gas emissions and economics of corn production. Environ. Sci. Technol. 42:6028–6033.

Loomis, RS, and WA Williams. 1969. Productivity and the morphology of crop stands: patterns with leaves. In JD Eastin et al. (eds.), Physiological Aspects of Crop Yield, pp. 27–47. American Society of Agronomy, Madison, WI.

Lovvorn, RL. 1945. The effect of defoliation, soil fertility, temperature, and length of day on the growth of some perennial grasses. J. Am. Soc. Agron. 37:570–582.

MacAdam, JW. 2009. Structure and Function of Plants. Wiley-Blackwell, Ames, IA.

MacAdam, JW, JJ Volenec, and CJ Nelson. 1989. Effects of nitrogen on mesophyll cell division and epidermal cell elongation in tall fescue leaf blades. Plant Physiol. 89:549–556.

Mousavi, SA, A Willems, X Nesme, P de Lajudie, and K Lindström. 2015. Revised phylogeny of *Rhizobiaceae*: Proposal of the delineation of *Pararhizobium* gen. nov., and 13 new species combinations. Syst. Appl. Microbiol. 38: 84–90.

Penning de Vries, FWT, HH VanLaar, and MCM Chardon. 1983. Bioenergetics of growth of seeds, fruits, and storage organs. In Potential Productivity of Field Crops Under Different Environments, pp. 37–59. International Rice Research Institute, Los Baños, Philippines.

Schwab, EC, CG Schwab, RD Shaver, CL Girard, DE Putnam, and NL Whitehouse. 2006. Dietary forage and nonfiber carbohydrate contents influence B-vitamin intake, duodenal flow, and apparent ruminal synthesis in lactating dairy cows. J. Dairy Sci. 89:174–187.

Singh, M, WL Ogren, and JM Widholm. 1974. Photosynthetic characteristics of several C_3 and C_4 plant species grown under different light intensities. Crop Sci. 14: 563–566.

Smith, D. 1962. Carbohydrate root reserves in alfalfa, red clover, and birdsfoot trefoil under several management schedules. Crop Sci. 2:75–78.

Smith, D, RJ Bula, and RP Walgenbach. 1986. Forage Management, 5th ed. Kendall Hunt Publishing Company, Dubuque, IA.

Thompson, P, CG Bowsher, and AK Tobin. 1998. Heterogeneity of mitochondrial protein biogenesis during primary leaf development in barley. Plant Physiol. 118:1089–1099.

US Environmental Protection Agency. 2016. Future Climate Change. https://www3.epa.gov/climatechange/science/future.html (accessed 20 May 2016).

Volenec, JJ, A Ourry, and BC Joern. 1996. A role for nitrogen reserves in forage regrowth and stress tolerance. Physiol. Plant. 97:185–193.

Walton, PD. 1983. Production and Management of Cultivated Forages. Pearson Education, Upper Saddle River, NJ.

Wilson, D. 1982. Response to selection for dark respiration rate in mature leaves in *Lolium perenne* and its effects on growth of young plants and simulated swards. Ann. Bot. 49:303–312.

Yamori, W, K Hikosaka, and DA Way. 2014. Temperature response of photosynthesis in C_3, C_4, and CAM plants: temperature acclimation and temperature adaptation. Photosynth. Res. 119:101–117.

CHAPTER

5

Environmental Aspects of Forage Management

Jeffrey J. Volenec and C. Jerry Nelson

The environment that plants experience depends on climate and the variation of weather events within the climate. **Climate** refers to the long-term history of temperature, precipitation, and radiation for a given region. Climate is the principal factor affecting adaptation of forage species or cultivars to a given location. In contrast, **weather** includes day-to-day and short-term extremes in temperature, precipitation (as rain, snow, or hail), relative humidity, wind, and solar radiation at a given site. Forage producers need to be aware of the climate (i.e., the average year) when selecting adapted species and cultivars to plant (see Chapter 15) and weather events that alter within-year productivity and influence day-to-day management decisions. These relationships are the focus of this chapter.

Effects of Climate

Plant geographers have long recognized the relationship between climate, primarily precipitation and temperature, and natural vegetation at a given site (see Chapter 14). **Grasslands** consisting of **grasses, legumes, forbs,** and shrubs compete naturally with trees and predominate in dry areas where trees are less adapted. Well-managed grasslands can restrict tree invasion because they limit establishment by competing effectively with seedlings for water, nutrients, and light. Conversely, if trees and other woody species become established, they shade the grasslands, are rarely grazed by domestic ruminants, and become dominant in the ecosystem. Invasion by woody species, such as mesquite in the southwestern grasslands and eastern red cedar in the Central Great Plains, causes severe problems. Likewise, deciduous trees are the natural vegetation and will encroach into pastures in high rainfall areas in the eastern USA.

Like trees, grasses exist naturally in distinct geographic regions. Grasses with C_4 photosynthesis predominate in the tall-grass prairie of the Central Great Plains, but shorter grasses that are more resistant to drought are dominant in short-grass prairies in more arid areas. Grasses in short-grass prairies are mainly C_4 types in southern latitudes of the USA, but are C_3 types in northern latitudes and at high altitudes in southern regions (Fig. 5.1). Some forage species, such as alfalfa, possess great genetic diversity that enables adaptation to a wide range of temperature and water-stress conditions, but most grasses and legumes are more restricted in adaptation. In the eastern USA, where moisture is adequate, temperature is the principal factor affecting the distribution of forage species.

Climate Continues to Change

Most scientists agree that the levels of carbon dioxide (CO_2), nitrous oxide, methane, and other greenhouse gases are increasing in the atmosphere and that they cause global warming. Global climate models indicate that in the future the earth's surface and air temperature will increase gradually, only slightly at the equator, but by 8–10°F at higher latitudes due to the way in which the air circulates in each hemisphere. Average air temperature in the central USA is expected to increase by 5–10°F by 2065. Annual precipitation will remain similar to present levels, but more will occur in winter, increasing the potential for soil erosion, and less in summer, increasing the potential for drought stress. The frost-free growing season will be longer, as

Forages: An Introduction to Grassland Agriculture, Seventh Edition. Edited by Michael Collins, C. Jerry Nelson, Kenneth J. Moore and Robert F Barnes.

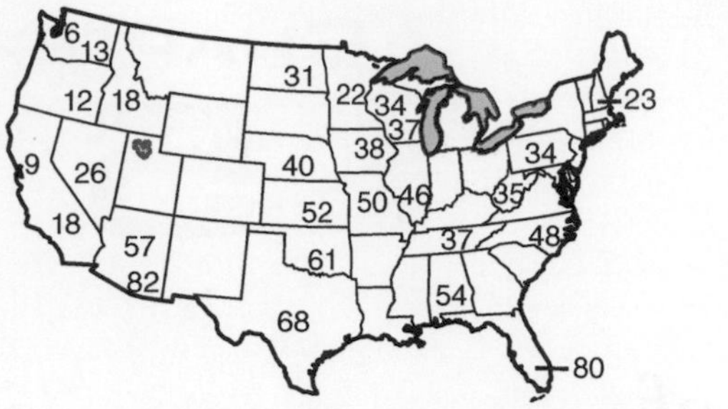

FIG. 5.1. Percentage of species with C_4 photosynthesis among naturally occurring grass floras in the USA. Species include annuals and perennials. No C_4 species were found on the arctic slopes of Alaska or in northern Manitoba, Canada. (Adapted from Teeri and Stowe, 1976.)

indicated by reduced snow cover in March, and the frequency of severe storms will increase. Current climates will be displaced northward to account for increased temperature and slightly eastward to account for changes in precipitation (Fig. 5.2). They will be displaced about 1500 ft (457 m) upward in elevation in mountainous regions. However, daylength will remain the same.

The higher CO_2 levels will benefit C_3 plants with increases in photosynthesis, yield, and, for leguminous species, greater N_2 fixation. The higher temperature may benefit growth of C_4 plants, but leaves and stems will have a reduced forage quality. Associated water deficit stress will reduce yield, but plants may be more deeply rooted. The impacts of climate change on forage-livestock agriculture are challenging to estimate with certainty because they will be location-specific and seasonal in nature. Present-day milk production is estimated to have been reduced by

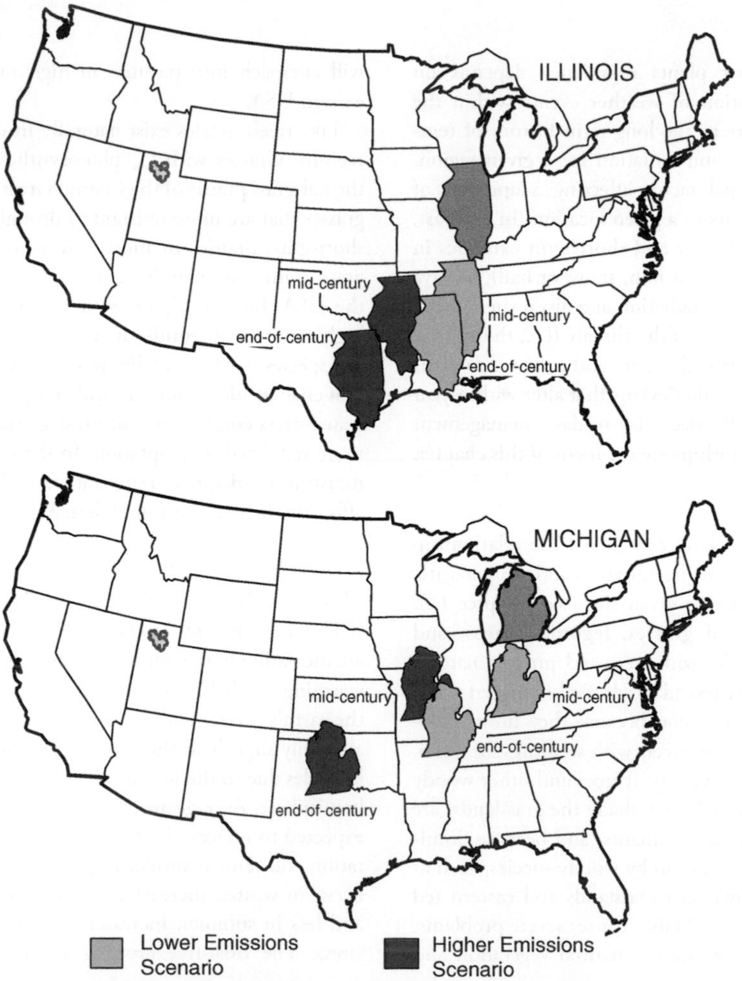

FIG. 5.2. Predicted changes in temperature and precipitation for North America in the latter part of the twenty-first century. (From Karl et al., 2009.)

1.9% by climate change to date, and this reduction is projected to increase to 6.3% by the end of the twenty-first century (Mauger et al., 2015). Using present-day prices, this corresponds to annual losses of US$ 670 million per year today, rising to US$ 2.2 billion per year by the end of the century. State-specific reductions in end-of-century milk production are predicted to vary from 0.4% in Washington State to a 25% loss in annual milk production in Florida.

Weather Within a Climate

Grasses and other forage species occur naturally in climates characterized by wide ranges in seasonal or daily temperature and moisture extremes. However, productivity may be low, and these ecosystems are more fragile and easily disrupted. Thus the area of consistent economic production of a given species is smaller than for the general area of climatic adaptation. As species are grown away from the regions to which they are best adapted, the need for careful management increases, in order to minimize the impacts of these stresses. For example, inconsistent winter survival of perennial, cool-season (C_3) forages often restricts their northern geographic limits, and high **respiration** rates restrict their adaptation to the South. Conversely, the northern distribution of many perennial warm-season (C_4) grasses is restricted by lack of winter hardiness and low minimum temperatures in July (Fig. 5.1). Temperature and drought stress often coincide and exacerbate the negative impacts of these individual stresses on adaptation of annuals and **perennation** and short-lived perennials that must flower and reseed naturally (Beuselinck et al., 1994).

Microclimate

Environmental conditions in and around the soil and plant **canopy** constitute the **microclimate**. Because the microclimate can be manipulated to a greater degree by crop management than can the climate, it is critical that we understand its role in the growth and development of forage plants. Temperature, radiation, moisture, CO_2, and wind speed near the soil surface are affected by the plant canopy (Fig. 5.3); the temperature and humidity within the canopy may be markedly different from that in a standard meteorological shelter 5 ft (1.5 m) above the ground. **Diurnal** (i.e., day vs. night) variations are also large in the microclimate zone (Fig. 5.4). Near the ground, the air is coolest just before sunrise and warmest near midday. At 5 ft (1.5 m) above the ground, the maximum temperature occurs about 1 to 2 hours later than it does immediately above the soil surface.

Since solar radiation strongly affects microclimate temperatures, diurnal temperature change is much greater on clear days, when incoming radiation from the sun greatly exceeds that re-radiated from the earth, with the result that surface temperatures increase rapidly. At night the reverse occurs, and surfaces cool rapidly. Clouds reduce both incoming and outgoing radiation, resulting in less diurnal temperature change and a smaller difference between the microclimate and the **macroclimate,** the environment measured at 5 ft (1.5 m). Relationships between incoming and outgoing radiation largely determine temperatures at plant and soil surfaces, which in turn affect plant growth and development, and ultimately agronomic performance. The impact of radiation balance on plant temperature is further modulated by transpirational cooling, where latent heat of vaporization consumes 540 cal/g of water evaporated from plant surfaces.

Management can markedly alter the microclimate. For example, on clear days the soil temperature increases rapidly by as much as 20°F after a forage is cut or grazed, especially when nearly all the canopy is removed (Fig. 5.4).

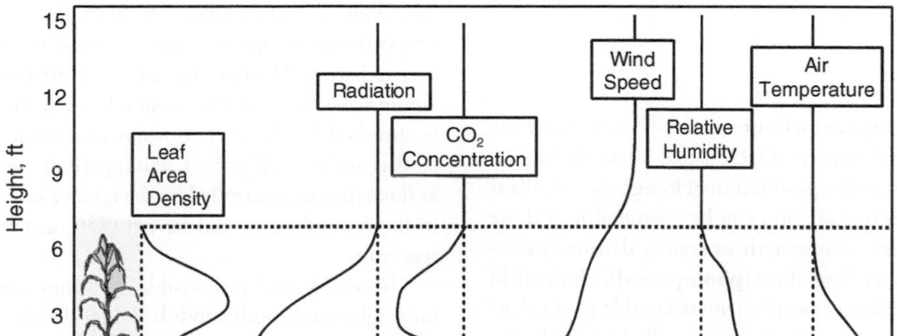

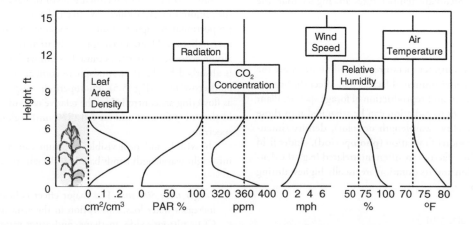

FIG. 5.3. Distribution of leaf area, photosynthetically active radiation (PAR), carbon dioxide concentration, wind speed in miles per hour (mph), relative humidity, and air temperature above and within a corn canopy. Relationships in shorter-growing forages would be similar, but compressed in height. (Adapted from Lemon, 1969.)

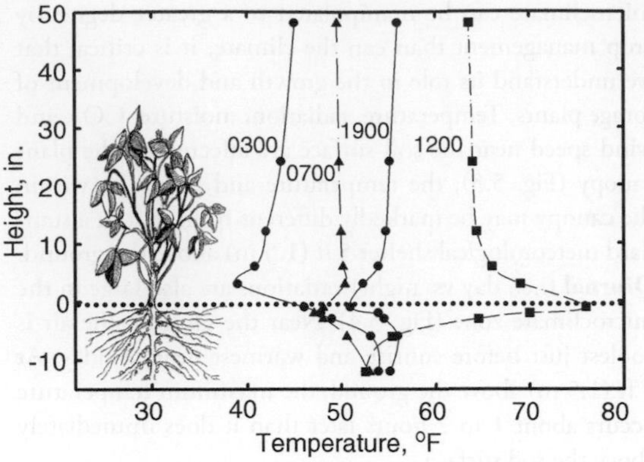

Fig. 5.4. Microclimate temperatures above and below the soil surface at different times of the day. Note that the soil surface temperature changes more during the day than does the air temperature a few feet above the soil, especially around noon when direct radiation hits the soil and at night when the soil is cooled by water evaporation and radiation back to the atmosphere. Soil temperatures below a plant canopy show less daily change than bare soil because the leaves intercept incoming radiation during the day and trap outgoing radiation and heat loss from the soil at night. (Adapted from Geiger, 1965.)

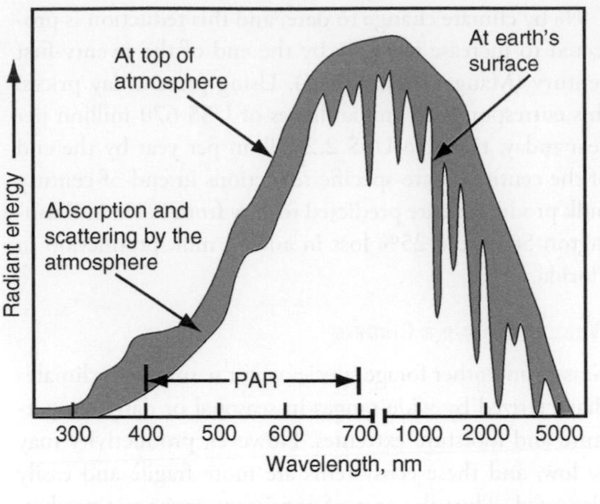

Fig. 5.5. Radiant energy at the top of the atmosphere and at the earth's surface as a function of wavelength. The atmosphere screens much of the ultraviolet radiation at wavelengths of less than 400 nm. Photosynthetically active radiation (PAR) includes wavelengths in the range 400–700 nm. Water vapor, nitrous oxide, and CO_2 are major absorbers in the atmosphere at long wavelengths (> 800 nm). Note the change in scale of the *X*-axis from linear to logarithmic.

This increases the temperature of buds located near the soil surface that develop into **shoots** and **tillers,** and can reduce shoot or tiller regrowth rate. Higher temperatures at soil level may be beneficial in cool climates or during cool parts of the growing season, but they are detrimental when microclimate temperatures are higher than the optimum.

Solar Radiation

Light (or radiation), temperature, and soil moisture are the three cardinal environmental factors that affect the vegetative development and reproduction of forage species. Plant growth responses to radiation can be separated into those due to the quality (wavelength or color), density (intensity), and duration of radiation (**photoperiod**). Under field conditions these factors are often interrelated (e.g., the density of radiation and its duration are usually highest during the same season).

Quality (Wavelength)

Quality refers to the wavelength of the rays that contribute to the radiation spectrum (Fig. 5.5). Plant development is better under the full spectrum of sunlight than under any single portion of the spectrum. In addition, radiation in the visible range (400–700 nanometers, nm) is most active in photosynthesis and is referred to as photosynthetically active radiation (PAR). Plants grown mainly under **infrared** wavelengths (> 700 nm) usually grow tall and thin, and are often fragile, mimicking plant growth in shade. Plants grown mainly under **ultraviolet** wavelengths (< 400 nm) may show retarded growth or have their tissues injured or even killed. Winter radiation tends to be proportionately higher in infrared radiation than is summer radiation. Plants often appear dwarfed or stunted in alpine ecosystems, partly because less ultraviolet radiation is absorbed by the thin atmosphere compared with lowland ecosystems (Fig. 5.5). Photoperiodic responses such as flowering are controlled by the relative ratio of radiation in the red (660 nm) and far-red (730 nm) regions of the spectrum.

The visible and near-visible spectrum can be divided into eight wavelength bands based on the effects on plants.

1. *Longer than 800 nm:* The major effect is heat, which increases water loss. Absorption in the atmosphere by CO_2, nitrous oxide, methane, and water vapor generates heat and causes global warming.
2. *Between 800 and 700 nm:* This promotes elongation growth and far-red effects on the **phytochrome** system.

3. *Between 700 and 610 nm:* There is peak chlorophyll absorption and maximum photosynthetic activity, as well as red effects on the phytochrome system.
4. *Between 610 and 510 nm:* Effects are minimal. Plants appear green because they reflect much of this radiation.
5. *Between 510 and 400 nm:* There is absorption by yellow pigments and chlorophyll for photosynthesis; **phototropism** responses occur.
6. *Between 400 and 320 nm (UV-A):* This affects leaf shape. Plants are shorter and leaves are thicker.
7. *Between 320 and 280 nm (UV-B)*: This is detrimental to most plants; damage is caused to DNA, and mutations occur.
8. *Shorter than 280 nm (UV-C):* This is highly deleterious, resulting in the rapid death of plants.

Fortunately, water vapor in the earth's atmosphere screens out much of the radiation with wavelengths above 700 nm. Carbon dioxide also absorbs radiation with wavelengths above 700 nm and contributes to global warming. Ozone absorbs much of the radiation with wavelengths below 320 nm. Recently the breakdown of ozone in the upper atmosphere has led to worldwide concern about increased transmission of UV-B radiation to the earth's surface, and the subsequent negative effects on plant growth as well as diseases such as cancer in animals and humans.

Density (Amount)

Radiation density is measured in energy units, and at noon on a clear day during summer it may be near 2000 μmol **photons** m^{-2} s^{-1}. When nutrient and water supplies are adequate, photosynthesis and growth rates of plants are a direct function of radiation density (Fig. 5.6). Reductions in light intensity (e.g., due to cloud cover) reduce photosynthesis and growth, especially of C_4 and sun-adapted C_3 plants. Shade-adapted C_3 plants (sciophytes) exhibit a smaller reduction in photosynthesis when shaded, and are well adapted to mixtures. Forage growth rate is related more to percentage radiation interception than to photosynthetic activity per unit of leaf area until extensive leaf area accumulates following cutting or grazing. A full canopy of leaf blades is needed to intercept the maximum amount of radiation, but the amount of leaf area required depends on the leaf angle (see Chapter 4).

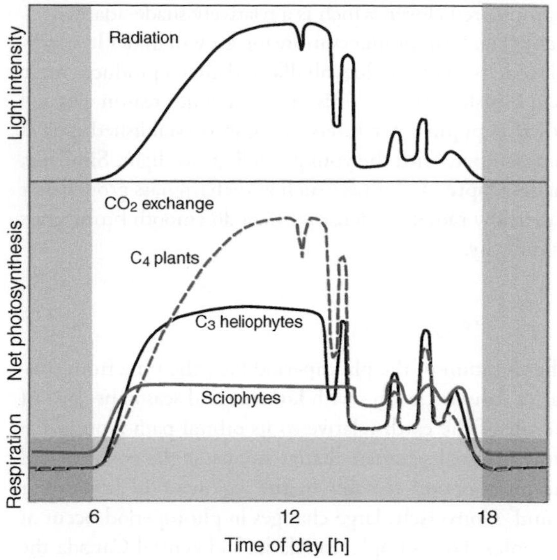

Fig. 5.6. Daily course of light intensity, net photosynthesis, and dark respiration for C_4 plants, C_3 sun-adapted plants (heliophytes), and C_3 shade-adapted plants (sciophytes). Cloud appearance from noon onward reduces the light intensity, and reduces photosynthesis of C_4 plants most and C_3 shade-adapted plants least. (From Larcher, 1995. Reproduced with permission of Springer.)

Family, Friend, or Foe?

Recently, scientists have learned that plants can "detect" their neighbors based on the spectral properties of the light reflected from the leaves and other tissues. About 10% of the light that strikes a leaf is reflected, and this reflected light from each plant species has a characteristic spectral signature (i.e., a well-defined proportion of wavelengths is reflected). As a result, some crop plants, such as corn and soybean, grow well in **monoculture** but can detect a weed nearby because of its characteristic reflectance spectrum, and can then change their growth habit or yield. Conversely, many forage grasses and legumes grow best in mixtures, and the compatibility of species may be determined, in part, by the spectra of reflected light. This light can signal the receiving plant to increase its leaf area, change its leaf or tiller angle, or alter other growth characteristics to either accommodate or interfere with the reflecting plant. As more is learned about this response, it may become feasible to genetically engineer cultivars for fighting weeds, for specific mixtures, or for growth conditions. This spectral signature is already utilized in remote-sensing applications for detecting nutrient or environmental stresses by using satellite-derived images.

Managing competition for light is especially important when forages are established with a grain companion crop that shades young seedlings, or when a grass and a legume are grown together in mixtures. Companion crops such as spring oat are often cut at an early stage of growth for hay or silage to remove this competitive canopy. Species respond differently to variations in radiation density. For example, red clover, which is a relatively shade-adapted C_3 plant (Fig. 5.6), produces more top growth under low radiation densities than does alfalfa, and alfalfa produces more than birdsfoot trefoil. This may be a major reason why red clover is popular for interseeding into established grasses that compete with the young seedlings for light. Similarly, shade-adapted C_3 grasses such as orchardgrass grow better under low radiation densities than do smooth bromegrass or timothy.

Duration (Photoperiod)

The duration of the photoperiod (i.e., the time from sunrise to sunset) changes with **latitude** and season because of the tilt of the earth relative to its orbital path around the sun. Minimal seasonal change occurs at the equator, and the photoperiod remains in the region of 12 hours year round. Conversely, large changes in photoperiod occur at the poles. For example, in Alaska and central Canada the photoperiod on the summer solstice (approximately June 21) can exceed 22 hours, but it may be only a few hours long on the winter solstice (approximately December 21). The range in photoperiod for a given location depends on its latitude. For example, the threefold range in the amount of solar radiation received per day at West Lafayette, Indiana, located at 41° N latitude (Fig. 5.7), is largely due to photoperiod differences during the year.

Day-to-day variation in solar radiation is mainly due to cloud cover, which can reduce radiation by as much as 90% and thus reduces photosynthesis, especially of C_4 plants (Fig. 5.6). Seasonal trends in solar radiation also drive changes in air and soil temperatures. However, soil temperatures are buffered from change, especially in winter, by the high heat capacity of the soil. The rates of photosynthesis and growth are generally highest during the longest days of summer, when the most radiation per day is received.

In addition to the photosynthetic role of light, photoperiod influences plant growth and development. Most temperate grasses and legumes flower during long photoperiods (Fig. 5.8), a response in perennial forage grasses that also requires **induction** caused by cold temperatures. Flowering of tropical species is less affected by photoperiod than is flowering of temperate and arctic species, because species of tropical origin have evolved in environments where photoperiod varies little.

Photoperiod also affects the vegetative growth form of many forage species. Leaf and stem growth in spring and summer is often erect under long photoperiods, but growth under short photoperiods in autumn tends to be prostrate and more branched. However, exceptions to this generality exist. For example, tall fescue produces long leaves under shortening photoperiods in early autumn, making this species particularly useful for **stockpiling** forage for winter grazing.

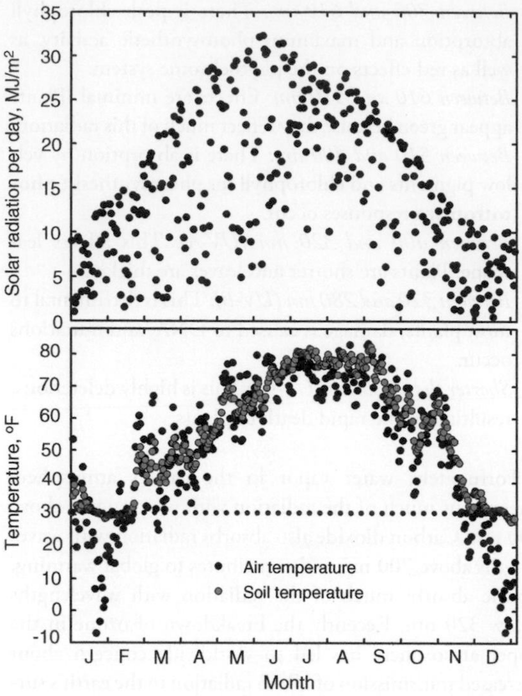

Fig. 5.7. Changes in daily solar radiation (upper panel), and daily mean soil (4 in. [10 cm] depth) and air temperature (5 ft [1.5 m] above soil) (lower panel) during the year 2000 at West Lafayette, Indiana. Note the day-to-day variation in radiation due to cloud cover, and the strongly buffered soil temperature in winter. The high specific heat of water, the heat of fusion for converting water to ice, and the mulch effects due to snow and the crop canopy help to keep soil temperatures near or slightly below freezing even when air temperatures are well below freezing. Plants with overwintering tissues in the soil can survive by becoming cold hardy.

Cultivars within a forage species also can differ in their response to photoperiod (Fig. 5.9). Fall-dormant alfalfa cultivars adapted to cold climates show reduced shoot growth in autumn. In contrast, non-dormant alfalfa cultivars show extensive shoot growth in autumn and winter; they are used in the southwestern USA where forage growth in winter is important but temperatures are not cold enough to injure or kill the plants. Knowledge of how photoperiod influences flowering and vegetative growth of forages facilitates the design of pasture and hay management systems that are optimally adapted to the different climatic regions.

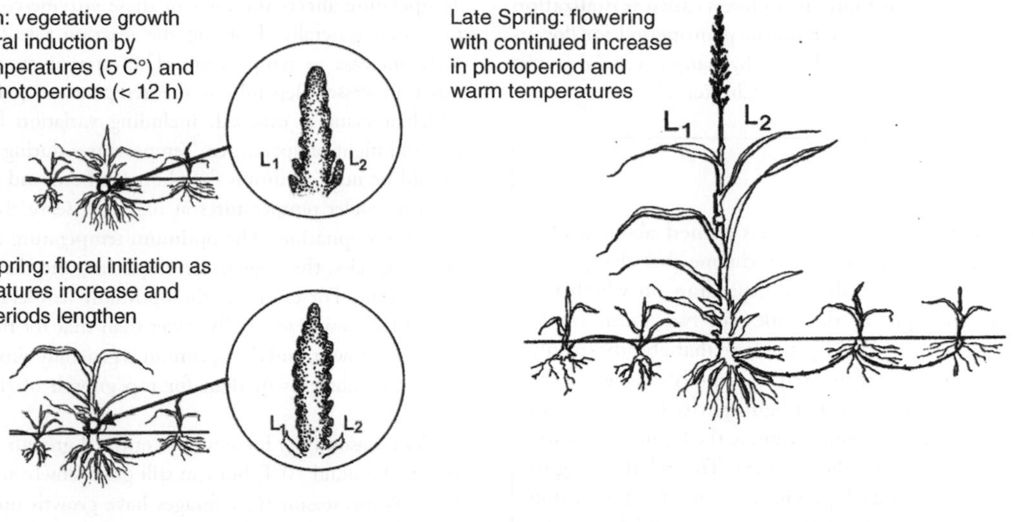

Fig. 5.8. Floral induction, initiation, and development of a typical forage grass exhibiting a long-day photoperiodic response. Low temperatures and short days in autumn are required for induction (the process of preparing the shoot apex for flowering). Floral initiation occurs in spring in response to lengthening days and warmer temperatures. Floral development in late spring completes differentiation of the vegetative apex into the inflorescence and its reproductive structures. Internode elongation elevates the developing inflorescence upward through the whorl of leaves to emerge at the top of the plant. Young tillers at the base of the elongating culm and those developed on stolons or rhizomes will go through the process the following year. (From Gardner and Loomis, 1953.)

Fig. 5.9. Differences in fall dormancy (FD) of alfalfa in November at West Lafayette, Indiana. The foreground cultivar has an FD rating of 1, whereas the cultivar in the background with an FD rating of 10 has extensive shoot growth that has been damaged by frost. (Photo courtesy of Jeffrey Volenec, Purdue University.)

Early studies emphasized the role of photoperiod in flowering and plant development, but it is now known that the photoperiodic responses are actually regulated by the length of the dark period (see below). Even so, the "daylength" terminology has remained. The effect of photoperiod on flowering has been studied intensively, as it influences both seed production and the natural geographical distribution of species.

Short-day plants flower only within a range of relatively short photoperiods (actually long nights), whereas **long-day plants** flower only within a range of relatively long photoperiods (actually short nights). For example, korean lespedeza, a short-day plant, produces little viable seed in states north of Missouri and central Indiana because the long days delay flowering such that mature seed is not produced before the frosts. Conversely, birdsfoot trefoil produces the highest seed yields in northern states because the long photoperiods exceed the minimum required for profuse flowering. Other plants, referred to as **day-neutral,** are capable of flowering under both long and short photoperiods.

The flowering response has been found to be more complicated than this classification, commonly known as **photoperiodism,** would indicate. For example, many forage species (especially perennial forage grasses) require exposure to low temperatures (less than 40°F) for an extended period (4 weeks or more) in a process called **vernalization** prior to exposure to the requisite photoperiod for flowering to occur (Fig. 5.8). This is why many perennial grasses flower only once each year (see Chapter 2).

Phytochrome, Flowering, and Biological Clocks

Several decades ago, scientists learned about a blue protein in leaves called phytochrome that changed its chemical structure slightly depending on whether it was in the light or the dark. The two forms would oscillate. During the day, the form that absorbs mainly red light (phytochrome red, P_r) rapidly changes to the form that absorbs far-red light (phytochrome far-red, P_{fr}) (Fig. 5.10). During darkness, the P_{fr} form is slowly converted back to the P_r form. The relative lengths of the light and dark period determine the proportion of each form of phytochrome present, which provides plants with a biological clock that detects daylength. This spawned the classification system for flowering behavior in plants, namely long-day, short-day, and day-neutral. Soon afterwards it was determined that the length of the night controlled flowering behavior, but by this stage it was too late to change the terminology that was already in widespread use. The amount of P_{fr} present at sunrise sends a signal that initiates the flowering response in long-day (short-night) plants and sends an inhibitory signal in short-day (long-night) plants. The mechanism whereby the phytochrome signal is transferred from the leaves to the flower bud remains a mystery.

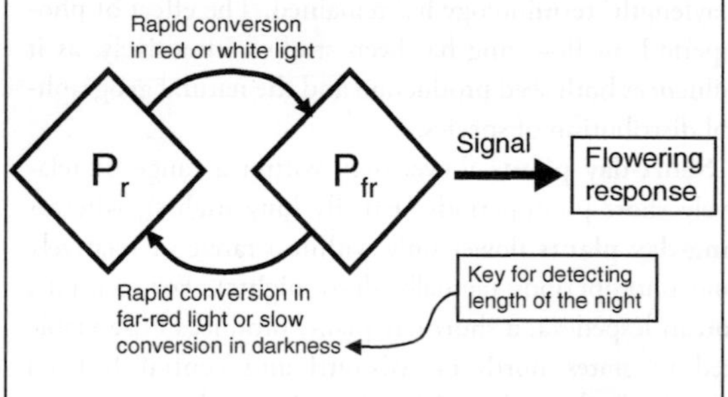

Fig. 5.10. Flowering of many forage plants depends on daylength and its influence on the relative amounts of phytochrome P_r and phytochrome P_{fr}.

Temperature

Plant metabolic pathways (e.g., photosynthesis and respiration) and growth processes (e.g., cell division, cell expansion, and cell wall synthesis) are catalyzed by enzymes. Temperature affects the rates of these enzyme-controlled processes, generally doubling the reaction rate for each 18°F increase in temperature. The rates of growth and other processes depend on the temperature pattern to which a plant is exposed, including variation between day and night temperatures. Temperatures during the day should be near optimum for photosynthesis and growth, whereas cooler temperatures at night conserve energy by reducing respiration. The optimum temperature depends on the species, the stage of development, and the specific plant tissue. For example, the optimum temperature for vegetative growth is usually lower than that for flowering and fruit growth, and the optimum is generally about 10°F lower for root growth than for top growth of the same plant.

Cool-season (C_3) forages have optimal growth temperatures of around 70°F, but can still grow slowly at around 35°F. Warm-season (C_4) forages have growth optima of around 90°F, and grow little at temperatures below 50°F—hence the use of 50°F as the base temperature in growing degree day (GDD) models for C_4 plants such as corn. For example, the rates of photosynthesis and dry matter accumulation for seven C_3 grasses in one study were highest at 50–70°F. However, the rates for the C_4 species, bermudagrass and bahiagrass, were highest at 95°F and decreased rapidly at lower temperatures. In another study, the growth rates of several C_4 grasses were highest at 97°F day/88°F night temperatures, and were decreased by about 75% at 60°/50°F (Fig. 5.11). In contrast, the relative growth rates of C_3 grass species were highest at 81°/72°F and were decreased by about 40% at 97°/88°F. Switchgrass, a C_4

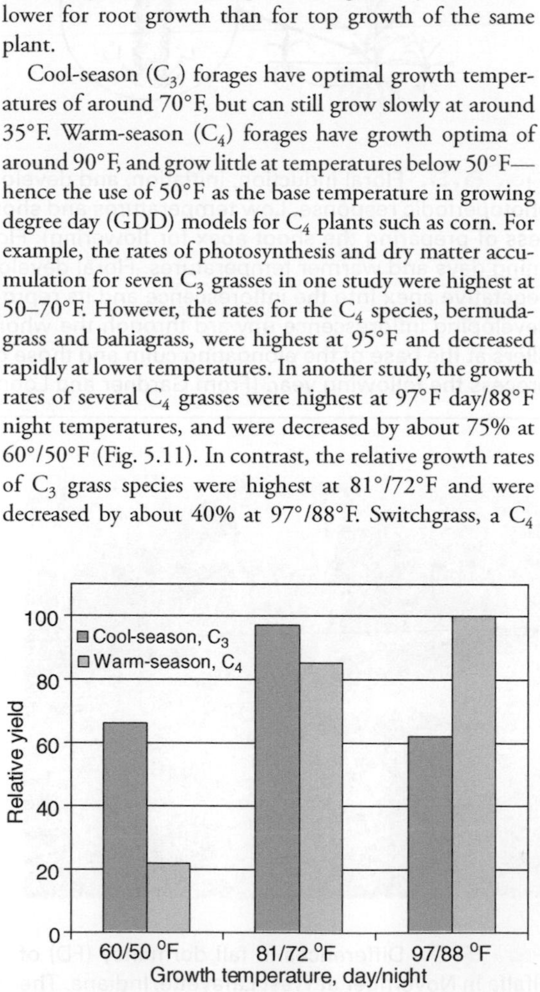

Fig. 5.11. Relative yield of cool-season (C_3) and warm-season (C_4) grasses grown in three day/night temperature regimes showing the cool-temperature sensitivity and warm-season preference of C_4 species. Data are averages of seven C_3 and eight C_4 grasses. (Adapted from Kawanabe, 1968.)

species, grows best and responds maximally to fertilizer nitrogen (N) at 90°/79°F, whereas rapid growth and maximum responses to N fertilizer of timothy, a C_3 species, occur at temperatures ranging from 60°/50°F to 70°/60°F.

Growth occurs as a result of cell division and cell expansion. The rates of cell division are closely related to the temperature of the **meristems**, where large amounts of energy are needed and the duration of the mitotic cycle is temperature dependent. The rates of cell expansion likewise depend on temperature and vary according to the species. For example, shoot growth rates of dicots, which generally have the shoot apex at the top of the canopy, are closely related to air temperature. In contrast, the shoot apex and leaf meristems of grasses are often near soil level, especially during the vegetative stages, so shoot or tiller growth depends on temperature conditions near the soil surface.

Air and soil temperatures often differ depending on the season (Fig. 5.7) and management, and different growth responses can result. In addition, adaptation to a given climate depends on temperature relationships and represents the major physiological basis for separating forage species into cool-season, warm-season, and tropical types. Interestingly, temperature responses of growth and photosynthesis are highly correlated within a species, illustrating the close association between source (photosynthesis) and **sink** (growth) activities. Growth responses are generally more sensitive to low temperature than is photosynthesis, allowing excess photosynthate to accumulate in storage organs when growth is slowed. High respiration rates at elevated temperatures can reduce carbohydrate storage, growth rates, and survival of forage species. This restricts adaptation of many perennial cool-season grasses to temperate latitudes and high altitude.

Higher temperatures increase the rate of plant development and decrease the time from seeding to flowering. In warm environments, vegetative plant development is accelerated—plants tend to be shorter and to bloom earlier than in cool environments. This response is one reason why the forage yields of cool-season species such as alfalfa and red clover are lower during hot summer periods. Earlier onset of flowering means that more frequent harvests may be needed in warm environments to maintain the desired forage quality. Conversely, tropical species are better adapted and most productive under hot, humid conditions.

When grown at high temperatures, most cool-season grasses and legumes produce smaller cells with thicker cell walls, have a lower leaf:stem ratio, store less **non-structural carbohydrate,** and produce herbage of lower digestibility. Thus the forage quality of cool-season grasses and legumes decreases during spring growth, due in part to increasing temperature, which alters metabolism, hastens stem development for flowering, and decreases the leaf:stem ratio. Conversely, the forage quality of cool-season grasses changes more slowly with time during autumn, because temperatures are decreasing, little stem growth occurs, and non-structural carbohydrates that are completely digestible accumulate in the leaves.

High-Temperature Stress

Plants are subjected to rapid daily changes in temperature, and stress occurs when temperatures are above or below the optimal range for that particular species. The severity of damage depends on the stage of plant development and on stress intensity and duration. High-temperature stress often occurs concurrently with moisture stress, making it difficult to separate the two effects. Limited water availability decreases transpiration and evaporative cooling of plant tissues, so high-temperature stress is exacerbated. Irrigation managers use this knowledge to measure canopy temperature with an infrared thermometer to help to determine when a field needs water.

High temperatures may lead to a number of metabolic disorders in plants, including elevated dark respiration, enzyme inactivation, imbalance among reaction rates, membrane dysfunction, and reduced synthesis of cellular constituents. Excessively high temperatures can induce flower sterility, especially pollen abortion, leading to poor seed production. High temperatures during late seed development can reduce the subsequent germination of seeds and seedling vigor.

Chilling Temperature Stress

Chilling injury can occur at temperatures above freezing in some plants, mainly by affecting either membrane structure and function or metabolism. As a group, C_3 plants are generally more tolerant of chilling stress than are C_4 plants, in which photosynthesis is impaired at temperatures below 50°F. However, certain C_3 plants, such as rice, are also very susceptible to chilling injury. **Translocation** of photosynthate from the chloroplasts and mesophyll cells (sources) to the meristems (sinks) requires metabolic energy to load sugars into the **phloem**. For example, following exposure to night-time temperatures below 60°F, the C_4 species digitgrass shows reduced translocation of photosynthate from the leaves at night. Starch is retained in the chloroplasts and photosynthesis is reduced the following day, even when daytime temperatures are near optimum.

Reasons for the lack of sensitivity to low, non-freezing temperatures include the ability of chilling-tolerant species to rapidly alter their membrane structure to maintain both fluidity and the unimpaired function of membrane-bound processes such as the light reactions of photosynthesis. In addition, most cool-season grasses accumulate **fructan,** a storage carbohydrate, in cell vacuoles, in contrast with legumes and warm-season grasses that store starch in the chloroplasts. At low temperatures both synthesis and breakdown of fructan occur more readily than is the case for starch.

Freezing Temperature Stress

Winter injury can be caused by excessively cold temperatures, **ice sheets,** or frost (Bowley and McKersie, 1990). The availability of new cultivars that are genetically capable of withstanding adverse winters has extended forage plant adaptation to colder climates. However, little progress has been made toward breeding forage cultivars to reduce heaving or ice sheet damage. Therefore management practices remain the major control mechanisms.

Frost Heaving

Plant heaving is a serious problem in regions where temperatures fluctuate around 32°F and forages are grown on imperfectly drained, fine-textured soils. Liquid water expands in volume by about 10% when it changes to ice. Especially during late winter (Fig. 5.7), the alternate freezing and thawing of the surface soil causes repeated cycles of soil expansion and relaxation. Plants are gripped by the freezing surface soil and lifted upward, but plants with taproots such as alfalfa and red clover may not settle back when the soil thaws (Portz, 1967). The next freezing event grips the plant lower down and moves the plant up even more and eventually, with repeated cycles, the taproot breaks. Plants die because the crown meristems that are normally protected by the soil are elevated above the ground and exposed to freezing temperatures and desiccation.

Plants with branched root systems survive heaving conditions better than plants with a taproot system because they tend to move up and down with the soil. Prostrate plants tend to have top growth produced near the soil surface, which acts like mulch that moderates temperature fluctuations and the freezing and thawing that lead to heaving (Table 5.1). Similarly, maintaining abundant stubble helps to trap snow that insulates the soil and plants from temperature change. Research has shown that removing snow from an alfalfa field decreases winter survival of adapted cultivars, whereas increasing the snow depth enhances the survival of cultivars with poor winter hardiness (Leep et al., 2001).

Ice Sheets

Encasement of plants in or under a layer of ice is a common problem in poorly drained soils in northern environments, and it can cause serious damage to forage plants. When the crown is completely encased in water or ice for several days or even weeks, exchange of oxygen and carbon dioxide is slowed and cells accumulate by-products of **anaerobic** dark respiration, usually alcohols and CO_2, and are injured or die. Winter-hardy species have low rates of respiration during winter, enabling them to survive ice encasement longer than non-hardy species with high dark respiration rates. Leaving tall stubble over winter may reduce ice-sheet damage as the protruding stubble provides a route for gas diffusion through the ice. Ice also has a high thermal conductivity, causing the temperature of ice-encased plants to decline to levels that cause intracellular ice formation and death, especially in marginally hardy species or cultivars. Ice sheets are less of a problem in southern latitudes because the ice does not remain for extended periods.

Cold Hardiness

Forage species differ widely in their ability to withstand cold temperatures. For example, freezing tests have shown that biennial sweetclover can withstand colder temperatures than 'Ranger' alfalfa, which in turn is more tolerant of cold than red clover. Differences within species also exist. For example, alfalfa cultivars range from non-tolerant to very cold tolerant (Fig. 5.9). Also, common white clover has greater **cold tolerance** than ladino clover, and it survives longer than ladino clover when both are encased in ice. Grasses also differ in cold tolerance. Seedlings of crested wheatgrass are more cold tolerant than those of western wheatgrass, which in turn are more cold tolerant than seedlings of smooth bromegrass. Similarly, timothy is more cold tolerant than smooth bromegrass, which in turn is more cold tolerant than orchardgrass.

Development and Maintenance of Cold Hardiness

Winter survival of a forage species depends upon the ability of the plant to make certain metabolic changes.

Table 5.1. Influence of snow and stubble residue on soil temperature

	Soil temperature, °F[a]			
Date in January	Air temperature	Bare soil	Snow	Snow and residue
15	30	30	32	35
20	18	29	30	35
25	20	25	31	34
30	2	16	28	33

Source: Adapted from Bouyoucos and McCool, 1916.
[a]Soil temperature measured 3 in. below the soil surface.

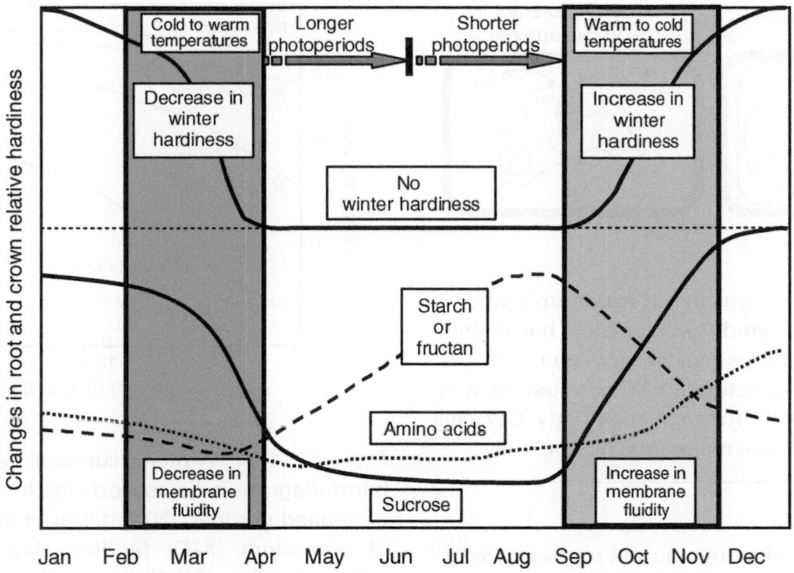

FIG. 5.12. Cold acclimation responses in overwintering structures of forages in autumn in response to cold temperatures and short photoperiods that gradually lead to increased winter hardiness. Deacclimation occurs rapidly in spring as temperatures warm and photoperiods lengthen. (Adapted from Larcher, 1995.)

Even grasses and legume species that are considered very cold hardy could not survive freezing temperatures if they occurred in summer. Within their inherent genetic capacity, overwintering forage species gradually develop **cold resistance** with the onset of the shorter days and colder temperatures of autumn (Fig. 5.12).

In the north central USA, perennial legumes begin to develop **winter hardiness** by mid-September, and this continues until late November or early December when the soil freezes. Cold resistance is highest shortly after the soil surface freezes, and is generally maintained from early December to mid-February. Cold resistance begins to decrease in mid-February with the onset of warmer temperatures and lengthening photoperiods as snow cover decreases and the soil surface thaws. Cold resistance is lost much faster than it is acquired. Extensive winter killing in the northern areas occurs during late winter and early spring when the snow cover has disappeared and plants are exposed to temperature fluctuations above and below freezing. By this time, plants have lost some cold resistance in response to warm temperatures, and when exposed to a sudden temperature decline they may not have time or the physiological capacity to reharden.

The Hardening Process

Under controlled conditions, cold hardening of alfalfa and other species is favored by short photoperiods (7–8 hours), alternating temperatures between warm (60°F) during the day and cold (32–40°F) at night, and adequate radiation for good photosynthetic activity. For example, alfalfa shows poor winter hardening at warm temperatures (75/61°F) regardless of photoperiod. In contrast, winter hardiness develops in cool environments (45°/36°F) better with short photoperiods (8 hours) than with long ones (16 hours). Although cold temperature appears to be the key to cold acclimation of forages, the metabolic processes associated with increased hardiness are also modified by the shortening photoperiod.

Winter hardening in autumn involves several physiological changes (Fig. 5.13). Cells of roots and crowns generally dehydrate, reducing the amount of tissue water present for ice formation. Ice formation inside cells is fatal. If ice formation must occur in plants, it is preferable for it to occur extracellularly in the cell wall than intracellularly. Inside the cell the relative abundance of **bound water,** which adheres to the surfaces of cell constituents such as proteins and carbohydrates and does not readily freeze, increases during cold acclimation. Thus the amount of **free water** (that which is not bound) available for ice formation is reduced and survival is enhanced. Solutes that act as **osmotica,** such as potassium ions, amino acids, and sucrose, accumulate in the free water and lower the freezing point of cellular water from 32°F to as low as 25°F, thereby preventing ice formation in cells in this temperature range. The concentrations

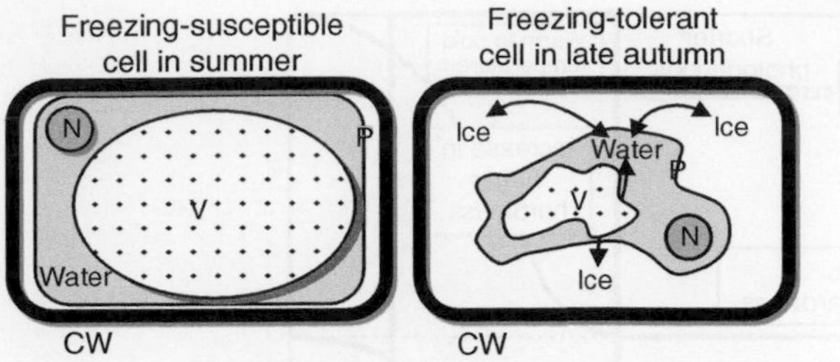

Fig. 5.13. Changes in cell water amount and distribution associated with winter hardening. Hardy cells contain less water, and much of it is bound to cellular constituents. If ice must form in tissues, it is best if it forms extracellularly. CW, cell wall; V, vacuole; P, plasmalemma; N, nucleus.

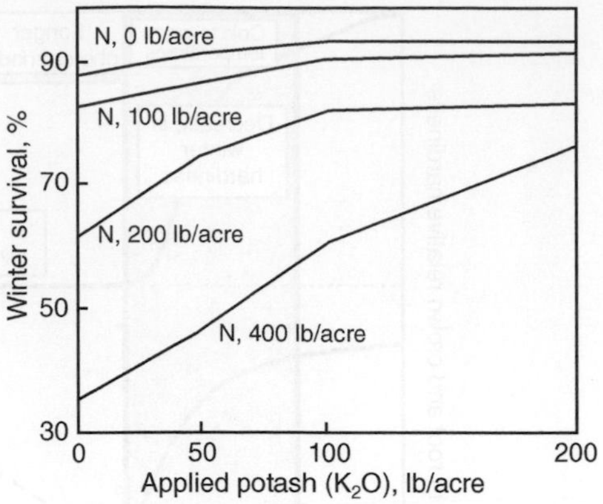

Fig. 5.14. Winter survival of "Coastal" bermudagrass is decreased with increasing levels of applied nitrogen (N) fertilizer at any given level of potassium (K_2O) fertilizer, but is improved with high rates of K_2O, especially at high N rates. (Adapted from Adams and Twersky, 1960.)

of polysaccharides such as fructans and starches stored in the cells of stem bases and roots decline during acclimation as they are hydrolyzed to serve as sources of sucrose. The structure of membrane lipids changes during cold acclimation, to ensure that the membranes remain fluid and to help to stabilize membrane-bound proteins.

Management to Improve Winter Hardiness

Conditions that result in active growth during hardening in autumn can hinder the development of cold resistance. For example, non-dormant alfalfa cultivars adapted to the southwest USA continue shoot growth into late autumn and often do not survive winter in the Midwest (Fig. 5.9). Cold resistance may be reduced or its development retarded by cutting or grazing of shoot growth, especially of legumes, during the hardening period in autumn.

Adequate soil fertility is as important for overwintering of plants as it is for their growth during summer. Two nutrients of major importance are potassium (K) and N. Stimulation of growth of grasses by fertilizing with N during the hardening period reduces cold resistance, an effect that can be negated, in part, with adequate K fertilizer management (Fig. 5.14). The role of K in winter hardiness is complex; possibly it serves as an osmoticum that depresses the freezing point of cell water. It remains clear that a high level of soil K is essential for development of maximum winter hardiness in both grasses and legumes. Cold resistance in grasses appears to be favored by a high K:N ratio.

The growing period prior to the first **killing frost** (24°F) is usually a critical time in forage management, particularly of forage legumes. Plants need leaf area during autumn to synthesize carbohydrates through photosynthesis and to accumulate **organic reserves** before winter. Accumulated carbohydrate and N reserves are needed for developing cold resistance, providing metabolic energy for dormant tissues during winter, and supporting new growth the following spring. The colder and longer the winter, the more imperative it becomes that plants enter it with a high level of organic reserves. This is best ensured by not harvesting or grazing upright legumes such as alfalfa and red clover during the cold-hardening period in autumn. In northern areas of the USA it is generally recommended that forage legumes are not cut or grazed during autumn, beginning approximately 6 weeks before the average date of the first killing frost. In southern areas the date of the first killing frost is less predictable, but plants still need time to harden.

The level of winter injury caused by untimely fall cutting depends upon several factors, including geographic location, the duration of autumn-like weather after harvest, and the severity of the winter. In Ontario, fall harvesting on September 10 caused the greatest reduction in alfalfa population for the northernmost site, but the greatest negative effect of fall harvesting occurred later, on September 27, for the southern site (Fig. 5.15). The greatest yield reduction occurred at the central site when plants were harvested on September 22, when 75% of the plants died over winter. Harvesting or grazing late in autumn, 2–3 weeks after frost has killed the top growth, is less hazardous because organic reserves have accumulated, the plants have become cold resistant, and they are generally dormant and will not regrow.

Although top growth can be grazed or harvested after the plants become dormant, removal of herbage at this time decreases the above-ground residue that traps snow, maintains high soil temperatures, and reduces the soil freeze–thaw cycles that cause heaving (Table 5.1). In northern and

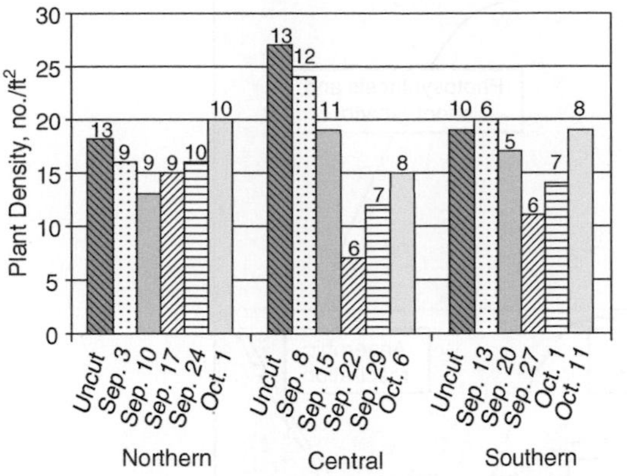

Fig. 5.15. Plant population and shoot height (in.), shown in numerals over bars, of alfalfa in May. A range of final cutting dates was applied the previous fall at three locations in Ontario, Canada. Note the improved survival achieved by leaving plants uncut, cutting early enough to allow the plants to harden properly before a killing frost, or cutting after they were hardy (in October at this location). (Adapted from Fulkerson, 1970.)

southern latitudes, cool-season grasses are less sensitive to fall grazing or cutting than are legumes because their leafy canopy near the soil surface moderates changes in soil temperature.

Management of Weakened Stands

Stands that have been weakened or injured by a severe winter usually need careful management to regain productivity. Damaged legumes may appear weakened and yellow in early spring, with only a few shoots per plant, but can recover with proper weed control and delayed cutting. The plants are not vigorous, so weed control in spring is important to reduce competition for light and nutrients, especially K. In the case of alfalfa, cutting winter-injured stands before buds appear on the shoots may kill already weakened plants or slow the rate and extent of their recovery from injury. Delaying the first cutting until most shoots have flowers gives plants more time to heal injured tissue and develop a high level of organic reserves in the storage organs (see Chapter 4). The herbage harvested may be lower yielding and weedy, but with delayed cutting the regrowth and yield of the subsequent harvests may return to near-normal productivity. Live plants that are frost-heaved less than 1 in. above the soil will often recover by settling back into the soil, but if they have heaved further the **crowns** may be damaged by mowing or grazing, and the weakened plants will probably die.

Water Relations

Seasonal distribution patterns of precipitation, total quantity of precipitation, and **evapotranspiration** demands affect water availability and adaptation of forage species. In addition, available water in the soil depends on soil texture and plant rooting depth. Water loss from plant surfaces (**transpiration**) is driven largely by solar radiation. Stomata (small pores in the leaf surfaces) open during the day to permit CO_2 diffusion into the leaves for photosynthesis. However, most transpired water also passes through the stomata, which close in darkness, reducing water loss from plants. Thus water loss from the soil by direct soil evaporation and by transpiration through plants depends largely upon the amount of water available in the soil and upon incident radiation. This is why plants in dry soil on cloudy days may be less stressed than plants in well-watered soil on bright, sunny days. Air temperature, relative humidity, and wind speed have less effect than radiation on plant water use.

Drought Stress

Adequate soil moisture is essential for maintaining the cell **turgor** that drives normal growth. Turgor is related to water status within the plant tissues, which is commonly called the **water potential** (the force with which water is held in a cell). The balance between soil moisture availability and the transpiration rate from plant surfaces influences plant water potential and turgor pressure. A decrease in tissue water potential affects some plant processes more than others. Cell division and cell enlargement are reduced by even a slight decrease in water potential, and directly reduce shoot growth (Fig. 5.13). Shoot growth slows down well before water stress becomes severe enough to cause stomatal closure and a decline in photosynthesis.

Sugars from photosynthesis often accumulate during mild to moderate drought stress because growth is slowed. Some forage species gain **drought tolerance** by accumulating osmotically active solutes such as sugars, amino acids, and ions that attract water into cells and thus prevent injury (Fig. 5.13). Solutes can also aid in **drought resistance** by contributing osmotically in meristematic areas to sustain turgor and growth during drought, or solutes can be stored and used for growth later on, after the drought has been alleviated. Some cool-season grasses that experience drought stress during summer may show enhanced growth during autumn and the following spring. In grasses such as tall fescue, this enhanced growth is due to faster leaf elongation and greater weight per tiller compared with plants that were well watered through the summer. Leaf growth, which is largely due to cell enlargement, recovers quickly when plants are re-watered after a short stress, but the onset of recovery is delayed if the drought is prolonged.

During severe drought, leaf growth ceases, the leaves curl to reduce the exposed surface area, the stomata

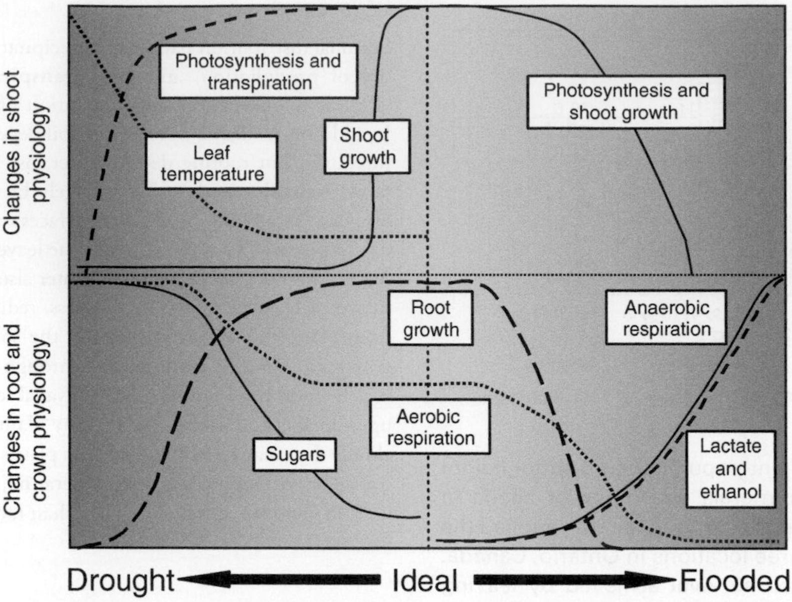

FIG. 5.16. Effects of drought and flooding on growth and physiology of forage plants. Note the difference in relative root and shoot growth when plants are subjected to drought compared with flooding.

remain closed, and photosynthesis and transpiration decline markedly (Fig. 5.16). Evaporative cooling of plant tissue stops and the tissue temperature increases. This can increase shoot dark respiration rates, and with extended drought, consume carbohydrate reserves that are needed later for plant growth and tissue maintenance. Forage species differ in their response to drought stress. Many plants reduce shoot growth but maintain root growth under moderate drought stress conditions. Legumes such as alfalfa are relatively drought tolerant compared with grasses, because the long taproots access water from deeper regions of the soil profile (Carter and Sheaffer, 1983). This type of drought resistance has been termed **drought avoidance**, as the plants can still obtain an adequate water supply, whereas other shallow-rooted species cannot. Grass species also differ with regard to drought avoidance. Smooth bromegrass tolerates drought better than orchardgrass and kentucky bluegrass because it is deep rooted and thus accesses a larger soil volume.

Even with abundant soil moisture, atmospheric evaporative demand is often high enough during sunny days with high transpiration to cause shoot water potentials to decrease to levels that reduce turgor and prevent rapid growth during the daytime. Consequently, cell expansion may be rapid during the night when the stomata are closed, transpiration rates are very low, and water potentials and turgor are increasing as plants absorb water from the soil faster than it is lost through transpiration. If a drought has been mild in terms of intensity or duration, cessation of cell enlargement and similar plant responses are not permanent, and re-watering allows cell enlargement and related developmental processes to resume.

Excessive Moisture Conditions

Poorly drained soils in high-rainfall climates provide an unfavorable environment for growth of many forage species, especially legumes. Forages such as reed canarygrass, a cool-season grass, and paragrass, a warm-season grass, tolerate long-term waterlogged conditions and often are the only forage species that can be grown in such areas. **Flooding tolerance** may be related to the ability of the roots to withstand low oxygen supply, or to anatomical changes in the stems, leaf sheaths, and roots, with the formation of channels for internal diffusion of oxygen to the root system. In addition, the roots of flooding-tolerant species can be less sensitive to toxic levels of soil minerals such as manganese (Mn^{2+}) that become abundant in many flooded soils. On poorly drained soils, serious soil compaction and plant damage caused by the hooves of livestock, especially after excessive rainfall, contribute further to poor aeration and reduced plant growth.

Water fills air spaces in flooded soils, reduces soil oxygen concentration, and quickly limits normal **aerobic respiration** of roots (Fig. 5.16). As a result, **anaerobic respiration** increases, but it is very inefficient in terms of the number of ATP molecules produced per glucose molecule consumed (see Chapter 4). Low availability of ATP quickly limits root growth, ion uptake, and other key root processes. In addition, anaerobic respiration produces large amounts of lactic acid and ethanol that gradually accumulate and can be toxic to many forage species. For most forages, the negative effects of 1 or 2 days of flooding on shoot growth and photosynthesis are minimal, but both growth and photosynthesis decline with prolonged flooding.

Flooding can also exacerbate problems caused by root and crown diseases. Stand losses of alfalfa growing in poorly drained soils during periods of high precipitation or excessive irrigation have often been attributed to diseases such as *Phytophthora* root rot. **Damping-off** of seedlings, caused by such soil fungi as *Pythium* species, *Rhizoctonia* species, or *Fusarium* species, can be a serious factor in stand establishment, particularly if periods of high rainfall follow seeding. Invariably, fungal diseases are most serious during periods of excessive rainfall and high humidity. In some cases, pathogens may only weaken plants, which subsequently fail to regrow after harvest or are killed by winter or other **abiotic** stresses.

Dormant plants are generally more tolerant of flooding than are actively growing plants. A major factor is the lack of oxygen, so plant survival is better if the water is cool (to reduce respiration) and flowing (to aid oxygenation). Thus spring floods caused by melting snow are less injurious than inundation that results from excessive rainfall in summer. Defoliated plants that are actively regrowing are more susceptible to flooding-induced injury than are plants at mature stages where the shoots are left intact (Fig. 5.17). Even short periods of flooding reduce dinitrogen (N_2) fixation by legumes, because this process requires oxygen to produce enough ATP to reduce N_2 to plant-useable N.

Forage species differ in their ability to survive flooding and water temperature, both of which can influence survival because of their impact on dark respiration rates. For example, in Canada most cool-season grasses survived spring flooding with cold water for up to 35 days, and timothy and reed canarygrass survived for up to 49 days (McKenzie, 1951). However, legumes survived for only 14–21 days. In warm environments, hardinggrass and bermudagrass are flooding tolerant, but dallisgrass is very sensitive to inundation when it is actively growing (Coleman and Wilson, 1960).

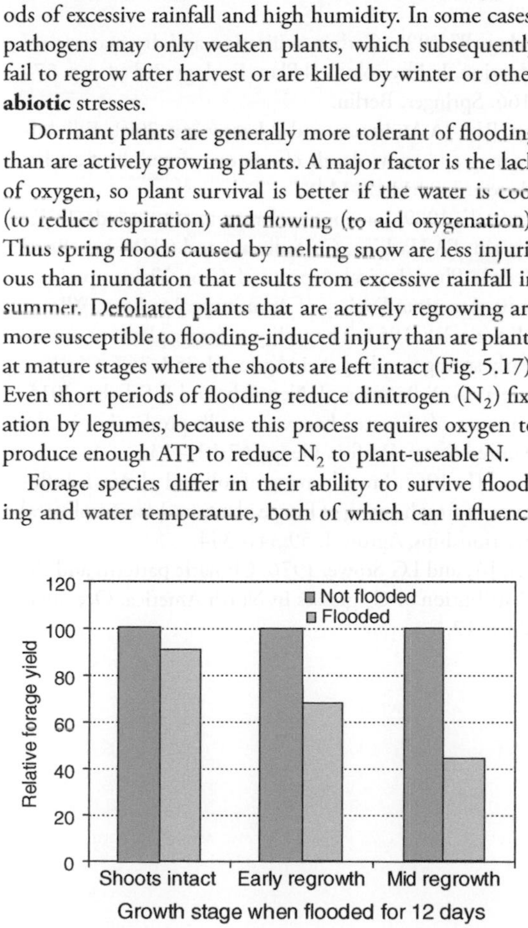

Fig. 5.17. Effect of flooding at different stages of shoot regrowth on yield of alfalfa. Note that plants flooded during active regrowth were more sensitive than those that were not harvested at the time of flooding. (Adapted from Barta, 1988.)

Summary

The environment in which plants grow is described by the prevailing climate and abundance of natural resources such as soil properties for minerals and water storage, solar radiation for heat and photosynthesis, and the frequency and amounts of precipitation. The major goal of forage managers is to select species that are adapted, and then to manage them in ways that optimize the use of the environment and help to offset negative events. This usually involves compromises with regard to factors such as harvest timing to protect forage yield or retain enough leaf area to reduce soil temperature in summer and retain soil heat in winter. Solar radiation received at any given location is affected by cloud cover and latitude, which alters the seasonable photoperiod and regulates flowering in many species.

Weather is experienced on a day-to-day basis and illustrates the extremes that can occur. In contrast, climate is the long-term average of weather conditions, and provides a more general evaluation of growing season and overwintering conditions. With pending global climate change, it is likely that the climatic zones will move northward in the northern hemisphere to have temperatures similar to those that prevail today, but the photoperiod will not change. This has implications for species that replenish the seed bank every year. Each species of forage plant has an ideal climate for growth and reproduction. These characteristics are well known for most species, and should be used to determine what species to plant and then how to manage it to meet the priority needs for the livestock producer.

Questions

1. What climatic factors determine whether the natural vegetation will be forest or grassland?
2. Which region of the solar spectrum is the most important for plant growth? Why is this?
3. Why is plant growth maximized when at least 95% of incoming solar energy is intercepted by plant leaves?
4. How would a change in latitude influence the flowering and potential seed production of a long-day grass species? Would the responses be similar for a day-neutral species?
5. List some physiological differences between temperate and tropical grasses. How do these grasses differ in seasons of productivity?

6. Differentiate between climatic factors involved in chilling injury and those involved in plant survival over the winter.
7. Why is knowledge of the growth stages when minimum and maximum organic reserves occur important to the management of perennial or biennial species?
8. What climatic and soil conditions favor the development of cold resistance? What conditions favor the loss of cold resistance?
9. What atmospheric constituents are driving global warming? What are the implications for cool-season forages and for warm-season forages?
10. What is the diurnal pattern of growth of forage plants? How is it affected by mild drought stress?

References

Adams, WE, and M Twersky. 1960. Effect of soil fertility on winter-killing of Coastal bermudagrass. Agron. J. 52:325–326.

Barta, AL. 1988. Response of field-grown alfalfa to root waterlogging and shoot removal. I. Plant injury and carbohydrate and mineral content of roots. Agron. J. 80:889–892.

Beuselinck, PR, JH Bouton, WO Lamp et al. 1994. Improving legume persistence in forage crop systems. J. Prod. Agric. 7:311–322.

Bouyoucos, GJ, and MM McCool. 1916. The freezing point method as a new means of determining the concentration of the soil solution directly in the soil. Mich. Agr. Expt. Sta. Tech. Bull. 24.

Bowley, SR, and BD McKersie. 1990. Relationships among freezing, low temperature flooding, and ice encasement tolerance in alfalfa. Can. J. Plant Sci. 70: 227–235.

Carter, PR, and CC Sheaffer. 1983. Alfalfa response to soil water deficits. I. Growth, forage quality, yield, water use, and water-use efficiency. Crop Sci. 23:669–675.

Coleman, RL, and GPM Wilson. 1960. The effect of floods on pasture plants. Agric. Gaz., July, 337–347.

Fulkerson, RS. 1970. Location and fall harvest effects in Ontario on food reserve storage in alfalfa *(Medicago sativa* L.). In Proceedings of the 11th International Grassland Congress, Surfers Paradise, Queensland, Australia, pp. 555–559.

Gardner, FP, and WE Loomis. 1953. Floral induction and development in orchardgrass. Plant Physiol. 28:201–217.

Geiger, R. 1965. The Climate Near the Ground, rev. ed. Harvard University Press, Cambridge, MA.

Karl, TR, JM Melillo, and TC Peterson (eds.). 2009. Global Climate Change Impacts in the United States. Cambridge University Press, New York.

Kawanabe, S. 1968. Temperature responses and systematics of the Gramineae. Proc. Jap. Soc. Plant Taxon. 2:17–20.

Larcher, W. 1995. Carbon utilization and dry matter production. In Physiological Plant Ecology. 3rd ed., pp. 57–166. Springer, Berlin.

Leep, RH, JA Andresen, and P Jeranyama. 2001. Fall dormancy and snow depth effects on winterkill of alfalfa. Agron. J. 93:1142–1148.

Lemon, E. 1969. Gaseous exchange in crop stands. In JD Eastin, FA Haskins, CY Sullivan, and CHM Van Bavel (eds.), Physiological Aspects of Crop Yield, pp. 117–140. American Society of Agronomy, Madison, WI.

McKenzie, RE. 1951. The ability of forage plants to survive early spring flooding. Sci. Agric. 31:358–367.

Mauger, G, Y Bauman, T Nennich, and E Salathé. 2015. Impacts of climate change on milk production in the United States. Profess. Geogr. 67:121–131.

Portz, HL. 1967. Frost heaving of soil and plants. I. Influence of frost heaving of forage plants and meteorological relationships. Agron. J. 59:341–344.

Teeri, JA, and LG Stowe. 1976. Climatic patterns and the distribution of C_4 grasses in North America. Oecologia 23:1–12.

CHAPTER 6

Grasses for Northern Areas

Michael Collins and C. Jerry Nelson

Cool-season perennial grasses provide most of the forage consumed by beef cattle and sheep in temperate areas of the USA and around the world. The production and utilization of these grasses are often improved by adding a legume. Dairy rations typically include higher percentages of legumes to increase the protein concentration, and corn or sorghum silage to increase digestible energy levels. Perennial warm-season grasses and summer annual grasses are also used to provide summer grazing in pasture systems for northern areas. **Crop residues,** especially corn stover, which consists of stalks, leaves, and husks, are often used to support or extend the grazing season.

Annual cereals such as spring oat, spring barley, winter rye, winter and spring wheats, and winter and spring triticales are also used as forage. Cool-season grasses are used as turf, conservation, and wildlife plants; they also add to human well-being by contributing to the aesthetics of the landscape. Because managers must select from among the available forage species and cultivars to fit their soils, climatic conditions, and livestock systems, our goals are to describe and contrast characteristics that affect use and management of the major grasses for temperate climates of North America. Further details are provided by Moser et al. (1996) and Barnes et al. (1995a, 1995b).

Cool-Season Grasses

Except for reed canarygrass and perhaps kentucky bluegrass, all of the cool-season grasses used on significant acreages in the eastern USA were introduced from other parts of the world. There is uncertainty as to the exact origin of the bluegrasses, but colonists probably introduced some seed. Cool-season wheatgrasses and wildryes, a complex of many native and introduced genera and species (Asay, 1995), are cultivated in sub-humid to arid parts of the northern **prairie** and western states.

Many native and introduced cool-season grasses occur in hay, **pasture,** and **rangeland.** Other cool-season grasses may have fewer desirable agronomic characteristics or may be limited with regard to their area of adaptation, yet they contribute to total forage production, particularly in marginal land areas.

Regions of Origin

Smooth bromegrass and some **cultivars** of kentucky bluegrass were introduced from Eurasia, tall fescue from the Mediterranean area and North Africa, reed canarygrass from temperate areas of Northern Europe, and orchardgrass from Western Europe and North Africa. These introduced grasses replaced less productive native grasses such as redtop, povertygrass, velvetgrass, and tall oatgrass in the northeast, and the tallgrass prairie species in the north central region. The wheatgrasses and wildryes that were introduced to the West came from Eurasia.

The exponential increase in tall fescue acreage in the USA, from a few thousand acres in 1940 to more than 35 million acres in 1990, indicates how quickly acreage can expand for a well-adapted, introduced species. This expansion rate exceeded that of alfalfa a few years earlier, probably due to advances in seed production and seed distribution systems. Since being introduced in 1906, crested wheatgrass has become a major component of temperate and semiarid rangeland.

Growth Habits

Bunch-type grasses that produce mainly vertical **tillers** include timothy, orchardgrass, the ryegrasses, most of the wheatgrasses and wildryes, and tall fescue when it

Forages: An Introduction to Grassland Agriculture, Seventh Edition. Edited by Michael Collins, C. Jerry Nelson, Kenneth J. Moore and Robert F Barnes.

is cut or grazed infrequently. Sod-forming grasses produce elongated **rhizomes** that spread laterally underground to develop a rather uniform sod and tiller density. They include the bluegrasses, reed canarygrass, smooth bromegrass, and tall fescue when it is cut or grazed frequently. Many wheatgrasses produce short rhizomes, but they generally appear to have a bunch-type habit because plant density is low in semiarid regions.

Except for timothy, the major introduced cool-season species all require **vernalization** (see Chapter 5). For these species, short days and cold temperatures in autumn and winter followed by long days in spring are needed for stem elongation and formation of the **inflorescence.** Wheatgrasses, wildryes, ryegrasses, and the small grains wheat, barley, rye, and triticale have **spike** inflorescences. Timothy has a compact **panicle** with single-seeded **spikelets** borne on very short branches. The other species have open and spreading panicles. Species that require vernalization produce stems and flower only once each year, during spring. Because it does not require vernalization, timothy can flower more than once each season.

Growth Distribution

Seasonal distribution of dry matter production is very important in grazing systems because animals require a consistent supply of forage. Cool-season grasses are bimodal in seasonal distribution of growth (Fig. 6.1A). Spring yield is high due to rapid growth of both stems and leaves. Yields in summer are low because the spring reproductive growth that is harvested or senesces is replaced by vegetative growth made up of leaves. Non-vernalized tillers that overwinter and tillers produced in spring are needed to produce summer and fall growth. Each new tiller must develop and elongate its own roots during late spring and summer to access water and nutrients.

The bimodal distribution of seasonal growth is partly due to the fact that the vegetative canopy is shorter than the

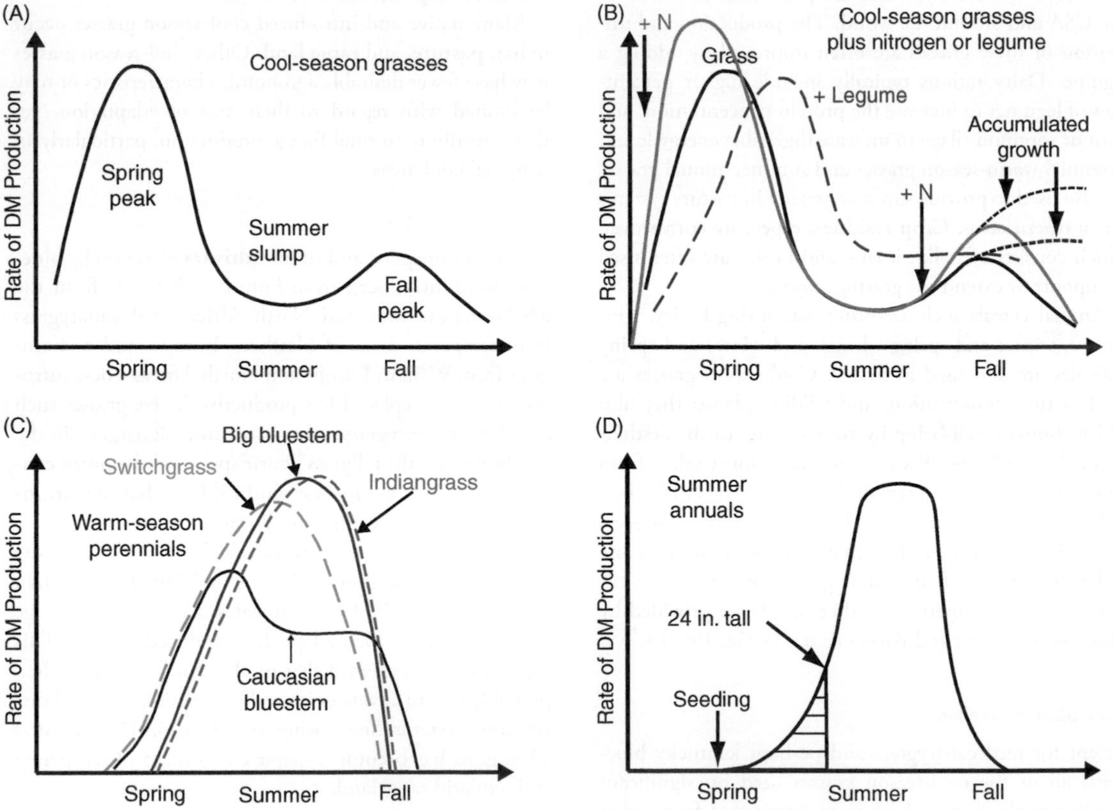

FIG. 6.1. Seasonal production profiles for (A) cool-season grasses, (B) cool-season grasses with N added in spring and late summer and with an added legume, (C) warm-season perennial grasses, and (D) summer annual grasses. Note that N applied to cool-season grasses or addition of a legume affects the growth distribution. Sudangrass or sorghum × sudangrass hybrids are seeded in spring and should not be grazed until the plants are 24 in. tall, due to the high concentrations of prussic acid in young leaves and tissues.

reproductive canopy and is not as efficient in terms of light interception and **photosynthesis**. Summer temperatures are above optimum for C_3 photosynthesis, and the associated drought stress reduces growth (see Chapters 2 and 4). Growth rates improve again in autumn when drought and high-temperature stress are relieved.

Some grasses are negatively affected by cutting or grazing in late spring during the transition from reproductive to vegetative growth. Species that tiller less actively in spring, primarily timothy and smooth bromegrass, are especially sensitive. They have neither adequate organic reserves nor young tillers ready for regrowth when cut early (see Chapter 4). The "critical stage" usually occurs between initiation of stem elongation and emergence of the inflorescence. This is one of the weak points in the **perennation** process, mainly because this stage occurs at the optimum time for the first cutting or grazing. The critical-stage problem is less severe with grazing, because all tillers and plants are not defoliated uniformly.

Adaptation of Cool-Season Grasses

Smooth bromegrass and timothy are more winter hardy than are tall fescue and orchardgrass (Table 6.1). Smooth bromegrass survived during the severe and extended droughts of the 1930s, probably because its roots extend deeper into the soil than did the roots of other cool-season grasses (Table 6.2). It is best adapted to the fertile, well-drained, deep calcareous soils of the northern prairie states, the northern Corn Belt states, and in some areas of the East. However, most soils in the East are too acidic, shallow, and deficient in basic nutrients for good productivity and **persistence** of smooth bromegrass. A close relative, prairie grass or "Matua" bromegrass, has recently been promoted for pasture during winter and early spring in northern areas. However, its disease susceptibility, inconsistent winter survival, and high management requirements have limited its use.

Timothy matures later than smooth bromegrass and most other cool-season grasses and legumes, such that it is often at the critical growth stage when alfalfa reaches first flower and should be cut for hay (see Chapter 4). When timothy is cut at this stage, regrowth is very slow and weak, it is quickly shaded by alfalfa, and the weakened plants may not survive summer (Table 6.3). Red clover and birdsfoot trefoil mature later than alfalfa, so mixtures with these species are cut after the critical stage for timothy. When cut later, timothy is ready to regrow, compete for resources, and persist.

Tall fescue and orchardgrass are relatively heat tolerant, so their adaptation extends southward into the transition zone between cool and warm climates of the USA. Tall fescue, especially that infected with the endophytic fungus, tolerates heat and drought better than orchardgrass.

Tall fescue quickly gained favor as a well-adapted grass on marginal soils in the transition zone of the USA. Plants form a dense **canopy** and an extensive root system that holds soil to reduce erosion. The leaves grow into late fall, and they resist frost injury better than the leaves of most other grasses, so they retain good quality into the winter to extend grazing of accumulated forage.

Tall fescue tillers profusely, has vegetative tillers at several developmental stages, and has a less well-defined critical growth stage in spring. It is quick to recover after cutting and is more tolerant of **continuous stocking** than are orchardgrass, smooth bromegrass, and timothy. Tall fescue and orchardgrass mature earlier, and regrowth comprises a greater percentage of the total seasonal yield than is the case for smooth bromegrass, timothy, kentucky bluegrass, and perennial ryegrass.

Orchardgrass matures early and its life cycle matches well with that of alfalfa in mixtures for hay. However, orchardgrass that is managed for hay or pasture frequently thins in the North because of its susceptibility to winterkill, and in the transition zone because of late spring frost damage, diseases, and susceptibility to heat and drought. Tall fescue does not persist well in droughty, light soils, especially where nematode populations are high.

Reed canarygrass is adapted to low, wet areas as well as to drier upland areas. Among the major species, it is the most tolerant of flooding (see Chapter 5). Regrowth is better than that of most cool-season grasses due to its drought and heat tolerance. Although reed canarygrass is more winter hardy than tall fescue and orchardgrass, its leaves are very susceptible to frost damage, so it does not retain forage quality into late fall and winter. Its growth habit and adaptation characteristics make it difficult to grow in mixtures with legumes.

The bluegrasses and ryegrasses are shorter and usually yield less than the tall-growing grasses. They are most productive during spring, and are least productive in summer due to low heat tolerance and shallow root systems (Table 6.2). Ryegrasses are the least drought tolerant and the least winter hardy of the major cool-season grasses. Kentucky bluegrass is very tolerant of close continuous grazing because of its extensive system of rhizomes and ability to maintain leaf area near the soil surface, and it is a preferred species for horse pastures. Perennial ryegrass persists best when it is stocked rotationally. Both are compatible with legumes such as white clover and birdsfoot trefoil that are low growing and tolerate grazing.

The most drought and heat tolerant of the cool-season grasses are the wheatgrasses and wildryes, but they do not compete well in high-rainfall areas. These species occur west of the Mississippi River and are abundant in the Northern Plains and intermontane states, where it is too dry for more productive grasses. Western wheatgrass and thickspike wheatgrass, the major native wheatgrasses in the Northern Plains, grow in association with needlegrasses, grama grasses, shrubs, and **forbs.** Crested wheatgrass and intermediate wheatgrass are the most abundant

Table 6.1. Adaptation and management characteristics of perennial cool-season grasses adapted to northern areas

Common name	Soil adaptation	Ease of establishment	Winter hardiness	Drought tolerance	Tolerance of wet soils	Primary use	Forage quality	Grazing tolerance	Competitiveness in mixtures
Northeast and humid north central states									
Timothy	Well to somewhat poorly drained, fine-textured soils	Easy	Excellent	Poor	Fair	Hay or silage	Very good	Low	Fair
Smooth bromegrass	Well-drained, deep, fertile soils	Slow to establish	Excellent	Excellent	Poor	Hay or silage	Very good	Low	Fair
Orchardgrass	Well-drained loams and silt loams	Easy	Good	Good	Fair	Hay or silage	Good	Fair	Very good
Reed canarygrass	Moist or wet lowlands to dry upland fertile soils	Problematic	Very good	Very good	Excellent	Hay or silage	Poor to good (alkaloid dependent)	Good	Good
Tall fescue	Tolerant of both wet and dry soils	Easy	Good	Very good	Good	Pasture or hay	Poor to good (endophyte dependent)	Good	Very good
Kentucky bluegrass	Fine-textured, well-drained to somewhat poorly drained soils	Easy	Excellent	Poor	Good	Pasture	Excellent	Very good	Very good
Perennial ryegrass	Most heavy, fertile soils	Establishes rapidly	Fair	Poor	Good	Pasture	Excellent	Very good	Poor
Annual ryegrass	Most heavy, fertile soils	Establishes rapidly	Fair	Poor	Good	Pasture	Excellent	Good	Poor
Semiarid and arid areas									
Western wheatgrass	Heavy alkaline soils; many soil types	Slow to establish	Excellent	Excellent	Good	Grazing	Good	Good	Good
Thickspike wheatgrass	Upland light-textured soils	Easy	Excellent	Excellent	Poor	Grazing	Good	Good	Good
Bluebunch wheatgrass	Most fertile soils, light sandy loams to heavy clays	Easy	Excellent	Very good	Fair	Grazing	Excellent	Poor	Good

Slender wheatgrass	Loams and sandy loams; dry to moderately wet	Easy	Excellent	Very good	Fair	Grazing or hay	Good	Fair	Fair
Crested wheatgrass	Fertile, light sandy loams to heavy clays	Easy	Excellent	Excellent	Poor	Grazing or hay	Good	Excellent	Excellent
Intermediate wheatgrass	Well-drained, fertile soils; tolerates moderate alkalinity	Easy	Good	Fair	Poor	Grazing or hay	Good	Fair	Good
Tall wheatgrass	Persists in soils that are too alkaline for other wheatgrasses	Easy	Excellent	Fair	Fair	Grazing or hay	Fair	Good	Good
Canada wildrye	Adapted to a wide range of soil types	Moderately easy	Excellent	Good	Fair	Grazing or hay	Fair	Fair	Fair
Russian wildrye	Loam and clay-loam soils; tolerant of moderate alkalinity	Slow	Excellent	Good	Poor	Grazing	Excellent	Good	Good
Great Basin wildrye	Broad soil texture adaptation; deep soils	Difficult	Excellent	Good	Fair	Grazing	Good	Poor	Fair

Source: Input from Drs. AI Frank, John Berdahl, John Hendrickson, and Tom Jones on arid and semiarid grasses is gratefully acknowledged.

Table 6.2. Distribution of root mass of some common perennial cool-season grasses

Grass	Soil depth (in.)					Total mass
	0–3	3–6	6–9	9–12	9–12	
	(lb dry matter/acre in.)					
Smooth bromegrass	718	201	140	124	106	4185
Orchardgrass	1247	124	77	51	30	4677
Kentucky bluegrass	1126	85	53	6	3	3828
Timothy	680	41	20	12	6	2295

Source: Adapted from Gist and Smith, 1948.

Table 6.3. Stand condition for several grasses grown in mixtures with alfalfa; data were obtained following 2 years of three different harvest schedules, each with two cutting heights

Grass	Pasture, 4 cuts		Early hay, 3 cuts		Late hay, 2 cuts	
	1.5 in.	4 in.	1.5 in.	4 in.	1.5 in.	4 in.
	(percentage of full stand)					
Orchardgrass	100	100	97	100	95	97
Reed canarygrass	87	100	95	99	100	100
Tall fescue	99	100	89	100	17	89
Smooth bromegrass	36	91	10	70	99	92
Timothy, early type	39	91	22	80	69	74
Timothy, medium type	65	97	7	50	69	75
Timothy, late type	75	96	2	55	39	46
LSD 0.05[a]		10		15		23

Source: Adapted from Smith et al., 1973.

[a]Data within a harvest schedule differing by more than the least significant difference (LSD) are statistically significantly different.

wheatgrasses in the drier western rangelands. Salt tolerance of grasses adapted to the western rangelands varies among species, and is much greater than that for temperate-origin species.

Establishment

The important cool-season grasses germinate and become established readily when high-quality seed is planted into firm seedbeds at the correct times of the year and good management practices are used (see Chapter 11). Common seeding rates for forages, which are inversely related to seed/lb, are given in the Compendium of Common Forages. When seeding a mixture of grasses and legumes, the seeding rate of each component is usually reduced in proportion to the number of species, so that the number of seed planted per unit area is similar.

The ryegrasses establish quickly and are frequently included in seed mixtures to give rapid short-term cover for turf and conservation plantings (Table 6.1). Timothy, kentucky bluegrass, and orchardgrass also establish easily. However, freshly harvested seed of orchardgrass and kentucky bluegrass contain germination inhibitors. These dissipate if newly harvested orchardgrass seed is aged for at least 1 month and kentucky bluegrass seed is aged for about 6 months prior to planting. With good management, ryegrasses, bluegrasses, timothy, and orchardgrass can be established successfully in spring or fall in most areas.

What's in a Name?

Timothy growing somewhat as a weed in England was known as "meadow cat's tail" (perhaps you can imagine why) and was probably brought to the Colonies as a contaminant. John Herd found it growing in New Hampshire, and it became known as Herd's grass. Then, in Maryland, Timothy Hanson promoted the grass for use in North Carolina, Virginia, and Pennsylvania. Benjamin Franklin "clarified the issue" when he received a sample of Herd's grass and responded, stating that the new grass was "mere timothy."

Orchardgrass is called cocksfoot in England because the inflorescence resembles a "cock's foot," but why did the name change? Orchardgrass is commonly found growing in waste areas and in shady spots. In about 1830, Philip Henshaw collected seed from plants growing in his father's orchard in Virginia, took them to Kentucky, and planted them there. Another Kentucky farmer, perhaps knowing the original source, collected seed from Henshaw's farm and produced "orchardgrass" seed for sale. Henshaw thus lost his chance to be immortalized!

Then there was Colonel William Johnson, who brought "johnsongrass" to the South around 1840, and promoted it as a vigorous, high-yielding perennial forage grass. Although it is still used to some extent today as a forage, it has become a major weed problem. Unfortunately for Mr. Johnson, the name stuck!

When seeded in spring, most grass seedlings remain vegetative during the summer and fall, are not very competitive, and are not fully productive until the following year. Smooth bromegrass establishes slowly, so it is preferentially seeded in spring in low-rainfall areas to allow development of a good root system to minimize low-temperature damage in fall and winter. In higher-rainfall areas, smooth bromegrass is best seeded in late summer or early fall to avoid the flush of weeds that commonly occurs in late spring and early summer and shades the short, vegetative seedlings.

Reed canarygrass is rather difficult to establish because its seed germinate slowly and at different times following planting. Both reed canarygrass and tall fescue can be established in marginal soils. Tall fescue is slower to germinate and develop seedlings than is orchardgrass, but where adapted, fully productive stands of tall fescue usually develop within 1 year.

Most of the rhizomatous wheatgrasses are poor seed producers. They are usually planted in spring.

Forage Quality of Grasses

Forage quality can be defined as the physical and chemical characteristics of a forage that make it valuable to animals as a source of **nutrients** and well-being (see Chapters 14, 15, and 16). Although the chemical composition is similar, the cool-season grasses differ in **palatability** and physical texture, which affect **intake.** The soft-textured leaves and fine stems of the bluegrasses and ryegrasses are consumed more readily than the stiffer leaves and thicker stems of other species. Forage that is slow to digest or that is low in final **digestibility** resides in the **rumen** and gut for a long time, and reduces appetite and intake. Thus intake rate often depends on rate of digestibility.

The developmental stage of growth is the dominant factor that affects digestibility of grasses (Table 6.4). In general, species of cool-season grasses differ very little in digestibility when they are harvested at the same developmental stage (Table 6.5). The leaf:stem ratio decreases as grasses develop in spring, and in mature grasses the stems are more lignified and less digestible than the leaves. Forage digestibility in cool-season grasses decreases by about 0.3–0.5 percentage units per day between the vegetative stage and **anthesis**.

Lignin has a strong influence on digestibility (Fahey et al., 1994). Its concentration increases rapidly from about 2% to as much as 8% of dry weight as grasses mature during the jointing and heading stages. Lignin adds rigidity to cell walls to help to prevent lodging, and provides resistance to diseases and insect pests, but it also reduces digestibility. The lignin molecules become intermeshed with the **cellulose** and **hemicelluloses,** making the entire cell wall less digestible. At post-anthesis stages, less than 33% of stem tissue in mature grasses is digestible, compared with 70% or more in leaves.

As a group, the ryegrasses have very high forage quality (Table 6.1), but are productive only for a short time during cool, moist periods of the year in most regions of the USA. Smooth bromegrass, kentucky bluegrass, timothy, and

Table 6.4. Effect of harvesting at different growth stages on first harvest yield, forage quality, and ground cover for timothy in the northeastern USA

Stage at harvest	Time between stages (days)	Relative yield (%)	*In-vitro* digestibility (%)	Ground cover[a] (%)
Pre-joint[b]	–	19	80.6	56
Early head[c]	23	72	68.3	42
Early anthesis[d]	16	100	63.9	58
Post anthesis[e]	14	110	55.8	66

Source: Adapted from Brown et al., 1968.

[a]Ground cover, a measure of persistence, was determined in spring following the 3 harvest years.

[b]Pre-joint = shoot apex less than 2.5 in. above ground level.

[c]Early head = heads emerged on less than 10% of reproductive tillers.

[d]Early anthesis = anthers on less than 10% of reproductive tillers.

[e]Post anthesis = 2 weeks after early anthesis.

Table 6.5. Effect of growth stage on *in-vitro* digestible dry matter and digestible crude protein of timothy, smooth bromegrass, orchardgrass, and reed canarygrass in the northeastern USA (growth stages as in Table 6.4)

	Growth stage at first harvest			
Species	Pre-joint	Early head	Early anthesis	Past anthesis
	(% of dry matter)			
In-vitro *digestible dry matter*				
Timothy	80.6	68.3	63.9	55.8
Smooth bromegrass	79.9	70.1	61.9	52.6
Orchardgrass	82.0	71.3	62.3	54.8
Reed canarygrass	82.4	72.1	70.0	59.8
Digestible crude protein				
Timothy	24.6	12.0	7.8	4.6
Smooth bromegrass	23.8	14.7	9.0	5.4
Orchardgrass	24.7	17.2	10.6	8.3
Reed canarygrass	23.0	11.5	9.6	6.7

Sources: Adapted from Brown et al., 1968, Decker et al., 1967, Washko et al., 1967, and Wright et al., 1967.

orchardgrass usually have higher forage quality than tall fescue and reed canarygrass, which often contain alkaloids that reduce animal performance and health (see Chapter 16). Due largely to temperature, the digestibility of leaves of the same age is lower in summer than in spring or fall.

Anti-Quality Factors in Grasses

Some chemical constituents, such as alkaloids, have large effects on forage quality even when present in small amounts (see Chapter 16). For example, a class of alkaloids called tremorgens in the indole diterpene group occurs in perennial ryegrass and results in "ryegrass staggers" in Australia and New Zealand. Sheep are more sensitive to tremorgens than are cattle.

Palatability and weight gain in both sheep and cattle are negatively correlated with the concentration of indole alkaloids synthesized in reed canarygrass. Concentrations are higher in leaf blades than in leaf sheaths and stems. Regrowth of leaves at warmer temperatures contains higher concentrations than does first-growth forage. Sheep are more sensitive than cattle, being affected by concentrations in excess of 0.2% of dry weight. Animal symptoms include diarrhea and rough hair coat. **Cultivars** such a "Vantage" and "Palaton" have low concentrations of alkaloids and are associated with improved forage intake and weight gain.

Interestingly, tall fescue and ryegrass have some alkaloids in common, but concentrations are low in ryegrass and high in tall fescue. Tall fescue harbors the endophytic fungus *Neotyphodium coenophialum* (Morgan-Jones & W. Gams) Glenn, C.W. Bacon& Hanlin (syn. *Acremonium coenophialum* Morgan-Jones & W Gams), which lives mainly in the leaf sheaths and cannot be recognized by visual observation. As the plant grows, the fungus systemically infects each new sheath in an intercellular manner, and eventually ascends the elongating stem to the panicle and seed. The fungus, in association with the host plant, synthesizes harmful alkaloids that are transported from the sheaths to the leaf blades.

The Best of Both Worlds?

Tall fescue and perennial ryegrass are closely related genetically. Furthermore, fertile crosses have been made between plants of the two species, but what is the value of this?

Scientists in the UK are attempting to improve the grazing persistence and drought tolerance of perennial ryegrass by incorporating tall fescue genes, and scientists in France, Poland, and the USA are trying to improve tall fescue forage quality by incorporating perennial ryegrass genes. But how would this work when perennial ryegrass is a diploid with two sets of seven chromosomes each, and tall fescue is a hexaploid with six sets of seven chromosomes each?

Based on cytogenetics and molecular mapping, the scientists learned that two sets of the chromosomes in tall fescue are very similar to those of perennial ryegrass, which means that perennial ryegrass is very probably a progenitor of tall fescue. When × "*Festulolium*" crosses were made, by crossing meadow fescue and tall fescue with perennial ryegrass and annual ryegrass, the hybrids show improved winter hardiness and drought tolerance compared with the ryegrass parents, and improved forage quality compared with the tall fescue parents.

Several Festulolium varieties are now available.

The fungus in tall fescue is transmitted from infected seed to seedlings. Fungus-free cultivars of tall fescue have been released, with less than 5% infected seed, with reduced detrimental effects of the fungus–alkaloid association. The fungus cannot survive in seed for long periods of storage at warm temperatures. Seed stored for 12 months at 70°F has few live fungi yet retains acceptable germination and vigor.

The fungus and its associated alkaloids provide benefits to the host plants. Infected plants are less affected by insects, diseases, and nematodes, and they show reduced effects from drought, N deficiency, low soil pH, and soil mineral imbalances. These positive effects are more important in the warmer southern USA than in more northern areas, where tall fescue is better adapted. Fungal infection is preferred in turf cultivars of tall fescue because it increases **persistence.**

Harvesting for Hay or Silage

Average yields of cool-season grass species are generally in the range of 1.5–2.5 tons/acre in the northern states. They are usually about 1 ton less than those reported for alfalfa and mixed grass/legume stands. Improved cutting and fertilization management of cool-season grasses can increase average dry matter yields by at least two- to threefold. Tall-growing grass and legume combinations are usually preferred and recommended for pasture and hay production in the USA because they produce more dry matter, have a higher pasture **carrying capacity,** and are more nutritious than grasses grown alone.

Due to stem production in spring, maximum seasonal yields of cool-season grasses are achieved by taking the first harvest at more advanced stages of development for most species (Table 6.6). However, in all species, forage quality is highest during the vegetative stages and declines as the shoots mature (Table 6.5). Forage quality is higher for leaves than for stems, and the quality of the leaf component does not decrease as rapidly with age.

In contrast, stand persistence is best achieved for most species when harvests are delayed until the mid- to late-heading stage, after the critical growth stage has passed (Table 6.4). Cutting the first growth of cool-season grasses

Table 6.6. Effect of first-cut growth stage on dry matter yield of timothy, smooth bromegrass, orchardgrass, and reed canarygrass in the northeastern USA

Growth stage	Dry matter yield		
	Total	Regrowth	Regrowth
	(tons/acre)		(% of total)
Timothy[a]			
Pre-joint	2.46	0.75	30
Early head	2.54	0.75	30
Early anthesis	3.48	0.82	24
Post anthesis	3.67	0.77	21
Smooth bromegrass[b]			
Pre-joint	2.38	1.22	51
Early head	2.70	1.24	49
Early anthesis	3.58	1.10	31
Post anthesis	3.94	0.98	25
Orchardgrass[c]			
Pre-joint	3.21	1.96	61
Early head	3.40	2.22	65
Early anthesis	3.98	1.88	47
Post anthesis	4.05	1.64	40
Reed canary grass[d]			
Pre-joint	2.77	1.30	47
Early head	3.30	1.77	54
Early anthesis	3.42	1.55	45
Post anthesis	3.46	1.17	34

Sources: Adapted from Brown et al., 1968, Decker et al., 1967, Washko et al., 1967, and Wright et al., 1967.

[a]Three-year averages from five locations.

[b]Three-year averages from three locations and 2-year averages from two other locations.

[c]Three-year averages from six locations.

[d]Three-year averages from four locations.

between the **boot stage** and the early anthesis developmental stage provides a good compromise for yield and persistence for most species, while providing quality forage for most classes of livestock. Cutting captures the quality and yield of forage at the time, but both are reduced by leaf loss during handling and respiration losses during storage (see Chapter 17). Cutting later gives lower forage quality (Table 6.5) and less aftermath production (Table 6.6). Plants that are cut before they reach the critical stage make better-quality forage, but yield is low and the plants are less persistent.

Planting species with different maturity dates helps to schedule optimal harvests and grazing times. Orchardgrass reaches anthesis first, followed in order by tall fescue, smooth bromegrass, and timothy. In addition, cultivars of timothy differ widely in maturity (Table 6.3), allowing for even more flexibility in management. Adequate soil moisture and moderate summer temperatures favor regrowth of both timothy and smooth bromegrass. Low-yielding regrowth of these species is often grazed rather than cut for hay. Regrowth of wheatgrasses is also very dependent on rainfall amounts after reproductive growth has ceased. Wild ryes produce little regrowth.

Species that are sensitive to cutting at a critical growth stage are also more sensitive to frequent cutting or grazing. In mixtures, they are often displaced by the grasses that tolerate more frequent defoliation. Grasses that have a low proportion (less than 30%) of vegetative tillers in spring usually do not persist well under grazing because there are few vegetative tillers that retain their shoot apex near soil level to produce leaves for regrowth. In contrast, species such as orchardgrass, tall fescue, and kentucky bluegrass tiller profusely, have a higher percentage of vegetative tillers in spring (more than 50%), and persist well under grazing.

Managing Regrowth

In northern areas of the USA, from one-half to two-thirds of the annual production of cool-season grasses occurs during reproductive growth in the months of April, May, and June (Fig. 6.1). Studies in the Northeast indicate that regrowth yield represents a higher percentage of the total seasonal yield for orchardgrass and reed canarygrass than for smooth bromegrass and timothy (Table 6.6). Tall fescue and orchardgrass are quick to recover after cutting.

The stems of the vegetative tillers of smooth bromegrass and reed canarygrass do elongate to some extent in second and subsequent growths, but the internodes are short and no inflorescence is produced. This causes the elongated shoots to develop new leaves further above the soil, producing an elevated canopy, and allowing more light to penetrate. The elongated stems of smooth bromegrass and reed canarygrass regrowth will usually maintain up to six live leaves per tiller. Most other species maintain about three leaves per vegetative tiller. Once this number is achieved, an older leaf undergoes **senescence** as each new leaf develops.

Fertilizer Management

Nitrogen (N) is typically the most limiting **nutrient** element for grass production. Phosphorus (P) and potassium (K) are also required to maintain production levels. Minerals are removed with the hay and silage, so in order to keep the grass productive, regular applications of lime, P, and K are required. Most P and K consumed by grazing animals is returned to the soil in excreted feces and urine, but these recycled nutrients are not uniformly distributed over the pasture.

Grasses respond to N, often increasing their yield by up to threefold. Nitrogen can be provided as fertilizer, as manure, or from fixation by an associated legume (see Chapter 4). Uptake and use of N are most efficient when grasses are actively growing and near-term conditions for growth are ideal. During these periods, plants produce 20–30 lb of dry matter for every 1 lb of applied N (Wedin, 1974), due to increases both in the number of tillers and in leaf size.

Species and species mixtures differ with regard to time of growth initiation, amplitude, and duration of the spring and fall peaks of production. Application times of fertilizer N can also affect seasonal distribution (Fig. 6.1B). For example, applying N early, just as the leaves begin to grow in spring, accelerates the plants for early grazing and stimulates spring growth of leaves and stems to increase yield at the desirable hay-cut stage.

Cutting infrequently and fertilizing heavily with N can reduce stand density, increase lodging, decrease tillering as a result of shading and interplant competition, and produce a clumpy stand. Nitrogen fertilization increases the protein concentration in the leaves, but has little effect on forage intake or digestibility of grasses.

Application of N in late summer or early fall promotes vegetative growth that consists mainly of leaves, which can be deferred for grazing in late fall and early winter (Fig. 6.1B). This works well with tall fescue to extend the grazing season in the transition zone. However, in more northern areas, high rates of N applied in late fall increase the susceptibility of plants to winterkill (see Chapter 5). For example, when high rates of N were applied in the fall in Iowa and in Pennsylvania, winter injury of species was most severe in orchardgrass, followed in order by tall fescue, reed canarygrass, and smooth bromegrass. Thus the use of N to manage tall fescue for fall-saved forage and winter grazing requires consideration of persistence in northern areas.

Timothy is an effective scavenger of available soil N. In regions where it is adapted, timothy is usually more productive than other grasses under low N conditions. Orchardgrass is noted for its efficient use of N fertilizer when compared with other tall-growing cool-season

Table 6.7. Adaptation and management characteristics of major perennial warm-season grasses used in northern areas

Common name	Soil adaptation	Ease of establishment	Winter hardiness	Drought tolerance	Tolerance of wet soils	Primary use	Forage quality	Grazing tolerance	Competitiveness in mixtures
Big bluestem	Moderately to excessively well drained soils	Weak seedling vigor; slow to establish	Excellent	Good	Poor	Pasture or hay	Good	Poor	Poor
Little bluestem	Well-drained, medium to dry; infertile soils	Establishes slowly	Excellent	Excellent	Poor	Pasture or hay	Fair	Fair	Fair
Indiangrass	Deep, well-drained sandy to clay soils	Moderate rate of establishment	Good	Good	Poor	Pasture or hay	Good	Fair	Fair
Switchgrass	Moderately deep to deep, dry to poorly drained soils	Moderate rate of establishment	Good	Good	Good	Pasture or hay	Fair	Fair	Fair
Caucasian bluestem	Moderate or better drainage, low fertility	Slow germination and establishment	Fair	Excellent	Poor	Pasture	Poor	Good	Good

grasses. A **sodbound** condition can develop in some rhizomatous species (e.g., smooth bromegrass) due to the accumulation and sequestration of N in organic matter, which renders it unavailable to the plant. This undesirable condition can be corrected by applying N to encourage microbial activity that breaks down the thatch and reduces the C:N ratio.

Use of Legumes

Legumes have a high protein content and high forage quality. As a group, white clover, alfalfa, red clover, birdsfoot trefoil, and annual lespedezas are usually digested more rapidly and to a greater extent than cool-season grasses. Thus both intake and digestibility of the mixture are increased. The carrying capacity of the grass/legume mixture is higher, especially in summer, and animal gain is often improved by 35% or more compared with grass alone.

Adding legumes to form mixtures extends the spring growth period and improves summer yield (Fig. 6.1B). Legumes usually initiate growth later in spring than grasses, and are more productive in summer. Legumes have less effect than N on fall growth of grasses, and they need to be managed carefully during fall to maintain the stands (see Chapter 8). Adding a legume also helps to offset the negative effects of alkaloids in grasses.

Warm-Season Perennial Grasses

Whereas cool-season grasses usually have low productivity during summer, warm-season grass species such as switchgrass, big and little bluestem, Caucasian bluestem, and indiangrass grow rapidly (Fig. 6.1C). Little bluestem is very drought hardy but is not very productive, whereas Caucasian bluestem is low growing and can tolerate near continuous stocking but is not very winter hardy (Table 6.7). Local populations of big bluestem, indiangrass, and switchgrass overlap adjacent adaptation regions to extend each species from Mexico into Canada (Moser and Vogel, 1995). They are used as pasture and range in the prairie states.

As with cool-season grasses, the quality of warm-season perennial grasses decreases with maturity. The quality of warm-season grasses is generally lower than that of cool-season grasses at similar growth stages because the protein concentration is lower, the leaf:stem ratio is lower, and there is more structural tissue in the leaves of warm-season grasses (see Chapter 4).

Cool-season grasses and legumes, with the exception of birdsfoot trefoil and annual lespedeza, initiate spring growth about 20–30 days earlier than warm-season grasses, and are very competitive with them. Maintaining separate pastures of warm-season grasses can provide grazing in summer when cool-season pastures are low in productivity and benefit from a rest period (Fig. 6.1A, C). The warm-season grasses differ with regard to periods of growth and time of heading, and thus fit into a system differently (Table 6.8). Switchgrass and Caucasian bluestem begin growth and development early, and they reach grazing height while the cool-season species are still productive.

Switchgrass grows more rapidly in late spring than is needed by grazing animals, so harvesting a portion of it for hay can be useful. Caucasian bluestem grows more slowly, has many prostrate stems, and tolerates near-continuous stocking during summer, but the forage is usually of lower quality. Both species show good regrowth and grow into early fall to bridge with the recovery of cool-season pastures. In the central USA about 2 acres of cool-season pasture per acre of warm-season pasture are required for a full-season grazing system.

Big bluestem begins spring growth later and matures later than switchgrass and Caucasian bluestem, so it matches better with mixtures of cool-season grasses and legumes. Indiangrass is the latest to mature among the five warm-season grasses, and has the highest forage quality. Forage quality is generally lowest for Caucasian bluestem, intermediate for switchgrass and little bluestem, and highest for big bluestem and indiangrass (Table 6.7). None of

Table 6.8. Origins and botanical characteristics of perennial warm-season grasses adapted to northern areas

						Heading dates		
Common name	Native or introduced	Mature height	Growth habit	Seed/lb	Inflorescence type	Relative	Range	Regrowth potential
Big bluestem	Native	Tall	Bunch	165,000	Panicle	Medium	Wide	Fair
Little bluestem	Native	Medium	Bunch	260,000	Panicle	Medium	Wide	Poor
Indiangrass	Native	Tall	Bunch	175,000	Panicle	Late	Wide	Poor
Switchgrass	Native	Tall	Bunch	300,000	Panicle	Early	Wide	Good
Caucasian bluestem	Introduced from Eurasia	Medium	Bunch	860,000	Panicle	Early	Wide	Good

the warm-season grasses perform well in autumn, and all of them need to be rested in winter.

The cool-season grasses, especially tall fescue, that were rested during summer can be fertilized with N in early August to stimulate fall growth, and will then be ready to be grazed in mid-September (Fig. 6.1B). This rest and growth period allows animals to be moved back to cool-season pastures. Some of the fall growth may be **stockpiled** for winter grazing after growth has slowed down or stopped. In more northern areas, corn stover is often abundant and can be used to extend the grazing season.

Overgrazing and row cropping have depleted many natural populations of warm-season perennial grasses, so new stands must be established by seeding, nearly always in spring. Seedling vigor is low in comparison with that of cool-season grasses, and careful management is needed to avoid weed competition and invasion by cool-season species (Table 6.7). Warm-season pastures should periodically receive weed control measures such as burning in spring to reduce invasion by woody and cool-season species. In grazing systems, the sequencing of separate cool-season and warm-season pastures works well for beef herds because abundant forage of adequate quality is available for much of the year.

Annual Grasses for Forage

Sudangrass, sorghum × sudangrass hybrids, and pearlmillet are warm-season annual grasses that produce abundant medium- to high-quality forage during summer (Fig. 6.1D). These annuals, which are established in prepared seedbeds when soil temperatures are 60–65°F for the sorghums and 70°F for pearlmillet, can be grazed or harvested for hay or silage (Fribourg, 1995). The crop can be harvested as hay, but the thick stems do not dry rapidly, so use of the crop for grazing or making silage is preferred. None of the species are frost tolerant, so the leaves and stems are killed by early frosts.

The quality of forage is good, and better than that of warm-season perennial grasses, but the leaves and stems of sorghums produce prussic acid that can lead to cyanide poisoning of animals that consume the forage (see Chapter 18). Prussic acid levels are highest in young leaves and tissues that have been frosted. Therefore forage of sorghum hybrids should not be cut or grazed until it is 24 in. tall. Animals should be removed from pastures at a stubble height of about 8 in. to reduce grazing of young basal leaves on new tillers. Grazing should be stopped for a few days after a killing frost to allow dissipation of the prussic acid. Ensiling and hay making both reduce prussic acid levels.

Sudangrass is lower yielding than the sorghums or hybrids, but the prussic acid levels in this species are lower, making it less risky to feed or graze. Pearlmillet does not produce prussic acid, matures later than the sorghums, and produces grain in the head that helps to support daily gain. It exhibits less regrowth than the sudangrass and sorghum × sudangrass hybrids. Summer annual forages fit very well into dairy systems, especially on good soils, but establishment costs each year and the need for N inputs need to be taken into consideration. Summer annuals are also popular as emergency forages.

Crop Residues

Residues left in the field after corn and sorghum grain harvest can be used to supplement the forage supply and extend grazing into late fall and winter in northern areas. Corn stover is the most abundant residue. The forage quality of stover is high when the grain is fully developed, but decreases as the grain dries. When harvested early for high-moisture grain, the quality of the stover is still good. If harvested after the grain has dried, the stalk has lower quality but the digestibility of the husks is still around 60%. Two acres of cornstalk residue will provide grazing for a beef cow for about 80 days (Wedin and Klopfenstein, 1995).

Grain sorghum residue is used in the areas where it is too dry for corn production or for extended growth of tall fescue. Grain sorghum is a perennial but acts as an annual in northern areas of the USA. The leaves do not die at grain maturity, and forage quality is retained. However, as with other sorghums, the leaves may accumulate high levels of prussic acid, especially after a frost. Corn and sorghum residues are used most efficiently during late fall and winter on well-drained land that does not collect water. Manure deposited during grazing of residues contributes to the production of succeeding crops.

Residues from oat and barley are of better quality than wheat residues, but all are of lower quality than corn stover and are available at times when perennial forages are growing actively. Thus, like corn and sorghum residues, small-grain residues can be baled, treated with ammonia to increase their crude protein concentration and preservation qualities, and fed to livestock at times when pastures are not growing.

Non-Forage Uses

In addition to providing high-quality pasture, range, hay, and silage for ruminant livestock, cool-season grasses contribute greatly to soil conservation and aesthetic value. Some species, such as tall fescue, perennial ryegrass, and kentucky bluegrass, are used for turf. Straw remaining after seed harvest is used as mulch, as livestock bedding, and also as fuel. Other species, such as switchgrass, reed canarygrass, and tall fescue, are used as biological filters to remove nutrients from industrial and/or municipal wastewater and to protect streams. Switchgrass is being investigated intensively for use as a biofuel crop.

Sod-forming grasses such as smooth bromegrass, reed canarygrass, bluegrass, tall fescue, and the rhizomatous wheatgrasses are best for site stabilization (Ditsch and Collins, 2000). They reduce soil erosion more effectively than bunchgrasses, legumes, and row crops, and are seeded to prevent erosion from road banks, surface-mined areas, stream banks, the edges of ponds, and sod waterways. Tall fescue and reed canarygrass are commonly used in sod waterways. They withstand machinery traffic and animal treading better than bunch-type grasses and shallow-rooted sod-forming grasses.

Summary

Grasses adapted to areas north of Tennessee and Oklahoma include a range of cool-season species, nearly all introduced from Europe. Each has its own region of adaptation that is associated primarily with tolerance of heat or cold temperatures and long periods of drought. In most cases the preferred grasses are grown in mixtures with legumes that increase forage and pasture quality and fix atmospheric nitrogen to reduce input costs. For example, timothy, bromegrass, and orchardgrass tend to grow upright and are popular for hay and silage for dairy herds in the north, especially when grown with alfalfa.

Due to its drought tolerance and ability to grow well into winter, tall fescue is a popular species for grazing for beef cows in the transition zone. It is long-lived and usually grown with red or ladino clover in continuous pastures, with the first cutting used for hay or seed production. The clovers offset some of the quality problems associated with the fungal endophyte, but they are also used with tall fescue cultivars that are endophyte-free. Annual sorghums are used for emergency forages and grazing in summer when cool-season grasses grow slowly. Non-grain residues of corn are also grazed to lengthen the grazing season for beef cows, or packaged in large round bales for winter feed.

Questions

1. Why are almost all of the cool-season grasses that are currently used in more humid areas introduced species?
2. Describe why sod-forming grasses are often better than bunch grasses for continuous stocking.
3. What are the main reasons why forage quality decreases during the spring growth of cool-season grasses?
4. What are the advantages of adding a legume to stimulate production of cool-season grasses compared with applying N fertilizer in spring? Would the advantages be the same compared with applying N fertilizer in late summer?
5. What is the critical stage of grasses? What features of the plant determine its importance for cutting management in spring? Why is there less concern about critical stages during summer?
6. Why do grasses planted in spring remain vegetative and low in competitiveness during the first year?
7. What are the advantages and disadvantages of using summer annual grasses or warm-season perennial grasses for summer pastures in sequence with mixtures of legume and cool-season grasses?
8. What characteristics of cool-season grasses make them well adapted for soil conservation uses?
9. Why are warm-season perennial grasses less adapted than cool-season grasses for planting in mixtures with perennial legumes such as alfalfa or ladino clover?
10. How can problems with prussic acid be reduced? How can problems with plant alkaloids be reduced?

References

Asay, KH. 1995. Wheatgrasses and wildryes: the perennial Triticeae. In RF Barnes, DA Miller, and CJ Nelson (eds.), Forages: An Introduction to Grassland Agriculture, 5th ed., Vol. I, pp. 373–394. Iowa State University Press, Ames, IA.

Barnes, RF, DA Miller, and CJ Nelson. 1995a. Forages: An Introduction to Grassland Agriculture, 5th ed., Vol. I. Iowa State University Press, Ames, IA.

Barnes, RF, DA Miller, and CJ Nelson. 1995b. Forages: The Science of Grassland Agriculture, 5th ed., Vol. II. Iowa State University Press, Ames, IA.

Brown, CS, GA Jung, KE Varney, RC Wakefield, and JB Washko. 1968. Management and productivity of perennial grasses in the Northeast. IV. Timothy. West Virginia Univ. Ag. Exp. Sta. Bull. 570T. Agricultural Experiment Station, West Virginia University, Morgantown, WV.

Decker, AM, GA Jung, JB Washko, DD Wolf, and MJ Wright. 1967. Management and productivity of perennial grasses in the Northeast. I. Reed canarygrass. West Virginia Univ. Ag. Exp. Sta. Bull. 550T. Agricultural Experiment Station, West Virginia University, Morgantown, WV.

Ditsch, DC, and M Collins. 2000. Reclamation considerations for pasture and hay lands receiving 66 centimeters or more precipitation annually. In RI Barnhisel et al. (eds.), Reclamation of Drastically Disturbed Lands, pp. 241–271. Agronomy Monograph 41. American Society of Agronomy, Madison, WI.

Fahey, GC, M Collins, DR Mertens, and LE Moser. 1994. Forage Quality, Evaluation and Utilization. American Society of Agronomy, Madison, WI.

Fribourg, HA. 1995. Summer annual grasses. In RF Barnes, DA Miller, and CJ Nelson (eds.), Forages: An Introduction to Grassland Agriculture, 5th ed., Vol. I, pp. 463–472. Iowa State University Press, Ames, IA.

Gist, GR, and RM Smith. 1948. Root development of several common forage grasses to a depth of eighteen inches. J. Am. Soc. Agron. 40:1036–1042.

Moser, LE, and KP Vogel. 1995. Switchgrass, big bluestem, and indiangrass. In RF Barnes, DA Miller, and CJ Nelson (eds.), Forages: An Introduction to Grassland

Agriculture, 5th ed., Vol. I, pp. 409–420. Iowa State University Press, Ames, IA.

Moser, LE, DR Buxton, and MD Casler. 1996. Cool-Season Forage Grasses. Agronomy Monograph 34. American Society of Agronomy, Madison, WI.

Smith, D, AVA Jacques, and JA Balasko. 1973. Persistence of several temperate grasses grown with alfalfa and harvested two, three, or four times annually at two stubble heights. Crop Sci. 13:553–556.

Washko, JB, GA Jung, AM Decker; RC Wakefield, DD Wolf, and MJ Wright. 1967. Management and Productivity of Perennial Grasses in the Northeast. III. Orchardgrass. West Virginia Univ. Ag. Exp. Sta. Bull. 557T. Agricultural Experiment Station, West Virginia University, Morgantown, WV.

Wedin, WE. 1974. Fertilization of cool-season grasses. In DA Mays (ed.), Forage Fertilization, pp. 95–118. American Society of Agronomy, Madison, WI.

Wedin, WF, and TJ Klopfenstein. 1995. Cropland pastures and crop residues. In RF Barnes, DA Miller, and CJ Nelson (eds.), Forages: The Science of Grassland Agriculture, 5th ed., Vol. II, pp. 193–206. Iowa State University Press, Ames, IA.

Wright, MJ, GA Jung, CS Brown, AM Decker, KE Varney, and RC Wakefield. 1967. Management and Productivity of Perennial Grasses in the Northeast. II. Smooth Bromegrass. West Virginia Univ. Agric. Exp. Sta. Bull. 554T. Agricultural Experiment Station, West Virginia University, Morgantown, WV.

Chapter 7

Grasses for Southern Areas

Daren D. Redfearn and C. Jerry Nelson

The southern area of the USA, below about 36° N latitude, has long frost-free growing seasons and warm summer temperatures. Annual rainfall ranges from sparse in western areas, where dry conditions limit production, to more than 35 in. (0.9 m) annually eastward from eastern Texas and Oklahoma. Areas along the Gulf Coast, including Florida, have semi-tropical climates with abundant rainfall, often exceeding 60 in. (1.5 m) annually. Rainfall is fairly uniformly distributed throughout the year in the southern USA. Soil types range from well drained with low organic matter content to those that are poorly drained and difficult to manage. Most southeastern soils have low fertility due to nutrient leaching.

The major domestic livestock enterprises are beef (*Bos taurus*) cow-calf and stocker operations. These livestock systems fit the climate as there is minimal need for winter feeding. Some areas around major markets have dairy farms with more intensive forage management. There are few sheep (*Ovis aries*) and goats (*Capra hircus*), but pleasure horse (*Equus caballus*) numbers are increasing rapidly. In western areas, ruminant wildlife such as deer (*Odocoileus* species) and pronghorn antelope (*Antilocapra americana*), and other herbivores such as rabbits (*Sylvilagus* species) and other rodents are part of the grassland ecosystem. Some pastures and rangeland areas are operated for fee-based hunting and other outdoor activities.

Because of the temperate growing conditions, the southern USA offers good potential for developing year-round forage systems. These systems use the valuable soil and grassland resources in an economic and environmentally friendly manner. Our goal is to introduce the major forage grasses used in the South and to discuss their use, management, and potential. Some of the species mentioned are covered in the compendium.

Warm-Season Perennial Grasses

The perennial warm-season grasses that form the base species for pastures and forage systems in the South (Table 7.1) have the C_4 photosynthetic pathway that contributes to their adaptation to the hot and sometimes dry conditions (see Chapter 4). Frequently the dormant warm-season grasses are overseeded with a winter-annual grass such as annual ryegrass or a small grain for winter and early spring grazing. The addition of a winter annual legume adds nitrogen (N) to the soil and provides higher-quality pasture or forage. With a mix of cool- and warm-season species and good management, it is possible to have 10 months of available forage for grazing.

With the exception of eastern gamagrass, big bluestem, indiangrass, and switchgrass (Moser and Vogel, 1995), the major forage grasses used for managed pasture and hay production in the southern USA are introduced. Most of the perennial warm-season grasses used in the southern region came from Africa or South and Central America. Warm-season perennial grasses are very productive in warm weather but, due to the lower protein content and higher fiber content in the leaves, are frequently of lower forage quality than cool-season grasses.

Adaptation, Productivity, and Use

Adapted grasses exhibit a wide range of morphology, from very prostrate and low-growing species such as bermudagrass to tall upright grasses such as elephantgrass and big bluestem (Table 7.1). Each has a unique growth pattern

Forages: An Introduction to Grassland Agriculture, Seventh Edition. Edited by Michael Collins, C. Jerry Nelson, Kenneth J. Moore and Robert F Barnes.

Table 7.1. Characteristics of warm-season perennial grasses adapted to humid areas of the southern USA (below 36° N latitude)

Species	Growth habit	Major states[a]	Major Use	Forage quality	Ease of establishment	Perennation strategy
Bermudagrass	Prostrate	All	Pasture, hay	Good	Easy	Rhizomes and stolons
Bahiagrass	Decumbent	South	Pasture	Fair	Slow	Rhizomes
Buffalograss	Decumbent	Southwest	Range	Excellent	Slow	Rhizomes
Digitgrass	Decumbent	Florida	Pasture	Fair	Moderate	Stolons
Elephantgrass	Upright	Florida	Pasture	Fair	Difficult	Tillers
Limpograss	Erect	Florida	Pasture	Good	Slow	Stolons
Dallisgrass	Decumbent	All	Pasture	Good	Slow	Rhizomes
Big bluestem	Upright	Southwest	Range	Very good	Slow	Tillers
Indiangrass	Upright	Southwest	Range	Very good	Slow	Tillers
Switchgrass	Upright	Southwest	Grazing, hay	Good	Moderate	Rhizomes
Caucasian bluestem	Decumbent	North	Grazing	Fair	Moderate	Tillers
Yellow bluestem	Semi-erect	Southwest	Grazing	Good	Moderate	Tillers
Eastern gamagrass	Upright	North	Grazing	Excellent	Difficult	Tillers

[a]States or regions of the southern USA.

that needs to be understood in order to optimize its management. With the exception of white clover, there are few perennial **herbaceous** legumes that are widely adapted for use in the South. Annual legumes are important, and some legume trees and shrubs provide **browse** (see Chapter 9).

Bermudagrass

Bermudagrass was introduced from Africa and is adapted in the USA as far north as southern Kansas and eastward to the Atlantic Coast. Today it is the most widely used perennial forage grass in the southern USA, mainly because it is high yielding, persistent, and well adapted for grazing due to its extensive production of **rhizomes** and **stolons**. New cultivars selected for increased cold tolerance have extended its area of adaptation northward. Although the emphasis is often on yield, factors such as **winter hardiness**, disease resistance, and forage quality are equally important.

> **Small Changes Make Big Differences**
>
> Although early bermudagrass cultivars were well adapted to the South and almost weed-like, their forage quality was low. Some states even passed laws that prohibited the planting of bermudagrass. However, the true value of this species was in preventing soil erosion.
>
> Through the years, winter hardiness was improved and new methods for measuring digestibility became available. Plant breeders (Burton et al., 1967) located a bermudagrass germplasm with better digestibility and crossed it with "Coastal", which had become the cultivar of choice. The progeny ("Coastcross-l") had 12.3% higher digestibility (53.5% vs. 60.1%), which probably also increased intake. The net energy for maintenance requirement (31 units) of animals consuming "Coastal" would use 58% of the digestible forage (31/53.5), but those consuming Coastcross-l would use only 52% (31/60.1). That remaining to support the net energy for growth requirement (22.5 vs. 29.1) was about 30% higher in Coastcross-l. Animal gains that directly compared the two forages verified these results.
>
> Today, it is well known that a small increase in forage digestibility that also increases intake is more important than a proportional increase in forage yield. The increase in digestibility of warm-season grasses is usually due to changes in cell wall properties, rather than in protein content. Improved bermudagrass cultivars have had a large positive effect on the ruminant livestock industry in the South.

Bermudagrass is deep-rooted and more drought tolerant than most warm-season forages, but it is not very productive on waterlogged soils (Table 7.2). It prefers an acidic soil pH of 5.0–5.5, and is very responsive to N fertilization. Bermudagrass produces abundant rhizomes and stolons, making it tolerant of grazing, and it has few diseases or insect pests. It is productive and has good quality when properly managed (Table 7.1). Due to its aggressive vegetative reproduction, bermudagrass can be a difficult weed species. especially in areas of cotton (*Gossypium hirsutum* L.) and corn production.

Some cultivars can be planted using seed, but most require vegetative propagation. Hybrid cultivars are

Table 7.2. Adaptation characteristics of perennial warm-season grasses for the southern USA (below 36° N latitude)

			Tolerance			
Species	Frost[a]	Winter[b]	Acid soil	Wet soil	Low fertility	Grazing
Bermudagrass	Poor	Good	Good	Fair	Fair	Excellent
Bahiagrass	Poor	Fair	Good	Good	Good	Excellent
Buffalograss	Poor	Very good	Poor	Poor	Good	Excellent
Digitgrass	Poor	Poor	Good	Fair	Fair	Fair
Elephantgrass	Poor	Poor	Good	Poor	Good	Good
Limpograss	Poor	Poor	Good	Excellent	Fair	Good
Dallisgrass	Poor	Fair	Good	Good	Good	Excellent
Big bluestem	Poor	Very good	Poor	Poor	Poor	Fair
Indiangrass	Poor	Very good	Poor	Poor	Poor	Fair
Switchgrass	Poor	Very good	Poor	Poor	Poor	Fair
Caucasian bluestem	Poor	Good	Good	Fair	Fair	Excellent
Yellow bluestem	Poor	Good	Good	Fair	Fair	Excellent
Eastern gamagrass	Poor	Very good	Poor	Poor	Poor	Poor

[a]Ability to remain green after night temperatures of 28°F.
[b]Ability to survive over winter.

sterile, produce little or no viable seed, and must be vegetatively propagated. With adequate rainfall and proper fertility, bermudagrass yields range from up to 5 tons per acre in northern parts of Oklahoma and Arkansas to over 10 tons per acre in southern parts of Georgia and Mississippi. Bermudagrass is also grown in western states under irrigation.

Selection of the best cultivar is important, as cultivars differ in important characteristics, including establishment, coverage rate, persistence, forage yield, and forage quality. Bermudagrass cultivars are broadly classified as either a "grazing type" or a "hay type." Grazing types have short stature, typically spread faster during establishment, and achieve a more complete cover than do hay-type cultivars. However, both types can be used for either purpose. Some grazing-type cultivars form a very dense stand, reducing weed competition to a greater extent than the hay-type cultivars. However, once established, hay-type cultivars have greater yield potential than do grazing-type cultivars. Established bermudagrass is difficult and expensive to destroy for replacement with an improved cultivar (Redfearn and Wu, 2013).

All bermudagrass cultivars respond to a good fertility program that supplies adequate amounts of nitrogen (N), phosphorus (P), and potassium (K). The soil should be tested before preparation of the seedbed, to determine the P, K, and lime requirements. Incorporation of these elements 4–6 in. into the surface soil is more efficient than broadcast applications after planting. Nitrogen is not normally applied at planting, to lessen the effects of weed competition, but is applied later when the established stolons begin to elongate.

Early sprigging (February 1 to April 10) is important for successful stand establishment. Sprigs (crowns, rhizomes, and stolons) should be dug and planted in spring before the plants break winter dormancy and initiate growth. This growth utilizes the carbohydrate reserves from the rhizomes. Furthermore, the sprigs to be planted will be mostly rhizomes and root crowns, which are also the sources of buds. Sprigs that are dug after spring growth has begun will contain top growth, which has a very short life and seldom contributes to establishment success.

Sprigs should be planted in the prepared "seedbed" as soon as possible after digging. They should be watered and covered with a tarp during transport to the field to avoid desiccation. Planting in the early spring means that temperatures are cooler, the humidity may be higher, and it is easier to keep the sprigs from drying out. In the range of 15–30 bushels (1.1 m^3) of sprigs per acre should be used for rapid establishment. The sprigs need to be disked into the soil, followed by a heavy roller to firm the soil.

During the establishment year, grassy weeds and palatable broad-leaved weeds can be suppressed by **mob grazing** when necessary. More severe weed problems may require chemical control. Mowing of broadleaf weeds is helpful, but by the time the weeds are tall enough to mow, they have already competed severely with the bermudagrass propagules for light. With adequate fertility and moisture, the new plantings can be cut or grazed late in the establishment year.

Bahiagrass

Bahiagrass is native to South America and is widely distributed in Argentina, Uruguay, Paraguay, Brazil, and the West Indies. In the USA it is adapted from east Texas to the Carolinas to as far north as northern Arkansas and central Tennessee. Because bahiagrass is less cold tolerant than bermudagrass (Table 7.2), it is primarily used along the coastal area in the southern USA. It is adapted to a wide range of coastal plain soils, but performs best on sandy soils with a pH of 5.5–6.5. Bahiagrass produces short rhizomes and develops a dense sod. It grows better than most other pasture grasses on sandy, drought-prone soils with relatively low fertility.

Bahiagrass is popular in the South because it is more persistent and produces higher yields on low-fertility soils than either bermudagrass or dallisgrass (Burson and Watson, 1995). It also is more competitive with weeds, is established by seed, is relatively disease- and insect-free, and withstands close defoliation. Bahiagrass is used primarily for permanent pastures, although some is harvested for hay (Table 7.1). Because it is a low-growing plant, as much as 60% of the forage may remain uncut when harvested as hay. However, bahiagrass hay is leafy and has few stems and seedheads.

Bahiagrass is generally seeded in early spring after the last killing frost. Later plantings in summer usually have severe competition from weeds, which makes establishment difficult. Young, small bahiagrass seedlings are weak competitors with weeds (Table 7.1), so some form of weed control is often necessary for successful establishment. Grazing during establishment is not recommended because many seedlings will be trampled and damaged. Fertilization is necessary for successful establishment, and should be based on a soil test. Because bahiagrass is extremely slow to establish, it sometimes takes up to 3 years for a full stand to be obtained. However, with good management and growing conditions, grazing can begin as early as the second year following establishment.

The dense, compact sod of bahiagrass generally limits the success of associated legumes. However, white clover, or winter annuals such as crimson or arrowleaf clover, can be established and grown in a bahiagrass pasture or hayfield. The soil needs to be well fertilized with P and K, and the grass kept short to reduce competition. The P and K should be applied prior to planting winter annual legumes. If the legume growth is satisfactory, the normal spring topdressing of N is not needed. To maintain high grass production through the summer, N should be topdressed in June and again in late July (Hanna and Sollenberger, 2007).

Dallisgrass

Common dallisgrass is distributed throughout the southeastern USA and is widely used for permanent pasture (Table 7.1). It is best adapted to areas that receive at least 35 in. of annual rainfall, and it grows best on clay or loam soils that are moist but not wet (Table 7.2). It initiates spring growth earlier than most warm-season perennial grasses, and generally continues to grow later into the fall.

Fig. 7.1. Common dallisgrass can withstand close and frequent grazing because it has numerous short, compact rhizomes that produce many tillers. Each tiller produces several leaves, many of which are near the base of the canopy. The horizontal lines on the board are about 13 in. apart. (Photo courtesy of Byron Burson.)

The plant has a large amount of basal leaf area (Fig. 7.1), survives well under high stocking rates, and has excellent forage quality when properly managed.

The use of dallisgrass is limited by several factors. Establishment can be slow due to poor seed quality and slow germination (Hanna and Sollenberger, 2007). Commercial seed production is hindered by low seed set, low germination rates, and contamination of the seed lots with sclerotia, the fruiting bodies of the ergot fungus (*Claviceps* species) that form in the seedhead. Ergot produces the powerful alkaloid ergovaline, which can cause serious animal health problems if hay or pasture has mature seedheads with sclerotia. Low forage production limits the usefulness of dallisgrass in some regions.

Elephantgrass

Elephantgrass (also called napiergrass) is an upright grass that is adapted to nearly all tropical and subtropical regions of the world, from sea level to 6500 ft (1980 m), that receive at least 40 in. (1.02 m) of annual rainfall. After being introduced from Africa, it has seen only limited use as a forage plant in the USA (Table 7.1). It grows best when temperatures are very hot (85–95°F), yet it tolerates cool temperatures down to 50°F before growth ceases (Ocumpaugh and Sollenberger, 1995). Frost will kill the top growth, but the roots will remain unharmed (Table 7.2). However, if the soil freezes, root damage and plant loss will occur.

Periods of drought restrict plant growth, but plants resume rapid growth with the onset of rain. Elephantgrass does not tolerate waterlogged or flooded conditions, and is not adapted to wet sites. It can persist in soils that tend to be very well drained, but it does not do well in heavy clay soils. The soil pH should be maintained above 5.5 (Sollenberger et al., 1988).

Many cultivars of tall elephantgrass are used in the tropics, but the only cultivar used in US forage systems is "Mott" dwarf elephantgrass, which has been selected for improved quality. Elephantgrass is normally propagated from stem cuttings consisting of a few nodes, internodes, and axillary buds. It is very responsive to fertilization, and is one of the fastest-growing, highest-yielding grasses. This high growth rate makes grazing management difficult, but it means that the plant is highly suitable for silage production.

Native Bluestems, Indiangrass, and Switchgrass

Throughout much of the South, many acres of native grasslands have been converted and replaced with introduced forages. However, the native prairie grasses are still important grassland components (see Chapter 6). In the South they are important forages in low-rainfall areas of Oklahoma and parts of Texas and eastward. Switchgrass greens up and matures earlier than big bluestem and indiangrass, but has lower forage quality (see Chapter 6). Little bluestem is more drought tolerant than big bluestem, and is used in semi-arid areas or sites.

These native perennials are seeded in spring when soil temperatures reach 60°F, yet are slow to establish as they invest most of the early growth in the root system (Table 7.1). This allows the root to occupy a large soil volume, but the small seedlings are subject to shading by other species. Fertilization with P and K to meet the requirements of the soil test is needed. Nitrogen is used sparingly because it stimulates the competitive growth of weeds and other species. Management is covered in Chapter 6.

Old World Bluestems

The Old World bluestems are part of a complex of species in the genera *Bothriocola*, *Capillipedium*, and *Dichanthium* that are native to Africa, the Middle East, and southern Asia. They are best adapted to fine-textured soils. Yellow bluestem is the most important of these in the USA because it is the easiest to establish (Table 7.1), and has good drought tolerance and winter hardiness (Sims and Dewald, 1982). Caucasian bluestem is slightly more winter hardy than yellow bluestem, but may be less drought tolerant. Caucasian bluestem is best adapted to the southern Corn Belt, where it provides high yields of medium-quality forage for midsummer grazing (see Chapter 6).

Recently released cultivars of yellow bluestem have supplanted "King Ranch" bluestem because of their greater winter hardiness and resistance to leaf rust. Seed production is predominantly **apomictic**. The weed-like nature of Caucasian and yellow bluestems attests to their aggressiveness and ability to establish (Table 7.1).

Yellow bluestem is more erect than Caucasian bluestem, yet both often form large, saucer-shaped clumps with the stems curving upward. Caucasian bluestem appears more decumbent. Because of their unique crown structure, they tolerate close, continuous stocking at high stocking rates (Voigt and Sharp, 1995). Somewhat like dallisgrass (Fig. 7.1), their crowns consist of densely packed stem bases that contain numerous sites for tiller initiation, and basal leaf material that escapes defoliation. After fertilizing with P and K, these grasses are seeded in late spring when soil temperatures reach 50°F. Forage quality is generally lower than for the native bluestems and indiangrass (Table 7.1).

Eastern Gamagrass

Eastern gamagrass has long had a reputation as a valuable forage grass. In the USA, eastern gamagrass is found from Texas northward to Kansas and eastward to Massachusetts. Believed to be one of the progenitors of corn, this tall grass has high photosynthetic rates and water use efficiency. It produces forage of very high quality, and is sought out by grazing ruminants, but is damaged by close and frequent grazing (Table 7.2). Difficulties with regard to seed production, seed quality, and establishment (Table 7.1) have limited both the amount of research on and the commercial use of this species until recently.

Warm-Season Annual Grasses

Summer annual forages for the South include forage sorghums, sorghum x sudangrass hybrids, pearl millet, and crabgrass (Table 7.3). These grasses contribute to systems that need emergency forage or high-quality forage. Pearl millet and sorghum × sudangrass hybrids are typically used for grazing or are harvested mechanically for hay or silage. Crabgrass is nearly always grazed.

These forage crops are usually planted in early summer, following winter annual forages, and are highly productive over a relatively short growing season. Crabgrass will naturally reseed to perennate, and if grazed early provides abundant forage of good quality. Summer annuals are best grown on good soils, and are very responsive to N fertilization (Table 7.4). Nitrogen fertilization is needed because legumes compete poorly with these grasses. With the exception of crabgrass, a disadvantage is the extra cost of establishing them each year. The sorghum family also poses a risk of prussic acid poisoning of livestock (see Chapter 6).

Management practices for pearl millet and sorghum × sudangrass production are similar. Either can be planted any time between April 15 and August 1. In the South, seeding rates for pearl millet are 25 lb of seed per acre (28 kg/ha) if drilled, and 30 lb of seed per acre (34 kg/ha) if

Table 7.3. Characteristics of cool-season perennial and annual grasses adapted to the southern USA (below 36° N latitude)

Species	Growth habit	Major states	Major use	Forage quality	Ease of establishment	Perennation strategy
Perennial grasses						
Tall fescue[a]	Upright	North	Pasture	Good	Moderate	Tillers
Smooth bromegrass	Upright	Northwest	Hay, pasture	Very good	Moderate	Rhizomes
Orchardgrass	Upright	Northeast	Hay, pasture	Very good	Easy	Tillers
Annual grasses						
Annual ryegrass	Upright	South	Pasture	Excellent	Very easy	None
Cereals	Upright	Southwest	Pasture	Excellent	Very easy	None
Crabgrass	Decumbent	All	Pasture	Fair	Moderate	Reseeding
Pearl millet	Tall, upright	All	Pasture, silage	Excellent	Easy	None
Sorghum × sudangrass hybrids	Tall, upright	All	Pasture, silage	Excellent	Very easy	None

[a]Endophyte infected.

broadcast. The corresponding rates for sorghum × sudangrass hybrids are 30 lb of seed per acre (34 kg/ha) and 35 lb of seed per acre (39 kg/ha), respectively. Grazing on the sorghum × sudangrass hybrids needs to be controlled, as young leaves and frosted plants can contain high levels of prussic acid that can kill ruminants (see Chapter 6).

Cool-Season Perennial Grasses

Although these grasses have C_3 photosynthesis and grow well in cool weather (see Chapter 4), they are not as well adapted to the South as they are to the North. They still fill important roles in southern forage systems, but need a higher level of management to offset environmental stresses. They are used for pasture in spring and fall, and for hay production in the northern part of the southern USA. Persistence through the summer is a major problem due to competition stress from weeds and warm-season grasses.

Tall fescue is the most important cool-season perennial in the South, where it plays an important role as a spring, fall, and winter pasture. Climatic factors (rainfall and temperature), edaphic factors (soil texture and moisture), geographic factors (latitude and elevation), and the presence of

Table 7.4. Adaptation characteristics of perennial cool-season grasses and annual grasses in the southern USA (below 36° N latitude)

Species	Frost[a]	Winter[b]	Tolerance: Acid soil	Tolerance: Wet soil	Tolerance: Low fertility	Grazing
Perennial grasses						
Tall fescue[c]	Very good	Good	Good	Good	Good	Excellent
Smooth bromegrass	Poor	Very good	Poor	Poor	Poor	Good
Orchardgrass	Good	Good	Fair	Poor	Poor	Good
Annual grasses						
Annual ryegrass	Very good	Good	Fair	Fair	Poor	Excellent
Cereals	Good	Very good	Fair	Fair	Poor	Excellent
Crabgrass	Very poor	None	Fair	Poor	Fair	Very good
Pearl millet	Poor	None	Poor	Poor	Poor	Good
Sorghum × sudangrass hybrids	Poor	None	Poor	Poor	Poor	Very good

[a]Ability to remain green after night temperatures of 28°F.
[b]Ability to survive over winter.
[c]Endophyte infected.

the endophytic fungus are primarily responsible for determining the distribution of tall fescue (Table 7.4). Generally, tall fescue is best adapted to the humid, temperate areas of the USA.

Although tall fescue grows best on deep, moist soils with a heavy to medium texture and high organic matter content (Table 7.4), it can grow on soils that range in pH from 4.7 to 9.5. It persists and helps to conserve thin topsoil on droughty slopes, yet it also forms dense sods and produces excellent growth on poorly drained soils where few other cool-season grasses persist. The massive root system of tall fescue and the presence of the endophytic fungus (see Chapter 6) are frequently cited as factors associated with its wide adaptation.

Are Some Endophytes Friendly?

Tall fescue and some of the ryegrasses are known to contain endophytic fungi that grow mycelia in the intercellular spaces of the leaf sheaths. This mutualistic association produces ergot-like alkaloids and other compounds that are transported in the plant and negatively affect animal performance, especially in summer when temperatures are warm. However, the same fungus confers resistance in the host plant to insects, nematodes, and some diseases. Endophyte-infected tall fescue also has better seedling growth and better drought resistance.

Scientists have now learned that the causes may be mutually exclusive, such that the fungus can be genetically altered so that it allows improved animal performance yet retains genes for enhancing resistance to pests and drought. Recently, cultivars with this friendly endophyte have become available.

Why did this association evolve in the first place? Tall fescue originated in the Atlas mountain regions of Morocco and Algeria, where the Mediterranean climate (winter rains and dry summers) was combined with intensive sheep and goat grazing. A mutualistic association of tall fescue and the endophytic fungus developed that improved pest resistance and root growth, but at the same time discouraged summer grazing. Both features aided plant survival during the long, dry summers.

Tall fescue produces approximately two-thirds of its annual growth in the spring during **culm** elongation (reproductive growth). In the South, the vegetative regrowth slows down and then ceases during midsummer under heat and drought stress, but resumes in the fall when temperature and moisture conditions are more favorable for growth. Tall fescue is renowned among cool-season grasses for its active fall growth, with high forage quality, that remains green after frost (Table 7.4) for winter grazing.

Nearly all of the tall fescue that is grown in the South is infected with the endophytic fungus (see Chapter 6), as the latter confers better drought and insect resistance. The presence of the fungus reduces the quality of tall fescue, and hence animal performance. Legumes such as white clover should be grown in the mixture with infected tall fescue because they help to offset these negative effects.

Cool-Season Annual Grasses

Annual ryegrass is widely used because of its ease of establishment, high-quality forage, high yields, later spring growth than cereal grains, good reseeding ability, and adaptation to a wide range of soil types (Table 7.4). Annual ryegrass is adapted to all soils, from sand to clay, with soil pH in the range 5.5–8.0. However, growth is optimal at soil pH values above 5.7 because nutrient availability is improved and the risk of aluminum (Al) toxicity, which occurs at low soil pH values, is reduced.

Annual ryegrass is adapted to poorly drained soils because of its ability to produce adventitious roots on or near the soil surface. However, the highest yields still occur on fertile, well-drained soils. Annual ryegrass yields are greatest along the Gulf Coast, where winters are milder. Growth during the winter months is suppressed by cold weather further north, so the yield is reduced. However, cereal rye and wheat, which are more cold hardy than annual ryegrass, can be included in seeding blends of winter annuals to improve forage production during cold months (Fig. 7.2).

Much of the annual ryegrass planted in the USA is used for winter pasture in the South. The most common

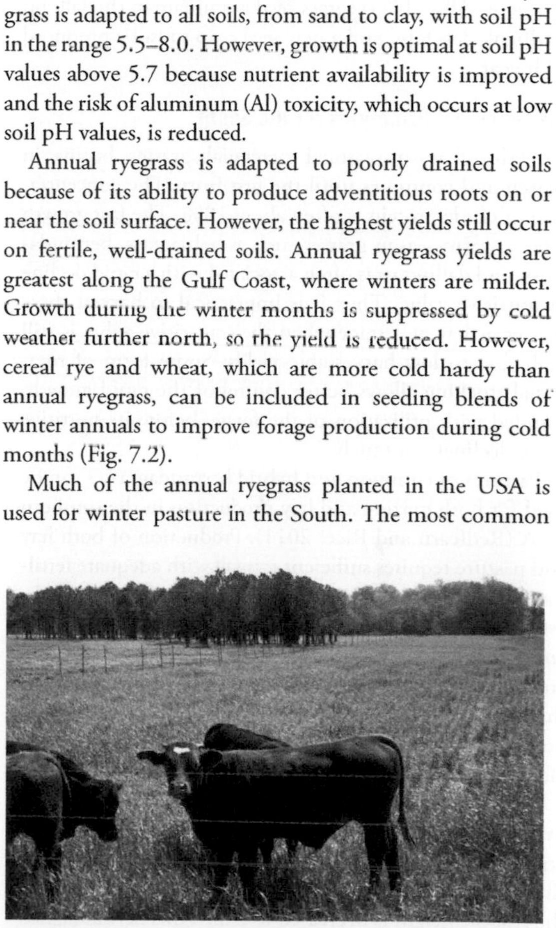

Fig. 7.2. In many areas of the South, warm-season perennial grasses are overseeded with winter annual cereal crops such as cereal rye or winter wheat. These grasses grow when the summer grasses are dormant to provide good animal performance with minimal supplementation. (Photo courtesy of Wayne Coblentz. Reproduced with permission from the Samuel Roberts Nobel Foundation.)

practice is to overseed into warm-season grasses to extend the grazing season and to produce high-quality forage. Although annual ryegrass is easy to establish, the sod of the perennial warm-season grass is usually cut or grazed closely prior to seeding. This reduces plant competition from the perennial, and hastens establishment.

Annual ryegrass will reseed itself if managed properly. Grazing should be interrupted when the inflorescences first begin to appear, which is usually in April in the lower South. After the seed shatters, livestock can be returned to graze the ryegrass stubble and new growth of the warm-season perennial grass. A light disking just prior to the time when annual ryegrass germinates will improve stand establishment. All viable ryegrass seed germinate in the fall, no seedbank develops, and a new seed crop must be produced each year.

Management Principles for the South

Growth of warm-seasonal perennial grasses begins in March and continues until the first frost. The concentration of crude protein is often about 10%. The fast growth rate of warm-season grasses such as bahiagrass, bermudagrass, and dallisgrass is often associated with a rapid decline in nutritive value. Thus it is impractical to harvest these grasses as hay at a stage when their nutritive value is still high due to low harvestable yields. Some form of rotational stocking allows better control of the rapid growth, coupled with utilization of the forage before its nutritive value declines too much.

Pastures of common and hybrid bermudagrasses can be used for both grazing and hay production in the southern USA (Redfearn and Rice, 2011). Production of both hay and pasture requires sufficient rainfall with adequate fertilization for high yields and high nutritive value. The highest bermudagrass yields are obtained on good soils with ample water-holding capacity. The strong system of stolons and rhizomes helps bermudagrass to form a dense canopy that is not very hospitable for the growth of legumes. Since bermudagrass responds well, N fertilizer is applied annually to stimulate growth.

Due to its growth characteristics, bermudagrass is extremely grazing tolerant and can withstand frequent relatively severe defoliations. Grazing should not begin until the bermudagrass is 6 in. tall, and should be stopped when the stubble height is decreased to 2 in. Thus bermudagrass pastures can often be grazed every 3–4 weeks as long as approximately 2 in. of residual height remains following each grazing period. Bermudagrass should be harvested as hay every 4–5 weeks to optimize yield and nutritive value.

Grazing in the Shade: Silvopastoral Systems

According to the Natural Resource Conservation Service, there are nearly 250 million acres in 14 southeastern states that have potential for using a silvopastoral system (i.e., a mixture of trees and pasture). The system works with evergreen, hardwood, and nut trees (Clason and Sharrow, 2000). Trees are generally planted in rows with an adapted grass species seeded between the rows to prevent soil erosion. The forage receives a modest level of inputs, mainly overseeding with a winter annual legume or a low rate of fertilizer. With legumes, the hope is that N added to the system in this way will reduce the need for commercial fertilizer.

Forage yields are close to 80% of optimum when trees are small, but this figure decreases as the tree canopy closes and the forages become shaded. Often there is less than 15% damage to the trees from grazing. The grazing animals reduce the forage canopy, thereby conserving water and increasing nutrient cycling in the forage–tree–animal ecosystem, and so improving the availability of nutrients for the trees.

Recent research suggests that trees can be harvested systematically to provide a diverse mixture of tree sizes that can maintain both tree and forage production for several years or even decades. This multi-use practice helps to optimize economic returns while providing resource conservation and a sustainable, diversified cropping system.

Dallisgrass can also tolerate close defoliation, and is used for both grazing and hay production. The productivity of dallisgrass is lower than that of bermudagrass, and can be approximately 1–3 tons of forage per acre depending on soil fertility and soil moisture. The nutritive value and palatability of dallisgrass are usually greater than those of either bermudagrass or bahiagrass.

In contrast with bermudagrass, dallisgrass forms an open sod, allowing white clover to grow well in a mixture and facilitating establishment of winter annual legumes. Thus a longer grazing season of improved pasture quality may be possible. However, seedheads of dallisgrass contain ergot bodies that can be toxic if consumed in large quantities. Therefore grazing must be frequent enough to limit seedhead production.

Management Systems

Based on the growth and quality characteristics of the forage species and the management objectives of the producer, a forage system is developed that integrates economic return and environmental concerns. Invariably, the base pasture of the selected perennial warm-season grass can be improved by managing inputs such as fertilizer to improve production, or by adding winter annual grasses or legumes to extend the grazing period.

Increasing Carrying Capacity

Warm-season grasses are very responsive to N fertilizer and to rainfall (Fig. 7.3). Being C_4 species, they are efficient at capturing CO_2 and fixing it into sugars. This allows the

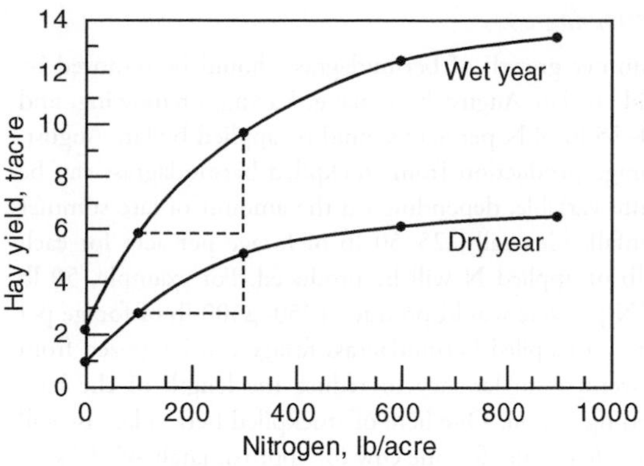

Fig. 7.3. Bermudagrass responds to nitrogen fertilizer in both a wet year and a dry year, but the nitrogen-use efficiency decreases as the rate increases. For example, the increase in N from 100 lb/acre to 300 lb/acre increased yield at the rate of 23 lb/acre/lb N with low rainfall and 38 lb/acre/lb N with high rainfall. The rainfall between April 1 and November 1 was 40 in. for the wet year and only 14 in. for the dry year. (Adapted from Burton and Hanna, 1995.)

leaves to function very efficiently with low levels of protein (enzymes contain about 16% N). The protein content of warm-season grasses is often near 10%. In contrast, the protein content of cool-season grasses is often close to 15%, and that of winter annual legumes can be even higher. The lower protein content of warm-season grasses and their efficient photosynthesis allow rapid accumulation of carbon and hence growth.

In a study by Burton and Hanna (1995), when bermudagrass was cut every 5 weeks in South Georgia, N fertilization increased yields in a curvilinear manner in both a wet year and a dry year (Fig. 7.3). However, the years differed in rainfall and efficiency of N use (lb dry weight per 1 lb of N applied). In the dry year, the efficiency of the first 100 lb of N was 49 lb forage/lb N, whereas in the wet year it was 73 lb forage/lb N. The efficiency for the second increment of N (300 − 100 = 200 lb more) was only 27 lb forage/lb N for the dry year and 50 lb forage/lb N for the wet year. The efficiency decreases with each added increment. Cool-season grasses under good growing conditions have a protein content that is about 50% higher than that of warm-season grasses, and typically have N-use efficiencies of about 30 lb forage/lb N.

Because rainfall cannot be predicted, it is recommended that bermudagrass is fertilized at a rate of 50–60 lb N per acre for each ton of desired forage production. In addition to yield increase, the protein content also increases slightly (Snyder and Leep, 2007), but as the N rates are increased and N-use efficiency decreases, the potential for N moving downward and escaping the root system is increased. This could lead to contamination of groundwater. Most N is very water-soluble and can move with run-off to reach streams and lakes.

Moderate rates of N are used more efficiently, so splitting large amounts into multiple applications a few weeks apart will reduce the potential for both groundwater and surface water contamination. In grasslands on non-hilly topography, a good guideline is to apply up to 150 lb of actual N per acre (112 kg/ha) in a single application. If more is needed, or if the landscape is hilly, the N should be applied in smaller increments spaced some time apart. Increased efficiency helps to offset the extra cost of multiple applications (Redfearn et al., 2010).

The primary benefits of N are that it increases yield and, to a lesser degree, the protein content of the forage, thus increasing both the **carrying capacity** and average daily gain. Nitrogen fertilization alone does not lengthen the productive period of grasses. Other steps should be taken to extend the grazing season.

Extending the Grazing Season

Cool-season annual grasses or legumes can be overseeded into warm-season grass pastures to extend production on the same land to provide a longer-term grazing system. Some producers extend the forage production season by overseeding bermudagrass or other warm-season perennial grass sods with a winter annual grass such as annual ryegrass, cereal rye, or winter wheat. However, this system requires added N.

Winter annual legumes such as arrowleaf clover, red clover, crimson clover, or hairy vetch can also be used. The choice of a particular grass or winter annual legume depends on the soil, the management objectives, and the adaptive characteristics of the forage. Since bermudagrass is only productive during the summer months, overseeding a winter annual will often increase the forage production season from 4–5 months for bermudagrass alone up to 8–9 months (Fig. 7.4). The inputs and management levels required will also increase, but the economic returns are nearly always positive.

Bermudagrass overseeded with a winter annual is probably the least expensive type of winter and early spring pasture to produce. However, the total amount of forage produced by the winter annual will be less than if the winter annual was planted into a prepared seedbed. This is primarily due to competition from the warm-season perennial that slows establishment and early growth of the winter annual. However, the establishment cost is kept low, and erosion is markedly lower.

Winter-Feeding Options

The most common winter-feeding option is hay. It is less risky than the other available options and requires the least amount of planning, but it is also the most expensive. **Stockpiled forage** offers an alternative that can reduce the

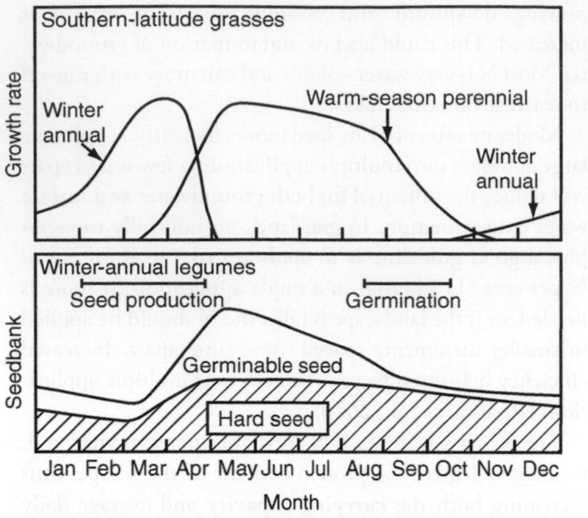

FIG. 7.4. Upper panel: Winter annual grasses or legumes can be established in the sod of warm-season perennial grasses to extend the grazing season. Lower panel: The legumes can develop a seedbank for re-establishment in the fall if they are allowed to produce seed in the spring. Legumes differ with regard to seedling competition and time of seed production. (Adapted from Beuselinck et al., 1994.)

associated costs of extended hay feeding during the winter. Pasture preparation is key to producing quality fall-grown forages.

Stockpiled Forages

The grazing of stockpiled forages is not a new concept. In several areas of the USA, livestock producers have grazed dormant native grass pastures for many years. Hay is generally not fed on native grass pastures during the winter, except during periods of heavy snow cover. Stockpiling native grass pastures is different from stockpiling introduced grass pastures such as bermudagrass and tall fescue. Stockpiling fall growth of bermudagrass and tall fescue requires N fertilization, whereas stockpiling native grass pastures is from non-fertilized summer growth.

The stockpiled bermudagrass and tall fescue forage is allowed to accumulate in the pasture for grazing when production is slow during the winter months. These perennial pastures are already established, so the only additional cost associated with growing stockpiled forages is the fertilizer cost plus any application cost. The primary disadvantage is that the level of success depends almost entirely on rainfall. When successful, the cost of grazing stockpiled forage per animal unit is lower than the cost of feeding hay or the cost of supplemental feeding on dry grass (Lalman et al., 2000).

Bermudagrass

Summer growth of bermudagrass should be removed by mid- to late August by grazing, haying, or mowing, and 50–75 lb of N per acre should be applied by late August. Forage production from stockpiled bermudagrass can be quite variable, depending on the amount of late summer rainfall. Generally, 25–50 lb of forage per acre for each 1 lb of applied N will be produced. For example, 50 lb of N per acre would produce 1250–2500 lb of forage per acre. Stockpiled bermudagrass forage can be grazed from November to January to reduce the length of the hay-feeding season. One acre of stockpiled bermudagrass will provide grazing for one cow for approximately 45 days.

Tall Fescue

Management of stockpiled tall fescue is similar to that of bermudagrass (Teutsch et al., 2005). The two main differences are the amount of N fertilizer applied and the grazing period. With regard to fertilizer, 75–100 lb of N per acre (112 kg/ha) should be applied by early September to take advantage of late-summer rainfall. As with bermudagrass, fall production is highly dependent on rainfall. Stockpiled tall fescue can be grazed from late December through February, with 1 acre of fall-fertilized tall fescue providing grazing for one cow for approximately 45 days.

Managing Perennial Species

Harrowing or light disking may be necessary to open the sod so that the seed can reach the soil. Tilling the soil surface lightly exposes the soil, ensuring that the seed has much better soil contact, and establishment is quicker for the winter annuals. However, heavy tillage weakens the perennial plants too much, such that spring growth of the perennials the following year is retarded, even after the winter annuals have stopped growing.

Establishing Winter Annual Grasses

Planting in mid-September is preferred, as the warm-season grass is slowing growth and has reduced forage quality. Seeding at this time allows the cool-season species to have good growth in the fall for grazing and maintaining plant vigor over winter. However, there should be adequate soil moisture present before overseeding. In most of the South, there is a low probability of getting enough rain to saturate the dry soil profiles during September. In addition, at least 1 in. of rain is needed after overseeding to support good germination and seedling vigor.

If sufficient rainfall occurs before mid-September, the summer growth of the warm-season perennial will be extended and will need to be suppressed by disking or spraying with an herbicide to minimize competition. If the rain occurs after mid-September, the growth of bermudagrass will not normally be a problem, due to the fact that cooler weather slows the growth (Fig. 7.4).

Establishing Winter Annual Legumes

Winter annual legumes take longer to become established than winter annual grasses, and need more management to ensure a good stand. Correct soil pH and adequate P are key factors for ensuring optimum legume growth. Soil samples should be taken a few months prior to planting the legume, so that soil pH and nutrient deficiencies can be corrected well before planting.

Legume seedlings are more sensitive to shading than are grass seedlings. Vegetative growth of the perennial should be slowed by burning, grazing, haying, or mowing by early September. The grass should be kept less than 2–3 in. tall. This reduces shading of the legume seedlings and decreases water use by the perennial, so more light and water are available for the seedlings. Both of these factors greatly increase the likelihood of success.

The best method of establishing fall-seeded forages is to drill the seed into areas that were recently grazed closely or cut for hay. The second best method is to broadcast seed after the perennial species has been lightly disked. After broadcasting the seed it is beneficial to firm the soil with a cultipacker to improve soil-to-seed contact.

Managing Winter Annual Species

Overseeded forages should not be grazed until the roots are well established, usually when the new grass plants have initiated tillering (i.e., when they are 10–12 in. [25–30 cm] tall or have 5–6 tillers per plant). With good establishment techniques and adequate fall moisture, cereal rye or wheat will provide potential for some late-fall or early-winter grazing. When annual ryegrass, perennial cool-season grasses, or legumes are overseeded, grazing should not be anticipated until spring.

Several factors are extremely important for successful overseeding. The timing of each event is very critical, as there will only be small windows of opportunity. If a soaking rain occurs before early September, there is a strong possibility of both fall and spring grazing from overseeding. However, if it has not rained by mid-October, the plants do not establish early enough to produce growth during early winter, and overseeding may not be a good option.

Annual Crabgrass as the Base

Systems that use a combination of both winter annual cool-season and summer annual warm-season grasses, such as pearl millet, sorghum × sudangrass hybrids, and crabgrass, have also been successful. However, these systems typically require more labor and equipment, as tillage and seeding must be performed on an annual basis. Fertilizer N also

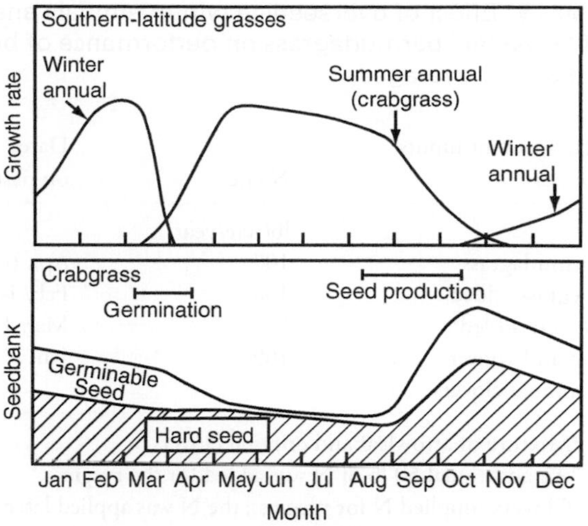

Fig. 7.5. Upper panel: Cereal rye, a winter annual, can be grown in sequence with crabgrass to provide a long grazing season. Lower panel: The cereals do not form a seedbank, but one will develop for crabgrass if it is managed. Grazing the winter annual in spring is important to open the canopy to enable the crabgrass to become established. Annual ryegrass and most legumes are too competitive to have dependable seedling emergence and stand establishment of crabgrass.

needs to be used in larger amounts because available forage legumes are not compatible with crabgrass.

In areas where winter annual grasses such as cereal rye, winter wheat, and annual ryegrass are grown, crabgrass can be an alternative for summer (Fig. 7.5). Crabgrass can be managed as a reseeding summer annual for summer production without the need to replant each year. To facilitate natural reseeding, cattle should be removed from the crabgrass pasture 3–4 weeks prior to the first frost. This will allow seed set to occur and a seedbank to develop. When production of the winter annual ceases the following spring, the soil can be lightly tilled using a disk or harrow to stir the soil slightly to improve soil-to-seed contact for the crabgrass. Crabgrass productivity depends on moisture being available for seed germination and development of a good seedling root system.

In this system, it is best to replant the winter annual grasses each year, as little seedbank will develop for these species. As with perennial grasses, adequate soil moisture and minimized vegetative growth of the crabgrass in early autumn are critical for establishment of the cool-season species and the success of this system. An advantage is the crabgrass seedbank, which reduces costs and management needed to maintain the system.

Table 7.5. Effect of overseeding winter annuals and nitrogen fertilization as inputs to a base pasture of "Coastal" bermudagrass on performance of beef cows and calves in southern Alabama; data are 3-year averages

Management inputs		Dates	Animal gain		
Species	N rate	on pasture	Cows	Calves	Calves
	lb/acre/year		lb/day	lb/day	lb/acre
Bermudagrass	100	Apr. 6 – Oct. 5	0.49	1.59	293
Ryegrass added	150[b]	Feb. 14 – Oct. 5	0.81	1.76	422
Clovers added[a]	0	Mar. 11 – Oct. 5	1.37	1.94	410
Rye and winter annual clovers	100[c]	Jan. 8 – Oct. 5	0.90	1.91	511

[a]Mixture of arrowleaf and crimson clovers.
[b]An additional 50 lb of N was added for the ryegrass.
[c]Clovers supplied N for ryegrass; the N was applied later for bermudagrass.
Source: Data from Ball et al., 1991.

An Example of Progress

In studies on bermudagrass receiving 100 lb of N per year (a low rate), gain of beef calves was 1.59 lb per day and seasonal gain was 293 lb per acre (Table 7.5). Using overseeded legumes to extend the grazing season by nearly 2 months increased average daily gain by 11% and the gain per acre by 44%, but it required more N. Using winter annual legumes extended the grazing season less than for grasses, but daily gain was increased by 22%, indicating the added quality of legumes. Gain per acre was increased by 40%, and the N requirement was met by the legume.

Using a mix of cereal rye and clover had the largest effect, due to the early grazing from the cereal and the advantage of the legume. The N used for the bermudagrass was the same as the control, but calf daily gain for the system was improved by 20% and gain per acre was improved by 74%. Cow performance was also improved. These examples clearly illustrate the way that forage species and management systems interact.

Summary

Perennial warm-season grasses form the primary base for pastures in the southern region of the USA. Primarily temperate conditions offer a long growing season and, by using multiple species, the opportunity exists to provide year-round grazing in this region. Bermudagrass is high yielding, persistent, and well adapted for grazing, and is the most widely used perennial forage grass in most of the southern USA. The cool-season grass tall fescue is also very persistent and tolerant of grazing, and forms the primary base for pastures in the northern part of the region, also called the transition zone.

Numerous other warm- and cool-season annual and perennial grasses are also important components of forage systems across the southern USA. Bahiagrass, dallisgrass, pearl millet, ryegrass, and many other species are versatile components of systems that fit specific soil or climatic conditions or nutritional needs of different livestock species and classes.

Questions

1. Why are warm-season grasses the base of a pasture system in the South, whereas cool-season grasses are the base in the North?
2. What characteristics make bermudagrass better adapted than elephantgrass for grazing?
3. Why is it easier to use winter annual legumes than perennial legumes in pasture systems in the South?
4. What factors affect the nitrogen-use efficiency of grasses? Why is the efficiency higher for C_4 grasses than for C_3 grasses?
5. Why are endophyte-free cultivars of tall fescue preferred in the North, whereas endophyte-infected cultivars are preferred in the South?
6. How are improved bermudagrasses propagated? Why are they propagated in this way?
7. What characteristics of dallisgrass make it a desirable species of perennial forage for the South?
8. Why is it easier to grow legumes with dallisgrass than with bahiagrass?
9. What is ergot, and why is it important with regard to the management of some grasses?
10. Why is vegetation control of the perennial grass more critical when overseeding a winter annual legume than when overseeding a winter annual grass?

References

Ball, DM, CS Hoveland, and GL Lacefield. 1991. Southern Forages. Potash and Phosphate Institute, Atlanta, GA.

Beuselinck, PR, JH Bouton, WO Lamp, AG Matches, MH McCaslin, CJ Nelson, LH Rhodes, CC Sheaffer, and JJ Volenec. 1994. Improving legume persistence in forage crop systems. J. Prod. Agric. 7:311–322.

Burson, BL, and VH Watson. 1995. Bahiagrass, dallisgrass, and other *Paspalum* species. In RF Barnes, DA Miller, and CJ Nelson (eds.), Forages: An Introduction to Grassland Agriculture, 5th ed., pp. 431–440. Iowa State University Press, Ames, IA.

Burton, GW, and WW Hanna. 1995. Bermudagrass. In RF Barnes, DA Miller, and CJ Nelson (eds.), Forages: An Introduction to Grassland Agriculture, 5th ed., pp. 421–429. Iowa State University Press, Ames, IA.

Clason, TR, and SH Sharrow. 2000. Silvopastoral practices. In HE Garrett, WJ Rietvald, and RF Fisher (eds.), North American Agroforestry: An Integrated Science and Practice, pp. 119–147. American Society of Agronomy, Madison, WI.

Hanna, WW, and LE Sollenberger. 2007. Tropical and subtropical grasses. In RF Barnes, CJ Nelson, KJ Moore, and M Collins (eds.), Forages: The Science of Grassland Agriculture, 6th ed., Vol. II, pp. 245–255. Blackwell Publishing, Ames, IA.

Lalman, DL, CM Taliaferro, FM Epplin, CR Johnson, and JS Wheeler. 2000. Review: Grazing stockpiled bermudagrass as an alternative to feeding harvested forage. J. Anim. Sci. 79:1–8.

Moser, LE, and KP Vogel. 1995. Switchgrass, big bluestem, and indiangrass. In RF Barnes, DA Miller, and CJ Nelson (eds.), Forages: An Introduction to Grassland Agriculture, 5th ed., pp. 409–420. Iowa State University Press, Ames, IA.

Ocumpaugh, WR, and LE Sollenberger. 1995. Other grasses for the Humid South. In RF Barnes, DA Miller, and CJ Nelson (eds.), Forages: An Introduction to Grassland Agriculture, 5th ed., pp. 441–449. Iowa State University Press, Ames, IA.

Redfearn, D, and C Rice. 2011. Bermudagrass Pasture Management. Oklahoma Cooperative Extension Service PSS-2591. Division of Agricultural Sciences and Natural Resources, Oklahoma State University, Stillwater, OK.

Redfearn, D, and Y Wu. 2013. Choosing, Establishing, and Managing Bermudagrass Varieties in Oklahoma. Oklahoma Cooperative Extension Service PSS-2263. Division of Agricultural Sciences and Natural Resources, Oklahoma State University, Stillwater, OK.

Redfearn, D, B Arnall, H Zhang, and C Rice. 2010. Fertilizing Bermudagrass Hay and Pasture. Oklahoma Cooperative Extension Service PSS-2263. Division of Agricultural Sciences and Natural Resources, Oklahoma State University, Stillwater, OK.

Sims, PL, and CL Dewald. 1982. Old World Bluestems and Their Forage Potential for the Southern Great Plains: A Review of Early Studies. USDA-ARS Agric. Rev. Man. ARM-S-28. Agricultural Research Service, US Department of Agriculture, Washington, DC.

Snyder, CS, and RH Leep. 2007. Fertilization. In RF Barnes, CJ Nelson, KJ Moore, and M Collins (eds.), Forages: The Science of Grassland Agriculture, 6thed., Vol. II, pp. 355–377. Blackwell Publishing, Ames, IA.

Sollenberger, LE, GM Prine, WR Ocumpaugh, WW Hanna, CS Jones Jr., SC Schank, and RS Kalmbacher. 1988. 'Mott' Dwarf Elephantgrass: A High Quality Forage for the Subtropics and Tropics. Univ. Fla. Agric. Exp. Stn. Circ. S-356. Agricultural Experiment Station, University of Florida, Gainsville, FL.

Teutsch, CD, JH Fike, GE Groover, and S Aref. 2005. Nitrogen rate and source effects on the yield and nutritive value of tall fescue stockpiled for winter grazing. Forage and Grazinglands. doi:10.1094/FG-2005-1220-01-RS.

Voigt, PW, and WC Sharp. 1995. Grasses of the Plains and Southwest. In RF Barnes, DA Miller, and CJ Nelson (eds.), Forages: An Introduction to Grassland Agriculture, 5th ed., pp. 395–408. Iowa State University Press, Ames, IA.

CHAPTER

8

Legumes for Northern Areas

Craig C. Sheaffer, M. Scott Wells, and C. Jerry Nelson

Forage legumes contribute significantly to agriculture in the northern USA. They provide forage that is a nutrient and fiber source for a diversity of livestock. Legumes also fix atmospheric nitrogen (N) that can be a substitute for fertilizer N, and provide carbon and organic matter to the soil. In addition, perennial legumes such as alfalfa conserve soil by providing year-round ground cover. Most legumes are less tolerant of harsh soil and climatic environments than are cool-season grasses, and require a higher level of fertility and harvest management, but the savings in fertilizer N costs and the superior feeding value make legumes valuable components of most forage systems.

This chapter focuses on important legumes that are harvested for **hay** or **silage**, grazed in **pastures** and **rangelands**, or used as winter cover crops in the northern USA. Most economically significant legumes are perennial, persisting for 3 years or more. Alfalfa, harvested for hay and haylage, and white clover, grazed in **permanent pasture** and **rotational pasture**, are the most widely grown legumes. Other perennials such as birdsfoot trefoil, red clover, alsike clover, crownvetch, sainfoin, cicer milkvetch, and kura clover occupy fewer acres and are adapted to specific uses and environments. Biennials such as yellow and white sweetclover, and some annual reseeding legumes such as hairy vetch, and striate and korean lespedezas occupy fewer acres, but are useful for specific environments or production goals. Several other legumes exist in pastures and rangelands of the northern USA or are grown for specific purposes. Some of these, such as annual clovers, are important in the South and their use extends northward into the transition zone.

Description, Adaptation, and Distribution

Legumes used in northern regions have distinct morphological features similar to those of other members of the legume family. These include compound leaves, showy flowers, seeds borne in pods, and the presence of N-fixing nodules on roots. Perennial, annual, and biennial legumes are used for forage in the north. Perennials are the economically most important group. Perennials such as alfalfa live for 3 years or more and can set seed each year. Annual plants such as hairy vetch and berseem clover germinate, flower, set seed, and then die within one growing season. Hairy vetch, when used as a winter annual cover crop, is planted in the late summer, overwinters, and flowers the following spring. Sweetclover, the sole biennial legume, produces only stems and leaves in the first year, and flowers and dies in the second year. The lifespan of individual plants in a perennial stand may be from one to several years, but stands can persist through vegetative reproduction and natural reseeding (Table 8.1). Crown formers, such as alfalfa and red clover, have well-developed **crowns** with many **axillary buds** for regrowth; they depend largely on individual plant survival for **perennation**. Clone formers, such as white clover and kura clover, establish new plants by vegetative reproduction via stolons and **rhizomes**, respectively, to ultimately replace the parent plants. Annual legumes must be reseeded each year or depend on natural reseeding. Most crown and clone formers also reseed naturally if managed with that objective. Natural reseeding is very common in white clover and birdsfoot trefoil. Compared with most perennial grasses, **stand persistence** of perennial legumes is less dependable and

Forages: An Introduction to Grassland Agriculture, Seventh Edition. Edited by Michael Collins, C. Jerry Nelson, Kenneth J. Moore and Robert F Barnes.

Table 8.1. Description of plant lifespan, growth habit, and persistence strategies of forage legumes

Legume	Plant lifespan[a]	Persistence strategy	Stem origin	Growth habit
Alfalfa	Long-lived perennial	Crown former	Crown bud	Upright
Red clover	Short-lived perennial	Crown former, reseeder	Crown bud	Upright
White clover	Long-lived perennial	Clone former, reseeder	Stolon bud	Prostrate
Birdsfoot trefoil	Short-lived perennial	Crown former, reseeder	Crown bud	Decumbent
Alsike clover	Short-lived perennial	Crown former, reseeder	Crown bud	Upright
Kura clover	Long-lived perennial	Clone former	Rhizome bud	Plastic[b]
Crownvetch	Long-lived perennial	Clone former	Buds from roots	Decumbent
Sainfoin	Long-lived perennial	Crown former	Crown bud	Upright
Cicer milkvetch	Long-lived perennial	Crown former, clone former	Rhizome bud	Decumbent
Sweetclover	True biennial	Crown former, reseeder	Crown bud	Upright
Annual lespedeza	Summer annuals	Reseeder	Buds from leaf axils	Upright

[a]Plant lifespan is the relative longevity of an individual plant of a crown former or a stem segment of a clone former.
[b]Depends on grazing pressure; it is upright with rotational stocking, and more decumbent with close grazing and continuous stocking.

is an important management objective (Beuselinck et al., 1994).

Winter survival is an important issue that affects production and profitability of perennial legumes in the north. Perennials naturally prepare for winter by becoming dormant in the fall. Dormancy is promoted by short days and low temperatures in the fall. However, because of their morphological and physiological features, perennials vary greatly in winter hardiness and winter survival. In addition to cold tolerance per se, the most winter-hardy plants remain dormant from late fall to spring. Less winter-hardy plants will break dormancy during periods of warm weather in the winter and early spring, and are then susceptible to winterkill due to resumption of cold temperatures. There are many environmental factors, such as depth of snow cover and temperature extremes, that interact with legume genetics and morphology to affect legume winter survival (see Chapter 5).

All legumes of agronomic significance were introduced from other parts of the world. For example, alfalfa originated in central Asia, red clover in the Middle East and southeastern Europe, and white clover and birdsfoot trefoil in the Mediterranean region. Legumes native to North America, such as purple prairie clover and leadplant, are an important component of prairies and rangelands, and are seeded in prairie restoration, but are not a primary source of forage in **managed agroecosystems**. Legumes of economic importance have been improved by plant breeding. Selected traits vary depending on the plant life cycle, but include forage yield, adaptation to environmental stress, and insect and disease resistance.

Alfalfa

Alfalfa is the economically most important perennial forage legume. It is harvested for hay and silage and sometimes grazed. Alfalfa is grown in pure stands and in mixtures with regionally adapted grasses. It has a pronounced crown that is the source of new upright stems in the spring and following cutting. The stems vary in number from 10–20 per plant, and can be up to 2.9 ft (90 cm) tall. Stem numbers vary with plant populations. Alfalfa also has a prominent **taproot** that can be extensively branched and grows several feet deep into the soil (Table 8.1). Some alfalfa types have stems originating from rhizomes and adventitious roots. Flowers are borne in an elongated raceme inflorescence and are typically purple, but can also be white, yellow, green, or blue (Fig. 8.1). Alfalfa leaves typically contain three leaflets in a pinnately compound arrangement; however, specialized multifoliolate alfalfa varieties with 4–5 leaflets per leaf are marketed.

Region-specific **cultivars** with high yield and adaptation features for **persistence**, such as resistance to specific diseases and insects, and with appropriate cold hardiness have

Fig. 8.1. Flowering alfalfa. (Photo courtesy of David L Hansen, University of Minnesota.)

Table 8.2. Adaptation characteristics, yield, and bloat-causing potential of forage legumes

Legume	Tolerance of low soil fertility	Tolerance of soil acidity	Tolerance of drought	Tolerance of poor drainage	Cold hardiness	Yield potential	Bloat potential
Alfalfa	Poor	Sensitive	Excellent	Poor	Excellent	Very high	Yes
Red clover	Fair	Moderate	Good	Fair	Very good	High	Yes
White clover	Fair	Moderate	Poor	Good	Very good	Medium	Yes
Birdsfoot trefoil	Good	Good	Good	Good	Good	Medium	No
Alsike clover	Fair	Moderate	Fair	Excellent	Very good	Medium	Yes
Kura clover	Fair	Moderate	Good	Good	Excellent	Low	Yes
Crownvetch	Good	Moderate	Good	Fair	Good	Medium	No
Sainfoin	Good	Low	Excellent	Poor	Very good	Medium	No
Cicer milkvetch	Good	Moderate	Very good	Good	Excellent	High	No
Sweetclover	Good	Sensitive	Excellent	Poor	Excellent	High	Yes
Annual lespedeza	Excellent	Tolerant	Good	Good	None	Low	No

been developed. Cultivars with low-set crowns have also been developed for grazing (Brummer and Bouton, 1992). This provides a range of adapted cultivars that allows alfalfa to be grown throughout the USA, northern Mexico, and southern Canada.

Genetically modified cultivars with resistance to the herbicide **glyphosate** have been developed to facilitate weed control. Some cultivars have been genetically modified to have 12–18% lower levels of the **anti-quality factor** lignin. Reduction in lignin content improves the digestibility of the forage and provides producers with harvest-timing flexibility (see Chapter 14).

Alfalfa performs best on well-drained soils that allow oxygen (O_2) movement to the roots and that reduce disease problems (Table 8.2). It does not tolerate wet soils. The large, long taproot extracts water from a large soil volume compared with shallower-rooted legumes, so it does not become drought stressed as easily.

Forage yields and quality at any harvest are affected by stage of development, which is described based on development from vegetative to floral stages. In regrowth following dormancy or harvest, alfalfa yields increase to early flowering and then decline due to leaf loss. Depending on the climatic conditions within a northern region, alfalfa will have from two to as many as five harvests per season, with total seasonal yields in the range of 3–10 tons/acre (6.7–22.4 mt/ha). Alfalfa is typically harvested from the bud stage to the flowering stage. Stand productivity is typically greatest in the first 2 years following seeding, and then declines due to disease and winter injury. Consequently, stands are typically maintained for 3–4 years following seeding in the north before reseeding is necessary. When grown in mixtures with perennial grasses, stands are often maintained for more than 4 years; however, the contribution of alfalfa to yield will decline greatly over time. Mixtures with perennial grasses are used to reduce the incidence of **bloat** in grazing ruminants (see Chapter 16).

Alfalfa seed planted throughout the USA is mostly produced in California, Washington, Idaho, and Oregon. This is a specialized process requiring pollination by bees, irrigation, and timely harvests to maximize seed yield.

Medics are yellow-flowered, prostrate annuals related to alfalfa. In winter rainfall regions of the USA, such as northern California, medics are used for winter forage, and they have shown potential for use as summer annuals in the north. These include burr medic, barrel medic, and snail medic. In Minnesota, spring-seeded annual medics produced up to 2.2 ton/acre (4.9 mt/ha) of forage and fixed over 90 lb/acre (100 kg/ha) of N (Zhu et al., 1998). Black medic frequently volunteers in **permanent pastures**. This species has been grown as a replacement for summer fallow in moisture-limited regions of the central Great Plains.

Alfalfa Roots in Time

Alfalfa probably originated in central Asia, north of present-day Iraq and Iran, in a region with cold winters and hot dry summers (Michaud et al., 1988). Its use as a livestock feed dates back to as early as 1400 BC, by the Hittites, in present-day Turkey. It was subsequently used for livestock feed by armies of Medians, Greeks, and Romans. The use of alfalfa for horse and livestock feeding by conquering armies spread its cultivation throughout Europe and into China.

Alfalfa was brought to the Americas first by the Spanish in the 1500s and later by North American colonists in the 1700s. Many of these early alfalfas were not well adapted to northern US climates. One of the first winter-hardy alfalfas, "Grimm" alfalfa, was selected by Wendelin Grimm, a farmer who brought alfalfa to Minnesota from Germany. Other winter-hardy alfalfas were brought to the USA by plant explorers such as Niels Hansen, a USDA plant explorer. Early alfalfas lacked the necessary disease resistance to allow them to persist under frequent cutting management, and to survive. In the early years after its introduction, alfalfa was often cut only twice per growing season. Beginning in the 1940s, alfalfa breeding programs of the USDA-ARS, state universities, and private companies developed varieties with resistance to important diseases (e.g., bacterial wilt), insects (e.g., potato leafhopper), and nematodes. Efforts to improve the crop have now led to faster regrowth following cutting, increased winter survival, and improved **forage quality**. Today, alfalfa is grown on over 5 million acres in the northeast and Midwest regions of the USA.

White Clover

White clover is the most widely distributed true clover (*Trifolium* species). It has palmately trifoliate leaves, and is a short-lived perennial that is used mostly as a pasture legume in mixtures with grasses. Seedlings have a short taproot, a short main stem, and prostrate stolons (horizontal aboveground stems) that radiate from the crown of the plant and produce adventitious roots at the nodes (Fig. 8.2). White clover derives its name from its white inflorescences or blossoms that are born on long **peduncles** originating from stolons (Fig. 8.3). The taproot and main stem die after 1 or 2 years, leaving a network of disconnected stolons with shallow roots at the nodes. Perennial stands are maintained from plants generated from buds on stolons and from volunteer seedings generated from shattered seed (Fig. 8.4). Plants produce many **hard seed** that can remain viable in the soil for up to 30 years. These perennation strategies cause white clover to "leave" during years with dry weather and "come back" during wet years.

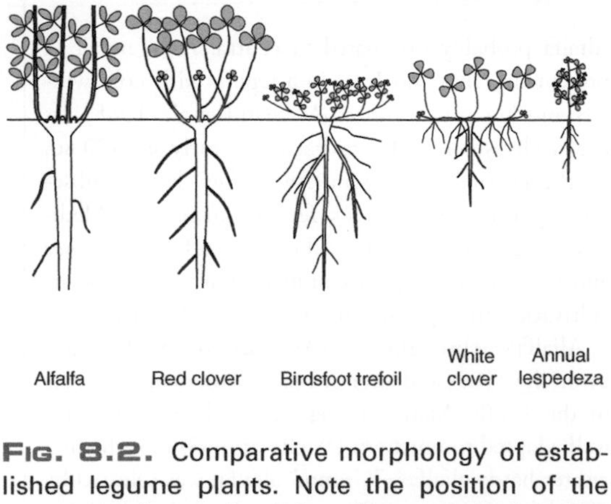

Fig. 8.2. Comparative morphology of established legume plants. Note the position of the crown relative to the soil surface and the location of the shoot apices where the young leaves are developing. Shoot apices of alfalfa are not shown; like those of annual lespedeza, they are at the top of the canopy.

Fig. 8.3. White clover. The individual flowers of white clover are arranged together in a round infloresence. (Photo courtesy of David L Hansen, University of Minnesota.)

There are three types of white clover—small, intermediate, and large—that can interpollinate. The small type is very tolerant of defoliation and is the most winter hardy (Sheaffer et al., 2003). It occurs in many **naturalized** pastures and home lawns, but it has low yield so is rarely sown (Table 8.1). Plants in these naturalized populations have gradually shifted genetically to be able to survive freezing temperatures in the North or hot summers in the South. The large type has the highest yield but is less winter hardy and does not persist as well as the shorter types. It is used in managed intensive grazing systems. White clover yields vary with plant type, but for large types are similar to red clover. White clover has very high forage quality because of its leafiness, but it can cause bloat.

White clover is a popular pasture legume due to its good grazing tolerance. However, its use as a sole forage source can cause bloat in ruminants. Having 50% or more grass in the mixture significantly reduces the probability of bloat. White clover is compatible with most common cool-season grass species. Leaves of white clover contain low

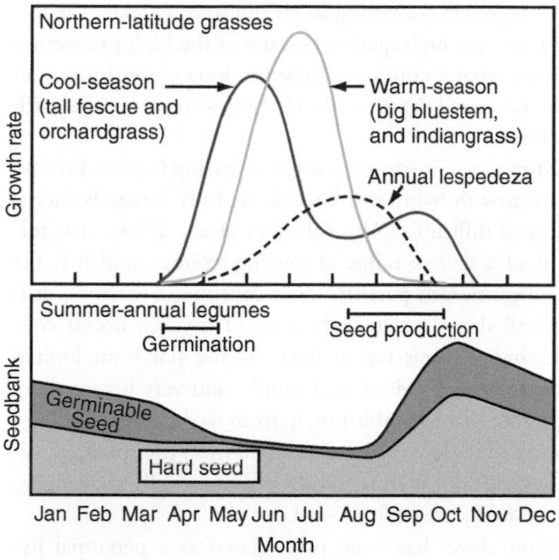

Fig. 8.4. Upper panel: The grass canopy needs to be managed in spring to encourage survival of germinated seed, and in autumn for flowering and seed development. Lower panel: Maintaining a seedbank is important for annual and other short-lived legumes. Seed of annual lespedeza are produced in autumn to increase the size of the seedbank of both hard and germinable seed. The seed provide a source of food for wildlife over winter, and some hard seed become germinable. The seedbank is reduced when seed germinate in spring. (Adapted from Beuselinck et al., 1994.)

amounts of cyanogenic glucosides and an estrogen, coumestrol, but these chemicals are not known to reduce animal production.

White clover grows best at cooler times of the year and on moist soils. It is not productive in droughty soils or poorly drained soils. It tolerates a soil pH in the range 4.5–8.0, but it gives optimal yields and persists best in soils with a pH range of 6.0–6.5.

Red Clover

Red clover is an upright-growing true clover with stems that terminate in a red inflorescence (Fig. 8.5). Its stems and leaves are pubescent (hairy), with green leaves usually distinctly marked with a white inverted "V", and it has a prominent taproot that is more extensively branched than that of alfalfa. Red clover is primarily used for hay and silage production, but is also used in mixtures with perennial grasses for pasture. Most modern varieties are the "medium" or multiple harvest type. A late-blooming type, referred to as mammoth red clover, is used in more northern climates and produces a single hay harvest annually.

Fig. 8.5. Red clover. (Photo courtesy of David L Hansen, University of Minnesota.)

Changes in red clover yield and quality have a similar relationship to changes in crop maturity as in alfalfa. However, long-term yields of red clover are substantially less than those of alfalfa, and because red clover is less persistent then alfalfa, it is typically harvested less each growing season. Red clover forage is wetter and harder to dry than alfalfa forage.

Red clover is most productive when it is growing in relatively cool temperatures during spring and fall. It is susceptible to crown- and root-rotting disease complexes, especially in regions with warm summer temperatures. Plants usually die 2–3 years after establishment due to winter injury or disease. In pastures, red clover stands can be maintained for several years by allowing natural reseeding to form a seed bank, or by **overseeding** every 2–3 years in late winter or early spring (see Chapter 11). Red clover has relatively large seed and good seedling vigor, making it one of the easiest forage legumes to establish.

Red clover can cause ruminant bloat when grazed. Reproductive problems for animals on red clover pastures have been associated with the estrogenic activity of the isoflavones, especially formononetin, biochanin A, and to a lesser extent daidzein and genistein. Grazing red clover can cause excessive slobbering in cattle, and especially in horses, due to the presence of a fungal toxin.

Red clover prefers well-drained soils with a high moisture-holding capacity. It is less sensitive to soil acidity than alfalfa, being productive at pH 6.0 or above (Table 8.2). Therefore it is often used as a hay substitute for alfalfa in regions with acidic soils.

Birdsfoot Trefoil

Birdsfoot trefoil has bright yellow flowers born in an umbel inflorescence (Fig. 8.6). The name "birdsfoot" is derived from the claw-like arrangement of seed pods that develop following pollination of the flowers. The plant has semi-prostrate stems that develop from the crown and from

FIG. 8.6. Birdsfoot trefoil flowers and seed pod arrangement that resembles a bird's foot. (Photo courtesy of David L Hansen, University of Minnesota.)

axillary buds. Birdsfoot trefoil has a well-developed taproot that is even more extensively branched than that of red clover (Fig. 8.2). It is a winter-hardy perennial, and some cultivars are adapted well into Canada. In northern areas, individual plants can survive for 3–5 years if infrequently cut. However, in warmer areas such as Missouri, the plants die due to root- and crown- rotting diseases during summer, and 90% of the plants die within 2 years (Beuselinck et al., 1994). In these areas, perennation depends on natural reseeding. Plants store only small amounts of reserves in the roots during summer, so regrowth is dependent on residual leaf area and active axillary buds on the stems.

Birdsfoot trefoil is most frequently used in pasture systems in mixtures with perennial grasses, but can be harvested for hay. It is well adapted for pasture if not grazed closely and if the plants are managed to produce seed. It can be "stockpiled" in place while retaining forage yield and quality because it retains its leaves at maturity. It does not cause bloat like clovers and alfalfa because of the presence of tannins in its forage.

Birdsfoot trefoil has been planted along roadsides because of its value as a soil stabilizer. However, because of its prolific seed production there is concern about its volunteer spread into natural areas.

Birdsfoot trefoil is unique as a legume in that it is tolerant of waterlogged soils and can survive prolonged flooding. It is also more tolerant of acid soils (pH 5.0) compared with most clovers and alfalfa.

Kura Clover

Kura clover is a long-lived perennial clover (Table 8.1). It has a deep crown and taproot, and produces rhizomes that give rise to new plants and enable it to spread vegetatively. In the spring it produces **inflorescences** that are grouped into large pink heads, borne on stems that after harvest show regrowth consisting solely of leaves. Overall, the forage is of very high quality because of the high proportion of leaves that it contains. However, kura clover forage can cause bloat and therefore should be grown in mixtures with grasses.

Kura clover is mostly used as a grazing legume because of its growth habit and because its leafy forage is succulent and difficult to dry (Sheaffer et al., 2003). It is tolerant of a diverse range of grazing regimes, and it is the only legume that persists under continuous stocking. It is also called honey clover because of its high nectar content that is attractive to bees. Its use has been limited because of the limited seed supply, and very low seedling vigor that makes establishment from seed challenging. Productive fields have been established from transplanted rhizomes, but this is more costly and logistically challenging than establishment from seed.

Kura clover has been investigated as a perennial living mulch for providing both forage and perennial cover (Zemenchik et al., 2000) (Fig. 8.7). In this system, established stands of kura clover are suppressed using herbicides in the spring, and corn is interseeded. The kura clover supplies some of the nitrogen required by the corn (Seguin et al., 2000). After the corn has been harvested the kura clover regrows, and by the following season it can be used as a pasture.

Kura clover has excellent tolerance of stressful climatic conditions and defoliation strategies, and is the only perennial clover that has consistently tolerated extreme winter conditions in the north. It tolerates soils that have low

FIG. 8.7. Later summer corn growing in a kura clover living mulch system. Corn is seeded into an established stand of kura clover in the spring using no-till seeder and herbicides for suppression of the kura clover. The kura clover regrows to suppress weeds and provide year-round ground cover. (Photo courtesy of Ken Albrecht, University of Wisconsin.)

fertility or are poorly drained, and it survives hot and dry periods by going dormant (Table 8.2). It has been used on a limited scale in the North where summer moisture is more abundant.

Alsike Clover

Alsike clover is a short-lived, crown-forming perennial with a small taproot (Table 8.1). It has weak upright stems that produce white or pink flowers in a small round inflorescence. Its flowering habit is **indeterminate**, and the stems tend to lodge or fall over with crop maturity. It is used mainly for pasture in the Great Lakes region and for pasture and hay in parts of the Pacific Northwest. It is not sown in monoculture, but is usually grown as a component of a mixture with grasses and other legumes. Alsike clover will reseed itself if harvest is delayed to allow seeds to develop.

Alsike clover is adapted to cooler, wetter climates than red clover, and can tolerate waterlogged, acidic soils better than other clovers (Table 8.2). It has poor drought tolerance.

Annual Clovers

There are several annual clovers, such as crimson clover, berseem clover, and arrowleaf clover, that are commonly grown as winter annuals in the southern USA, and that are sometimes grown as emergency summer annuals or as **green manures** for producing forage or N for subsequent crops in rotation. They can also be sown in late summer and used as forage and cover crops in the fall. Growth of annual clovers sown in the spring is limited by summer temperatures, and when sown in later summer, by fall frosts.

Crownvetch

Crownvetch is a perennial that spreads by means of **adventitious buds** that arise from lateral roots (Table 8.1). It has trailing prostrate stems that produce pinnately compound leaves with multiple leaflets. Its white-pink flowers are borne in a crown-like, umbel arrangement.

Crownvetch is a popular legume for planting in disturbed areas such as roadsides, and for revegetation of strip mines. Crownvetch is not widely used as a forage crop because of its intolerance of frequent defoliation and its lack of **palatability**, which is caused by anti-quality components, primarily the glucose esters of 3-nitropropanoic acid (NPA). It can be harvested as hay or grazed, but is less tolerant of frequent harvesting than alfalfa. Crownvetch is best adapted to well-drained, fertile soils with a pH higher than 6.0, but will tolerate somewhat acid and infertile soils (Table 8.2).

Sainfoin

Sainfoin is a perennial that is used mainly in the western USA, where it is adapted to dry calcareous soils (i.e., with high calcium levels and high pH). It is also adapted to soils low in P. It is less persistent than alfalfa, and provides perennial cover in humid regions (Sheaffer et al., 2003). Nitrogen may be limiting because of poor nodulation and inadequate N_2 fixation. Sainfoin has deep, branched taproots and hollow stems that originate from a crown. The hay yield from sainfoin is less than that from alfalfa, but forage quality and animal performance are similar for the two species. Sainfoin does not cause bloat and has no major insect pests. It is popular for early-season pasture, but it does not tolerate close and continuous defoliation.

Cicer Milkvetch

This very winter-hardy legume is produced mainly in areas of lower rainfall in the Great Plains and western USA, and in adjacent areas of Canada. Cicer milkvetch is a long-lived perennial, has a branched taproot, and spreads by means of rhizomes. It is adapted to dry areas but also grows well in humid areas or when irrigated. Cicer milkvetch tolerates a range of soil pH, from slightly acid to moderately alkaline (Table 8.2). Yield is about 75% of that for alfalfa under similar growing conditions. Forage composition and quality are similar to those for alfalfa, but low palatability and intake often limit animal performance. Cicer milkvetch does not cause bloat, and it tolerates grazing if sufficient leaf area is maintained to support regrowth. As in birdsfoot trefoil, little carbohydrate is stored in the roots and rhizomes during summer. Grazing cicer milkvetch can cause skin sensitivity of animals to solar radiation, which suggests the possibility that the plant may contain a phytotoxin that causes **photosensitization** of animals.

Sweetclover

Sweetclovers have woody stems that grow up to 4 ft tall. Yellow or white flowers are borne on elongated raceme inflorescences. Sweetclovers are not true clovers but have a pinnately compound leaf arrangement similar to that of alfalfa. Sweet clover is a biennial with single-stemmed, vegetative growth in the seeding year and multiple stems from crown buds in the second year. If left uncut, plants flower in early summer of the second year, produce seed (much of which is hard), and then die. The name "sweetclover" is derived from the presence of **coumarin** in the forage, which when cut emits a sweet vanilla-like odor, and from the use of the plant as a honey crop. Sweetclovers are adapted to low-rainfall areas mainly in the Great Plains region. Sweetclover has excellent drought tolerance and winter hardiness, and is best adapted to soils with a pH higher than 6.5. Most cultivars are true biennials, but annual cultivars are also available. All cultivars depend on seed production for long-term stand maintenance. Sweetclovers have deep taproots and the crop is very drought tolerant.

Sweetclover was a traditional green manure crop that was plowed down to add N in crop rotations, but it is currently not an important forage crop and is seldom sown.

In many of the humid northern areas, sweetclover still volunteers on roadsides and wasteland. In drier regions, the sweetclovers volunteer with grasses for pasture. When used for grazing, optimum quality can be achieved when plants are about 24 in. in height, before the stems become too woody. However, the sweet smell of coumarin is misleading because its taste is bitter and its presence reduces the palatability of the forage. Sweetclover can be cut for hay; first-year plants should be cut or grazed to leave a 10 in. stubble, as the regrowth comes from axillary buds located along the main stem. Making sweetclover hay can be challenging because the forage is very moist, dries slowly, and is prone to molding. During molding of sweetclover hay or silage, coumarin is converted to dicoumarol, which interferes with blood clotting and can cause livestock death from internal bleeding.

From Coumarin to Warfarin

When extensive acreages of sweetclover existed, it was known that cattle which consumed moldy sweetclover hay could bleed excessively, or even die, if they were scratched or injured. The cause was traced to coumarin, a chemical in the plant that is converted to dicoumarol. Dicoumarol thins the blood and interferes with blood clotting. With high levels of dicoumarol, the blood thins so much that it oozes out of the capillaries and causes death by internal bleeding.

A biochemist at the University of Wisconsin evaluated dicoumarol as a means of controlling mouse and rat populations. He mixed dicoumarol from moldy sweetclover with a cereal food source and placed it in rodent-infested areas. It was extremely effective, and a patent was processed through the Wisconsin Alumni Research Foundation. The patented compound was named warfarin, and several decades later it is still being used extensively to control rodents.

Warfarin is now manufactured using laboratory processes. It is used in smaller doses to thin the blood and thereby reduce the risk of clot formation in heart-attack victims, and it also has many other medical uses.

Would you call warfarin a "rat poison"? Did you catch the origin of the name?

The Annual Lespedezas

Korean and striate lespedezas are warm-season annual legumes that are usually grown with cool-season grasses for pasture. Annual lespedezas are grown in the transition zone between the northern and southern USA. The western boundary is limited by lack of moisture and the northern boundary is limited by the plant's inability to flower and form mature seed before a killing frost. The growth habits of korean and striate lespedeza are similar—the plants have a shallow taproot, are fine stemmed, and are most productive during summer when cool-season grasses grow slowly.

Cultivars of korean lespedeza tend to mature earlier than those of striate lespedeza, and are grown in the upper two-thirds of the lespedeza region. Striate lespedeza matures later and is more important in the southern part of the region. Korean lespedeza can be distinguished from striate lespedeza by its broader leaflets, larger stipules, and stem pubescence that is appressed upward. Annual lespedezas grow better than other legumes on eroded, acidic, and low-P soils (Table 8.2), due in part to low competition from other species.

Stands of annual lespedeza perennate by means of seed production in autumn and establishment of new seedlings each spring. Plants should not be grazed or cut from early September until after a killing frost, to allow for adequate flower and seed production. A seedbank gradually develops to perpetuate the stand even when seed production has been low.

Hairy Vetch

Hairy vetch is one of over 150 *Vicia* species that are grown worldwide. Vetches typically have trailing viny stems that can be up to 6.6 ft (2 m) long. Their leaves are pinnately compound with 8–10 leaflets terminating in a tendril that allows the plant to attach to and climb associated plants grown in mixtures. The flowers, which are born in clusters of 10–12, are purple and showy. Vetches are adapted to a diverse range of soils. In the north, hairy vetch can be grown as a summer or winter annual. When grown as a winter annual cover crop for soil conservation and green manure, hairy vetch is sown in late summer, grows in the fall and early spring, and is incorporated prior to crop planting (Clark, 2007). Hairy vetch has sufficient winter hardiness to overwinter in most northern regions, but during winters with extreme cold and lack of snow cover it can suffer winter injury. It can also be used as a forage crop.

Food for Bees and Honey for Humans

Clovers have long been important sources of nectar for pollinators. Nectar, the liquid secreted by flowers, is collected by pollinators and taken to hives for conversion to honey, which is then used as an energy source for the hive. The sweetness of honey is due to the high glucose and fructose content of the nectar. Legumes vary in their nectar production and in the quality of the honey they produce.

Clover honey is light and mild and is preferred by some consumers. Populations of honey bees and other pollinators are in decline, and greater use of legumes on landscapes will provide an increased food supply for bees. Bees also aid legume seed production by ensuring flower pollination (Fig. 8.8).

Fig. 8.8. Honey bee visiting individual flowers on a white clover blossom. (Photo courtesy of Michael Collins, University of Missouri.)

Soil Fertility Requirements

Legumes vary in their tolerance of poor soil fertility, and some can persist in soils with low levels of **nutrients** and below neutral pH. To achieve maximum legume yield and persistence, these plants often benefit from applications of **macronutrients** such as phosphorus (P) and potassium (K), and micronutrients such as sulfur (S) and boron (B). A soil pH of 6.5–7.0 is usually required for maximum forage yield. Fertilizer nutrient application is based on soil nutrient status and crop nutrient removal. For example, for alfalfa that typically contains about 1.5% K and 0.2% P, removal of each ton of hay also removes 60 lb of K and 14 lb of P per ton of forage removed. Under conditions of adequate biological dinitrogen fixation, legumes seldom respond to N fertilizers, and their use is not recommended. Fertilizers that are added to meet nutrient needs and lime to adjust soil pH are typically applied prior to seeding and incorporated into the soil to allow mixing throughout the root zone.

Legume–Grass Mixtures

For alfalfa production systems that rely on herbicides for weed control during establishment, alfalfa is grown in monoculture. However, alfalfa and other legumes are often grown in mixtures with perennial grasses, especially for grazing (Sheaffer et al., 2009). Diversification of functional groups of forage species within a field is an effective approach for enhancing resource utilization and ensuring productivity in the event that one of the species fails to perform or dies. Inclusion of grasses with legumes reduces weed invasion, increases the drying rate of cut forage, decreases the incidence of legume bloat in pasture, and reduces the incidence of winter injury to legume stands. Some upright grasses also reduce the incidence of lodging of weak-stemmed legumes such as alsike clover and birdsfoot trefoil. Achieving the correct balance of legume and grasses in mixtures is challenging and becomes more difficult as the stand ages. Legumes vary in their ability to compete with grasses in mixtures. Red clover is noted for its high seedling vigor and competitiveness. To minimize negative competition effects on the legume, grasses should be selected that mature with or after the legume, and harvest should be timed to favor the legume component. The recommended seeding rates for legumes and grasses grown in mixtures vary according to the region.

Nitrogen Fixation

Forage legumes can form a symbiotic relationship with bacteria that are capable of biological dinitrogen (N_2) fixation—that is, the conversion of inert atmospheric nitrogen into nutritionally valuable plant protein for growth and development. Biological N_2 fixation involves a symbiotic partnership between the legume plant and a host specific rhizobium bacterium that is either present in the soil or supplied via inoculum to the seed. Rhizobium bacteria invade the roots of legumes and form nodules that are the site of conversion of N from the soil air into protein (Fig. 8.9).

This symbiotic relationship is facilitated by having healthy plants and adequate soil pH and soil aeration to facilitate rhizobium growth. However, legumes typically utilize some available soil N when it is present in addition to that produced by fixation. Thus less N_2 fixation occurs in legumes on soils with high N availability compared with soils that have low levels of N.

Alfalfa has among the highest rates of N_2 fixation for forage legumes, which is consistent with its high forage yield (Table 8.3). The rates for red clover, white clover, and birdsfoot trefoil are usually lower than that for alfalfa. Fixation of N_2 is dependent on photosynthesis of the legume plant (see Chapter 4), so less fixation occurs under conditions of climatic stress, such as high or low temperatures and drought stress, soil nutrient limitations, or when grasses or weeds occupy space and shade some leaves.

In addition to independently providing N that is required for their own growth, legumes can also supply N to grasses grown in association, and to subsequent crops in rotations. For example, for alfalfa that is often grown in rotation with corn, alfalfa can supply the equivalent of 100 lb N/acre (112 kg/ha) for a subsequent corn crop. An additional 50 lb/acre (56 kg/ha) may carry over to the second year (Yost et al., 2015). Legumes can also provide N to grasses grown in association with them. In a mixed pasture composed of about 30% legumes, the legumes provide up to 50% of the N used by the grasses (Heichel and

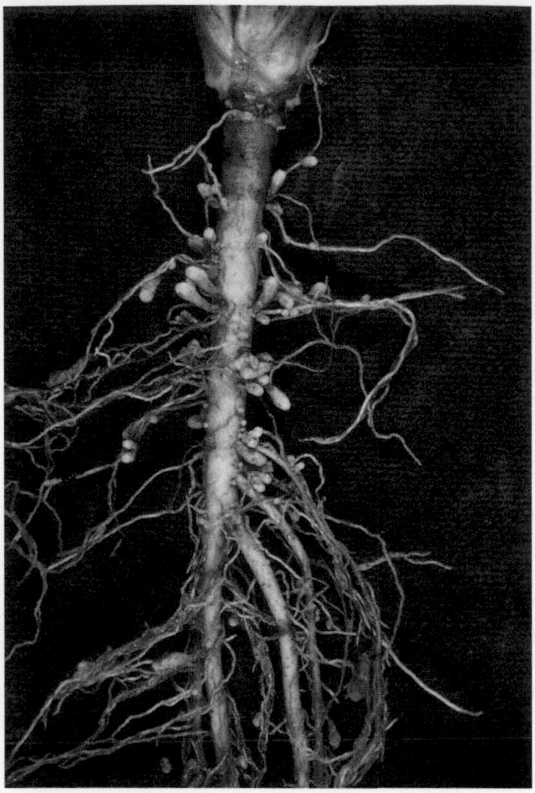

FIG. 8.9. Nodules on lateral root of red clover formed by rhizobium bacteria. (Photo courtesy of Michael Collins, University of Missouri.)

Henjum, 1991). Nitrogen fertilizer is rarely required for properly nodulated legumes.

Insect and Disease Problems

Many diseases and insect pests adversely affect forage legumes. For alfalfa, the most economically valuable legume, disease-resistant cultivars have been developed and seed treatments are used to protect against fungi (Samac et al., 2015). Fungicides have also been shown to reduce the incidence of foliar diseases. For other legumes with less economic value, efforts have not been made to protect plants against diseases. Helpful cultural practices include maintaining good soil fertility, and using proper harvest or grazing management (Marten et al., 1989).

Alfalfa weevil (*Hypera postica* Gyll.) and potato leafhopper (*Empoasca fabae* Harris) are two important insect pests of alfalfa (see Chapter 13), which affect forage yield and quality. They are controlled by insecticides, and potato leafhopper-resistant varieties have been developed.

Glandular Hairs Deter Leafhopper

The nymph stage of the potato leafhopper inserts its stylet into an alfalfa leaf and feeds on the proteins and sugars within the cells. However, the nymph also inserts chemicals that limit plant growth and that may plug the vascular tissue, thus reducing yield. Accurate methods of evaluating leafhopper populations, as well as effective insecticides, are available.

Some alfalfa germplasms have glandular hairs on the leaves (actually visible extensions of the epidermal cells) that exude a sticky mixture of substances from near their tip. The hairs do not reduce animal acceptance,

Table 8.3. Amount of N fixation for forage legumes and forage legume–grass communities during a growing season

Species	Total N_2 fixation (lb N/acre)	Location
Alfalfa		
Grown alone	189	Lexington, KY
Grown alone	102–200	Rosemount, MN
With orchardgrass mixture	13–121	Lucas County, IA
With reed canarygrass mixture	73–227	Rosemount, MN
Red clover		
Grown alone	62–101	Rosemount, MN
With reed canarygrass mixture	4–136	Rosemount, MN
Birdsfoot trefoil		
Grown alone	44–100	Rosemount, MN
With reed canarygrass mixture	27–116	Rosemount, MN
White clover		
Grown alone	114	Lexington, KY

Source: Adapted from Heichel, 1987 and Heichel and Henjum, 1991.

but they do deter some insects and limit their feeding and consequent damage.

This hairy trait has been bred into alfalfa cultivars. Erect hairs are physically more challenging for young nymphs to maneuver, and those with secretions actually "entrap" the nymphs so that they cannot feed (Ranger and Hower, 2001). Non-erect hairs that bend until they are parallel with the leaf surface are less effective. Several field experiments have shown the advantage of "glandular-haired" cultivars with regard to reducing damage caused by potato leafhopper.

Establishment

Legumes are seeded into tilled seedbeds or no-till seeded into crop residue or killed pastures (see Chapter 11). Similar strategies for achieving productive stands apply to both approaches. These include adjusting seeding rates, seeding dates, seedbed preparation, soil fertilization, and control of competition from weeds or grasses in mixtures. Because of the challenges with regard to establishment, high seeding rates are typically used to achieve target seeding year stands of 10–25 seedlings/ft^2 (108–270/m^2) by autumn of the seeding year (Table 8.4). For example, birdsfoot trefoil has about 375,000 seed/lb (827,000 seed/kg), so seeding at the recommended rates of 6–8 lb/acre (6.7–9 kg/ha) provides about 70 seed/ft^2. Likewise, alfalfa has about 200,000 seed/lb (440,000 seed/kg), so seeding at rates of 12–15 lb/acre (13–17 kg/ha) provides at least 55 seed/ft^2, more than three times the number needed if each seed produced a seedling. Seeding rates vary because of differences in seed size and seedling vigor.

Legumes are seeded at a depth of 0.25–0.5 in. (0.6–1.3 cm). Small-seeded legumes with small seeds, such as birdsfoot trefoil, white clover, kura clover, and alsike clover, are sown at shallow depths, whereas larger seeds such as those of alfalfa can be sown deeper. A firm seedbed that allows for shallow seeding and good contact between soil and seed is important for successful seeding. Firm seedbeds can be achieved by compaction with corrugated rollers or harrowing before seeding. Seeding with grain drills that utilize press wheels or Brillion^{TM} seeders with corrugated rollers will provide compaction following seeding.

Legumes are most often seeded either in early spring or in late summer. Spring seedings take advantage of soil moisture and favorable temperatures to promote establishment, and they allow two or more harvests during the seeding year. Late-summer seeding following harvest of a spring-seeded small grain or pea crop can also be successful if rainfall is timely. For conventional seedings of pure stands of legumes into tilled seedbeds and for no-till seedings, herbicides are used for control of broadleaf and grass weeds (see Chapter 11). The more traditional approach with regard to spring seedings involves the use of oat, barley, or wheat **companion crops**. Companion crops are more frequently used to established mixtures of legumes with grasses. Fast-growing companion crops reduce soil erosion and suppress weeds, but if they are not managed they can compete with legumes. Recommendations for reducing companion crop competitiveness include reduction of seeding rates, early companion crop harvest to decrease the duration of competition, and removal of straw from the field.

Legumes such as alfalfa, red clover, ladino clover, birdsfoot trefoil, and annual lespedeza are sometimes seeded into established grass pastures using "minimum tillage" practices to improve forage quality, yield, and seasonal distribution of growth of the pasture (Table 8.5; see also

Table 8.4. Establishment and seed characteristics of forage legumes used in the north

Legume	Ease to establish	Recommended seeding rate[a] (lb/acre)	Seed weight[b] (number/lb)	Seeding depth (in.)
Alfalfa	Easy	12–15	200,000	0.25
Red clover	Very easy	8–12	275,000	0.25–0.5
White clover	Easy	1–3	800,000	0.25
Birdsfoot trefoil	Fair	6–8	375,000	0.125–0.25
Alsike clover	Easy	4–6	700,000	0.25
Kura clover	Slow	4–8	300,000	0.25
Crownvetch	Fair	15–20	110,000	0.25
Sainfoin	Fair	30–35	30,000	0.25
Cicer milkvetch	Fair	15–20	130,000	0.25
Sweetclover	Easy	10–15	269,000	0.25
Annual lespedeza	Very easy	10–15[c]	200,000[c]	0.25

[a] Pure live seed.

[b] Adapted from US Department of Agriculture, 1948.

[c] With hull attached.

Table 8.5. Some examples of legume–grass mixtures for northern regions.

Alfalfa, 10 lb, with one of the following: orchardgrass, 3 lb; tall fescue, 5 lb; perennial ryegrass, 5 lb.
Birdsfoot trefoil, 6 lb, with one of the following: orchardgrass, 3 lb; smooth bromegrass, 6 lb; "Climax" timothy, 2 lb; perennial ryegrass, 6 lb.
Red clover, 6 lb, with one of the following: timothy, 4 lb; orchardgrass, 34 lb, perennial ryegrass, 5 lb; tall fescue, 12 lb.
Ladino clover, 1 lb, red clover, 2 lb, and alsike clover, 1 lb, with one of the following: orchardgrass, 4 lb; timothy, 4 lb; reed canarygrass, 8 lb; smooth bromegrass, 8 lb; perennial ryegrass, 5 lb.

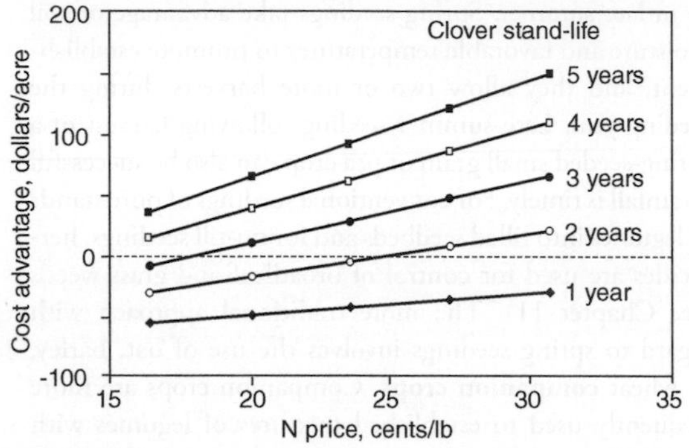

Fig. 8.10. Net returns for renovating a tall fescue pasture by introducing white clover compared with adding 180 lb N/acre. Note that returns from renovation and legume seeding were dependent on annual costs for N. Managing for long stand-life of the clover was beneficial because it delayed the cost of re-establishment of the clover. (Adapted from Burns and Standaert, 1985.)

Chapter 11). The addition of legumes can also eliminate the need to fertilize grass with N fertilizer (Fig. 8.10). The existing grasses must be managed in order to reduce competition. Depending on the seeding equipment available, this is accomplished by application of herbicides or tillage to open the sod and disrupt the existing grass. For "frost seeding" during the winter and early spring with white and red clover, the sod can be suppressed by grazing during the previous fall.

Harvest Management for Hay or Silage

Harvest timing involves balancing the relative needs for forage yield and forage quality with those of stand persistence. Because of differences in growth habit, persistence, and winter hardiness, management strategies for achieving a balance between yield, quality, and persistence are not consistent among legumes.

Plant Population and Persistence

The plant density of crown-forming legumes declines with the age of the stand (Fig. 8.11). However, stands of

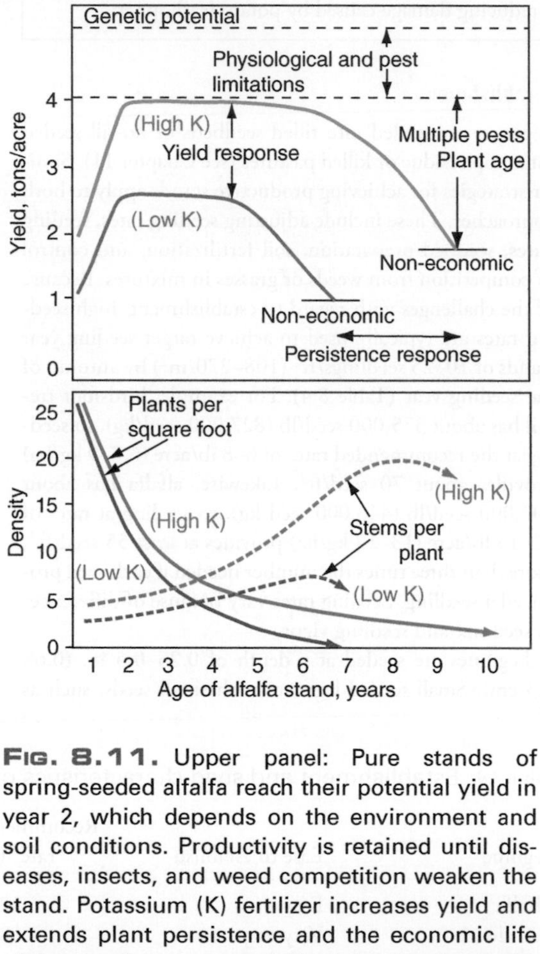

Fig. 8.11. Upper panel: Pure stands of spring-seeded alfalfa reach their potential yield in year 2, which depends on the environment and soil conditions. Productivity is retained until diseases, insects, and weed competition weaken the stand. Potassium (K) fertilizer increases yield and extends plant persistence and the economic life of the stand. Lower panel: Plant density decreases rapidly during the early stand-life, mainly due to plant–plant competition. Annual applications of K help plants to maintain vigor, so the crown increases production of new shoots to offset the loss of plants. The added vigor helps to increase the lifespan of plants by making them more resistant to disease and insect damage, and by reducing weed encroachment and competition. (Adapted from Moore and Nelson, 1995.)

clone-forming species such as kura clover and cicer milkvetch may actually increase with stand age. Stand dynamics are a function of harvest management and soil fertility. Management for stand persistence of crown formers such as alfalfa centers on keeping the existing plants alive. As plant density declines, yield is maintained through an increase in the number of shoots per plant until a minimal plant density is reached, below which the remaining plants cannot produce enough shoots. This allows weed encroachment. For example, for alfalfa, environments that support yields of 5–6 tons/acre (11.2–13.4 mg/ha) may require 50 or more shoots/ft^2 (540/m^2), whereas yields of 3–4 tons/acre may require 35 shoots/ft^2 (Undersander et al., 2011).

Harvest schedules that allow flowering of legumes promote storage of root reserves and improve stand persistence. The optimum number of harvests per growing season for persistence varies with the legume. For alfalfa, a crown-forming perennial, three harvests per growing season at early flowering provide for a high level of persistence in much of the Midwest. Increasing the number of harvests during the growing season beyond three cuttings generally reduces stand persistence. Maintaining good soil fertility, especially K (Fig. 8.11), also has a major effect on stand persistence.

Alfalfa and other legumes that are mechanically harvested for hay or silage are cut to leave a stubble height of 2–3 in. (5–7.6 cm). This provides for the greatest yield and facilitates regrowth from buds located on the stem bases. A higher cutting height is useful if legume stands have suffered winter injury.

In northern regions, cutting management of legumes in the fall has had a significant impact on winter survival. Although all legumes which depend on storage of energy reserves for winter survival are affected by fall cutting, these relationships have been documented for alfalfa in particular. To minimize the risk of winter injury in alfalfa, it has traditionally been recommended that cutting should not take place for 4–6 weeks before the last killing frost. For example, the "no-cut" window in much of Wisconsin and Minnesota has been from September 1 to the killing frost (25°F) that generally occurs in mid-October. If alfalfa is cut during this period, it will often regrow and exhaust its root reserves. However, research has shown that calendar date is less important than air temperature when describing the parameters of fall cutting (Dhont et al., 2004). Therefore, instead of defining the no-fall-cut period based on calendar date, Dhont and colleagues proposed that alfalfa needs 500 **growing degree days** (GDD, base 41°F) to be accumulated until a killing frost after the last summer cutting to regrow sufficiently for good winter survival. However, interactive factors such as the winter hardiness level of the cultivar, soil K levels, and the total season harvest regime of the alfalfa combine to interact with winter weather to influence the outcome of fall cutting (Undersander et al., 2011).

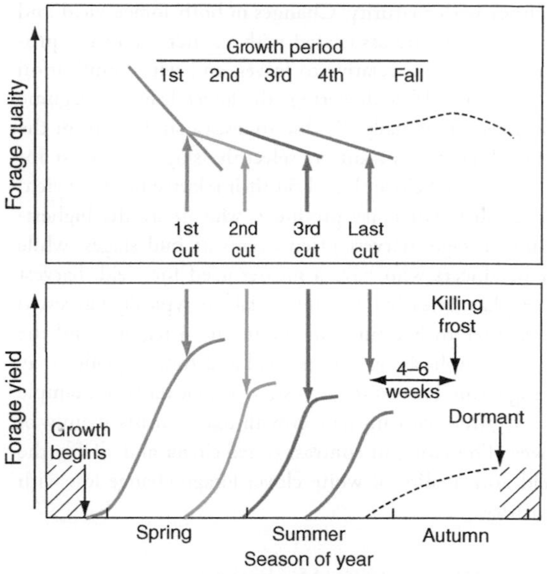

FIG. 8.12. Effect of time and season on growth and quality of alfalfa forage. Timing of the first harvest is critical because forage quality is highest then but decreases fastest during this growth period. Alfalfa regrowth changes less in quality, as air temperatures are high and daylength is long. Quality during autumn remains high as air temperatures are decreasing and daylength is shortening. Timing of the fall cut before winter hardening begins is important for persistence.

Alfalfa can be safely cut or grazed in late autumn, after growth has stopped and the plants are dormant and will not regrow. This occurs around mid-October in Wisconsin and around November in Missouri. This forage is of very high quality (Fig. 8.12) because it was produced under cool temperatures, but it is difficult to field dry in autumn.

A stubble of about 6-in. (15 cm) should be left to catch snow for insulation and reduce damage from ice sheeting. In milder winters, especially on poorly drained soils, the stubble helps to prevent the alternate freezing and thawing of the soil that causes plant heaving. Planting legumes in mixtures with grasses reduces heaving because the fibrous roots of grasses are resistant to heaving, and also because grasses trap snow and thus help to insulate the soil.

Forage Yield and Quality

For all legumes, forage yield and forage quality are affected by crop maturity. Yield and quality patterns change somewhat depending on the growth habit of the legume. For species with upright stems, such as alfalfa and red clover, yield increases until flowering and then plateaus or declines. In contrast to yield, forage quality typically declines with maturity. Changes in both forage yield and forage quality are associated with an increase in the proportion of stems relative to leaves, and with lignification of the stem. After flowering, the lower leaves in legume canopies can actually die due to disease and fall from the plant. Therefore producers select the stage of harvest for alfalfa and red clover based on their relative need for yield and quality. For dairy producers who desire the highest-quality forage, harvest often occurs at bud stages, while for producers who have a greater need for yield, harvest takes place at early flowering. Alfalfa is typically harvested from three to five times in the northern regions, and the relative emphasis on forage yield and forage quality may change from harvest to harvest. White clover is an example of a legume with upright growth that consists mainly of leaves. Therefore, in contrast to red clover and alfalfa, the yield and quality of white clover forage change less with maturity.

Management of Legumes for Pasture

Legumes are most often used in pastures as components of mixtures with grasses. Use of legumes in pasture with grasses is beneficial in terms of reduced cost of N, improved seasonal productivity (Fig. 8.10), and improved animal performance. Burns and Standaert (1985) evaluated 24 experiments and found that adding a legume to a grass pasture increased average daily gain of steers by 0.31 lb/day (0.14 kg/day) and that of beef calves by 0.35 lb/day (0.15 kg/day), compared with fertilizing the same pastures with N. However, sustaining legumes in pastures is challenging because of competition with grass components. Therefore management of soil fertility and grazing so as to achieve the desired balance is critical.

Grazing and Rest Periods

Continuous and rotational stocking are two stocking strategies (see Chapter 18). With continuous stocking, livestock are placed on a large pasture and allowed to graze freely. Under continuous stocking, livestock graze unevenly and can select and often re-graze more palatable legumes and grasses. Therefore few legumes persist under continuous grazing except for types of prostrate white clover that can escape defoliation.

Rotational stocking is a more intensive grazing system that moves or rotates livestock among paddocks or areas of the pasture. The number of paddocks used can vary with the goals of the producers, but most importantly, rotational stocking allows for managed grazing and efficient utilization of pastures. Rotational stocking allows for initiation of grazing at specific stages of legume development and specific periods of rest that are important for persistence of most legumes. Grazing of legumes is usually initiated before flowering. Alfalfa, red clover, and birdsfoot trefoil grazing should begin when growth is 15–20 in. (38–51 cm) and white clover and kura clover grazing should be initiated when herbage is 6–10 in. (15–25 cm) in height. Most legumes should be grazed to leave 2–4 in. (5–10 cm) of stubble height, because a higher stubble height promotes persistence. Legumes need sufficient rest (about 20–30 days between grazing periods) to allow reserves in the roots to be replenished. However, rest periods are likely to be shorter during favorable growth periods in the spring, and longer in the summer when the rate of regrowth decreases.

Summary

Legumes, especially perennials, play major roles in forage and pasture production over a range of soil resources and management practices. Alfalfa is the most important legume, due to its superior yield potential on good soils, its drought hardiness, and its high rates of N fixation that benefit other crops in the rotation. When grown alone or with adapted grasses, alfalfa provides high-quality hay or silage and can be harvested several times in a season. Some cultivars are grazing tolerant. Alfalfa is managed for winter survival and persistence, as autotoxicity restricts reseeding.

On poorly drained or low-fertility soils, which are common in long-term pastures, red clover and white clover are adapted and grown in mixtures with perennial grasses. Both species fix N, tolerate grazing, and improve productivity and animal performance on tall fescue pastures, especially those infected by the endophyte. Red clover is short-lived and must be reseeded every 2 years. White clover spreads by means of stolons and seed production, and combines well with low-growing grasses. Bloat can occur when white clover is the major species. Annual lespedezas, birdsfoot trefoil, and other perennial legumes offer options for special environment or management goals.

Questions

1. What characteristics distinguish crown-forming legumes from clone-forming legumes?
2. How does the root system of white clover compare with that of alfalfa?
3. What are the advantages of growing legumes in mixtures with grasses compared with growing legumes alone?
4. What is the importance of the soil seedbank for managing white clover or annual legumes?
5. Why is stage of growth critical for achieving high forage quality of alfalfa?
6. Why is the timing of fall cutting critical for perennial legumes?
7. Why does the quality of white clover and kura clover remain relatively high as the plants mature, compared with that of other legumes?
8. Why are forage legumes important for honey bees?

9. What is a living mulch system and what legume has been used in this system?
10. Why is rotational grazing important for legume persistence in grazing systems?

References

Beuselinck, PR, JH Bouton, WO Lamp, AG Matches, MH McCaslin, CJ Nelson, LH Rhodes, CC Sheaffer, and JJ Volenec. 1994. Improving legume persistence in forage crop systems. J. Prod. Agric. 7:311–322.

Brummer, EC, and JH Bouton. 1992. Physiological traits associated with grazing-tolerant alfalfa. Agron. J. 84:138–143.

Burns, JC, and JE Standaert. 1985. Productivity and economics of legume-based vs. nitrogen-fertilized grass-based pastures in the United States. In RF Barnes et al. (eds.), Forage Legumes for Energy-Efficient Animal Production, pp. 56–71. US Department of Agriculture Agricultural Research Service, Springfield, VA.

Clark, A. 2007. Managing Cover Crops Profitably, 3rd ed. USDA-Sustainable Agriculture Research & Education, College Park, MD.

Dhont, C, Y Castonguay, P Nadeau, G Bélanger, R Drapeau, and FP Chalifour. 2004. Untimely fall harvest affects dry matter yield and root organic reserves in field-grown alfalfa. Crop Sci. 44:144–157.

Heichel, GH. 1987. Legume nitrogen: symbiotic fixation and recovery by subsequent crops. In Z. Helsel (ed.), Energy and World Agriculture Handbook, Vol. 2. Energy in Plant Nutrition and Pest Control, pp. 63–80. Elsevier Science Publishing, Amsterdam, Netherlands.

Heichel, GH, and KI Henjum. 1991. Dinitrogen fixation, nitrogen transfer, and productivity of forage legume-grass communities. Crop Sci. 31:202–208.

Marten, GC, AG Matches, RF Barnes, RW Brougham, RJ Clements, and GW Sheath. 1989. Persistence of Forage Legumes. American Society of Agronomy, Madison, WI.

Michaud, R, WF Lehman, and MD Rumbauch. 1988. World distribution and historical development. In AA Hanson, DK Barnes, and RR Hill (eds.), Alfalfa and Alfalfa Improvement, pp. 25–91. Agronomy Monograph 29. American Society of Agronomy, Crop Science Society of America, and Soil Science Society of America, Madison, WI.

Moore, KC, and CJ Nelson. 1995. Economics of forage production and utilization. In RF Barnes, DA Miller, and CJ Nelson (eds.), Forages: An Introduction to Grassland Agriculture, 5th ed., pp. 189–202. Iowa State University Press, Ames, IA.

Ranger, CM, and AA Hower. 2001. Glandular morphology from a perennial alfalfa clone resistant to the potato leafhopper. Crop Sci. 41:1427–1434.

Samac, DA, LH Rhodes, and WO Lamp. (eds.) 2015. Compendium of Alfalfa Diseases and Pests, 3rd ed. American Phytopathological Society Press, St. Paul, MN.

Seguin, P, MP Russelle, CC Sheaffer, NJ Ehlke, and PH Graham. 2000. Dinitrogen fixation in kura clover and birdsfoot trefoil. Agron. J. 92:1216–1220.

Sheaffer, CC, NJ Ehlke, KA Albrecht, and PR Peterson. 2003. Forage Legumes, 2nd ed. Minnesota Agricultural Experiment Station, St. Paul, MN.

Sheaffer, CC, LE Sollenberger, MH Hall, CP West, and DB Hannaway. 2009. Grazinglands, forages and livestock in humid regions. In WF Wedin and SL Fales (eds.), Grassland, pp. 95 118. American Society of Agronomy, Crop Science Society of America, and Soil Science Society of America, Madison, WI.

Undersander, DJ, D Cosgrove, E Cullen, C Grau, ME Rice, M Renz, C Sheaffer, G Shewmaker, and M Sulc. 2011. Alfalfa Management Guide. American Society of Agronomy, Crop Science Society of America, and Soil Science Society of America, Madison, WI.

US Department of Agriculture. 1948. Grass: The Yearbook of Agriculture. US Government Printing Office, Washington, DC.

Yost, MA, JA Coulter, and MP Russelle. 2015. Managing the Rotation from Alfalfa to Corn. University of Minnesota Extension Bulletin. University of Minnesota, St. Paul, MN.

Zemenchik, RA, KA Albrecht, CM Boerboom, and JG Lauer. 2000. Corn production with kura clover as a living mulch. Crop Sci. 92:698–705.

Zhu, Y, CC Sheaffer, MP Russelle, and CP Vance. 1998. Dry matter accumulation and dinitrogen fixation of annual *Medicago* species. Agron. J. 90:103–108.

CHAPTER 9

Legumes for Southern Areas

Lynn E. Sollenberger and Michael Collins

Globally, forage legumes have contributed less to livestock production in warm-climate than in temperate-climate areas. In the southern USA, a relatively large number of legumes play valuable but minor roles. None of them approach the importance that alfalfa has in cooler temperate climates.

Why have legumes been less successful in warmer areas of the USA? Factors vary regionally but include vast areas of acid, infertile soils, soils with inadequate or excessive drainage, longer and more intense pressure from pests, and wide ranges in seasonal **weather** conditions. In addition, in the lower South, C_3 legumes must often compete directly with vigorous C_4 **perennial** grasses (see "Where Are the Warm-season Legumes?"). Furthermore, many **warm-season** legumes are not tolerant of continuous stocking to short stubble height (Table 9.1).

Legume contributions in the South include providing higher-quality forage than **grasses** for livestock and wildlife, increasing the length of the grazing season, reducing the need for N fertilizer inputs in grass-based forage systems, serving as rotation crops, supplying N for subsequent row crops, and providing numerous ecosystem services (see "Legumes Serving the Ecosystem").

In this chapter, the most important legumes for the South will be presented in groups based on their season of production (cool vs. warm), location of use (Gulf Coast vs. intermediate South), and primary role in production systems (livestock feed vs. multipurpose legumes).

Warm-Season Forage Legumes for the Gulf Coast Region

Warm-season forage legumes originated in the subtropics and tropics. Their tolerance of cold primarily determines where they are adapted in the Gulf Coast region (Fig. 9.1), but within a temperature zone, soil characteristics, especially drainage, have the greatest impact on species adaptation. Warm-season legumes are generally planted in mixtures with warm-season grasses. These legumes are grown on less than 100,000 acres (40,500 ha), a very small percentage of the millions of forage acres in the Gulf Coast region.

Where Are the Warm-Season Legumes?

Productive and long-lived grass–legume mixtures are the exception rather than the rule in warm climates. Two important reasons for this are differences in the physiology and morphology of grasses and legumes.

Tropical and subtropical grasses are C_4 plants, but both temperate and tropical legumes are C_3 plants. In C_4 grasses, carbon from CO_2 is first fixed into carbohydrate by phosphoenolpyruvate carboxylase (PEP), a very efficient **enzyme** that contributes to rapid growth rates, especially when temperatures are high. The enzyme that fixes carbon in legumes and other C_3 plants is ribulose bisphosphate carboxylase-oxygenase (rubisco). This enzyme also fixes O_2 in a process called **photorespiration** that wastes carbon and **ATP**. Photorespiration increases as the temperature rises, so legumes are disadvantaged at high temperatures compared with C_4 grasses. The more efficient photosynthetic system of tropical and subtropical grasses makes them very competitive with legumes.

The morphology of C_4 grasses also increases their competitiveness. Most perennial C_4 grasses originated in Africa and evolved under grazing by large herbivores, whereas warm-season legumes originated mainly in the

Forages: An Introduction to Grassland Agriculture, Seventh Edition. Edited by Michael Collins, C. Jerry Nelson, Kenneth J. Moore and Robert F Barnes.

Table 9.1. Traits that affect the competitiveness of warm-season grasses and legumes

Trait	Warm-season grasses	Warm-season legumes
Photosynthesis	C_4	C_3
First CO_2-fixing enzyme	PEP	Rubisco
Photorespiration	No	Yes
Growth rate at high temperatures	High	Lower
Growing points	Protected	Elevated
Grazing tolerance	Generally high	Lower
Treading tolerance	Generally high	Lower

PEP, phosphoenolpyruvate carboxylase; rubisco, ribulose bisphosphate carboxylase-oxygenase.

American tropics, where large herbivores were absent until a few hundred years ago. Grasses generally have growing points near or below the soil surface, whereas legume growing points are often elevated well above the soil surface, where they are more likely to be removed by grazing or cutting.

Warm-season grasses have a well-distributed, fibrous root system that allows them to take up more P and/or K than can legumes when these elements are in short supply. The C_4 grasses also produce more forage per unit of water and N used than do C_3 legumes. Thus there are numerous factors that favor grasses, and management of C_4 grass–C_3 legume associations must favor the legumes if they are to remain productive over time.

Adaptation, Productivity, and Use

In the cooler parts of this region (e.g., southeastern Texas, northern Florida, and southern Louisiana, Mississippi, Alabama, and Georgia), only the most cold tolerant of the

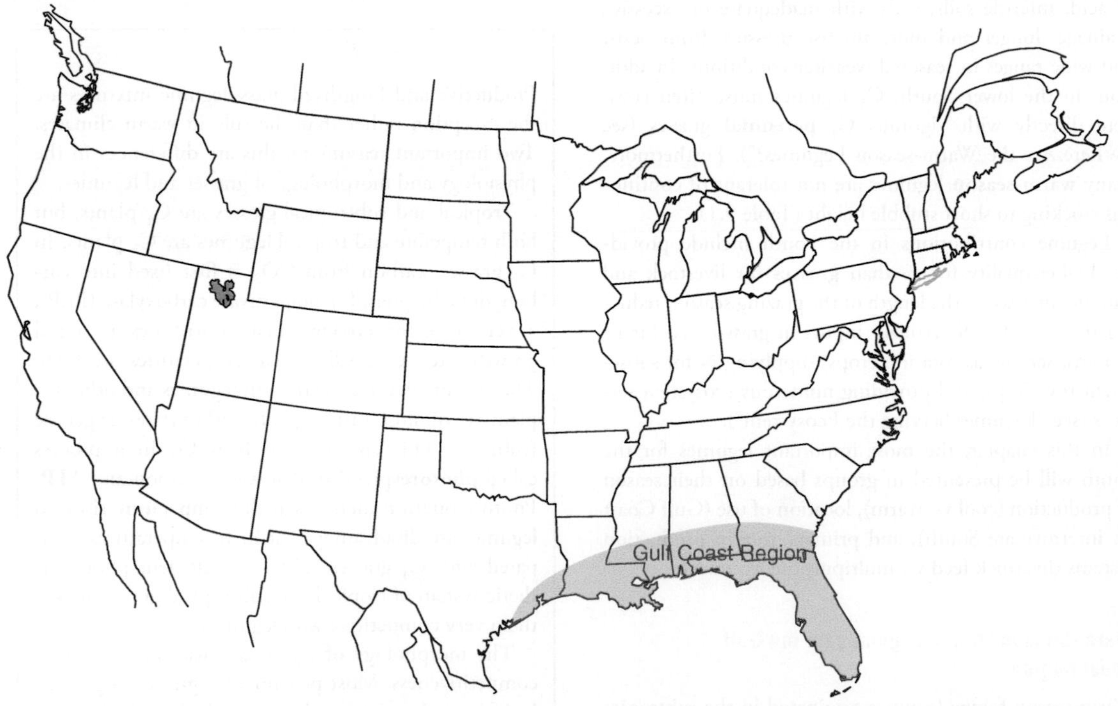

Fig. 9.1. Gulf Coast region of the USA where both tropical perennial legumes and cool-season annual legumes are grown. Winter annual legumes are used, but winter growth is reduced in regions north of the Gulf Coast.

warm-season legumes will perennate. These include rhizoma peanut and leucaena. The less cold-tolerant perennials, carpon desmodium and stylo, will consistently survive winter only in south Florida. Aeschynomene and alyceclover are summer **annuals** and are adapted throughout the region where soil conditions are suitable (Table 9.2).

Of this group, the most commercially important is rhizoma peanut. It is herbaceous and strongly rhizomatous, and is grown on well-drained soils primarily as a high-value hay crop. Total use is about 50,000 acres (20,250 ha), mainly in southern Georgia and northern Florida. Yield is in the range of 3–6 tons/acre (6.7–13.4 mt/ha), with most growth occurring from April through September. The **nutritive value** of rhizoma peanut hay is comparable to that of alfalfa, and it is sought after by dairies and by horse (*Equus caballus*) owners. **Crude protein** content is in the range of 13–18%, and **digestibility** is 60–70%. Rhizoma peanut is increasing in use and importance, but more rhizoma peanut hay could be sold if it were consistently available. Pinto perennial peanut is a seed-propagated relative of rhizoma peanut that is widely used in tropical environments. Pinto peanut is currently being evaluated to determine whether it is sufficiently cold tolerant to persist in the Gulf Coast region.

Rhizoma peanut is also used for **pasture**. Due to the fact that it produces a dense rhizome network in the upper 4 in. of the soil, it persists very well under grazing. When grazed, it is often grown in mixtures with bahiagrass or bermudagrass. There are rhizoma peanut stands in Florida that have been grazed or hayed annually for more than 30 years without any decline in legume stand density. The main factors limiting use of rhizoma peanut are that costly **vegetative propagation** is required to establish it and that stand establishment requires at least 2 years on non-irrigated sites. The other cold-tolerant perennial legume for this region is leucaena, which is grown on a few thousand acres, mostly in southern Texas. It is a tree legume that is managed to maintain a height of 6–10 ft, and is used exclusively for grazing. Cattle push the woody branches down so that they can graze the leaves. Branches may break, but the effects of this on plant persistence are minimal. Although it is well adapted to the climate of Florida, leucaena does not grow well on areas of poorly drained, acid soils that are common in south Florida. Approximately 1 year after planting, utilization of leucaena can begin. Well-established leucaena is very grazing tolerant and cold tolerant, and produces **palatable** forage of very high quality. The leaves and small twigs that animals select often contain 16–20% protein and have *in-vitro* digestibility of 65–70%. The limitations to leucaena use include the psyllid (*Heteropsylla cubana* Crawford) insect and the toxic amino acid **mimosine**. These will be discussed later (see also Chapter 16). In addition, some leucaena ecotypes are invasive, so it is important to know the characteristics of a particular cultivar or ecotype before planting it in a given environment.

Stylo is an erect to semi-erect herbaceous plant that grows to a height of 2 ft (60 cm) to 4 ft (1.2 m). It is adapted to sites with good surface drainage, but can tolerate short periods of saturated soil conditions. Even where it is well adapted, stylo behaves like a short-lived perennial. It can be used as a **reseeding annual** where the first frost occurs after the middle of December, because there is sufficient time for **seed** set after late-fall flowering. Early-season growth is slower than that of alyceclover or aeschynomene, but unlike those legumes stylo grows well into fall until frost occurs. Thus its primary use is for late summer to fall grazing for animals with high nutrient requirements. Yields of 3.5 tons/acre (3.9 mt/ha) with 10–17% crude protein and *in-vitro* digestibility of 55–70% have been obtained. Stylo is currently grown on less than 1000 acres (405 ha).

Carpon desmodium is a perennial herbaceous legume with a growth habit that varies from semi-erect (up to 3 ft [0.9 m] tall) to prostrate, depending upon **grazing management**. About 15,000 (6070 ha) acres are grown in mixtures with bahiagrass on the poorly drained soils of south Florida. When grazed at moderate stocking rates in bahiagrass pasture, carpon desmodium can contribute 30–75 lb (13.6–34 kg) of N per acre per year. About 65% of seasonal production occurs during midsummer through fall. The forage contains 12–20% crude protein, but *in-vitro* digestibility has been low, at 45– 60%, perhaps due in part to the presence of **tannin**.

The advantages of carpon desmodium over aeschynomene are that it is a perennial and it is more tolerant of grazing. The disadvantages are its slower establishment and its much lower digestibility. The two legumes can be successfully overseeded together into bahiagrass. Aeschynomene is productive during the first year, and carpon becomes prominent in subsequent years (Aiken et al., 1991a).

Aeschynomene is an erect-growing annual with a maximum height of about 5 ft (1.5 m) if it is not defoliated. Also known as American jointvetch or deervetch, it is adapted to poorly drained soils, and is best used for grazing. Its shatter-prone leaflets and woody branches limit its use for hay. It is most widely planted in south Florida, where it occurs on approximately 20,000 acres (8,100 ha), but it has been used on wet sites in Louisiana. Although more widely planted, aeschynomene is found mainly in scattered, poorly drained patches within pastures.

Aeschynomene overseeding increases the nutritive value of perennial grass pastures. The crude protein content of young leaves and fine stems can exceed 20%, and *in-vitro* digestibility is often higher than 60%. Adequate soil moisture and sustained control of grass competition with the aeschynomene seedlings are required for initial establishment. Exacting management of fall grazing is also needed to allow sufficient seed set for re-establishment the following year. Another disadvantage of aeschynomene is its short

Table 9.2. Characteristics of warm-season legumes adapted to the Gulf Coast region

Legume	Drought tolerance	Tolerance of acid soils	Tolerance of wet soils	Tolerance of low fertility	Tolerance of grazing	Grazing/hay management[a]	Typical establishment method	Ease of establishment
Aeschynomene	Low	Moderate	High	Moderate	Moderate	R: Enter at 18 in., exit at 8–10 in. C ≥ 12 in.	Seed in spring to early summer at 10–15 lb/acre	Moderate
Alyceclover	Moderate	Moderate	Low	Low	Moderate	R: Enter at 12 in., exit at 6–8 in. H18–24 in.	Seed in spring to early summer at 15–20 lb/acre	Moderate
Carpon desmodium	Low	Moderate	Moderate	Moderate	High	R: In summer/fall, enter every 3–6 weeks, exit at 6–8 in. C: In spring to 3–4 in.	Seed in spring to summer at 3–10 lb/acre	Difficult
Leucaena	High	Low	Low	Moderate	High	C: Leaf/small twig mass of ≥1000 lb/acre	Seed in early summer at 2–3 lb/acre; rows *c.* 15 ft apart	Difficult
Rhizoma peanut	High	Moderate	Low	Moderate	High	R: Enter every 6 weeks, and exit at 4 in. C ≥ 6 in. Every 6–8 weeks	Vegetative during Jan. to Mar. at 80 bushels of rhizomes/acre	Difficult
Stylo	Moderate	Moderate	Moderate	Moderate (high for P)	Moderate	R: Enter at 18–24 in., exit at 9–12 in. C > 8 in.	Seed in spring to early summer at 10–12 lb/acre	Moderate

Sources; Kretschmer and Pitman, 1995; Sollenberger and Kalmbacher, 2005.

[a]R, rotational stocking; C, continuous stocking; number refers to recommended pasture height. H, hay; recommended heights or frequencies of cutting are indicated.

productive season, which is generally from June to September. It flowers in mid- to late September and produces little leaf growth thereafter.

Alyceclover is a warm-season annual legume that is adapted to fertile, well-drained soils, and it can be grazed or used for hay production, especially for horses. A study by Bagley et al. (1985) showed that if alyceclover was harvested for hay when it was 22–24 in. (0.6 m) tall, yields were 1.0–1.25 tons/acre (2.2–2.8 mt/ha). Although not as tall, its growth habit and seasonal distribution of dry matter are similar to those of aeschynomene. Alyceclover has poor seedling vigor, which often results in severe weed competition early in its growing season. One of its advantages is that it maintains excellent **forage quality** in later summer and autumn when grass forage quality declines sharply. In Louisiana, alyceclover is recommended for late summer through fall grazing by beef steers. Gains of 700 lb (350 kg) (steers were gaining 1.7 lb/day [0.77 kg/day]) were obtained from July to September, and these were higher than gains on pearl millet (1.3 lb [0.6 kg]) and sorghum × sudangrass hybrids (1.2 lb [0.6 kg]) (Bagley et al., 1985). The acreage of alyceclover in the region is small.

Establishment

The legumes described above are all seed propagated except for rhizoma peanut, which must be propagated vegetatively using rhizomes. To ensure effective nodulation and N fixation, **inoculation** of legume seed or rhizomes is recommended. All legumes in this group except for leucaena are inoculated with *Bradyrhizobium* bacteria from the cowpea group. Leucaena must be inoculated with a specific *Rhizobium* strain.

A completely prepared, firm seedbed (see Chapter 11) is generally recommended for rhizoma peanut, although no-till planting is an option. Rhizomes are dug from existing fields and planted from January to March. Rhizome yield and vigor are greatest when dug from **swards** that were defoliated once or not at all during the previous growing season. After digging, rhizomes are spread across the surface at a rate of 80 bushels/acre and disked in or are planted using a bermudagrass sprig planter. Rhizomes should be planted approximately 1 in. (2.5 cm) deep, and the soil should be firmed using a roller or cultipacker after planting. To reduce establishment cost, techniques have been developed to strip plant the legume into existing perennial grass pastures (Castillo et al., 2014). Whatever planting method is used, minimization of weed competition after planting is important because it often takes two growing seasons for rhizoma peanut to achieve complete ground cover. Slow establishment occurs because a large amount of photosynthate is partitioned underground during the early stages of plant development to help to form the dense rhizome network. Irrigation can speed up establishment and result in complete ground coverage by the end of the first growing season.

The other legumes in this group are seed propagated. Alyceclover and leucaena are generally planted into a prepared seedbed, whereas carpon desmodium, aeschynomene, and stylo are generally overseeded into warm-season grass sods. Seedbeds should be firm to achieve a shallow planting depth, and then rolled or cultipacked to achieve good soil–seed contact. For all but leucaena, a good technique on sandy soils is to roll the seedbed, broadcast the seed, and roll again. Unlike the other legumes, leucaena is most often planted in rows that are 15–20 ft (3.7–6.1 m) apart. Because it is slow to establish, weed control is critical. Special cultivators can be used to uproot small weed seedlings after leucaena emergence. When the trees are 6 ft (1.8 m) tall, an adapted C_4 grass should be planted between the rows.

Overseeding of warm-season legumes into an existing grass sod can be done in March or April in years with good spring rains, but in Florida it is often best to wait until June when the summer rainy season begins. Before planting, the **canopy** must be grazed closely or cut for hay to remove competition, and the field disked lightly to expose soil in order to obtain good seed–soil contact. If seed are broadcast, the pasture should be rolled after planting to ensure soil contact. The grass should not be fertilized with N, and must be grazed closely, to 4–6 in. (10–15 cm), periodically after planting to reduce competition until the legume seedlings have grown to the height of the grass canopy. For aeschynomene, this occurs approximately 4 weeks after planting. Some legume seedlings are lost due to trampling, but failure to control grass competition usually results in complete failure. For natural reseeding, canopy management in spring is similar to that for initial **sod seeding**.

Defoliation Management

Rhizoma peanut, carpon desmodium, and leucaena are relatively grazing tolerant, and management is flexible (Table 9.2). If stocked rotationally, rhizoma peanut should be grazed to a 4-in. (10-cm) stubble, followed by about 6 weeks of **rest**. If stocked continuously, stubble height should be maintained at 6 in. or more throughout the growing season. Rhizoma peanut for hay is generally cut every 6–8 weeks. Leucaena can readily support continuous but moderate stocking. Livestock strip the leaves and tender twigs from the trees. Long rest periods during the growing season should be avoided, to prevent the trees from growing beyond the reach of cattle. Carpon desmodium–bahiagrass mixtures should be stocked heavily and continuously in spring to reduce grass competition with the legume, but lighter **grazing pressure** and/or **rotational stocking** in late summer are beneficial to the legume.

Aeschynomene, stylo, and alyceclover are best managed using rotational stocking (Table 9.2). Aeschynomene should be grazed when plants reach a height of 18 in. (46 cm), leaving a stubble of 8–10 in. (20–25 cm). Alyceclover can be grazed to leave a stubble of 6–8 in.

(15–20 cm) when plants reach 12 in. (30 cm) in height. When used for hay, alyceclover should be cut when plants are 18–24 in. (46–61 cm) tall. Stylo should be grazed when it reaches a height of 18–24 in. (46–61 cm), and animals removed after about 50% of the original height has been grazed. Stylo grows well in fall, and can be **stockpiled** by removing animals for 2 months beginning around the first of August. To increase winter survival, a stubble of 6–8 in. (15–20 cm) should remain after fall grazing.

Forage Nutritive Value

With the exception of carpon desmodium, which has relatively low digestibility, the nutritive value of these legumes is high (Table 9.3). Yearling cattle can gain 2 lb/day (1 kg/day) on nearly pure stands of rhizoma peanut, and gains of 1.5 lb/day (0.7 kg/day) can be achieved for the other legumes in pure stands or in mixtures with grasses. Weaned beef heifers grazing bahiagrass but with access to leucaena were found to gain nearly 0.9 lb/day more than those grazing only bahiagrass (Kalmbacher et al., 2001). Adding aeschynomene to limpograss pastures increased gain to 1.54 lb/day, from 0.86 lb/day for limpograss alone (Rusland et al., 1988). Studies of bahiagrass and limpograss show that even small proportions of aeschynomene or carpon desmodium (as little as 10% of the forage dry matter) can have a major positive effect on animal performance (Aiken et al., 1991b). **Bloat** is not considered to be a problem even with pure stands of these species.

Anti-quality factors are common in subtropical and tropical legumes, are not detected by common measures of forage nutritive value, and include a range of secondary plant metabolites. For the species discussed in this section, the most widely known anti-quality factor is the amino acid mimosine found in leucaena. Ingested mimosine is metabolized in the rumen to 3-hydroxy-4(1H)-pyridone (commonly abbreviated to 3,4-DHP), which can cause goiters and which reduces feed intake when leucaena represents a large proportion of the diet. However, in some situations no negative effects are observed; it has been found that these animals possess a unique rumen bacterium that

Table 9.3. Crude protein (CP), *in-vitro* digestibility (IVDMD), and animal performance for warm-season legumes adapted to the Gulf Coast region

Legume	Crude protein (%)	Digestibility (%)	Average daily gain (lb)
Aeschynomene	Diet: 20–27[a]	Diet 62–76[a]	1.54[a]
	Leaf: 22–25[b]	Leaf: 72–78[b]	
	Stem: 7–8[b]	Stem: 39–45[b]	
Alyceclover	15–18[c,d]	61–63[c,d]	0.8–1.7[d]
Carpon desmodium	Leaf: 16–19[b]	Leaf: 44–5 1[b]	1.1–1.5[e]
	Stem: 5–7[b]	Stem: 31–37[b]	
Leucaena	Leaf: 24–32[f]	Leaf: 61–66[f]	1.53[g]
Rhizoma peanut	17–22[h]	70–73[h]	2.05[h]
Stylo	Hay: 16–17[c]	Hay: 69[c]	0.88[i]
	Total herbage: 11–12[i]	Total herbage: 48–50[i]	
	Diet: 14–16[i]	Diet: 56–66[i]	

[a]Rusland et al., 1988; rotational stocking with 35-day rest periods. CP and *in-vitro* digestibility data from samples taken to represent the aeschynomene herbage consumed. Gain data for aeschynomene–limpograss mixture (25% aeschynomene).

[b]Aiken et al., 1991a; continuously stocked pastures of aeschynomene–carpon desmodium–bahiagrass at several stocking rates.

[c]Williams et al., 1993; planted in mid- to late June and harvested 60–70 days later to a 4-in. stubble.

[d]Bagley et al., 1985. CP and IVDMD data from a hay harvest 8–9 weeks after planting. Daily gain data from 700–950 lb steers during July to October.

[e]Aiken et al., 1991b; continuously stocked pastures of aeschynomene–carpon desmodium–bahiagrass at several stocking rates.

[f]Dalzell et al., 1998; youngest fully expanded leaf of 21 accessions of *Leucaena leucocephala* var. *glabrata* near Brisbane, Australia.

[g]Jones et al., 1998; average across two cultivars under rotational stocking at Lansdown, Australia.

[h]Sollenberger et al., 1989; rotational stocking with 35-day rest period. CP and *in-vitro* digestibility data from samples taken to represent the diet consumed.

[i]RS Kalmbacher, 2001, personal communication; continuous stocking of stylo–bahiagrass mixture by yearling cattle at 1.2 head/acre from July to September. Total herbage = cut to a 2-in. stubble. Diet = samples taken to represent stylo herbage consumed.

degrades 3,4-DHP and renders it harmless. This bacterium can be transferred to other animals by ruminal or oral dosing or simply via saliva from casual contact among animals housed together. Thus **mimosine** need not be a problem if animals are inoculated.

Some species in the *Desmodium* and *Leucaena* genera have high tannin levels, but the particular species used in the Gulf Coast region have low tannin concentrations. Tannin binds to protein molecules, reducing the ability of digestive enzymes to attach to and digest them. Thus too much dietary tannin reduces the supply of protein available to the animal.

Environmental Stresses and Pests

Tolerance of the warm-season legumes to environmental stresses is summarized in Table 9.2. With the exception of leucaena, these legumes are better adapted to acid soils than most temperate legumes, but in general a pH of at least 5.5 is recommended. Stylo is tolerant of low-P soils, but in most situations the other legumes will require P and K fertilization for good yield and persistence.

Rhizoma peanut and leucaena are long-lived perennials, and established plants of these species have survived winter lows of 15°F or colder following regular defoliation during the previous growing season. Stylo and carpon desmodium are not tolerant of freezing conditions, and behave as perennials only in the warmest parts of the region, primarily south of Orlando, Florida. None of the warm-season legumes have frost-tolerant shoots, and leaf tissue is killed and falls to the ground soon after frost.

The pest that has caused most damage to this group of legumes on a worldwide basis has been the psyllid insect on leucaena. This sucking insect damages young leaves and can completely defoliate or even kill trees. The principal means of psyllid control are introduction of natural predators and the use of resistant **cultivars**. Carpon desmodium and alyceclover are susceptible to the root-knot nematode (*Meloidogyne* species), and they should not be planted after a vegetable crop in infested areas. Susceptibility to the disease anthracnose (*Colletotrichum gloeosporioides*) has limited the use of stylo on a worldwide basis. Possibly due to the low acreage in the region, this disease has not had a major effect on stylo in the Southeast. The peanut stunt cucumovirus has been observed in the last decade in commercial fields of "Florigraze" rhizoma peanut. A new cultivar, "UF Tito", has tested negative for stunt virus.

Warm-Season Forage Legumes for the Intermediate South

The intermediate South is the region north of the Gulf Coast, extending as far as parts of North Carolina, Tennessee, Kentucky, Arkansas, Missouri, and Oklahoma. Large areas of this region are not suited for alfalfa and red clover. Common constraints are soil related, including drainage, acidity, and low fertility. In this region, the greatest contribution by legumes is made by cool-season species during winter, but there are a limited number of important warm-season legumes.

Adaptation, Productivity, and Use

The primary warm-season forage legumes in the intermediate region are two annual lespedezas (striate lespedeza and korean lespedeza), perennial sericea lespedeza, and annual soybean. General information on the adaptation of these species is provided in Table 9.4.

The annual lespedezas are grown from the Atlantic Coast west to eastern Oklahoma and north to southern Iowa (Fig. 9.2). Stands sometimes regenerate through natural reseeding, but periodic re-establishment is usually required. The western boundary of their range is established by low precipitation and the northern boundary occurs where shorter days and early frost prevent natural reseeding (Ball et al., 2015). Within this region, they grow well on drought-prone, upland soils that may be too acidic and low in P to support other legumes. They do respond to fertilizer and lime, and grow best on fertile, well-drained soils.

Both striate and korean lespedezas are fine-stemmed and leafy, have a shallow taproot, and grow to 16–20 in. (40–50 cm) in height. They are used primarily as pasture. Yields are low, in the range of 1–2 tons/acre (2.2–4.4 mg/ha) under rain-fed conditions, but forage quality is high from midsummer to late summer. Prior to the 1940s they were widely used as rotation crops for soil improvement.

Annual lespedezas play an important role in low-input production systems. Their optimum use may be in mixtures with endophyte (*Neotyphodium coenophialum*)-infected tall fescue in fields that will be grazed during summer, when the quantity and quality of tall fescue are at their poorest. Naturalized stands are ubiquitous on large acreages of low-input grasslands in the intermediate South. "Marion", a cultivar of striate lespedeza, is well adapted throughout the range of this species. It has better resistance to leaf diseases and improved seed production, especially in the northern parts of the region.

Sericea lespedeza is adapted to the humid east, bounded by Missouri to the west, Ohio to the north, and the Gulf Coast to the south (Fig. 9.2). It is tolerant of acid soils, relatively high soil Al, and drought, making it useful in marginal areas and for revegetation of mine lands. It is best adapted to clay or loam soils, and shows little response to P or K fertilizer. It is estimated to occur on 30,000–40,000 acres (12,000–16,000 ha) in Alabama alone.

Sericea is an erect-growing, deep-rooted, very long-lived perennial that reaches a height of up to 4 ft (1.2 m) if not defoliated. Like the annual lespedezas, it was formerly widely used for soil conservation and in crop rotations. Its thick stems and high tannin content hinder the use of

Table 9.4. Characteristics of warm-season legumes adapted to the intermediate South region

Legume	Drought tolerance	Tolerance of acid soils	Tolerance of wet soils	Tolerance of low fertility	Tolerance of grazing	Grazing/hay management[a]	Typical establishment method	Ease of establishment
Annual lespedezas	Moderate	High	Low	High	Moderate	R or C: Relies on leaf area for regrowth H: 1–2 cuts/year at early bloom, leaving stubble with some leaf	Seed in early spring at 25–35 lb/acre	Easy
Sericea lespedeza	High	High	Low	High	Low (except for "AU Grazer" which is moderate)	R: Enter at 8–10 in., exit at 4 in. H: 15–18 in.; 2–3 cuts/year	Seed in spring at 20–30 lb/acre	Moderate
Soybean	High	Low	Low	Low	Low	C: Short season, no regrowth H: In summer when 75% of pods are filled, 1 cut/year	Seed in May at 60–100 lb/acre	Easy

Sources: McGraw and Hoveland, 1995; Ball et al., 2015.
[a]R, rotational stocking; C, continuous stocking; H, hay; recommended heights at cutting are indicated.

Fig. 9.2. Area of adaptation of lespedezas, warm-season legumes used in the intermediate South in the USA. (From McGraw and Hoveland, 1995.)

common sericea for forage. Improved cultivars that are fine-stemmed ("Serala"), low in tannin ("AU Lotan" and "AU Donnelly"), and grazing tolerant ("AU Grazer") are available. The cultivar "AU Grazer" combines fine stems and low tannin levels with tolerance of close grazing or **continuous stocking**. Long stand life and the ability to naturally reseed from a soil seed reserve have contributed to sericea being considered invasive in some ecosystems, particularly following a change of land use to less intensive management (Pitman, 2006). However, the spread of sericea is limited to several feet over 7 years, apparently due to lack of an effective seed dispersal mechanism.

Annual soybean is best adapted to well-drained soils, but does show some tolerance of drought. It is usually harvested once for hay or silage, and it can be used for late-summer grazing. The forage usually has high digestibility (54–60%) and protein levels (15–20%). Hay yields from the single cut are generally in the range of 2–3 tons/acre (4.5–6.7 mg/ha), but may be higher for types selected for forage. Soybeans with the long-juvenile trait (i.e., delayed flowering under short-day conditions) fit well into forage production systems. They mature later, grow taller, produce higher yields, have thinner stems, and exhibit a more vine-like growth habit than earlier-maturing types.

Establishment

The lespedezas are usually grown with cool-season grasses; tall fescue is a common choice. Often they cannot compete effectively with warm-season perennial grasses such as bermudagrass. Soybean is most often grown in pure stands, but it is sometimes planted with warm-season annual grasses such as pearlmillet. Legumes in this group are seed propagated and require inoculation at planting, especially in areas where they have not previously been grown. The lespedezas are inoculated with *Bradyrhizobium* bacteria from the cowpea group, whereas soybean requires use of the specific strain *Bradyrhizobium japonicum*.

Annual lespedezas establish readily when sown at seeding rates of 25–35 lb/acre (28–39 kg/ha) from midwinter to early spring. Broadcasting seed into a grass pasture in late winter in colder parts of the region allows frost heaving to bury the seed, and normally provides a good stand.

In contrast to the annual lespedezas, for sericea a well-prepared, firm seedbed is considered essential, and seeding into grass sods has generally been unsuccessful. Tall fescue can be planted into sericea pastures after the legume is established. Sericea is slow to establish and does not compete well with weeds. Cool temperatures reduce germination and emergence with March and April seeding. Seeding later in spring may be more successful, but only if weed competition is controlled. Seeding rates of 20–30 lb/acre (22–34 kg/ha) are recommended (Ball and Mosjidis, 2007).

Soybean can be planted into a prepared seedbed or no-till planted at seeding rates of 60–100 lb/acre (67–112 kg/ha) in May. High-density plant populations produce a smaller stem diameter and have improved hay drying and palatability.

Defoliation Management

Annual lespedeza is used primarily for pasture and to a limited extent for hay. It is most productive during the late summer period when cool-season grasses are growing relatively slowly and are of low quality. Annual lespedezas do not rely on stored carbohydrates for regrowth, so some leaf area should remain after grazing (McGraw and Hoveland, 1995). In addition, grazing pressure must be low if plants are to flower and produce sufficient seed for natural reseeding to occur (Ball and Mosjidis, 2007). Hay should be cut at the early **bloom** stage for optimum yield and quality, and can be harvested twice a year.

Sericea can be used for hay or grazing. For hay, it can be harvested two or three times a year when it is 15–18 in. (38–46 cm) tall, usually resulting in a total annual yield of 2–3 tons/acre (4.5–6.7 mg/ha). It is important to leave a stubble of 4–5 in. (10–13 cm), because new growth starts from **axillary buds** located above ground on the stubble (McGraw and Hoveland, 1995). The last cutting should be made no later than early September, to allow accumulation of reserves prior to frost. One of the major factors limiting the use of sericea lespedeza, except for "AU Grazer", is its lack of tolerance of close grazing and/or continuous stocking. In general, sericea pastures should be rotationally stocked, starting in spring when the canopy height is 8–10 in. (20–25 cm). A stubble height of at least 4 in. (10 cm) should be maintained to avoid stand loss.

Soybean can provide short-season grazing or be used for hay, but it does not regrow following defoliation. Yields increase with maturity until 75–90% pod fill, after which leaf loss decreases both yield and nutritive value. Soybean hay is difficult to cure, and requires a hay conditioner (Ball et al., 2015).

Forage Nutritive Value

The lespedezas do not cause bloat. During July and August in Missouri, crude protein of Marion and "Summit" lespedezas is approximately 14% and neutral detergent fiber is approximately 55% (McGraw and Hoveland, 1995). In contrast, common sericea lespedeza has coarse stems and high leaf tannin levels that can reduce intake. High tannin levels in sericea have been associated with a reduction in fiber and N digestibility. Field drying of sericea hay reduces tannin levels markedly and results in greater palatability compared with fresh forage. Use of low-tannin sericea cultivars has resulted in higher gain per animal than has been observed on high-tannin types. Season-long gains of 1.7 lb/day (0.8 kg/day) have been obtained with beef steers grazing pure stands of "AU Lotan" (Ball and Mosjidis, 2007). However, dry matter yield and vigor are slightly lower for low-tannin than for high-tannin cultivars. Tannins have some potentially beneficial effects, including suppression of ruminal methane production, increased ruminal bypass of protein to the lower tract, gastrointestinal nematode suppression, and depression of fly larval development in manure (Muir et al., 2014).

The crude protein concentration of soybean forage increases and digestibility remains almost constant as plants mature from 50% bloom (17% protein and 59% digestibility) to 90% pod fill (21% protein and 61% digestibility). This is a result of the high levels of protein and energy in the seed. In central Florida, late-maturing lines had 15.0–18.5% protein and 59–64% digestibility while yielding 3 tons/acre.

Environmental Stresses and Pests

The lespedezas as a group are considered to be highly tolerant of drought and of acid, infertile soils, even those with relatively high concentrations of Al (Table 9.4). They are not considered tolerant of wet soils or of soil pH above approximately 7.2. Soybean is also best adapted to well-drained soils, but it is not well suited to soils with low pH or low fertility. Sericea shoots are killed back to ground level by cold during winter, and regrowth is initiated from **crown** buds each spring.

Korean lespedeza is particularly susceptible to bacterial wilt (*Xanthomonas campestris* pv. Lespedezae [Ayers et al.] Dye) in the northern part of its adaptation zone (McGraw and Hoveland, 1995), while powdery mildew (*Microsphaera diffusa* C. & P.), *Rhizoctonia solani* Kuehn, and southern blight (*Sclerotium rolfsii* Sacc.) can occur on annual lespedezas in the southern part of their adaptation zone. "Marion" striate lespedeza has a high level of resistance to bacterial wilt, tar spot, and southern blight, resulting in greater leaf retention. Sericea is relatively free of disease problems, although southern blight and *Rhizoctonia* species are sometimes observed in older cultivars. Insect damage to both sericea and the annual lespedezas is considered to be minimal. Root-knot nematodes can be serious pests of the lespedezas and soybean on light-textured soils. Nematode-resistant cultivars of soybean and sericea lespedeza have been released to address this limitation.

Cool-Season Forage Legumes

In the South, winter and spring temperatures are sufficiently warm to allow growth of some temperate legumes, primarily the clovers. They are planted during autumn into prepared seedbeds, or overseeded into dormant warm-season perennial grass sods. The clovers produce 75–90% of their total forage yield from late winter to spring, with very little growth occurring from late December to mid-February in most of the region. They are usually seeded along with cool-season annual grasses, particularly with annual ryegrass, because these grasses provide more forage for grazing during winter and reduce the potential for bloat during spring. Cool-season legumes extend the spring grazing season with forage that is generally higher in digestibility and protein than is the case for grasses.

Arrowleaf and crimson clovers are the two most important annual species. Other annuals of less importance include ball, berseem, rose, and subterranean clovers. Perennials such as alfalfa, red clover, and white clover are used in the region, but they act as short-lived perennials (in the case of alfalfa) or as annuals (in the case of red and white clover), particularly in the lower South.

Adaptation, Productivity, and Use

Arrowleaf and crimson clover are used primarily in pasture systems based on warm-season perennial grasses. Both are grown from the Atlantic coast west to central Texas and throughout the southeastern USA (Fig. 9.3).

Among the annual clovers, arrowleaf is the latest maturing and often the highest yielding (Ball et al., 2015). It has low bloat potential because it has higher tannin concentrations than other annual clovers (Ball et al., 2005). If soil moisture levels are adequate, arrowleaf clover may continue growth into early July. Arrowleaf is not recommended for overseeding of warm-season perennial grass hayfields because this prolonged growing season restricts grass growth during early summer.

Crimson clover tolerates a broader range of soil conditions than arrowleaf clover (Table 9.5). Good seedling vigor and early maturity make crimson clover ideal for overseeding into warm-season perennial grasses used for hay and as a winter cover crop in no-till rotations with warm-season row crops (Evers, 2000).

Ball, berseem, rose, and subterranean clovers are other annuals that are used to a limited extent in the region. Ball clover is best adapted to loam and clay soils, especially where poor internal drainage is a problem (it is primarily grown from east Texas and Arkansas to Alabama). It is considered to be more tolerant of acidity than other clovers (Hoveland and Evers, 1995). It is a good reseeder and tolerates intensive grazing, but it has high bloat potential (Table 9.5).

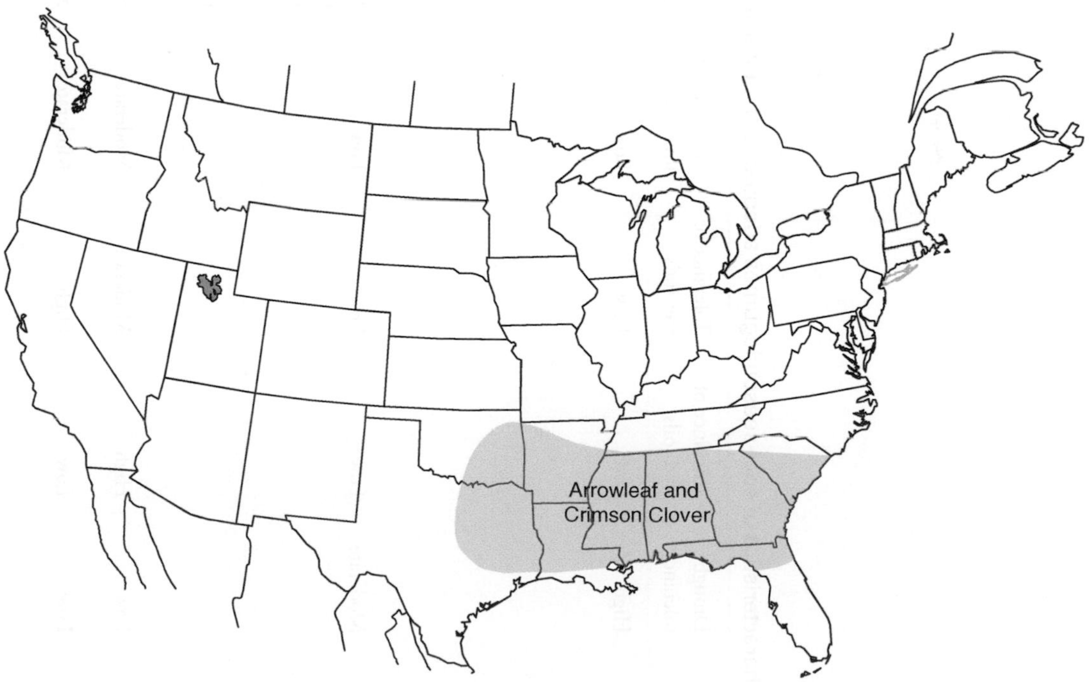

Fig. 9.3. Area of adaptation of arrowleaf and crimson clovers, cool-season legumes used in the intermediate South and the Gulf Coast region in the USA. (From Hoveland and Evers, 1995.)

Table 9.5. Characteristics of cool-season legumes adapted to the intermediate South and the Gulf Coast region

Legume	Drought tolerance	Tolerance of acid soils	Tolerance of wet soils	Tolerance of low fertility	Tolerance of grazing	Grazing/hay management[a]	Typical establishment method	Ease of establishment
Alfalfa	High	Low	Low	Low	Moderate	Bloat potential: high R: graze when vegetative, 4–6 weeks rest C: only with grazing-tolerant cultivars H: cut at first flower	Seed in late summer to autumn at 15–20 lb/acre	Easy
Arrowleaf clover	Moderate	Low	Low	Low	Moderate to high C: 3–4 in. R: avoid large accumulation of forage H: late-spring growth at early bloom	Bloat potential: low	Seed in autumn at 8–10 lb/acre	Moderate; excellent reseeding
Ball clover	Low	High	Moderate	Moderate	High	Bloat potential: moderate to high	Seed in autumn at 3 lb/acre	Easy; excellent reseeding
Berseem clover	Low	Low	High	Moderate	Moderate	Bloat potential: low C or R: enter at 6–8 in.	Seed in autumn at 20–25 lb/acre	Easy; poor reseeding

Crimson clover	Moderate	Moderate	Low	Moderate	Moderate to high C or R: begin grazing when 6–8 in. tall, graze no closer than 3 in. H: graze early growth, then harvest hay at early bloom	Bloat potential: low	Seed in autumn at 15–20 lb/acre	Easy; poor reseeding
Red clover	Moderate	Low	Low	Low	Moderate	Bloat potential: moderate R: leave more than 3-in. stubble H: cut at early bloom	Seed in autumn at 10–15 lb/acre	Easy
Rose clover	High	Moderate	Low	Moderate	Moderate	Bloat potential: low	Seed in autumn at 10–15 lb/acre	Difficult: excellent reseeding
Subterranean clover	Low	Moderate	Moderate	Moderate	High	Bloat potential: low C: 3–6 in.	Seed in autumn at 15–20 lb/acre	Easy; poor reseeding
White clover	Low	Moderate	High	Moderate	High	Bloat potential: high R: enter at 8–10 in., exit at 2 in. C: 3–4 in.	Seed in autumn at 2–3 lb/acre	Moderate

Sources: Ball and Lacefield, 2000; Evers, 2000; Lacefield and Ball, 2010; Ball et al., 2015.
[a]R, rotational stocking; C, continuous stocking; number refers to recommended pasture height. H, hay.

Berseem clover is less cold tolerant than the other annual clovers, and its use as a winter annual is limited to the lower South. Like crimson clover it has good early-season production, but it does not reseed well because its low percentage of **hard seed** leaves few viable seed in the soil to germinate later (Evers, 2000).

Rose clover has been used successfully in north central Texas and central Oklahoma, both of which are areas that are too dry for most clover species. It is deep rooted and an excellent reseeder, but poor nodulation and slow seedling growth are major limitations (Evers, 2000).

Subterranean clover or subclover is the common name for three *Trifolium* species, but only *T. subterraneum* will be discussed in this chapter. Subclover germination and seedling growth are favored by mild autumn temperatures, so early forage production is better than that for crimson clover near the Gulf Coast. Moving north, cooler temperatures reduce autumn and winter growth. Subclover has a prostrate growth habit and is tolerant of close grazing.

Red clover and white clover are productive during the spring in the lower South, but generally behave as annuals in this region because they do not persist during the hot summers. Moving north, the primary growing season is delayed, and red and white clover behave as short-lived perennials. This chapter will focus on their use as annuals in the lower South; for details of their use in temperate areas, see Chapter 8.

Red clover is adapted to moist, well-drained soils, but it is not as tolerant of wet soils as is white clover. In this region, red clover is generally seeded in autumn into a warm-season perennial grass sod, and it is often planted with annual ryegrass. It matures later than the annual clovers, except for arrowleaf, and will grow into May or June if moisture is available. Because of the differences in the timing of maturity of red and crimson clovers, they may be seeded together to extend the grazing season. The non-dormant red clover cultivars "Cherokee" and "Southern Belle", which were developed in Florida, begin growth earlier in spring than the other cultivars.

White clover grows best in moist, well-drained soils, but because it has a shallow root system it prefers wetter sites than red clover. Like red clover, it is seeded in autumn, most often in mixtures with annual ryegrass planted into warm-season perennial grass sods. White clover remains productive into the summer if soil moisture is adequate and grass competition is not too severe. It is a leafy plant that grows 8–12 in. (20–30 cm) tall and spreads by means of stolons. Because of its growth habit, it is best used for grazing; when it is harvested for hay or haylage, yields are lower than for red clover or alfalfa.

Alfalfa is not widely grown in the lower South because of its short stand life and high production costs. Soil constraints and pests (Table 9.5) are the major factors limiting its use and stand life. However, new cultivars that are better suited to the lower South have addressed some of the pest constraints. In the lower South, alfalfa is most often seeded in pure stands into a prepared seedbed in autumn, although seeding into existing bermudagrass is sometimes practiced. Alfalfa is most productive during spring and early summer. It is used primarily as hay for dairy cows and for horses, but grazing-tolerant cultivars such as "Bulldog 805" have increased its use in pastures in the South.

Establishment

Arrowleaf clover and crimson clover are planted in autumn, either by sod seeding or into prepared seedbeds. Autumn growth from sod-seeded stands is considerably less than that obtained using a prepared seedbed, resulting in lower total production and later onset of grazing for sod-seeded stands. Seed should be inoculated with the appropriate *Rhizobium* bacteria prior to planting. Arrowleaf clover has moderate seedling vigor and nodulation. If it is sod seeded, grass competition must be controlled by cutting or grazing to a short stubble height and then disking lightly prior to seeding. Ready establishment and early maturity make crimson clover ideal for overseeding into warm-season perennial grasses.

Arrowleaf clover stands can regenerate by natural reseeding. This species produces high seed yields that include up to 90% hard seed, so it develops a large soil seedbank. Planting 4–5 lb/acre (4.5–5.6 kg/ha) of **scarified** seed in the first reseeding year ensures a good stand. This practice is not necessary in subsequent years. Arrowleaf clover will produce seed in early summer, even when stocked continuously at a moderate stocking rate. If arrowleaf has been grown in a pasture for several years, its ability to reseed is severely limited by a *Fusarium* species disease complex (Hancock, 2014).

Crimson clover has poor reseeding ability because its hard seed percentage is only about 10% (Evers, 2000), and the seed do not survive for long in the soil. Its seed heads can also be damaged by clover head weevils (Ball and Lacefield, 2000).

Defoliation Management

Maximum forage yields of arrowleaf clover are usually obtained by frequent defoliation from the start of grazing until mid-April, followed by a single hay harvest at early bloom in late May (Ball et al., 2005). Total yields of 1–3 tons/acre (2.2–6.7 mg/ha) can be achieved. When grazed to a height of 3–4 in. (8–10 cm), arrowleaf continues to produce new leaves and has a longer growing season than when large amounts of forage accumulate. Under hay management, arrowleaf has few or no buds at the base of the plant in April or May, and therefore there is no regrowth after hay harvest during this time (Hoveland and Evers, 1995).

In February to March, crimson clover forage reaches a height of 6–8 in. (15–20 cm) and can be grazed for the first time. Total-season yields of 1–2 tons/acre are typical. The frequency of grazing does not markedly affect yield or persistence of crimson clover, but high grazing pressure during flowering will greatly reduce seed production.

Forage Nutritive Value

The nutritive value of clovers is generally quite high, with crude protein often in the range of 15–20% or higher, and digestibility in the range of 60–70%. Both decline as the plant matures. Arrowleaf clover has higher digestibility than crimson clover across a wide range of regrowth intervals. Winter pastures including arrowleaf or crimson clover in mixtures with cool-season annual grasses can support live-weight gains of cattle of 2–3 lb/day (0.9–1.4 kg/day).

The primary anti-quality factor associated with crimson clover is the risk of bloat, although the risk is much lower than for alfalfa and white clover (Table 9.5). Very mature seed heads of crimson clover contain barbed hairs that when fed in hay can harm horses, but this problem can be avoided by harvesting at early bloom (Ball and Lacefield, 2000).

Environmental Stresses and Pests

Tolerances of the cool-season legumes to a range of environmental stresses have been summarized in Table 9.5. The major pest problems in arrowleaf clover are viruses transmitted by aphids (*Aphidae* spp. J) (Hoveland and Evers, 1995) and root rots. The leaves of infected plants turn bright red to purplish red, and plants may undergo premature **senescence** (Hoveland and Evers, 1995). Viruses predispose plants to infection by *Phytophthora* root rot, and crown and stem rot (*Sclerotinia trifoliorum* Eriks.), which are fungal diseases that can occur if non-grazed forage accumulates during warm wet weather (Hoveland and Evers, 1995). Crimson clover is also susceptible to crown and stem rot in damp weather if there is significant forage accumulation (Ball and Lacefield, 2000). This problem can be avoided by grazing to remove excessive growth. On light-textured soils in the lower South, root-knot nematodes can also be a major pest of arrowleaf clover. Pest problems can be minimized by rotation of interseeded areas.

Multipurpose Forage Legumes

In addition to producing forage, legumes have other uses in the South. For example, they can be grown as green manure or rotation crops for soil enrichment, as well as being used for weed suppression, and as wildlife feed.

Soil Fertility Enhancement

The use of legumes in the South for soil conservation and to improve soil fertility peaked in the first half of the twentieth century. Species that were widely used included kudzu, sweetclover, cowpea, velvetbean, and lablab. Most of these were also used for forage. In 1919, 3.7 million acres (9.1 million ha) of velvetbean were grown in the southern USA for soil enhancement, as a smother crop, as forage, and as a source of beans or bean meal for cattle and swine (Eilittä and Sollenberger, 2001). Further details on the adaptation and use of these species have been provided elsewhere (Miller and Hoveland, 1995; Ball et al., 2015).

Wildlife Food

A number of legumes are planted as cover or as a food source for wildlife. Aeschynomene, also called deer vetch, has been planted by hunting clubs because it increases live weight and rack size of whitetail deer (*Odocoileus virginianus*). Crimson clover and white clover are widely used for deer plots. Bobwhite quail (*Colinus virginianus*) are attracted to annual lespedeza and partridge pea, and annual lespedeza is thought to be the most important winter food source for quail in the South (Ball et al., 2015). Cowpea and forage soybean can provide summer forage for deer, cover for quail, and seed for several types of wildlife (Ball et al., 2015).

Forbs for Winter Forage in the South

Forbs are herbaceous forage species other than grasses and legumes. Their primary use is as specialty crops or to fill a niche within a forage production system (Smith and Collins, 2003). In the South, forbs are an alternative to cool-season legumes and grasses to provide forage during the winter. The most prominent genus of forbs in the South is *Brassica*, to which turnip, kale, forage rape, and swede all belong. Brassicas establish rapidly from seed, tolerate frost, and their forage quality does not decline significantly with increasing maturity (Kalmbacher et al., 1982). They are similar in productivity and quality to small-grain cereals, annual ryegrass, and annual legumes when planted during the fall or early winter. However, because they are not legumes, they have relatively high N fertilizer requirements.

The energy value of brassica forages is high, and fiber and dry matter levels are particularly low. These traits make it difficult to utilize them for green chop or to preserve them as hay or silage, and they are much better suited for grazing (Smith and Collins, 2003). For swede and turnip, multiple harvests have been shown to increase leaf yield but decrease root yield, whereas single harvests increased root yield and decreased leaf yield (Kalmbacher et al., 1982). Limited observations show that winter gains of beef cattle grazing mixed pastures of forage rape and annual ryegrass are inferior to those of animals grazing on winter rye–annual ryegrass–crimson clover pastures. It has been suggested that the presence of anti-quality constituents may limit performance in some feeding situations. These constituents include glucosinolates that affect thyroid activity, especially of fetuses and younger, growing animals. Brassicas can also accumulate toxic levels of nitrate.

Summary

Numerous legumes are available for forage use in the southern USA, but far fewer acres are planted with legumes in southern than in northern regions. The reasons for this include lack of persistence, difficult establishment, and highly competitive warm-season grasses. Many cool-season legumes are used as winter annual species in the southern USA when conditions are suitable for their growth.

Questions

1. What factors have limited the success of legumes in the southern USA?
2. Why do annual legumes play a more prominent role in the South than in the northern USA?
3. Give two examples of anti-quality factors associated with legumes used during the warm season in the Gulf Coast region.
4. How do these anti-quality factors affect grazing livestock?
5. Why are annual lespedezas so well suited for overseeding into endophyte-infected tall fescue pastures?
6. What characteristics have been incorporated into new cultivars of sericea lespedeza that make it more attractive as a forage in the intermediate South?
7. Why is crimson clover a better alternative than arrowleaf clover for overseeding bermudagrass hayfields?
8. Describe the constraints on the use of alfalfa in the South.
9. What characteristics of new cultivars are increasing the opportunities for use of alfalfa in pastures in the South?
10. List some important non-forage uses for legumes in the South.

References

Aiken, GE, WD Pitman, CG Chambliss, and KM Portier. 1991a. Plant responses to stocking rate in a subtropical grass-legume pasture. Agron. J. 83:124–129.

Aiken, GE, WD Pitman, CG Chambliss, and KM Portier. 1991b. Responses of yearling steers to different stocking rates in a subtropical grass-legume pasture. J. Anim. Sci. 69:3340–3356.

Bagley, CP, IM Valencia, and DE Sanders. 1985. Alyceclover: a summer legume for grazing. Louisiana Agric. 28:16–17.

Ball, DM, and GD Lacefield. 2000. Crimson Clover. Oregon Clover Commission, Salem, OR.

Ball, DM, and JA Mosjidis. 2007. Sericea lespedeza: a pasture, hay, and conservation plant. Alabama Coop. Ext. Serv. Bull. ANR-1318. Alabama Cooperative Extension System, Auburn, AL.

Ball, DM, GD Lacefield, and CS Hoveland, 2005. Arrowleaf Clover. Oregon Clover Commission, Salem, OR.

Ball, DM, CS Hoveland, and GD Lacefield. 2015. Southern Forages: Modern Concepts for Forage Crop Management, 5th ed. International Plant Nutrition Institute, Peachtree Corners, GA.

Castillo, MS, LE Sollenberger, AR Blount, JA Ferrell, C Na, MJ Williams, and CL Mackowiak. 2014. Seedbed preparation techniques and weed control strategies for strip-planting rhizoma peanut into warm-season grass pastures. Crop Sci. 54:1868–1875.

Dalzell, SA, JL Stewart, A Tolera, and DM McNeill. 1998. Chemical composition of *Leucaena* and implications for forage quality. In HM Shelton et al. (eds.), Leucaena—Adaptation, Quality, and Farming Systems, pp. 227–246. ACIAR Proceedings No. 86. Australian Centre for International Agricultural Research, Canberra, ACT.

Eilittä, M, and LE Sollenberger. 2001. The many uses of mucuna: Velvetbean in the southern United States in the early 20th century. In M Flores et al. (eds.), Food and Feed from Mucuna: Current Uses and the Way Forward, pp. 73–100. CIDICCO, Honduras.

Evers, GW. 2000. Principles of Forage Legume Management. Texas A&M University Agricultural Research and Extension Center, Overton, TX.

Hancock, DW. 2014. Arrowleaf Clover. Georgia Agric. Exp. Stn. Pub. University of Georgia, Athens, GA.

Hoveland, CS, and GW Evers. 1995. Arrowleaf, crimson, and other annual clovers. In RF Barnes et al. (eds.), Forages: An Introduction to Grassland Agriculture, 5th ed., pp. 249–260. Iowa State University Press, Ames, IA.

Jones, RJ, KK Galgal, AC Castillo, B Palmer, A Deocareza, and M Bolam. 1998. Animal production from five species of leucaena. In HM Shelton et al. (eds.), Leucaena—Adaptation, Quality, and Farming Systems, pp. 247–252. ACIAR Proceedings No. 86. Australian Centre for International Agricultural Research, Canberra, ACT.

Kalmbacher, RS, PH Everett, FG Martin, and GA Jung. 1982. The management of *Brassica* for winter forage in the sub-tropics. Grass Forage Sci. 37:219–225.

Kalmbacher, RS, AC Hammond, FG Martin, FM Pate, and MJ Allison. 2001. Leucaena for weaned cattle in south Florida. Trop. Grassl. 35:1–10.

Kretschmer, AE Jr, and WD Pitman. 1995. Tropical and subtropical forages. In RF Barnes et al. (eds.), Forages: An Introduction to Grassland Agriculture, 5th ed., pp. 283–304. Iowa State University Press, Ames, IA.

Lacefield, GD, and DM Ball. 2010. Red Clover. Oregon Clover Commission, Salem, OR.

McGraw, RL, and CS Hoveland. 1995. Lespedezas. In RF Barnes et al. (eds.), Forages: An Introduction to Grassland Agriculture, 5th ed., pp. 261–271. Iowa State University Press, Ames, IA.

Miller, DA, and CS Hoveland. 1995. Other temperate legumes. In RF Barnes et al. (eds.), Forages: An Introduction to Grassland Agriculture, 5th ed., pp. 273–281. Iowa State University Press, Ames, IA.

Muir, JP, TH Terrill, NR Kamisetti, and JR Bow. 2014. Environment, harvest regimen, and ontogeny change *Lespedeza cuneata* condensed tannin and nitrogen. Crop Sci. 54:2903–2909.

Pitman, WD. 2006. Stand characteristics of sericea lespedeza on the Louisiana Coastal Plain. Agric. Ecosyst. Environ. 115:295–298.

Rusland, GA, LE Sollenberger, KA Albrecht, CS Jones Jr, and LV Crowder. 1988. Animal performance on limpograss-aeschynomene and nitrogen-fertilized limpograss pastures. Agron. J. 80:957–962.

Smith, DH, and M Collins. 2003. Forbs. In RF Barnes et al. (eds.), Forages: An Introduction to Grassland Agriculture, 6th ed., pp. 215–236. Iowa State University Press, Ames, IA.

Sollenberger, LE, and RS Kalmbacher. 2005. Aeschynomene and carpon desmodium: Legumes for bahiagrass pasture in Florida. In FP O'Mara et al. (eds.), Proceedings of the 20th International Grassland Congress, p. 334. Wageningen Academic Publishers, Wageningen, Netherlands.

Sollenberger, LE, CS Jones Jr, and GM Prine. 1989. Animal performance on dwarf elephantgrass and rhizoma peanut pastures. In R Desroches (ed.), Proceedings of the 16th International Grassland Congress, pp. 1189–1190. The French Grassland Society, Versailles Cedex, France.

Williams, MJ, CG Chambliss, and JB Brolmann. 1993. Potential of 'Savanna' stylo as a stockpiled forage for the subtropical USA. J. Prod. Agric. 6:553–556.

CHAPTER 10

Forage Crops for Bioenergy and Industrial Products

Kenneth J. Moore and Emily A. Heaton

Plants contain energy in the form of carbohydrates, proteins, and other compounds that can be used to meet our energy needs in the form of heat, liquid fuels, and other energy products. Their often fibrous root systems, perennial life cycle, symbiotic nitrogen (N) fixation, and efficient nutrient uptake make forage crops among the best candidates for bioenergy production. Federal policies designed to increase **biofuel** use have resulted in dramatic increases in renewable biofuel production in the USA during the past decade. A portfolio of federal legislation and regulations was developed that among other things provided financial incentives for fuel blenders to use ethanol in their products, provided tax incentives to develop ethanol production plants, and established production mandates for the manufacture and use of biofuels (Moore et al., 2013a).

The Renewable Fuel Standard (RFS) was developed to create a market for biofuels and help to ensure a reasonable return for those who invested in the infrastructure to produce fuel from biological materials. The RFS established mandates for the production of biofuels beginning in 2006 and continuing up to 2022, and is regulated by the US Environmental Protection Agency (EPA). Three primary types of biofuels are covered by the RFS, namely conventional biofuels, advanced biofuels, and cellulosic biofuels. Conventional biofuels include ethanol derived from grain. Advanced biofuels include any fuels that are not made from cereal starch, such as biodiesel and fuels derived from cellulosic materials. Cellulosic biofuels are advanced biofuels that are derived from fibrous materials, and include those made from energy crops (Council for Agricultural Science and Technology, 2007). It is this latter category that is relevant to forage crops, because many of the species that can be utilized to produce cellulosic fuels can also be used as forage in animal production (Moore et al., 2013a).

The RFS mandated the production of 9 billion gallons of biofuel by 2008, and is increasing the production goal over time to 36 billion gallons by 2022. Conventional biofuel production was capped at 15 billion gallons in 2015. Further increases are to be made up of advanced biofuels and primarily cellulosic biofuels (Fig. 10.1). The initial rapid expansion of the biofuel industry was based on construction of plants to produce conventional fuels. Future growth in renewable fuels is likely to consist almost entirely of cellulosic biofuels, and these will include crops that can also be grown as forages.

The feasibility of producing enough biomass to displace 30% of US petroleum consumption was assessed in a report released jointly by the US Department of Agriculture (USDA) and the US Department of Energy (DOE) in 2005 and updated in 2011 (Perlack et al., 2005; US Department of Energy, 2011). This report refined estimates of biomass availability by accounting for the influence of price on availability, and it predicts that energy crops could surpass crop residues as the largest source of biomass by 2022. Clearly, energy crops will become increasingly important over the next decade if the current trajectory for biofuel production is maintained. In this chapter you will learn about the crops that are being evaluated for energy production, the environmental benefits of

Forages: An Introduction to Grassland Agriculture, Seventh Edition. Edited by Michael Collins, C. Jerry Nelson, Kenneth J. Moore and Robert F Barnes.

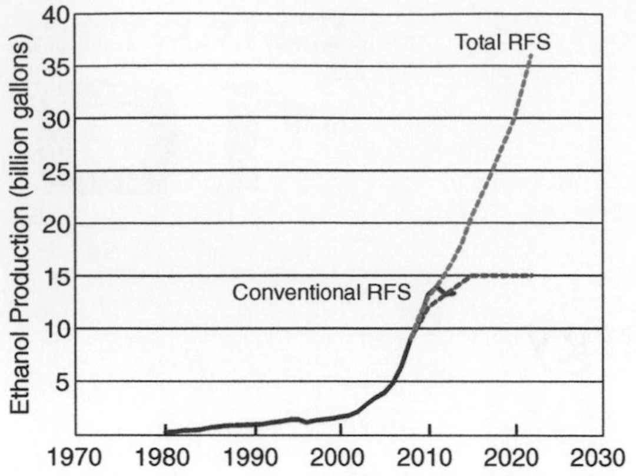

Fig. 10.1. US ethanol production since 1980 and Renewable Fuel Standard (RFS) mandates for the period 2008–2022. Conventional fuels produced from grain were capped at 15 billion gallons in 2016. The difference between the total RFS mandate and the conventional RFS mandate will mainly come from cellulosic biomass. (Adapted from Moore et al., 2013b.)

growing them, and the ways in which they can be converted to advanced biofuels.

Conversion Technologies

Forage crops have provided energy and other nutrients for traction and transportation since before the start of recorded history. Before the introduction of the modern tractor in the early twentieth century, draft animals provided nearly all of the traction required for plowing, cultivation, and other farm operations. A significant portion of the land owned or managed by most farmers had to be used to produce forage to feed these animals in addition to providing nutrients for food animals (Vogel, 1996). As farms became increasingly mechanized, land previously used to grow forage for draft animals was shifted to other uses, mostly cash crops or forages used for food animal production.

Thermochemical Processes

Combustion is the easiest and oldest method of converting biomass to heat energy. Simply burning plant material releases substantial energy that can be captured as heat and used in any number of processes, including electric power generation. Co-firing biomass with coal is one option that has been explored as a means of using biomass as a fuel (Van Loo and Koppejan, 2008). However, the current cost of biomass required to produce a given amount of heat energy is much higher than that of coal, and the process is not widely used commercially.

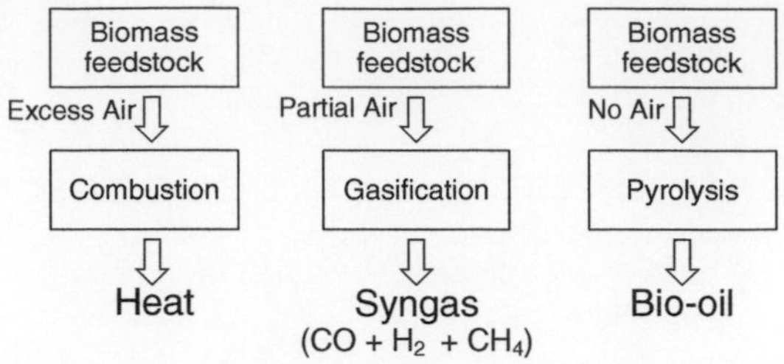

Fig. 10.2. Thermochemical conversion processes used to produce fuels and other products from plant biomass.

There are other thermochemical processes (Fig. 10.2) that involve the incomplete combustion of biomass under reduced oxygen to produce gaseous or liquid fuels (Brown, 2003). **Gasification** involves heating the biomass under conditions of high pressure and limited oxygen supply. Multiple chemical reactions occur under these conditions, converting the biomass to a flammable gas mixture called syngas. Syngas produced from biomass contains hydrogen, carbon monoxide, methane, and other gases that can be further combusted to release energy. It can be used for heat and power generation, and technologies are available for processing it into other chemicals, including liquid fuels. **Pyrolysis** uses even higher temperatures and pressures to combust biomass in the absence of oxygen. Under these conditions the reactions produce larger molecules that can be condensed into a liquid product referred to as bio-oil. This latter process is known as pyrolysis, and in addition to bio-oil it produces some syngas and a carbon-rich byproduct called biochar. Biochar is of great interest as a soil amendment because it is extremely resistant to breakdown and can increase soil organic matter content while sequestering substantial amounts of carbon that would otherwise be converted to CO_2 and released into the atmosphere (see "Biochar: Sequestering Carbon and Improving Soil Quality"). The oil produced by pyrolysis has properties that are similar to those of fossil oil, and can be refined into fuel products that are indistinguishable from gasoline and diesel derived from fossil oil.

Biochemical and Extraction Processes

All plants capture energy from sunlight, which fuels a vast array of reactions that produce the biochemicals required to support life. Many of these biochemicals are important nutrients for sustaining life in the organisms that consume them. As you will learn in Chapter 14, carbohydrates and proteins are the most important nutrients derived from forages, but they are also important sources of the vitamins and minerals that are necessary for growth and reproduction in livestock. From the standpoint of producing fuels, proteins and minerals are at best by-products, and at worst they are contaminants that can interfere with conversion processes (Brown, 2003).

The largest fraction of biochemicals produced by forages are carbohydrates. These include simple sugars involved in metabolism, storage polysaccharides such as starch and fructans, and structural polysaccharides including cellulose and hemicelluloses. Simple sugars and storage polysaccharides are relatively easy to extract and then convert to sugars that can be fermented to produce fuel. However, the cellulose and hemicelluloses that account for most of the carbohydrates in plants are much less easily extracted than simple sugars and storage polysaccharides.

The biochemical conversion of structural carbohydrates in a cellulosic ethanol refinery is roughly analogous to the process that occurs within the digestive system of a ruminant animal (Fig. 10.3). The plant biomass must be pretreated to reduce its particle size and render the structural polysaccharides more susceptible to enzymatic hydrolysis (US Department of Energy, 2006). However, the conversion efficiencies of complex carbohydrates in ruminants are far lower than the values that would be acceptable in a **biorefinery**. The digestibility of cellulose and hemicelluloses in forages consumed by ruminant livestock is generally only in the range of 40–60%, which means that a large proportion is excreted by the animal in an undigested form. Much higher conversion yields are necessary to economically convert these compounds to biofuels. Therefore

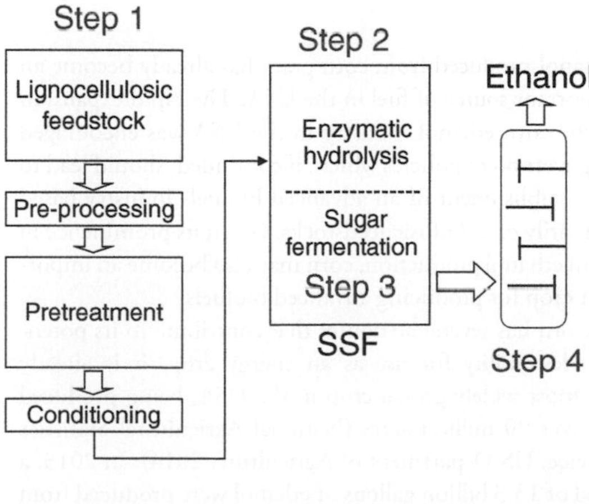

Fig. 10.3. Biomass conversion using biochemical extraction to release sugars which are then fermented to ethanol. The process involves four key steps: (1) pretreatment, (2) hydrolysis, (3) fermentation, and (4) distillation.

the biomass used in a biorefinery must be subjected to significant pretreatment to render the complex carbohydrates more susceptible to enzymatic hydrolysis.

Both pretreatment types, physical and chemical, are designed to increase the accessibility of glycosidic bonds to the hydrolytic enzymes that will cleave them. A relatively simple way to achieve this is to apply a dilute acid or base that will hydrolyze the molecular bonds between **lignin** and the polysaccharides that prevent access by the enzymes. Removing these barriers to hydrolysis can greatly enhance the yield of simple sugars.

After pretreatment, the biomass is subjected to enzymatic hydrolysis to release simple sugars from the material. This can be done either in a separate process from fermentation or simultaneously by combining the enzymes with the yeasts required to convert the sugars to alcohol in a single vessel. This latter method is often referred to as simultaneous scarification and fermentation (SSF). The alcohol produced must be separated from the remainder of the material and purified, usually by distillation.

Another extraction process that can be used to remove carbohydrates from biomass without using enzymes is solvolysis. In this process, the biomass is treated with a solvent under conditions of very high temperature and pressure to release the sugars from the complex carbohydrates. Superheated water is the most commonly used solvent for this process. Once the sugars have been extracted they can be converted to fuels or other biochemical products using biological or catalytic processes.

Integrated Processing

Production of fuels and other biochemical products from biomass is a relatively new development, and processes are still being developed and improved. There are many competing technologies, and it is likely that entirely new ones will be developed. Many of these processes will probably be integrated into modern biorefineries to improve their overall efficiency and increase the range of different products that can be formed. For example, many ethanol plants already use some type of thermochemical process to convert the residual materials remaining after extraction of carbohydrates to produce the heat required for the fermentation and distillation components of the process. This greatly improves the overall efficiency of conversion, and it reduces the input of fossil-fuel-derived energy required for the process.

It is not difficult to imagine biorefineries in the future that will replace the refineries currently used to convert fossil oil into fuels and other chemicals. Most of the technologies required to do this have already been developed. The remaining challenge will be to shift from an energy economy based on fossil fuels to one based on biorenewable materials. This transition has already begun and, given the finite availability of fossil fuels and the negative environmental consequences of their use, it is likely to continue and accelerate.

Biochar: Sequestering Carbon and Improving Soil Quality

One of the products of pyrolysis of biomass is biochar, a carbon-rich substance that is extremely resistant to chemical breakdown. Biochar has several properties that could make it an excellent amendment for improving soil health and productivity (Fig. 10.4).

Fig. 10.4. Biochar can be applied to agricultural fields to increase overall soil health and productivity by improving water-holding and cation-exchange capacity as well as other soil properties. (Photo courtesy of David Laird, Iowa State University.)

It has adsorptive properties and can increase the water-holding capacity of soil. It also has cation- and anion-exchange sites that can increase the capacity of the soil to retain and exchange nutrients in the soil environment. The anion-exchange capacity of biochar is of special interest because it may help to retain anions such as nitrate that are readily leached from the crop rooting zone. In fact researchers are exploring ways to increase the anion-exchange capacity of biochar for this very purpose. It may one day be used as the active component in bioreactors designed to purge effluent from crop fields of nitrate, thereby preventing nitrate pollution of surface waters.

Most of the benefits of biochar result from it being a stable form of organic matter in the soil. It has a very long half-life in the soil, of up to 1000 years. This makes biochar an excellent compound for sequestering carbon and keeping it out of the atmosphere. Biochar therefore has the potential to help to reduce greenhouse gas emissions from agriculture.

Energy Crops

Just about any forage crop could potentially be grown as a feedstock for biofuel production. However, there are some characteristics that make certain species more suitable for this purpose than others. The ideal energy crop would be high yielding, broadly adapted, have alternative uses and therefore market flexibility, be stable under environmental stresses, and be compatible with existing cropping systems (Moore et al., 2013a). Of the species that are most often grown for forage, those most likely to possess these traits are grasses with the C_4 photosynthetic pathway. These grasses generally use N and water more efficiently, are more stress tolerant, and produce higher yields than C_3 grasses and legumes in temperate and tropical environments (Moore et al., 2004). They also generally have much lower nutritive value and make poorer-quality forage. However, this is not a problem in relation to most of the technologies for producing biofuels, because they are better able to convert the complex polysaccharides, and in some cases phenolic compounds, which can be very recalcitrant to digestion by ruminants.

A host of different forage species have been evaluated as potential energy crops. Among these, several have been identified that appear to have excellent potential, and these will be addressed in this chapter. They include some annual grasses, such as corn and sorghum, and several perennial grasses, including switchgrass, miscanthus, and energy cane (Table 10.1). Others will certainly be developed, but for the time being these are the ones that are either currently being used or in planning for use in biofuel systems.

Annual Grasses

Corn

Ethanol produced from corn grain has already become an important source of fuel in the USA. The rapid expansion of the corn ethanol industry in the USA was encouraged by government policies which, if continued, should lead to the establishment of an advanced biofuels industry based primarily on cellulosic feedstocks. Given its prominence in grain ethanol production, corn may also become an important crop for producing advanced biofuels.

Corn has several attributes that contribute to its potential desirability for use as an energy crop. It is already the most widely grown crop in the USA, being produced on over 90 million acres (National Agricultural Statistics Service, US Department of Agriculture, 2014). In 2013, a total of 13.3 billion gallons of ethanol were produced from corn grain, just short of the RFS mandate of 13.8 billion gallons (Renewable Fuels Association, 2014, p. 2). Corn ethanol already provides 10% of the transportation fuel supply in the USA. Corn ethanol production consumes approximately 40% of the corn grain produced in the USA. The USDA and DOE estimate that the availability of crop

Table 10.1. Basic information on early and developing grass energy crop species in the USA

Crop	Establishment method	Life cycle	Established agronomics	Established markets	Typical biomass yield (tons DM/acre)
Maize (corn)	Seed	Annual	Yes	Food, feed, grain ethanol	5–10
Sorghum	Seed	Annual	Yes	Food, feed, forage	7–12
Sugarcane/Energy cane	Stem cuttings	Perennial	Yes	Sugar	23–52
Switchgrass	Seed	Perennial	Yes	Forage	3–10
Miscanthus	Rhizomes	Perennial	No	Not developed	10–15

Source: Bean et al., 2006; Pyter et al., 2007; Schmer et al., 2008.
DM, dry matter.

residue remaining from corn grain production that can be used sustainably under current production practices is over 100 million dry tons (US Department of Energy, 2011). This is only a fraction of what could be available if more sustainable production practices were developed.

Because of its historical and current prominence as a food, feed, and energy crop, the infrastructure for improving the performance of corn surpasses that for any of the other potential energy crops. More is known about the genetics, production, and use of corn than for any other commercial crop. Furthermore, the supporting knowledge and supply chain for inputs are highly developed. Many farmers are familiar with growing corn as a grain crop, and the changes in production practices needed to grow it for both grain and biomass are likely to be minor compared with growing a different crop species.

Mininum tillage or "no-tillage" cropping systems have developed that involve planting the corn or soybean directly into residues from the previous crop without plowing or major tillage (Moore and Karlen, 2013). These residues protect the soil from wind and water erosion and contribute to soil organic matter, and there is significant concern that removing corn crop residue as biomass could lead to environmental degradation. In addition to causing off-site concerns about air and water quality, removal of crop residues could lead to long-term losses in productivity due to soil losses through erosion and decreased soil organic matter. These negative impacts are likely to occur under current tillage and other management practices. However, in order to address these concerns, alternative management practices could be developed that would allow removal of crop residue with minimal environmental impacts (Moore et al., 2013b).

The facilities currently being developed to produce cellulosic ethanol plan to use corn crop residues as a feedstock. Therefore, at least for the foreseeable future, corn will continue to be an important bioenergy crop. Many believe that the initial development of plants for converting cellulosic biomass from corn will lead to the development of processes and markets for other energy crops.

Sorghum

Sorghum, like corn, is already an important food and forage crop, with different types of sorghum being used for different purposes. Milo types have a higher harvest index (ratio of grain yield to total biomass) than the others, and are used for grain production. A relatively small amount of sorghum grain is used to produce ethanol in regions where sorghum is better adapted than corn (generally areas with lower rainfall). There are also sweet sorghums that accumulate higher concentrations of sugars in their stalks, much like sugarcane. Sweet sorghums have traditionally been used to produce sorghum syrup, but have generated interest in their potential use as biomass crops because they provide a source of readily converted non-structural carbohydrates for ethanol production.

The forage sorghums include sudangrass, forage types, and hybrids between them. Because they are relatively high yielding, forage types and sorgum × sudangrass hybrids are of interest as potential bioenergy crops. Sorghums grown for food and forage are generally insensitive to photoperiod. Their growth and development proceed through the reproductive stages regardless of daylength as long as other conditions are favorable for growth. However, there are some sorghum varieties that are sensitive to daylength. These are sometimes called "tropical sorghums" because they generally flower only under very short daylengths. When these varieties are grown at higher latitudes they remain vegetative and continue to accumulate biomass until the end of the growing season. For this reason there is a great deal of interest in growing tropical sorghums for cellulosic energy production.

Some key attributes of sorghum may prove advantageous for its use as an energy crop. It is better adapted to hot dry conditions than corn, and it uses N and water more efficiently under these conditions. In those areas that become warmer and drier as a consequence of changes in the global climate, sorghum may prove to be a better alternative as an energy crop.

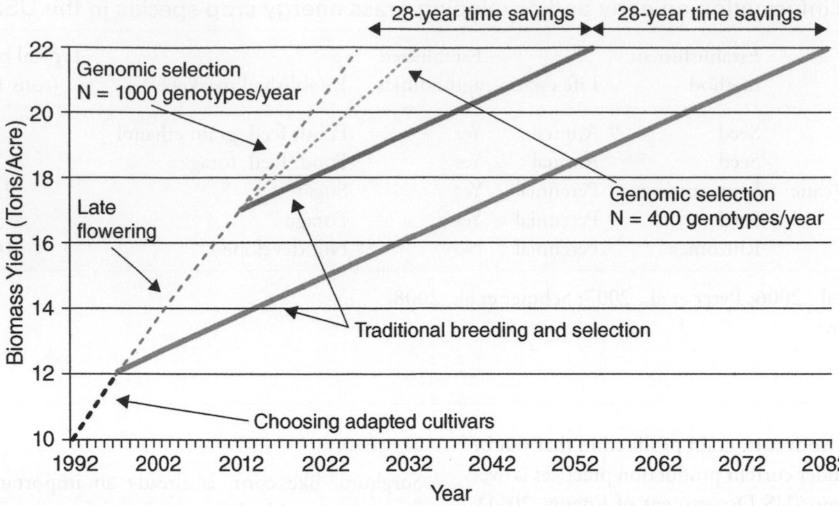

Fig. 10.5. Historical and projected timelines for genetic improvement of switchgrass biomass yield for USDA Hardiness Zones 3 to 6. The graph includes four phases of breeding: the appropriate choice of adapted cultivars for the correct hardiness zones (broken black line); traditional field-based breeding and selection (solid red line and solid purple line); incorporation of the late-flowering trait into winter-hardy plant germplasm (broken green line); and the use of genomic prediction methods based on DNA-sequence information (broken orange line and broken blue line). (Figure provided by Michael D. Casler, USDA-ARS, University of Wisconsin, and Great Lakes Bioenergy Research Center, Madison, WI. Historical and projected gains in biomass yield were taken from the following: Casler, 2010, 2014; Casler and Vogel, 2014; Ramstein et al., 2016.)

Perennial Grasses

Switchgrass

The most extensively studied and developed perennial grass species for use as an energy crop is switchgrass. It was identified early on as a promising energy crop because of its wide adaptation zone, relatively high productivity, efficient use of nutrients, especially N, and tolerance of environmental stresses (Wright and Turhollow, 2010). It is regarded as a model bioenergy crop, and is the only perennial grass species that has been improved specifically for this purpose. As discussed elsewhere (see Chapter 7 and the Compendium), switchgrass is native to the USA and has many positive agronomic characteristics. It has a perennial life cycle and the C_4 photosynthetic pathway, which imparts high N and water use efficiency.

"Liberty", the first switchgrass cultivar to be developed specifically for energy production, was released in 2014 by USDA-ARS and the University of Nebraska (Vogel et al., 2014). Switchgrass cultivars belong to one of two major ecotypes—upland and lowland. In general, upland cultivars are better adapted to areas that experience severe winters. Lowland cultivars are better adapted to areas that have mild winters, and they typically yield substantially more biomass. "Liberty" was developed by crossing a lowland cultivar with an upland one. The resulting population was selected to specifically yield more biomass and survive harsh winters. It is adapted in the USA as far north as Wisconsin, and has performed very well in experimental trials. It is anticipated that seed of "Liberty" will be available to farmers in 2017.

There are several possible approaches that can be utilized for improving and adapting switchgrass for use as an energy crop (Fig. 10.5). Synthetic cultivars of switchgrass have been traditionally developed by selecting superior genotypes from within an existing population or cultivar. Slow but steady progress has been made in improving switchgrass biomass yield using this approach. Switchgrass is sensitive to photoperiod, and the flowering date of any given cultivar is greatly affected by the latitude for which it was developed. Those that are adapted to more southern latitudes flower later when grown at more northern latitudes. One way to potentially increase the yield of a switchgrass cultivar would be to extend the period of vegetative growth. This can be accomplished by growing a forage cultivar that was developed at a lower latitude, but this approach often results in poor winter survival. However, by selecting the surviving plants for further development, it should be possible to produce new cultivars that are both later flowering and sufficiently winter hardy.

Additional improvements in switchgrass biomass production could be greatly accelerated by using modern genomic approaches. Genetic tools such as marker-assisted selection could be used to speed up the identification of superior genotypes. These methods are currently being used to develop new energy-crop cultivars of switchgrass, and should result in more frequent cultivar releases in the future.

Energy cane (*Saccharum* species)

Sugarcane is by far the most productive biomass crop grown in terms of dry matter yield per unit area (Somerville et al., 2010). Most commercial sugarcane varieties have been selected out of interspecific hybrids of *Saccharum officinarum* and *S. spontaneum* over several generations, emphasizing high sugar yield (Tew and Cobill, 2008). High sugar yield and minimal residue or bagasse are preferred for sugar production, to reduce the amount of material to be transported to the mill and the amount of residue remaining after the sugar extraction, thus increasing the efficiency of the process (Tew and Cobill, 2008). However, for producing bioenergy, varieties that produce both high yields of sugar and high yields of fibrous biomass might be superior. Type I energy canes for both high sugar and high fiber yields are typically produced from early-generation hybrids. Early-generation energy-cane hybrids produce higher biomass yields, exhibit greater persistence and therefore extended ratooning capacity, and are more tolerant of abiotic stresses. Type II energy canes are selected primarily for high fiber yield, and use less traditional sugarcane germplasm in their development. Consequently, they tend to be more tolerant of low temperature stress than other canes, and their production may be better adapted in more northern latitudes (Tew and Cobill, 2008).

Energy canes have excellent potential as bioenergy crops in the tropical and subtropical regions where they are best adapted. The technology and infrastructure for producing energy cane are already in place in many areas, and the genetics of the crop are well understood. Sugarcane is grown in about 80 countries worldwide, and produces about 70% of the global sugar supply (Tew and Cobill, 2008). It is likely that energy canes will become the dominant herbaceous biomass crop in the areas where they are adapted. Building off sugarcane infrastructure, energy-cane supply chains can be highly productive, especially when combined with heat, power, and high-value co-products in an integrated biorefinery.

Giant miscanthus (*Miscanthus* × *giganteus*)

Another perennial grass that is under development as a cellulosic biomass crop is the sterile hybrid *Miscanthus* × *giganteus*, often referred to as giant miscanthus. Although it has not been well studied in the USA, giant miscanthus has been assessed thoroughly in Europe in much the same way as switchgrass has been evaluated in North America. A hybrid between *M. sacchariflorus* and *M. sinensis*, this triploid does not produce fertile seed and is therefore typically planted using rhizome cuttings or live plants. Giant miscanthus was advanced as an energy crop in the EU in part because this sterility, coupled with a non-spreading growth habit, mitigated the risk of weediness or pollen outcrossing with compatible species. Following years of testing in multi-location trials around the EU, giant miscanthus was shown to produce consistently high biomass across a range of conditions with minimal inputs, and at temperatures and latitudes beyond the normal growing range of warm-season grasses. When evaluated in the USA, giant miscanthus produced record yields, and on average two to four times more biomass than switchgrass (Heaton et al., 2010).

The requirement for digging, sorting, transporting, and planting rhizomes dramatically increases the planting costs for giant miscanthus compared with traditional seed-based crops. However, the associated economic and environmental costs may be offset by the higher biomass yields and the low annual production inputs when the crop is grown on appropriate land. Like switchgrass, giant miscanthus has long stand lifetimes, low input requirements, and well-documented environmental benefits, but it is important that it is located and managed appropriately. For example, no-tillage methods do not exist for *M.* × *giganteus*, there is a "payback" time required to replace soil C, and it is less drought tolerant than North American prairie grasses, restricting current recommendations for its use to temperate, arable, rain-fed areas that are not highly erodible (Heaton et al., 2010). The management of and harvest practices for all energy crops can have a dramatic impact on their relative value for energy production and **greenhouse gas** abatement, highlighting a role for appropriate crop management (Davis et al., 2013), as will be discussed in the next section.

Management of Energy Crops

Management practices for forage species that are being grown for energy production are quite similar to those used when they are grown for forage. Seeding rate and depth, time and methods of planting, and pest control practices are virtually identical. However, fertilization requirements and optimum cutting management may differ. Forages are generally used near the area where they are produced, allowing nutrient recycling in livestock urine and manure, but biomass will probably be transported several miles to a central processing facility. In most cases the biomass will need to be pre-processed to achieve a physical form that is amenable to the conversion process. Thus the storage and handling methods that are used for biomass are likely to be very different to those used for forage crops.

Vegetative Propagation

Sterile crops, such as miscanthus, or those with slow seedling growth, such as energy cane, can be vegetatively propagated as shown in the examples of miscanthus propagation schemas in Fig. 10.6.

Fig. 10.6. The production of vegetative material for use in propagating miscanthus by stem propagation (left) and by rhizome production (right). (Photos courtesy of Emily Heaton, Nicholas Boersma, John Caveny, and Repreve Renewables.)

Stem propagation involves excising aerial stem nodes and then growing them in the greenhouse before transplanting them into the field. Rhizome propagation is more common, but requires more disturbance of parent fields as the rhizomes are lifted while they are dormant, and then processed to fit through transplant or custom planters. Both methods are more resource intensive than seed production and planting systems, which limits the adoption or expansion of vegetatively propagated systems in some regions.

Fertilization

The role of fertilization is to ensure that crop yield is not being limited by mineral nutrient availability. Fertilizer requirements depend on soil nutrient levels and the rate of removal by the crop. In general, macronutrients such as phosphorus (P) and potassium (K) should be applied at the rates needed to replace removal by the crop. As a rule of thumb, biomass crops remove about 5 lb of P and about 40 lb of K per ton of dry matter harvested, although these values can be much lower if the crop is harvested for biomass after frost to allow recycling of some of the nutrients. The growing of very high-yielding biomass crops therefore requires substantial inputs of these nutrients as fertilizer. Build-up nutrient applications may be beneficial for increasing production potential on low-fertility soils (see Chapter 12).

Unlike P and K, mineral N is not retained to any significant extent in the soil profile. If it escapes the root system of the crop it is then lost and can be leached into the groundwater, where it may cause health problems if it is consumed by humans and animals. Fortunately, most bioenergy crops are grasses with extensive fibrous root systems and high yield potential that increase their capacity to immobilize and recycle N.

Optimum fertilizer recommendations are not well defined for many bioenergy crops. Nitrogen fertilizer recommendations for switchgrass are in the range of 75–150 lb/acre. Lower optimum N rates of 50–75 lb/acre for miscanthus may result from free-living associative N-fixing bacteria in the root zone of this species, or from the extensive miscanthus root system that can explore larger soil volumes.

Harvest Management

Forage quality generally declines as the crop matures, while yield increases (see Chapter 14), so harvest decisions for animal feeding represent a compromise between these conflicting goals. Multiple cuttings of perennial forages are generally made to achieve an acceptable nutritive value. Harvest management of bioenergy crops is very different. In this case, the goal is generally to maximize biomass dry matter production. The elevated fiber levels that are found in very mature crops are not a negative factor in bioenergy conversion, and animal nutritive value is not a consideration. Most perennial grasses produce similar total seasonal yields regardless of the number of cuttings, so biomass systems often involve a single annual harvest.

The timing of cutting is important even with infrequent harvest. Cutting perennial crops within a month or so of the first killing frost can increase winter injury and lead to stand loss. Cutting too close to the frost date can stimulate new growth, lead to use of carbohydrate reserves that are needed for overwintering, and delay development of winter hardiness (see Chapter 5).

Nutrient cycling is an important concern for high-yielding biomass crops which require and remove large amounts of mineral nutrients. The minerals contained in biomass do not contribute to its biofuel value, and in fact high ash levels are detrimental to the conversion process. One benefit of delaying biomass harvest until after frost is the reduction in mineral concentrations. During the hardening period, perennial grasses remobilize proteins and other nutrients to below-ground storage organs. Leaves generally have much higher mineral concentrations than stems, and decay at a faster rate following frost, which also contributes to lower concentrations. The smaller leaf component reduces yield somewhat, but the recycling of nutrients and the reduced biomass ash concentrations that also occur are beneficial. If the cutting takes place before frost it should be made well before it, and ideally soon after the crop has achieved maximum biomass production.

Storage and Handling

Because of the larger scale, the logistics of harvesting, transportation, and storage of biomass are much more complicated than they are for the relatively small, distributed systems used for forages (Sokhansanj et al., 2009). Currently, the transportation of biomass crops harvested using conventional forage machinery is limited by volume rather than by weight. Bales of field-dried bioenergy crops have a density of 10–11 dry lb/ft^3, which is relatively low compared with other agricultural commodities. The transportation cost per unit of biomass can be decreased by densification of the biomass to maximize the weight transported on a truck or other vehicle. The biomass-processing systems envisioned currently involve particle size reduction, and it may be beneficial to undertake this near the site of production in order to increase the density of the material for transporting. Conversion methods and facility designs are still evolving, and will ultimately determine the optimum packaging form and storage methods for biomass (Mckendry, 2002).

Environmental Benefits of Energy Crops

Using renewable fuels produced from energy crops could help to alleviate many of the environmental concerns associated with fossil fuels. Because the carbon released during the combustion of fuels derived from biomass was only recently removed from the atmosphere, they do not contribute directly to the build-up of CO_2 in the atmosphere. Since some of the energy used in the production of energy crops is derived from fossil fuels, the offset is not entirely complete. However, in the case of thermochemical biofuels, a significant amount of carbon can be fixed as biochar, and incorporating it as a soil amendment could lead to biofuel production systems that actually fix more carbon than they release into the atmosphere.

Perennial energy crops can also help to address several other environmental problems associated with agricultural production. For example, perennial energy crops could be grown on sites with high erosion potential, thereby stabilizing the soil and reducing the amount of sediment and nutrients in waterways (Kort et al., 1998; Lemus and Lal, 2005). Perennial grass species generally partition a greater proportion of the carbon that is fixed through photosynthesis to below-ground organs than do annual species. This, together with the absence of tillage, leads to greater carbon sequestration potential than is possible in intensively managed row crops.

Many of these biomass crops are likely to be better adapted and relatively more productive on marginal land than grain crops. Land that is susceptible to flooding, with poor water-holding capacity, and other limitations to row crop production may be more suitable for biomass crop production with less risk of environmental problems.

Other Industrial Uses of Forage Crops

Chemicals produced by plants, including carbohydrates, proteins, phenolic compounds, and lipids, can be extracted and converted to any number of products. Plants were ultimately the source of many synthetic chemicals produced from fossil fuels. Such products produced from recently grown plants are considered to be renewable, as they do not rely on a finite resource such as fossil oil for their manufacture (Brown, 2003).

Chemicals extracted from plants can be converted into any number of other useful compounds through catalysis and/or fermentation methods. Some chemicals that are already commonly derived from plants include sugar alcohols, furfural, lactate, and other organic acids. Biodegradable plastics derived from plant starch have been developed and are used to manufacture bags, bottles, mulches, and a host of other products. Less degradable plastics can be made from cellulose derivatives, and can be used to produce more durable items.

Plants are a source of biorenewable fibers that can be used in a number of manufactured products. They can be subjected to a pulping process to produce paper and other consumer products, and they can be used as fillers and reinforcing matrices in plastic composites. Certain plant species, although technically not forage species, produce bast fibers that can be used to produce coarse textiles such as burlap. Natural fibers produced by plants could eventually be used to replace synthetic fibers that are manufactured from fossil oil.

Proteins are an important class of nutrients in the diet of all animals. Some forages, such as alfalfa and other legumes, produce very high concentrations of high-quality proteins in their leaves. Use of these proteins is generally restricted to the diets of ruminants and other herbivores that can digest fibrous plant materials. However, it is possible to extract food and feed-grade proteins from forages by a process called wet fractionation which involves expressing a protein-rich fraction from fresh plant material (Sanderson et al., 2007). The residue is high in fiber and can be used for a number of purposes, including cellulosic ethanol production. The wet fraction containing the protein can be further processed to produce high-quality food and feed-grade proteins that can be used in the diets of humans and monogastric livestock.

Summary

There has been much public concern about the cost of fossil fuels and the need to reduce greenhouse-gas emissions. Direct combustion of plant biomass to heat is not efficient, so high-yield crops are evaluated for their potential to recapture CO_2 via photosynthesis in biomass for ethanol, biodiesel, and other compounds. Government mandates, regulations, and policies are in place to increase the use of biofuels and their production. Plant biomass contains large amounts of sugars, starch, cellulose, and hemicelluloses that can be heated anaerobically to form high-energy fuels or biochar, a soil amendment.

Corn in particular, and also sorghum and several perennial grasses, including switchgrass, miscanthus, and energy cane, are considered most suitable. Perennials are preferred in order to reduce tillage and planting costs, and are usually harvested once in late fall to obtain the highest yield with low mineral content. Switchgrass appears promising due to its wide adaptation, high productivity, efficient N use, tolerance of environmental stresses, and positive environmental benefits. There is a need for improvement of the harvesting, packaging, and transport of the bulky biomass. The industrial production of fuels and other biochemicals from biomass is still relatively new, and its development is ongoing.

Questions

1. Explain the role of the Renewable Fuel Standard in the development of a market for advanced biofuels.
2. Briefly describe three thermochemical technologies that can be used to convert forages to fuel and heat.
3. In what ways is the biochemical conversion of biomass into fuel similar to forage utilization by ruminants?
4. What properties of biochar contribute to its potential value as a soil amendment?
5. List some of the characteristics of an ideal energy crop.
6. Describe two annual crop species that show excellent potential as energy crops, and explain why they are likely to be used for this purpose.
7. Which perennial grass has been the most extensively studied and developed bioenergy crop? What characteristics of the crop account for this status?
8. For a biomass crop that yields 5 tons/acre, how much phosphorus and potassium would need to be applied as fertilizer the following year to replace these nutrients?
9. Describe the potential environmental benefits that could be derived from growing energy crops and using them to produce fuels.
10. Apart from fuel, what are some other useful products that can be manufactured from biomass crops?

References

Bean, B, T McCollum, K McCuistion, J Robinson, B Villarreal, R VanMeter, and D Pietsch. 2006. Texas Panhandle Forage Sorghum Silage Trial. Texas Cooperative Extension and Texas Agricultural Experiment Station, Overton, TX.

Brown, RC. 2003. Biorenewable Resources: Engineering New Products from Agriculture. Iowa State Press, Ames, IA.

Casler, MD. 2010. Changes in mean and genetic variance during two cycles of within-family selection in switchgrass. BioEnergy Res. 3:47–54.

Casler, MD. 2014. Heterosis and reciprocal-cross effects in tetraploid switchgrass. Crop Sci. 54:2063–2069.

Casler, MD, and KP Vogel. 2014. Selection for biomass yield in upland, lowland, and hybrid switchgrass. Crop Sci. 54:626–636.

Council for Agricultural Science and Technology. 2007. Convergence of Agriculture and Energy: II. Producing Cellulosic Biomass for Biofuels. Council for Agricultural Science and Technology, Ames, IA.

Davis, SC, RM Boddey, BJR Alves, AL Cowie, BH George, SM Ogle, P Smith, M van Noordwijk, and MT van Wijk. 2013. Management swing potential for bioenergy crops. GCB Bioenergy 5:623–638.

Heaton, EA, FG Dohleman, FE Miguez, JA Juvik, V Lozovaya, JM Widholm, OA Zabotina, GF McIsaac, MB David, TB Voigt, NN Boersma, and SP Long. 2010. Miscanthus: A promising biomass crop. Adv. Bot. Res. 56:76–137.

Kort, J, M Collins, and D Ditsch. 1998. A review of soil erosion potential associated with biomass crops. Biomass Bioenerg. 14:351–359.

Lemus, R, and R Lal. 2005. Bioenergy crops and carbon sequestration. Crit. Rev. Plant Sci. 24:1–21.

Mckendry, P. 2002. Energy production from biomass (part 2): conversion technologies. Bioresource Technol. 83:47–54.

Moore, KJ and DL Karlen. 2013. Double cropping opportunities for biomass crops in the North Central USA. Biofuels 4:605–615.

Moore, KJ, KJ Boote, and MA Sanderson. 2004. Physiology and developmental morphology. In L Moser, B Burson, and LE Sollenberger (eds.), Warm-Season (C_4) Grasses, pp. 179–216. American Society of Agronomy, Crop Science Society of America, and Soil Science Society of America, Madison, WI.

Moore, KJ, EA Heaton, and SL Fales. 2013a. Use of grasses for biofuel. In: L Jank et al. (eds.), Forage Breeding and Technology, pp. 213–233. Embrapa, Campo Grande, Brazil.

Moore, KJ, DL Karlen, and KR Lamkey. 2013b. Future prospects for corn as a biofuel crop. In: SL Goldman and C Kole (eds.), Compendium of Bioenergy Plants: Corn, pp. 331–352. CRC Press, Taylor & Francis Group, Boca Raton, FL.

National Agricultural Statistics Service, US Department of Agriculture. 2014. Data and Statistics. www.nass.usda.gov/Data_and_Statistics/ (accessed May 30, 2014).

Perlack, RD, LL Wright, AF Turhollow et al. 2005. Biomass as Feedstock for a Bioenergy and Bioproducts Industry: The Technical Feasibility of a Billion-Ton Annual Supply. US Department of Energy and US Department of Agriculture, Washington, DC.

Pyter, R, TB Voigt, EA Heaton, FG Dohleman, and SP Long. 2007. Giant miscanthus: biomass crop for Illinois. In J Janick and A Whipkey (eds.), Issues in New Crops and New Uses, pp. 39–42. ASHS Press, Alexandria, VA.

Ramstein, GP, J Evans, SM Kaeppler, RB Mitchell, KP Vogel, CR Buell, and MD Casler. 2016. Accuracy of genomic prediction in switchgrass (*Panicum virgatum* L.) improved by accounting for linkage disequilibrium. Genes Genomes Genet. 6:1049–1062.

Renewable Fuels Association. 2014. Falling Walls & Rising Tides: 2014 Ethanol Industry Outlook. Renewable Fuels Association, Washington, DC.

Sanderson, MA, NP Martin, and P Adler. 2007. Biomass, energy, and industrial uses of forages. In: RF Barnes et al. (eds.), Forages: The Science of Grassland Agriculture, Volume II, 6th ed., pp. 635–647. Iowa State University Press, Ames, IA.

Schmer, MR, KP Vogel, RB Mitchell, and RK Perrin. 2008. Net energy of cellulosic ethanol from switchgrass. Proc. Natl Acad. Sci. USA 105:464–469.

Sokhansanj, S, S Mani, A Turhollow, A Kumar, D Bransby, L Lynd, and M Laser. 2009. Large-scale production, harvest and logistics of switchgrass (*Panicum virgatum* L.) – current technology and envisioning a mature technology. Biofuels Bioprod. Bioref. 3:124–141.

Somerville, C, H Youngs, C Taylor, S Davis, and SP Long. 2010. Feedstocks for lignocellulosic biofuels. Science 329:790–792.

Tew, TL, and RM Cobill. 2008. Genetic improvement of sugarcane (*Saccharum* spp.) as an energy crop. In W Vermerris (ed.), Genetic Improvement of Bioenergy Crops, pp. 250–272. Springer, New York.

US Department of Energy. 2006. Breaking the Biological Barriers to Cellulosic Ethanol: A Joint Research Agenda. DOE/SC-0095, US Department of Energy, Washington, DC.

US Department of Energy. 2011. US Billion-Ton Update: Biomass Supply for a Bioenergy and Bioproducts Industry. US Department of Energy, Oak Ridge National Laboratory, Oak Ridge, TN.

Van Loo, S, and J Koppejan. 2008. The Handbook of Biomass Combustion and Co-Firing. Earthscan, Abingdon, UK.

Vogel, KP. 1996. Energy production from forages (or American agriculture – back to the future). J. Soil Water Conserv. 51:137–139.

Vogel, KP, RB Mitchell, MD Casler, and G Sarath. 2014. Registration of 'Liberty' Switchgrass. J Plant Registr. 8:242–247.

Wright, LL, and A Turhollow. 2010. Switchgrass selection as a "model" bioenergy crop: A history of the process. Biomass Bioenerg. 34:851–868.

Compendium of Common Forages

Kenneth J. Moore

One of the most interesting and challenging aspects of working with forage crops is the diversity of species available. Each one has a unique set of characteristics and uses. Although most species are better adapted to certain areas and climates than others, within any geographical region there is an abundance of species that may be grown. The challenge is to find the right ones to use for the specific purposes intended by the producer.

This compendium provides brief descriptions of forage species commonly grown in the continental USA, including their adaptation and use, botanical characteristics, and management. It is designed to complement information presented within various chapters and to be a ready reference for students studying forages. Since the information on each species is necessarily brief, it should be regarded as a general overview. Readers intending to grow any of these forages should seek more comprehensive information about the culture and management of the species within their local area. This information usually can be obtained from the Cooperative Extension Service or local National Resource Conservation Service office.

Forages: An Introduction to Grassland Agriculture, Seventh Edition. Edited by Michael Collins, C. Jerry Nelson, Kenneth J. Moore and Robert F Barnes.

Courtesy of Michael Collins, University of Missouri; flower inset courtesy of Vivien Allen, Texas Tech University

ALFALFA

Alias: lucerne
Scientific name: *Medicago sativa* L.
Family: Fabaceae
Origin: Asia

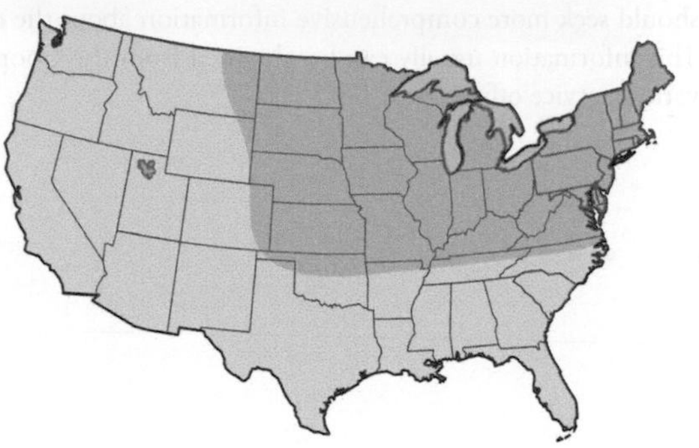

Adaptation and Use. This widely adapted perennial legume is grown in every state in the USA. Alfalfa does best on soils with fine to medium textures that are moderately to well drained and neutral or higher in pH. It tolerates drought but not prolonged flooding, and cultivars range from non-dormant to very winter hardy. Its major use is for hay, but it can be pastured, generally in mixtures with grasses to reduce bloat hazard and to extend useful stand life. Alfalfa forage is high in protein, vitamin A, minerals, and digestible energy, and is favored as a hay or haylage crop for dairy cattle and horses. It is also used in pet foods.

Plant Characteristics. Most alfalfa grows upright, but some cultivars are marketed as spreading or grazing types that are more decumbent. The leaves are pinnately compound. Most cultivars have trifoliolate leaves, but some that have more than three leaflets are marketed as multi-leaf with high forage quality. Leaflets are typically smooth on the surfaces and are oblong with the margin serrated close to the tip. Stipules are large and pointed. The raceme has typical legume flowers, which commonly are purple but may vary among shades of white, yellow, and purple, depending on the cultivar. The plant develops a deep taproot system.

Management. Alfalfa should be planted in early spring or late summer into a firm seedbed at a depth of 0.25–0.50 in. at rates in the range of 10–25 lb of inoculated seed/acre. Late summer seedings have less weed pressure, but they are more vulnerable to winter kill in colder climates. New stands often suffer from autotoxicity when alfalfa is planted immediately following alfalfa. Typical stand life is 3–5 years, but plants can persist for over a decade. Yield of digestible dry matter is highest when harvested at early bloom. Regrowth should be harvested at 4- to 6-week intervals. In northern areas, the final cut should be 3–4 weeks prior to frost to allow plants to harden for winter. Alfalfa does not persist well under continuous stocking. Bloat incidence can be markedly reduced by careful management.

Courtesy of Michael Collins, University of Missouri

ALSIKE CLOVER

Aliases: Swedish clover, hybrid clover
Scientific name: *Trifolium hybridum* L.
Family: Fabaceae
Origin: Eurasia

Adaptation and Use. A short-lived perennial adapted to southern Canada, northern US states, and higher elevations in the western USA, alsike clover prefers relatively cool and moist habitats. It tolerates acid, alkaline, low-fertility, and poorly drained soils, but not drought. It is used for hay and pasture, usually in mixtures with grasses and other legumes, such as red clover.

Plant Characteristics. Plants have an upright growth habit, branched stems that are often hollow, and a short taproot. Leaves are palmately trifoliolate with finely serrated leaf margins and prominent veins. The large stipules are sharply pointed. Growth is indeterminate. The inflorescence is a small head with white to pink flowers.

Management. Inoculated seed should be planted 0.125–0.25 in. deep in spring or late summer at 8–10 lb/acre (alone) or 4–6 lb/acre (in a mixture). The plant reseeds naturally under favorable conditions. Typically, only one hay crop is harvested near full bloom; alsike clover should not be harvested 4–6 weeks before frost. It can cause bloat and photosensitivity in cattle.

ALYCECLOVER

Aliases: buffalo clover, one-leaf clover
Scientific name: *Alysicarpus vaginalis* (L.) DC.
Family: Fabaceae
Origin: Asia

Adaptation and Use. Alyceclover is a summer annual legume adapted to the Gulf States. It prefers fertile, well-drained soils and will not tolerate flooding or poor drainage. It is used primarily for hay, pasture, and conservation.

Plant Characteristics. Plants have an upright herbaceous growth habit. Leaves are unifoliolate with a round to oblong shape. Stems are usually fine and are sometimes covered with fine hairs. Flowers are pink. Some varieties produce stolons that root at nodes.

Management. Alyceclover is often seeded into established grass. In this case, inoculated seed should be planted in late spring at a rate of 15–20 lb/acre and a depth of 0.25–0.5 in. Grazing should begin at 12–15 in. The optimum height for hay harvest is 18–24 in., and a second hay crop will be produced under favorable conditions. If allowed to mature, alyceclover reseeds naturally.

Courtesy of Lynn Sollenberger, University of Florida

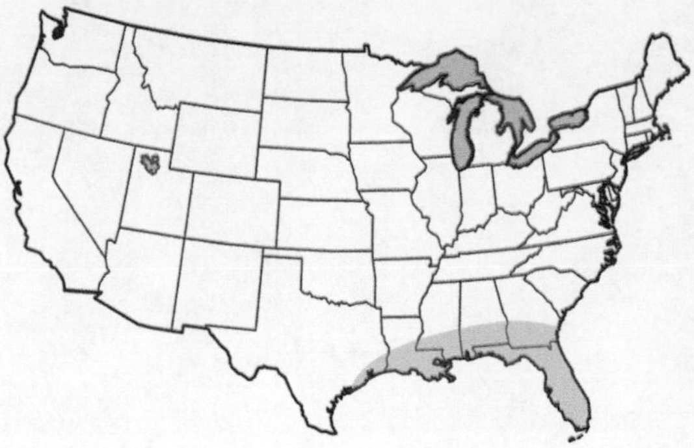

ANNUAL LESPEDEZAS

COMMON LESPEDEZA

Aliases: annual lespedeza, striate lespedeza, Japanese lespedeza
Scientific name: *Kummerowia striata* [Thunb.] Schindler

KOREAN LESPEDEZA

Scientific name: *Kummerowia stipulacea* (Maxim.) Makino
Family: Fabaceae
Origin: Asia

Adaptation and Use. Common and korean lespedezas are closely related short-day, warm-season annual legumes that are adapted to the humid transition zone and subtropical region of the USA. Both have become naturalized and reseed throughout their range of adaptation. Korean lespedeza can be grown as a summer annual further north, but it will not reseed naturally because it requires short days to flower. Adapted to a wide range of soils, annual lespedezas tolerate low fertility and acid soils, and are relatively tolerant of drought and poor drainage. They are used mostly for pasture in mixtures with grasses. Annual lespedezas are most compatible with bunch grasses, which offer less competition to seedlings that develop each year.

Plant Characteristics. Growth habit varies from upright to prostrate in both species, with common lespedeza being more prostrate. Stems are fine and hairy, with hairs angled toward the base in common lespedeza and upward in korean lespedeza. Leaves of common lespedeza are palmately trifoliolate with narrow, oblong leaflets and small stipules. Leaves of korean lespedeza are palmately trifoliolate with heart-shaped leaflets and large stipules, and leaflets near the end of stems have a scale-like appearance. Flowers of common lespedeza are clustered in short racemes that occur in leaf axils along the length of stems. Flowers of korean lespedeza are clustered in short racemes that occur in leaf axils near the end of stems. Two types of flowers occur in both species: some have blue to purple petals, while the others have no petals and are inconspicuous. Both species produce a shallow taproot system.

Management. Annual lespedezas are often seeded with grasses or into established grass. Inoculated seed should be planted in early spring at 10–25 lb/acre and 0.25–0.50 in. deep. Grazing frequency is generally determined by the companion grass, but minimum height is 3 in. Hay should be cut at the mid- to full-bloom stage. Both species reseed naturally if allowed to mature. They have low bloat potential.

Courtesy of Michael Collins, University of Missouri
COMMON LESPEDEZA

Courtesy of Michael Collins, University of Missouri
KOREAN LESPEDEZA

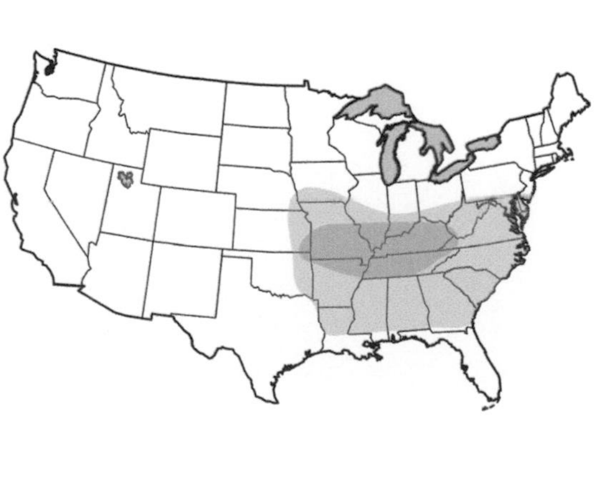

ANNUAL RYEGRASS

Alias: Italian ryegrass
Scientific name: *Lolium multiflorum* Lam.
Family: Poaceae
Tribe: Poeae
Origin: Europe

Adaptation and Use. Annual ryegrass is a cool-season annual grass that is well adapted for interseeding into warm-season grasses for winter pasture in the southeastern USA. It is also grown as a summer hay crop or for pasture in the Pacific Northwest and other northern states. Annual ryegrass is adapted to a wide range of soils, and will tolerate low fertility and poor drainage but not drought. It is not winter hardy. It is known for its ease of establishment, high nutritive value, and high yield.

Plant Characteristics. This bunch grass has fibrous roots and upright stems. Leaves are rolled in the sheath; blades are glossy, pointed, and keeled. The leaf sheath is split and overlapping. The collar is broad, the auricles are long and narrow, and the ligule is membranous and truncate to rounded. The long, narrow spike inflorescence has awned spikelets.

Management. Warm-season grass pastures should be grazed or cut closely before seeding annual ryegrass about 4 weeks before the first frost date. Seed should be planted at a depth of 0.25–0.50 in. and at a rate of 30–40 lb/acre alone or 15–30 lb/acre with an annual legume. Harrowing or light disking of the sod before broadcast seeding will improve establishment. Planting in a prepared seedbed should be done 8 weeks before frost. Annual ryegrass is usually planted in spring in cooler climates. Grazing should begin at 6–8 in. and stop at 2 in. Hay should be cut at boot stage. Multiple crops can be made under favorable conditions. If allowed to mature, the plant reseeds naturally.

Courtesy of Michael Collins, University of Missouri

ANNUAL RYEGRASS

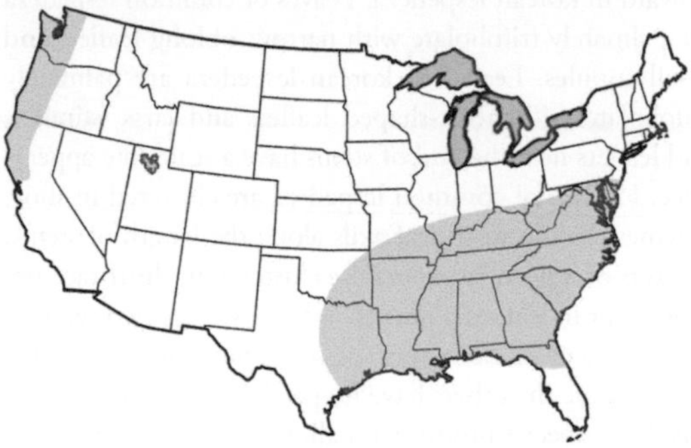

ANNUAL VETCHES

COMMON VETCH

Alias: garden vetch
Scientific name: *Vicia sativa* L.

HAIRY VETCH

Alias: winter vetch
Scientific name: *Vicia villosa* Roth
Family: Fabaceae
Origin: Eurasia

Adaptation and Use. Common vetch and hairy vetch are closely related winter annual legumes. Common vetch is adapted to regions with mild winters. Hairy vetch is more winter hardy and is widely adapted throughout the USA. Both species can be grown as a summer annual in areas with cool summer temperatures. Common vetch is adapted to well-drained, fertile soils. Hairy vetch is adapted to a wider range of soils; it is well suited to sandy soils, but it does not tolerate drought. Vetches are more acid tolerant than most legumes. They are often used to overseed warm-season grass pastures in the southern USA for winter grazing. Hairy vetch is also seeded with small grains as a hay or silage crop. It has very high nitrogen-fixing potential and is used routinely as a cover and manure crop.

Plant Characteristics. Growth of both species is prostrate when sown alone and viny in mixtures. Leaves are pinnately compound with multiple leaflets, and terminate in a tendril. The smooth, oblong leaflets of common vetch are tipped with a small, sharp point. The leaflets of hairy vetch are hairy, narrow, and tapered at both ends. Common vetch has smooth stems, and hairy vetch has hairy stems. Reddish-purple flowers occur singly or as pairs in the leaf axils of common vetch. The inflorescence of hairy vetch is a raceme with numerous blue-violet florets. Both species produce shallow taproot systems.

Management. Inoculated seed should be planted into a firm seedbed at a depth of 0.25–0.50 in. The seeding rate is 75–90 lb/acre for common vetch and 20–40 lb/acre for hairy vetch. Planting should be done during late summer when grown as a winter annual, and in early spring when grown as a summer annual. Annual vetches should not be grazed until they are 6 in. tall, to avoid removing the lower leaf axils from which branching occurs. Harvest for hay when the first seedpods form. Both species reseed naturally under favorable conditions. They can cause bloat.

COMMON VETCH

Courtesy of Jimmy Henning, University of Kentucky

HAIRY VETCH

Courtesy of Michael Collins, University of Missouri

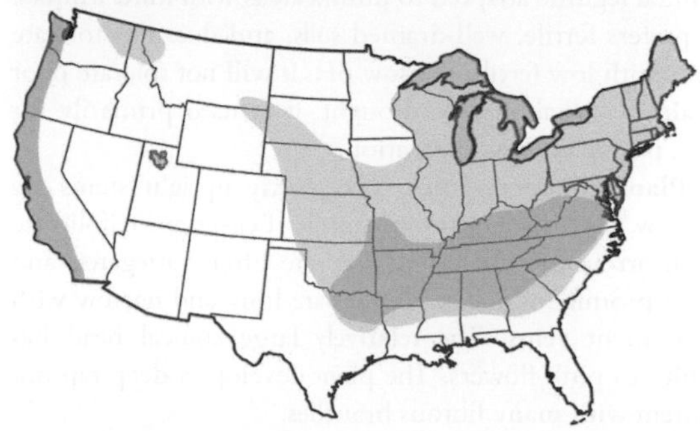

Courtesy of Gerald Evers, Texas A&M University

ARROWLEAF CLOVER

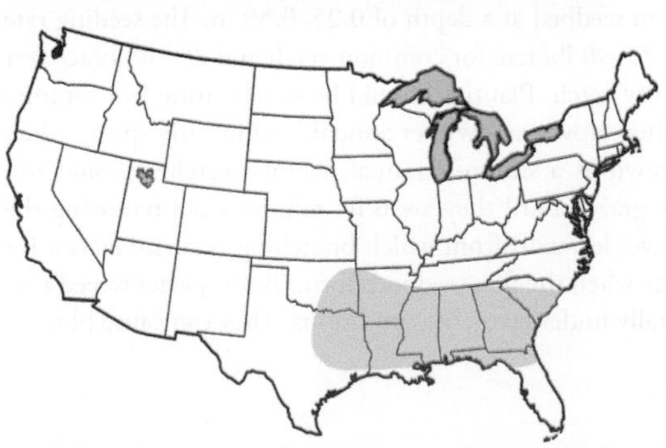

Scientific name: *Trifolium vesiculosum* Savi
Family: Fabaceae
Origin: Eurasia

Adaptation and Use. Arrowleaf clover is a winter annual legume adapted to humid areas with mild winters. It prefers fertile, well-drained soils, and does not tolerate soils with low fertility or low pH. It will not tolerate poor drainage, alkalinity, or drought. It is used primarily for hay, pasture, and conservation.

Plant Characteristics. The mostly upright stems are hollow, smooth, and often purple. Leaves are trifoliolate with arrow-shaped leaflets that are often variegated and have prominent veins. Stipules are long and narrow with prominent veins. The relatively large conical head has white to pink flowers. The plant develops a deep taproot system with many fibrous branches.

Management. Arrowleaf clover is often seeded into established grass. Inoculated seed should be planted in early fall at 10–15 lb/acre and 0.25–0.50 in. deep. Grazing should begin at 5–6 in. Hay should be cut at the early bloom stage. If allowed to mature, arrowleaf clover reseeds naturally. It has low bloat potential.

Courtesy of Michael Collins, University of Missouri, Lynn Sollenberger, University of Florida, and Vivien Allen, Texas Tech University

BAHIAGRASS

Scientific name: *Paspalum notatum* Flüggé
Family: Poaceae
Tribe: Paniceae
Origin: South America

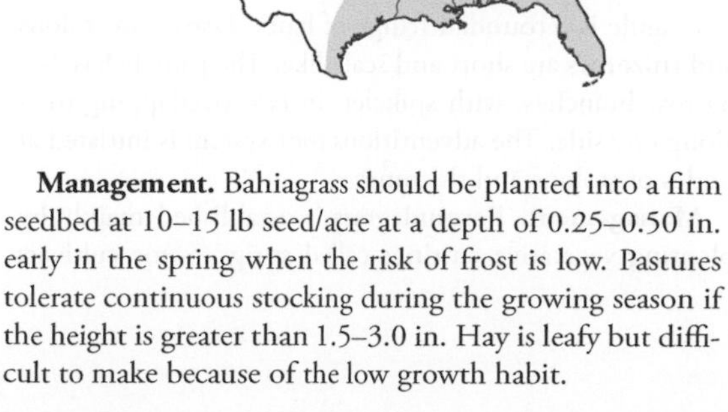

Adaptation and Use. This perennial warm-season grass is well adapted to coastal plains of the humid, subtropical region. Although adapted to a wide range of soils, bahiagrass is especially well adapted to sandy soils because of its tolerance of drought and low fertility. It is extremely competitive and difficult to grow in mixtures with legumes. It is used mostly for pasture because it tolerates close grazing and continuous stocking.

Plant Characteristics. Bahiagrass forms a dense sod from short, heavy rhizomes. Leaf blades are basal, flat, or folded, with hairy margins. The ligule is a fringe of hairs. Flowering stems grow upright and bear a racemose panicle with two or three racemes. The root system is deep and extensive with many highly branched fibrous roots.

Management. Bahiagrass should be planted into a firm seedbed at 10–15 lb seed/acre at a depth of 0.25–0.50 in. early in the spring when the risk of frost is low. Pastures tolerate continuous stocking during the growing season if the height is greater than 1.5–3.0 in. Hay is leafy but difficult to make because of the low growth habit.

Courtesy of Vivien Allen, Texas Tech University, and Jimmy Henning, University of Kentucky

BERMUDAGRASS

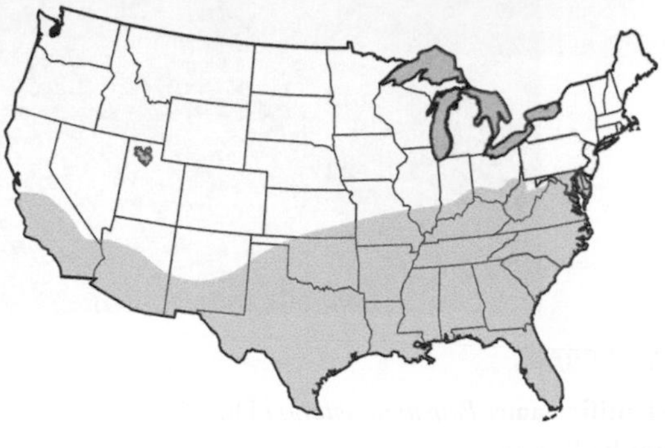

Scientific name: *Cynodon dactylon* (L.) Pers.
Family: Poaceae
Tribe: Chlorideae
Origin: Africa

Adaptation and Use. Bermudagrass is a highly productive warm-season grass that is adapted to warm climates with mild winters. Although it grows well on a wide range of soils and tolerates drought and short periods of flooding, it is sensitive to freezing temperatures. Bermudagrass is used mostly for pasture because it tolerates close grazing and continuous stocking. Bermudagrass pastures are often oversown in fall with cool-season annual grasses and legumes to extend the grazing season.

Plant Characteristics. Bermudagrass is a sod-forming perennial that spreads by means of stolons and rhizomes. Short, flat, narrow leaf blades occur on upright stem branches that arise from nodes of stolons and rhizomes. The ligule is a rounded fringe of hairs. Leaves on stolons and rhizomes are short and scale-like. The panicle has 3–7 narrow branches, with spikelets in two overlapping rows along one side. The adventitious root system is initiated at nodes of stolons and rhizomes.

Management. Bermudagrass is established mainly by planting vegetative cuttings called sprigs. Some cultivars are propagated by seed, but the seed is very small, making it difficult to plant. The seeding rate is 5–10 lb seed/acre. Many of the most productive cultivars of bermudagrass are sterile hybrids that must be propagated vegetatively. Specialized planting equipment is available for this purpose. Sprigs are planted into 40-in. rows, with one sprig every 2–3 ft at a depth of 1–2 in. Bermudagrass responds very well to N fertilization and is ideal for land application of livestock waste. It tolerates continuous stocking, but rotational stocking results in more uniform utilization. It can be an invasive weed (commonly known as wiregrass or devilgrass) in row crops because it spreads by means of rhizomes and is difficult to control.

BERSEEM CLOVER

Courtesy of Gerald Evers, Texas A&M University

Aliases: Egyptian clover
Scientific name: *Trifolium alexandrinum* L.
Family: Fabaceae
Origin: Middle East

Adaptation and Use. Berseem clover is grown primarily as a winter annual forage in the southeastern and south central states, and to a lesser extent under irrigation in the Southwest. It is also used as a summer annual in crop rotations with corn, soybean, and oat in the Midwest. It is adapted to medium-textured soils that are well to somewhat poorly drained. It does not tolerate acid soils, but does well on alkaline soils. It is sensitive to drought and freezing temperatures. Berseem clover is primarily used as a pasture crop, but may also be harvested for hay, silage, or green chop. It is sometimes grown in mixtures with annual grasses, such as oat. It is also used as a source of green manure. Berseem clover produces relatively high-quality forage with low bloat potential.

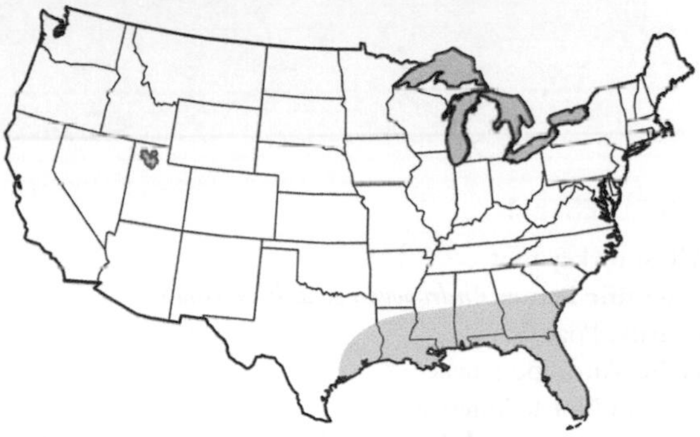

Plant Characteristics. Berseem clover has an upright herbaceous growth habit. Leaves are palmately trifoliolate with narrow, oblong leaflets that are toothed near the tip. Stipules are pointed with long hairs along the margin. The stems are hollow. The inflorescence is a small head with off-white flowers. The plant produces a short taproot that limits productivity on dry sites.

Management. Berseem clover should be seeded into a firm seedbed at a depth of 0.125–0.25 in. Seeding rates are 10–20 lb/acre. Planting should be done during late summer when used as a winter annual, and in early spring when used as a summer annual. Cutting or grazing should begin when the basal shoots initiate growth and before flowering. Delay of harvesting until after flowering results in significant stand loss. Berseem clover does not persist well through natural reseeding.

Courtesy of Michael Collins, University of Missouri

BIG BLUESTEM

Alias: turkey foot
Scientific name: *Andropogon gerardii* Vitman
Family: Poaceae
Tribe: Andropogoneae
Origin: North America

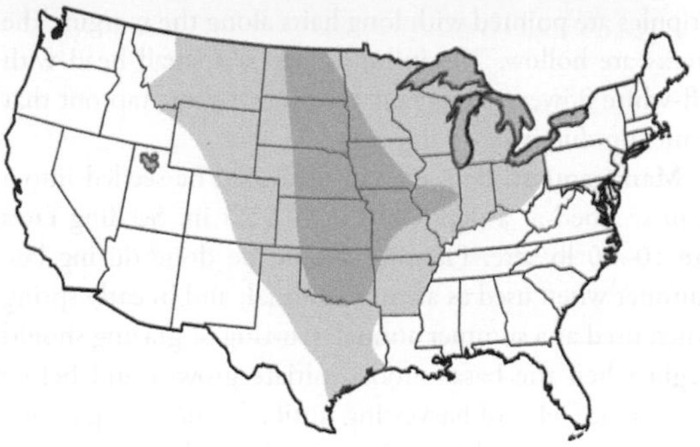

Adaptation and Use. Big bluestem is native to the tall-grass prairie and is adapted throughout much of the central USA. Although it tolerates poor soil drainage, moderate salinity, and acid to moderately alkaline soils, it prefers well-drained sites with high fertility. It has high drought tolerance and is winter hardy. Big bluestem is used primarily as a summer pasture or hay crop in temperate regions where introduced warm-season grasses are not well adapted. It is also used extensively in conservation and wildlife plantings.

Plant Characteristics. This perennial grass spreads by means of short rhizomes, yet usually displays its solid upright stems in a bunch-type growth habit. It grows rapidly during late spring and it flowers during summer. Leaf blades have white hairs near the base, and a short, blunt ligule with a fringe of hairs, but have no auricles. Sheaths are flattened, sometimes with long hairs, usually with a purple coloration near the base. The panicle often has three primary raceme-like branches that give the appearance of a turkey foot. The seed has a long, twisted awn and is often hairy. The fibrous root system is deep and extensive.

Management. Big bluestem is daylength sensitive, so cultivars are adapted to narrow latitudes. The chaffy seed are difficult to sow and may have a high degree of dormancy. Seedlings are slow to establish, so weed control is critical. Plant into a prepared seedbed or no-till at 5–10 lb seed/acre and 0.25–0.50 in. deep in late spring. A stand of 1–2 plants/ft^2 is adequate. Grazing should begin at 10–12 in. and not be closer than 6 in. in order to retain vigor and reduce weed invasion. Quality hay can be harvested up to boot stage, when quality decreases. Plants should not be cut or grazed during the 4–6 weeks prior to a killing frost.

Courtesy of Michael Collins, University of Missouri

BIRDSFOOT TREFOIL

Alias: birdsfoot deervetch
Scientific name: *Lotus corniculatus* L.
Family: Fabaceae
Origin: Mediterranean

Adaptation and Use. This short-lived perennial legume is adapted to humid, temperate regions. Adapted to a wide range of soils, birdsfoot trefoil tolerates acidity, alkalinity, low fertility, drought, and poor drainage. It is primarily used for pasture, and is usually grown in mixtures with grasses.

Plant Characteristics. The growth habit is prostrate or upright, depending on the cultivar. Types with prostrate stems are more winter hardy and are better adapted for grazing, whereas upright types make better hay. Leaves are pentafoliolate, with three leaflets attached at the end of the short petiole and two near its base. Leaflets are smooth and lance shaped. Leaves fold around the petiole at night. Stems are well branched and relatively fine. The umbel has 4–8 flowers that are generally yellow, sometimes with orange or red stripes. The plant forms a well-developed taproot with many branches. Some cultivars produce short rhizomes that can produce new plants. Birdsfoot trefoil needs long days to flower and reseed.

Management. Inoculated seed should be planted 0.125–0.25 in. deep in spring or late summer at 4–8 lb/acre. Seedlings are not very shade tolerant, so competition with other plants needs to be controlled. Grazing may begin at early flowering; it can be grazed frequently, but not closely. It does not cause bloat. Birdsfoot trefoil is well suited for stockpiling because it maintains quality after flowering. It should be cut for hay at the early flower stage, two to three times per year, but leaf shatter is a problem with field drying. It should not be harvested 4–6 weeks before frost. It must reseed naturally to perennate for more than 2–3 years.

Courtesy of Michael Collins, University of Missouri

BLUE GRAMA

Scientific name: *Bouteloua gracilis* (Kunth) Lag. ex Griffiths
Family: Poaceae
Tribe: Chlorideae
Origin: North America

Adaptation and Use. This warm-season perennial grass is native to the North American short-grass prairie. Although adapted to a wide range of soils, blue grama occurs more frequently on heavy upland soils. It tolerates drought and alkalinity but not wet or poorly drained soils. It is grown primarily for pasture and conservation in the semiarid region of the Great Plains.

Plant Characteristics. Plants have a prostrate growth habit. Flowering stems are fine, smooth, and erect. Leaf blades are narrow, tapered, and mostly basal. Sheaths are smooth and round with long, soft hairs at the collar. The ligule is truncated and has a fringe of hairs. Auricles are absent. The panicle has 1–3 spike-like branches with 40–90 spikelets per branch. It produces a dense mat of fine, fibrous roots.

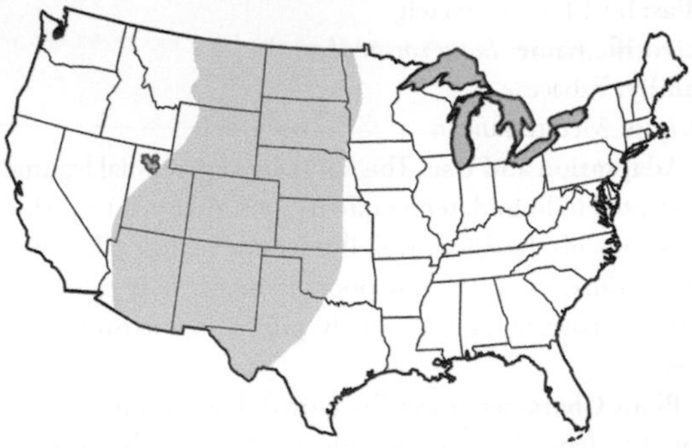

Management. Blue grama should be planted into a firm seedbed at 1–3 lb seed/acre and 0.25–0.50 in. deep in late spring or early summer. Locally adapted ecotypes perform best. Blue grama withstands close grazing and stockpiles well.

BRASSICAS

KALE

Scientific name: *Brassica oleracea* L.

RAPE

Alias: canola
Scientific name: *Brassica napus* L. subsp. *napus*

TURNIP

Scientific name: *Brassica rapa* L. subsp. *rapa*
Family: Brassicaceae
Origin: Eurasia

Adaptation and Use. These annual and biennial forbs belong to the mustard family. Used for summer or fall forage throughout the USA, brassicas are adapted to a wide range of soil conditions. They grow best on moderately to well-drained soils, and do not tolerate wet soils. They require medium soil fertility and have moderate drought and frost tolerance. Brassicas are used mostly for supplemental pasture.

Plant Characteristics. Kale resembles cabbage, but is leafier. It has a rosette growth habit while vegetative. Flowering stems are branched and bear many yellow flowers. Cultivars vary considerably in appearance, and some are stemless. Rape also resembles cabbage in appearance. Giant and dwarf cultivars are available. Turnip has an upright growth habit. It produces several large, erect leaves with sparse hairs. Flowering stems are branched. Flowers are yellow and borne in racemes. The plant produces a large taproot that is high in carbohydrates.

Management. Brassicas are sown in spring or summer depending on when they are to be used. Spring planting provides late summer forage, while late summer planting provides forage in the fall. Brassicas should be drilled in 6–8 in. rows at a depth of 0.25–0.50 in. Seeding rates are 3.5–4.5 lb/acre for kale and rape and 1.5–2.5 lb/acre for turnip. Rape and turnip are ready to graze in 80–100 days, and kale is ready in 150–180 days after seeding. The forage quality of brassicas is exceptionally high. Livestock grazing them can develop hemolytic anemia and goiter due to the presence of toxic compounds. Access to pasture should be limited by strip grazing and the diet supplemented with either hay or stockpiled grass. Brassicas can cause bloat.

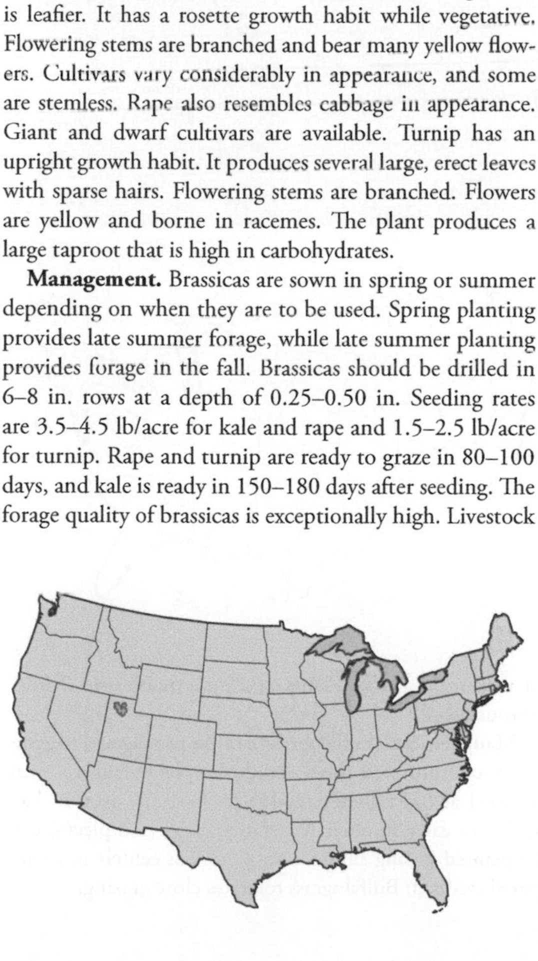

Courtesy of Michael Collins, University of Missouri

KALE

Courtesy of Michael Collins, University of Missouri

RAPE

Courtesy of Michael Collins, University of Missouri

TURNIP

Courtesy of Jennifer MacAdam, Utah State University

BUFFALOGRASS

Scientific name: *Buchloe dactyloides* (Nutt.) Columbus
Family: Poaceae
Tribe: Chlorideae
Origin: North America

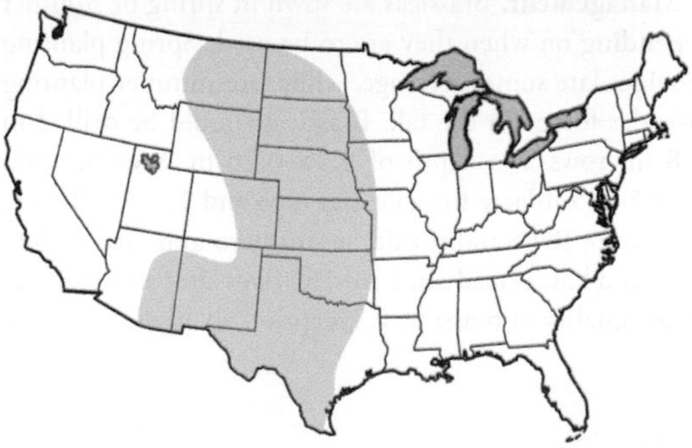

Adaptation and Use. Buffalograss, a warm-season perennial grass, is native to the North American short-grass prairie and is adapted to dry, medium to heavy upland soils. It tolerates drought and alkaline soils. It is grown primarily for pasture in semiarid regions of the Great Plains; some is used for soil conservation. Buffalograss is gaining popularity as a low-maintenance turf in drier areas of the USA.

Plant Characteristics. This sod-forming grass spreads by means of stolons that produce fine, erect, smooth stems at nodes. Leaf blades are narrow, relatively short, and curled. Sheaths are smooth with a few marginal hairs at the collar. The ligule is membranous and has a fringe of hairs. The panicle of male plants has 1–4 spike-like branches and 6–12 spikelets per branch. Female plants have bur-like clusters located near the middle of the stem or within the leaves. Plants develop a dense mat of fine, fibrous roots.

Management. Buffalograss can be propagated vegetatively or from seed. Seed should be planted into a firm seedbed at 8–15 lb/acre and 0.25–0.50 in. deep in late spring or early summer. Vegetative plugs (sod pieces) can be planted during summer on 3- to 4-ft centers in a prepared seedbed. Buffalograss tolerates close grazing.

BUFFELGRASS

Alias: African foxtail
Scientific name: *Pennisetum ciliare* (L.) Link
Family: Poaceae
Tribe: Paniceae
Origin: Africa, India

Adaptation and Use. This warm-season perennial grass is adapted to the southernmost region of the Gulf States and northern Mexico. Although adapted to a wide range of soils, buffelgrass prefers sandy soils and tolerates saline and alkaline soils. It is tolerant of drought and high temperatures and recovers quickly, but it does not tolerate flooding. It is grown primarily for pasture, soil conservation, and range improvement.

Plant Characteristics. The growth habit of buffelgrass varies from bunch- to sod-forming, spreading by means of short rhizomes. Leaf blades are flat and tapered with soft hairs on the upper side. The leaf sheath is open, mostly smooth, and the ligule is truncated and has a fringe of hairs. Stems grow erect, branching near the crown. The cylindrical, purple to black panicle has 2–4 spikelets per branch. The plant produces a dense fibrous root system.

Management. Buffelgrass can be planted in the spring or fall at a rate of 3–5 lb seed/acre at a depth of 0.25–0.75 in. The seed is light and difficult to plant without the use of specialized planters. Once established, it can withstand close grazing and can be grazed to a height of 2–4 in. Hay should be cut at the early head stage. To allow natural reseeding, it should be grazed or harvested after the seed have matured. It is considered to be an invasive weed species in some rangeland areas. Buffelgrass can accumulate high levels of oxalate.

Courtesy of Lynn Sollenberger, University of Florida

CAUCASIAN BLUESTEM

Alias: old world bluestem
Scientific name: *Bothriochloa bladhii* (Retz.) S.T. Blake
Family: Poaceae
Tribe: Andropogoneae
Origin: Russia

Adaptation and Use. Caucasian bluestem, a warm-season perennial grass, is adapted to the humid transition zone. It is adapted to medium-textured soils and tolerates low fertility and acid soils. However, it does not tolerate poor drainage or drought. It is grown for pasture and hay.

Plant Characteristics. The growth habit varies from bunch type in young stands to sod-forming in older ones. It produces fine, erect stems that are purple at the nodes. The smooth leaf blades are flat or folded and have a characteristic blue-green color. The ligule is membranous, and auricles are absent. The panicle produces numerous raceme-like branches. The plant has a deep, fibrous root system.

Management. Caucasian bluestem should be planted in the spring at a rate of 2–3 lb seed/acre at a depth of 0.25–0.75 in. The seed is chaffy, and the use of specialized planters is recommended. Weed control is critical, as seedlings grow slowly and do not compete well. Grazing of established stands should begin at 4–6 in., and although it is generally tolerant of close grazing, livestock should be removed at 2 in. Caucasian bluestem can produce two to three hay crops per year. Hay should be cut at the late boot stage. It should not be cut or grazed 2–4 weeks before frost.

Courtesy of Michael Collins, University of Missouri

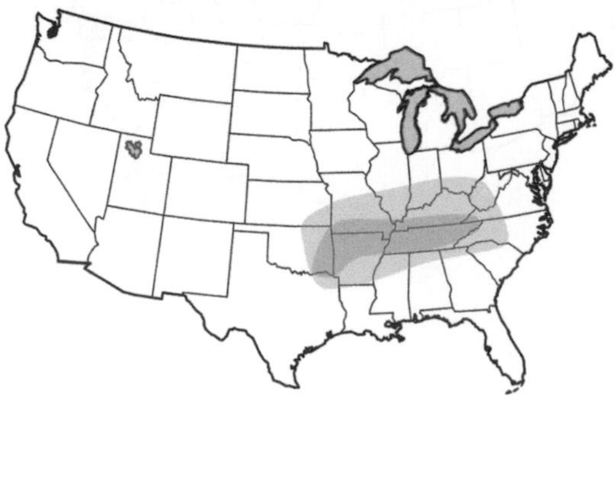

Courtesy of Michael Collins, University of Missouri

CHICORY

Aliases: coffeeweed, succory, blue sailors
Scientific name: *Cichorium intybus* L.
Family: Asteraceae
Origin: Mediterranean

Adaptation and Use. Chicory is a perennial forb that is widely adapted throughout the USA. It is adapted to a wide range of soil conditions but grows best on moderately or well-drained soils. It requires medium to high soil fertility. It has moderate drought tolerance, and it is used mostly for pasture.

Plant Characteristics. Chicory has a rosette growth habit during the cool season, and once established it produces many flowering stems in late spring and early summer. The basal leaves are large and lobed, while leaves growing on flowering stems are smaller and less lobed. Stems grow upright, and are smooth and branched. Blue to purple flower heads occur at the apex and upper leaf axils of the stem. The taproot system is large and deep.

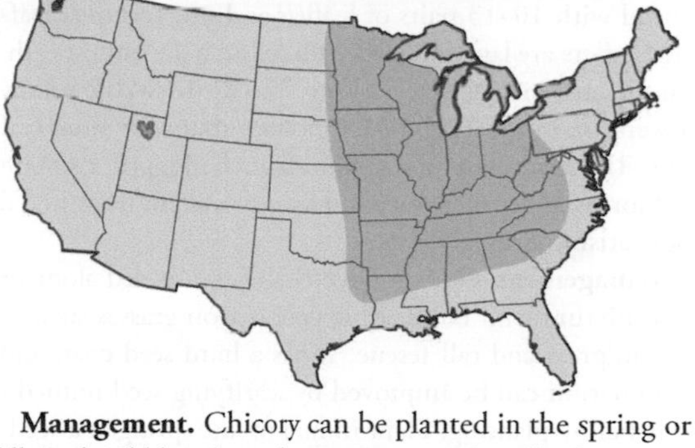

Management. Chicory can be planted in the spring or fall. It should be planted into a prepared seedbed at a rate of 3–4 lb seed/acre at a depth of 0.25–0.50 in. It is most productive under rotational stocking. Graze to a height of 2 in. and rest for at least 3–4 weeks. Prevention of stem growth by close grazing in late spring and early summer is critical to maintaining vigorous, productive growth later in the season. Chicory is a common roadside weed and is considered invasive in some states.

CICER MILKVETCH

Courtesy of Michael Collins, University of Missouri

Alias: chickpea milkvetch
Scientific name: *Astragalus cicer* L.
Family: Fabaceae
Origin: Eastern Europe

Adaptation and Use. This herbaceous perennial legume is well adapted to the Intermountain Region but also grows well in the Northern Great Plains and the northwestern part of the Corn Belt. It can be grown in most areas with cool and moist growing conditions, but often does not perform as well there as other, better adapted legumes. It grows best on moderately coarse soils with a pH in the range 6.0–8.1, but is adapted to a wide range of soils. It has moderate salt tolerance and can be grown on relatively poor and disturbed soils. It does not tolerate poor drainage or flooding, but tolerates drought. It is relatively frost tolerant and winter hardy. It is used primarily for pasture and conservation, but is sometimes harvested as hay.

Plant Characteristics. Cicer milkvetch is a long-lived perennial legume that spreads by means of rhizomes. It has an upright growth habit but tends to lodge as it matures. Stems are smooth and hollow, growing upright when young but later becoming decumbent and quite long, reaching lengths of 4–10 ft. Leaves are pinnately compound with 10–13 pairs of leaflets and one terminal leaflet. Leaflets are lance shaped with sparse hairs underneath. Stipules are triangular to oblong. The white to pale yellow flowers are borne in compact racemes that arise from leaf axils. The plant develops a short, branched taproot system and forms an extensive secondary root system from nodal roots arising from rhizomes.

Management. Cicer milkvetch may be seeded alone or in a mixture with compatible cool-season grasses such as orchardgrass and tall fescue. It has a hard seed coat, and germination can be improved by scarifying seed immediately before planting. Plant in mid to late spring at a seeding rate of 7 lb/acre when seeded alone or 5 lb/acre in a mixture. Drill or broadcast seed into a prepared seedbed at a depth of 0.5–0.75 in. Seedlings are slow to establish, so weed control is critical. This legume tolerates continuous grazing but should not be grazed closer than 4 in. It is a non-bloating legume, but development of photosensitivity has been observed in grazing livestock. Cicer milkvetch produces two cuttings of hay per season. The first should be made at 1/10 bloom and the second at the end of the growing season. More frequent cutting may reduce the total season yield.

Courtesy cf Michael Collins, University of Missouri

CORN

Alias: maize
Scientific name: *Zea mays* L. subsp. *mays*
Family: Poaceae
Tribe: Andropogoneae
Origin: Mesoamerica

Adaptation and Use. Corn is the most important row crop grown in North America. It is planted on more land than any other US crop. It is grown in almost every state, but its major production region is the North Central states. It is adapted to well-drained, fertile, medium-textured soils, and does not tolerate drought or flooding. Corn requires high fertility and moisture availability. As forage, it is used primarily as a silage crop but is sometimes used as an emergency pasture. Corn crop residues are often used as roughage for gestating beef cows.

Plant Characteristics. Corn has an upright growth habit and typically produces a single upright stem with solid internodes (stalk). Secondary tillering occurs in some hybrids, particularly at lower population densities. Leaf blades are long, wide, and tapered. The leaf sheath overlaps the stem. The ligule is membranous and blunt, or is absent in some hybrids. The male inflorescence (tassel), which occurs at the stem apex, is a panicle with paired spikelets containing paired florets. The female inflorescence (ear) is a thick spike with pairs of spikelets occurring in several rows, and is generally located six to seven nodes below the apex. Corn produces a dense, fibrous, adventitious root system and forms aerial (brace) roots at nodes near the soil surface.

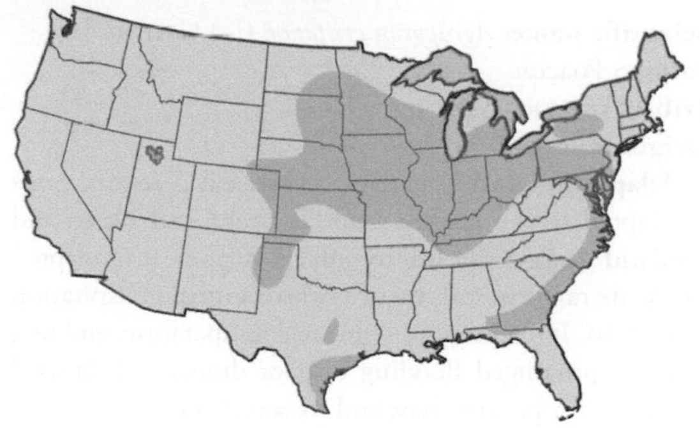

Management. High-yielding, full-season grain hybrids are generally grown for silage, although special-purpose silage hybrids are available. Corn is usually planted in the spring, when soil temperatures exceed 50°F, at population densities in the range of 25,000–35,000 plants/acre. Seeding is usually done in rows spaced 15–40 in. apart at a depth of 1.5–2.0 in. Optimum yield and quality of silage are obtained when corn is harvested between 80% milk line and 7–10 days after black layer formation. Residues remaining after grain harvest can be harvested for roughage or grazed.

Courtesy of Kenneth Moore, Iowa State University

CRESTED WHEATGRASS

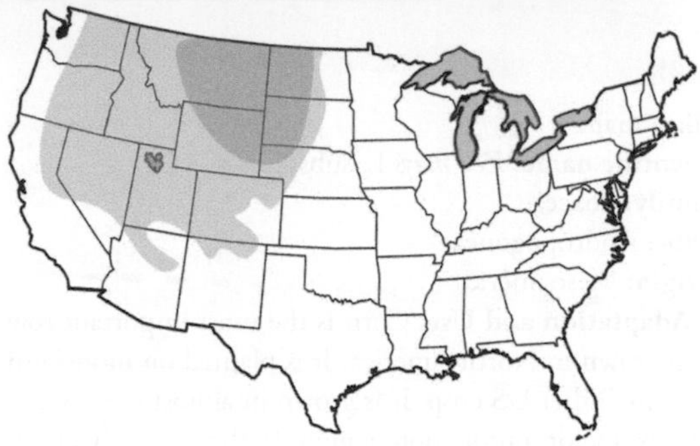

Scientific name: *Agropyron cristatum* (L.) Gaertn.
Family: Poaceae
Tribe: Triticeae
Origin: Eurasia

Adaptation and Use. This cool-season perennial grass is adapted to the northern and central Great Plains and semiarid areas of the Intermountain Region. It is adapted to a wide range of soils in areas where annual precipitation is 8–16 in. It tolerates drought, cold temperature, and fire, but not prolonged flooding or poor drainage. It is used primarily for pasture, hay, and conservation.

Plant Characteristics. Crested wheatgrass generally has a bunch growth habit, but some types produce rhizomes, particularly under wetter soil conditions. Leaf blades are flat, with a smooth surface below and slightly rough surface above. Leaves are rolled in the whorl. The leaf sheath is round and split with overlapping margins. The sheath is usually smooth but occasionally is hairy. The leaf collar is distinct and divided. The ligule is membrane-like with irregularly shaped edges. Auricles are slender and claw-like. Stems grow upright and are smooth. The spike is short, tapers at both ends, and is densely populated with awned spikelets containing 3–8 florets. The fibrous root system is extensive.

Management. Crested wheatgrass should be planted in early spring or late fall. Late summer seedings are not recommended unless the field is irrigated. The normal seeding rate is 5–7 lb seed/acre, but it should be increased to 10–12 lb/acre on poorer soils. Seeding depth is 0.25–0.50 in. Grazing should begin at 6 in. or later. This grass tolerates relatively close grazing but should not be grazed lower than 3 in. Hay should be cut at the late boot stage. Crested wheatgrass should not be cut or grazed 4 weeks before frost.

Courtesy of Michael Collins, University of Missouri

CRIMSON CLOVER

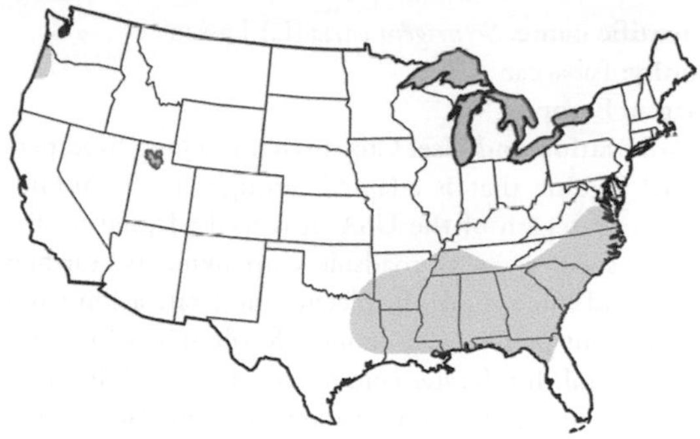

Aliases: scarlet clover, carnation clover
Scientific name: *Trifolium incarnatum* L.
Family: Fabaceae
Origin: Eurasia

Adaptation and Use. This winter annual legume is adapted to humid areas with mild winters. Crimson clover is grown to a lesser extent along the Pacific Coast and as a summer annual in northern states. It is adapted to a wide range of soil textures, but requires at least moderate drainage. It tolerates acid to moderately alkaline soils, but not salinity. It has low drought tolerance, and is sensitive to freezing or very high temperatures. It is used primarily for pasture, and is often seeded into warm-season grass pastures to provide winter and spring grazing. It can also be harvested for hay or silage, and it is used as both a cover and a manure crop in rotations.

Plant Characteristics. Growth is upright to ascending. The primary stem does not elongate, but produces many axillary branches at lower nodes. Leaves are palmately trifoliolate, with round to egg-shaped leaflets that are narrower near the base and finely toothed near the tip. Stipules are short, rounded, and hairy. Leaves and stems are very hairy. Flowering stems are produced in the spring. The inflorescence is a long, conical head with bright crimson flowers. The plant produces a taproot with many fibrous branches.

Management. Inoculated seed should be planted during late summer or early fall (when grown as a winter annual) or early spring (when grown as a summer annual) at 10–15 lb/acre and 0.125–0.25 in. deep into a firm seedbed. Warm-season pastures should be oversown with crimson clover later in the fall at a rate of 15–25 lb seed/acre. Grazing should begin and hay should be harvested at the early bloom stage. The minimum grazing height is 3–4 in. If allowed to mature, crimson clover reseeds naturally. It has low to moderate bloat potential.

Courtesy of Michael Collins, University of Missouri

CROWNVETCH

Scientific name: *Securigera varia* (L.) Lassen
Family: Fabaceae
Origin: Europe

Adaptation and Use. Crownvetch is a long-lived perennial legume that is adapted throughout the humid, temperate region of the USA. It is well adapted to disturbed soils, such as roadside embankments, earthen dams, and mine spoils. It tolerates moderate acidity and fertility, but prefers well-drained, fertile soils with a pH near neutral. It tolerates cold temperatures and drought. It is used mostly for conservation, but can also be used for pasture and hay.

Plant Characteristics. Crownvetch spreads by means of rhizomes. Stem growth is initially upright, but later reclines to form a dense mat. Leaves are pinnately compound with multiple leaflets, similar in appearance to vetch leaves, but do not terminate with a tendril. Leaflets are smooth and oblong. Stems are hollow, angular, and smooth. The inflorescence is an umbel with variegated white to purple flowers. The plant produces a well-developed taproot with many branches.

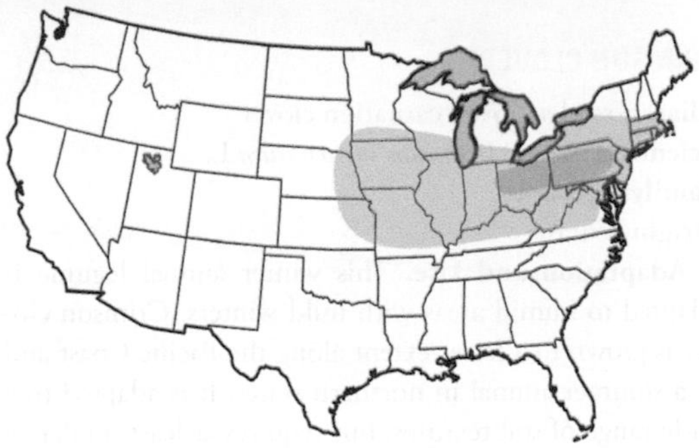

Management. Inoculated seed should be planted at 10–15 lb/acre into a firm seedbed at a depth of 0.125–0.25 in. The seed has a hard coat, and optimum results are obtained with scarified seed. Planting should be done during spring. Crownvetch is slow to establish due to poor seedling vigor. It does not tolerate continuous stocking and should not be grazed below 4 in. Although it does not cause bloat, it contains compounds that may be toxic to monogastric animals. Harvest for hay at full bloom stage for best regrowth, which is relatively slow. Crownvetch usually produces two cuttings in a season.

Courtesy of Vivien Allen, Texas Tech University

DALLISGRASS

Alias: paspalum
Scientific name: *Paspalum dilatatum* Poir.
Family: Poaceae
Tribe: Paniceae
Origin: South America

Adaptation and Use. Dallisgrass, a perennial warm-season grass, is well adapted to the southern coastal plains. Adapted to soils with fine to medium texture, it prefers moist lowland soils. It does not tolerate consistently wet or dry soils, but will tolerate periods of drought. It is used mostly for pasture because it tolerates close grazing and continuous stocking. It is compatible with legumes and other grasses when grown in mixtures.

Plant Characteristics. Dallisgrass has a bunch growth habit with tillers arising from a crown of short rhizomes. Leaf blades are basal; they are flat and smooth with few marginal hairs near the base. Leaf sheaths are flattened and smooth except near the base, where they are hairy. The ligule is membrane-like, and auricles are absent. Stem growth is upright to ascending. The inflorescence is a racemose panicle with 3–5 racemes. The root system is deep and extensive with many highly branched, fibrous roots.

Management. Dallisgrass is slow to germinate, and the seedlings are not very competitive. It should be planted into a firm seedbed at 8–15 lb seed/acre and 0.25–0.50 in. deep in early spring. Seeding with an annual companion crop is often beneficial for weed control. Pastures tolerate continuous stocking during the growing season if height is maintained above 3–4 in. Hay should be cut while in the vegetative stage. Seed heads are susceptible to an ergot-like fungus that can cause neurological symptoms when consumed by livestock.

EASTERN GAMAGRASS

Alias: ice cream grass
Scientific name: *Tripsacum dactyloides* (L.) L.
Family: Poaceae
Tribe: Andropogoneae
Origin: North America

Adaptation and Use. This perennial warm-season grass is native to the North American tall-grass prairie. It is adapted to a wide range of soils, but prefers moderately well-drained to poorly drained sites with high fertility. It does not tolerate drought, but will tolerate long periods of flooding. Eastern gamagrass is grown primarily as summer forage for pasture and hay in temperate regions where introduced warm-season grasses are not well adapted. It is also used extensively in conservation and wildlife plantings.

Plant Characteristics. This sod-forming grass spreads by means of rhizomes and produces a dense, rough crown. Most of its growth occurs during late spring and early summer, and it flowers during summer. Leaf blades are flat, wide, and rough. Leaf sheaths are round, smooth, and split with overlapping margins. The ligule is blunt with a fringe of hairs, and auricles are absent. Reproductive stems grow upright; they are solid in the center, slightly flattened, and smooth. The inflorescence is a panicle with 1–4 raceme-like branches. Eastern gamagrass is monoecious, with male spikelets located at the top of each panicle branch and female spikelets below. The fibrous root system is deep and extensive.

Management. Eastern gamagrass should be planted in the spring when the soil temperature reaches 55°F. Seed are enclosed in a cupule, which contributes to a high degree of seed dormancy. Optimum establishment occurs with stratified seed. Dormant planting of untreated seed in late fall, which allows for natural stratification, has been reported to be an effective alternative to spring seeding. Plant with a grain drill or a corn planter at a rate of 8–10 lb seed/acre at a depth of 0.5–1.5 in. Planting with corn as a companion crop reduces weed competition. Eastern gamagrass does not tolerate close grazing or continuous stocking. Grazing should begin at 14 in. and livestock should be removed at 8 in. It produces one to two hay cuttings in a season. Hay should be harvested at the boot stage, and it should not be cut or grazed 4–6 weeks before frost.

Courtesy of Michael Collins, University of Missouri

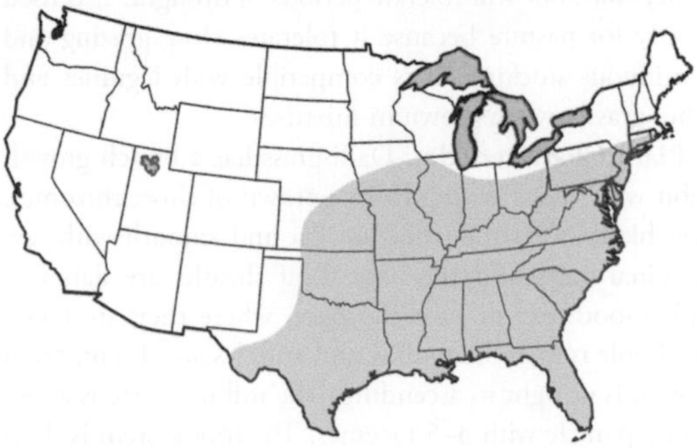

ELEPHANTGRASS

Courtesy of Lynn Sollenberger, University of Florida

Alias: napiergrass
Scientific name: *Pennisetum purpureum* Schumach.
Family: Poaceae
Tribe: Paniceae
Origin: Africa

Adaptation and Use. Elephantgrass, a perennial warm-season grass, is adapted to humid, tropical and subtropical regions, including the southernmost areas of the Gulf States. It is adapted to a wide range of soils but requires high fertility and good drainage for optimum growth. An extensive root system enables it to tolerate drought. It does not tolerate low temperatures or flooding. Elephantgrass is used mostly for pasture in the USA.

Plant Characteristics. Elephantgrass has a bunch growth habit but does produce short fleshy rhizomes. Leaf blades are wide and smooth with serrated margins and a prominent midrib on the lower side. Leaf sheaths are smooth. The ligule is a fringe of hairs, and auricles are absent. Stems are thick and grow upright. The inflorescence is a spike-like panicle that is usually yellow to brown but sometimes purple. The fibrous root system is extensive.

Management. Elephantgrass is usually propagated vegetatively from stem cuttings. Dwarf cultivars have shortened internode lengths and higher leaf-to-stem ratios; they are generally easier to manage for grazing than are other types. Stem cuttings should be planted at a depth of 1–2 in. during late summer at a rate of 1800–2000/acre. Dormant plantings can be made in late fall. Elephantgrass responds well to N fertilization. It does not tolerate continuous stocking. Minimum grazing height is 18–24 in. with 5-week rest periods. Hay or silage should be cut every 8–9 weeks at a height of 8 in.

FODDER BEET

Courtesy of Chris Benedict, Washington State University

Aliases: mangels, field beet, common beet
Scientific name: *Beta vulgaris* L.
Family: Chenopodiaceae
Origin: Europe

Adaptation and Use. Fodder beet is grown primarily as a summer annual in cool and moist temperate areas, but may be grown as a winter annual in areas with milder winters. It is widely adapted within the USA, but is grown as a forage crop mostly in the northeast. Fodder beet is adapted to a wide range of soils. It tolerates drought and can be grown on sandy and shallow soils, but does best on moist but well-drained fertile soils. It tolerates summer heat, but grows best at relatively cool summer temperatures. It is sensitive to cold temperature and prone to winterkill in areas with severe winters. The tops are generally harvested by grazing, but are sometimes ensiled. The roots are harvested and stored for later feeding.

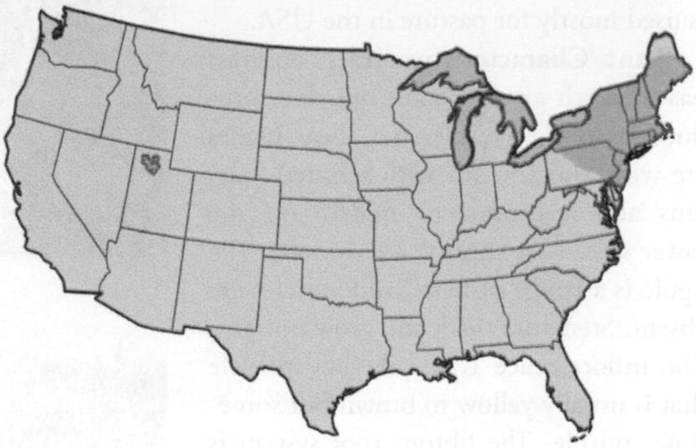

Plant Characteristics: Although typically grown as an annual, fodder beet has a biennial growth habit, producing vegetative growth in the first season, and flowering and producing seed (called bolting) in the second season. Leaves are egg shaped to heart shaped and grow in a rosette during the first season of growth. They are smooth and usually dark green. Stems are initiated following a period of vernalization, and elongate during the second season of growth. Leaves are borne on the elongated stem (stalk) in the second cycle of growth, and vary from being borne on a petiole near the base to being sessile near the top. The stem terminates in an elongated inflorescence with flowers borne individually or in clusters along its length. Flowers are sessile to the rachis, forming an irregular spike. Beet produces a large, fleshy taproot that is partially exposed in some cultivars.

Management: Fodder beet should be planted in rows 18–40 in. apart at a depth of 0.75–1.5 in. Seeding rate is in the range of 1–2 lb/acre. Planting should be done in early spring when grown as a summer annual, and during late summer when grown as a winter annual. Soaking seed in warm water for 12 hours followed by air drying may improve germination. Tops should be harvested prior to frost when grown as a summer annual, either by grazing or by chopping for silage. Cutting or grazing height is 2–4 in. Leaves have a laxative effect and contain high concentrations of oxalic acid, so should be fed with other forage to avoid problems with livestock health. Roots can be harvested by mechanical sugar beet harvesters and store well for later feeding if all the leaves are completely removed. Processing beets by mashing or slicing is recommended to improve animal performance.

INDIANGRASS

Courtesy of Kenneth Moore, Iowa State University

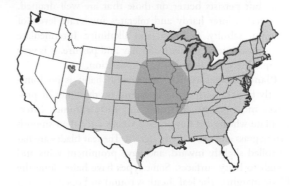

Scientific name: *Sorghastrum nutans* (L.) Nash
Family: Poaceae
Tribe: Andropogoneae
Origin: North America

Adaptation and Use. Indiangrass is a perennial warm-season grass adapted throughout much of the central and eastern USA. Adapted to a wide range of soils, it prefers well-drained and fertile lowland sites. However, it tolerates acid, alkaline, and sandy soils. It has moderate drought tolerance and is winter hardy. Indiangrass is grown primarily for summer pasture and hay in temperate regions, and is usually grown in mixtures with other warm-season grasses. It is also used in conservation and wildlife plantings.

Plant Characteristics. Indiangrass has a bunch growth habit but does develop short rhizomes. It produces most growth during summer, and flowers in late summer. Leaf blades are smooth with prominent veins, and are constricted near the base. Sheaths are usually smooth and round, but are sometimes flattened. The membrane-like ligule is deeply notched and the auricles are pointed. The yellow-brown panicle is plume-like with paired spikelets. The seed has a long, twisted awn. The fibrous root system is deep and extensive.

Management. Indiangrass is relatively difficult to establish. The chaffy seed are difficult to sow and may have a high degree of dormancy. In late spring, plant 0.25–0.50 in. deep into a prepared seedbed. Seeding rate is 6–8 lb/acre if drilled and 12–15 lb/acre when broadcast. Seedlings are slow to establish, so weed control is critical. A stand of 1–2 plants/ft^2 is adequate. Grazing should begin at 12–16 in. and should not be closer than 6 in. Hay should be harvested at early boot stage. Indiangrass has limited regrowth potential. It should not be cut or grazed during the 4–6 weeks prior to a killing frost, and a 6-in. stubble should be left for winter.

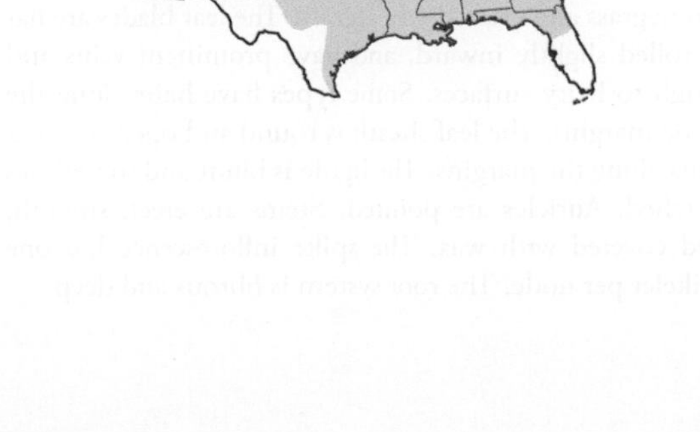

Courtesy of Kenneth Moore, Iowa State University

INTERMEDIATE WHEATGRASS

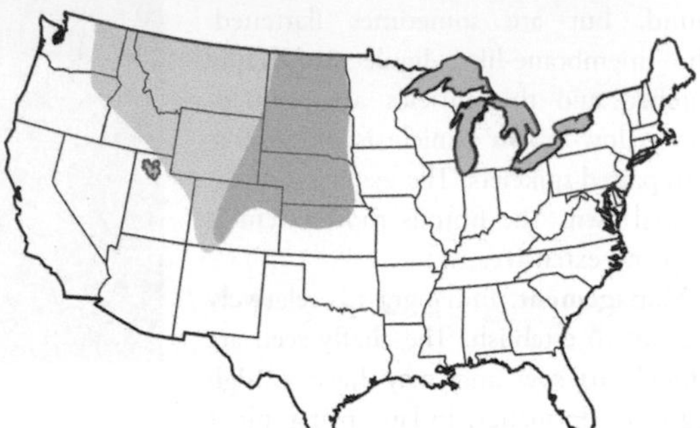

Alias: pubescent wheatgrass
Scientific name: *Thinopyrum intermedium* (Host) Barkworth & D.R. Dewey subsp. *intermedium*
Family: Poaceae
Tribe: Triticeae
Origin: Eurasia

Adaptation and Use. This cool-season perennial grass is well adapted to semiarid and sub-humid areas of the northern Great Plains. It is adapted to a wide range of soils, but persists better on those that are well drained. It is very winter hardy and tolerates moderate levels of drought, salinity, flooding, and alkalinity. Intermediate wheatgrass is used primarily for hay and pasture. It is very compatible with alfalfa for hay production.

Plant Characteristics. The growth habit is predominantly bunch, with some spreading by rhizomes. It produces most growth in spring and early summer. Intermediate wheatgrass flowers 1–2 weeks later than smooth bromegrass and crested wheatgrass. The leaf blades are flat to rolled slightly inward, and have prominent veins and rough to hairy surfaces. Some types have hairs along the blade margins. The leaf sheath is round and open, and has hairs along the margins. The ligule is blunt and sometimes notched. Auricles are pointed. Stems are erect, smooth, and covered with wax. The spike inflorescence has one spikelet per node. The root system is fibrous and deep.

Management. Intermediate wheatgrass has good seedling vigor and is relatively easy to establish. It should be planted into a firm, well-prepared seedbed in spring at 8–12 lb seed/acre when seeded alone, or 4–6 lb/acre when seeded with alfalfa. Seeding depth is 0.25–0.50 in. but may be increased to 1 in. on sandy soils. This grass does not tolerate close grazing or continuous stocking. Grazing should begin at a height of 10–12 in., and at 6 in. the livestock should be removed for at least 3–4 weeks. Hay should be cut at the early head stage. Regrowth is relatively slow, limiting the number of cuttings to one to two per year. Intermediate wheatgrass should not be cut or grazed 4–6 weeks prior to frost.

Courtesy of Michael Collins, University of Missouri

KENTUCKY BLUEGRASS

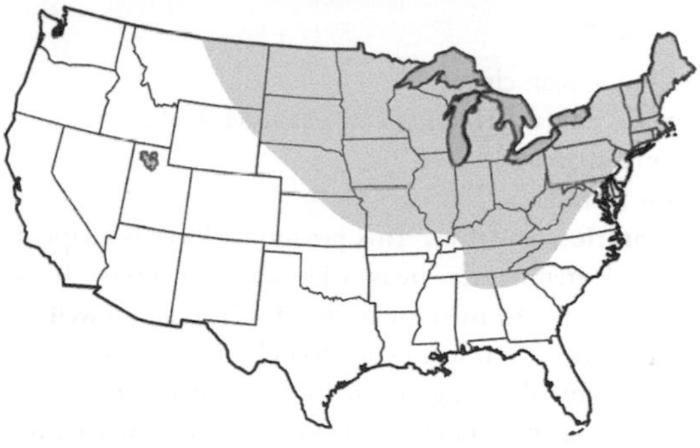

Scientific name: *Poa pratensis* L.
Family: Poaceae
Tribe: Poeae
Origin: Europe

Adaptation and Use. This cool-season perennial grass is well adapted to the humid, temperate region of the USA. Kentucky bluegrass has become naturalized throughout its area of adaptation. It is adapted to a wide range of soils, but persists better on well-drained, medium-textured soils. It has low tolerance of drought, but recovers well from it. Kentucky bluegrass will tolerate flooding, and is winter hardy. It is used primarily for pasture in association with other species, and is an important turf species.

Plant Characteristics. Kentucky bluegrass forms an extensive sod by means of rhizomes. It produces most growth in spring and fall. Leaf blades are smooth and form a keel that is pronounced at the tip. Leaves are folded in the whorl. The leaf sheath is flattened and smooth, and has overlapping margins. The collar is broad and divided, the ligule is blunt and membrane-like, and auricles are absent. Stems are mostly erect, fine, and slightly flattened. The inflorescence is an open panicle in the shape of a pyramid. The root system is fibrous and very shallow.

Management. Kentucky bluegrass occurs naturally in association with other species in pastures throughout its area of adaptation. It is less productive than taller grasses, and is rarely seeded as a pasture grass unless it is included in a mixture with other species. When seeded alone, it should be seeded at a rate of 6–12 lb/acre at a depth of 0.125–0.25 in. Because it tolerates close grazing it is often used for horse pasture. Grazing should begin at a height of 5 in. The minimum grazing height is 1.5–2.0 in.

Courtesy of Michael Collins, University of Missouri

KURA CLOVER

Alias: Caucasian clover
Scientific name: *Trifolium ambiguum* M. Bieb.
Family: Fabaceae
Origin: Eurasia

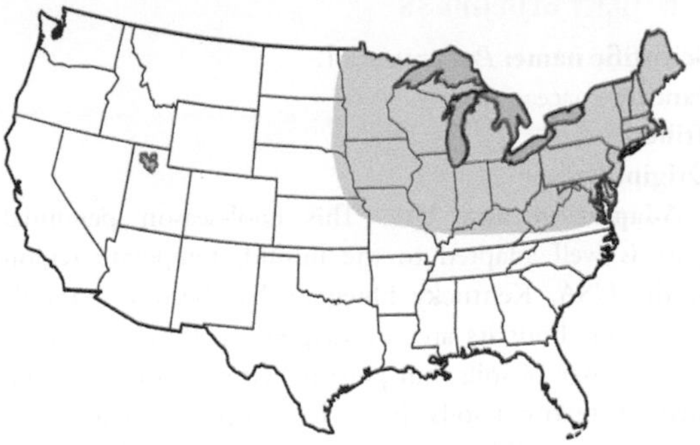

Adaptation and Use. This perennial clover is adapted to humid, temperate areas with severe winters. It is adapted to fine- to medium-textured soils that are well to poorly drained. It tolerates moderately acid and infertile soils and some flooding. There has been increasing interest in the use of kura clover as a persistent winter-hardy pasture legume. It is used mostly for permanent pasture in mixtures with grasses.

Plant Characteristics. Kura clover has a prostrate growth habit and spreads by means of rhizomes. Stems are short and smooth and arise from the crown and nodes of rhizomes. Leaves are palmately trifoliolate and borne on long, smooth petioles. The egg-shaped leaflets have serrated margins, are smooth, and often have a white "watermark." The round head has mostly white to pink flowers. The plant develops an extensive adventitious root system arising mainly from nodes of rhizomes.

Management. Kura clover has small seed and low seedling vigor, and generally requires 2–3 years to achieve a vigorous stand. Inoculated seed should be planted in early spring at a rate of 4–8 lb/acre when sown alone or 2–4 lb/acre when sown in a mixture. Seeding depth should be 0.25–0.50 in. Grazing management is generally determined by the companion grass. Kura clover tolerates close grazing and continuous stocking with a minimum grazing height of 2 in. It can cause bloat.

Courtesy of Michael Collins, University of Missouri

ORCHARDGRASS

Alias: cocksfoot
Scientific name: *Dactylis glomerata* L.
Family: Poaceae
Tribe: Poeae
Origin: Eurasia

Adaptation and Use. This perennial cool-season grass is well adapted to humid, temperate regions. Orchardgrass is adapted to medium-textured soils that are moderately to well drained, but it tolerates some flooding. It has moderate heat and drought tolerance, and is more shade tolerant than most forage grasses. Its moderate winter hardiness limits its northern range of adaptation. Orchardgrass is used primarily for hay and pasture. It is compatible with alfalfa for hay production and white clover for pasture.

Plant Characteristics. Orchardgrass has a bunch growth habit. Leaf blades are flat, have a smooth surface and rough edges, and are folded in the whorl. The leaf sheath is flat and usually smooth. The leaf collar is broad and divided. The leaves have a long membrane-like ligule that is pointed at the tip, and no auricles. Stems grow upright and are smooth. The panicle has dense clusters of spikelets borne on short branches. The fibrous root system is moderately extensive and deep.

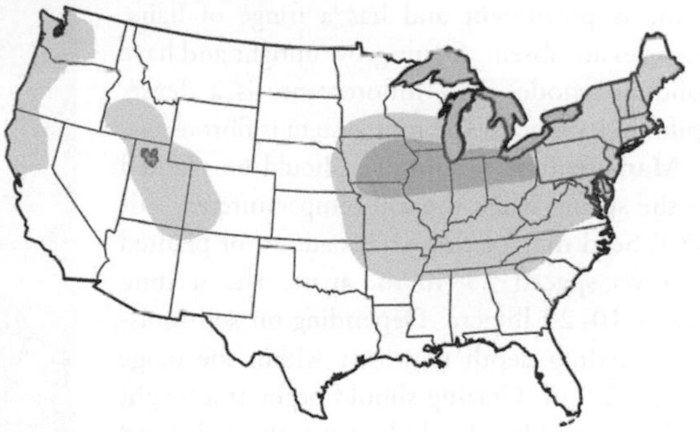

Management. Orchardgrass is relatively easy to establish. It should be planted into a firm, well-prepared seedbed in spring or late summer. Seed should be planted at a depth of 0.25–0.50 in. at a rate of 8–12 lb/acre when seeded alone or 4–6 lb/acre when seeded with a legume. It responds well to N fertilization, although high rates can reduce persistence. Orchardgrass should be stocked rotationally because continuous stocking weakens the stand. Grazing should begin at 6–8 in., and the livestock should be removed at 4 in. Hay should be made between boot and early head stages. Subsequent cuttings should be made every 4–6 weeks, but not during the 4- to 6-week period before frost.

PEARLMILLET

Aliases: American fountaingrass, cattail millet
Scientific name: *Pennisetum glaucum* (L.) R. Br.
Family: Poaceae
Tribe: Paniceae
Origin: Africa

Adaptation and Use. Pearlmillet is an important cereal crop in the semiarid tropics. In North America, it is grown primarily as a summer annual forage. It can be grown in almost every US state, but its major production region is in the southeastern coastal plains. It is adapted to a wide range of soils and is more tolerant of adverse soil conditions than most other cereal grains. Pearlmillet tolerates low pH and fertility, although it responds well to fertilization. It has moderate drought tolerance, but does not tolerate flooding or poor drainage. Although used primarily for pasture, it can also be used as a hay or silage crop.

Plant Characteristics. Pearlmillet has a bunch growth habit with many upright tillers. Some cultivars produce aerial tillers. Leaf blades are long and wide, and have rough margins and a prominent midrib. The leaf sheath is split. The ligule is prominent and has a fringe of hairs. Auricles are absent. Stems grow upright and have solid internodes. The inflorescence is a dense, spike-like panicle. The root system is fibrous.

Management. Pearlmillet should be planted in the spring when the soil temperature exceeds 65°F. Seed may be drilled, broadcast, or planted in rows spaced 15–40 in. apart. The seeding rate is 10–20 lb/acre. Depending on soil moisture, seeding depth may vary within the range of 0.5–2.0 in. Grazing should begin at a height of 20–25 in. Hay should be cut at the early head stage, and a 4- to 6-in. stubble should be left to ensure adequate regrowth.

Courtesy of Vivien Allen, Texas Tech University

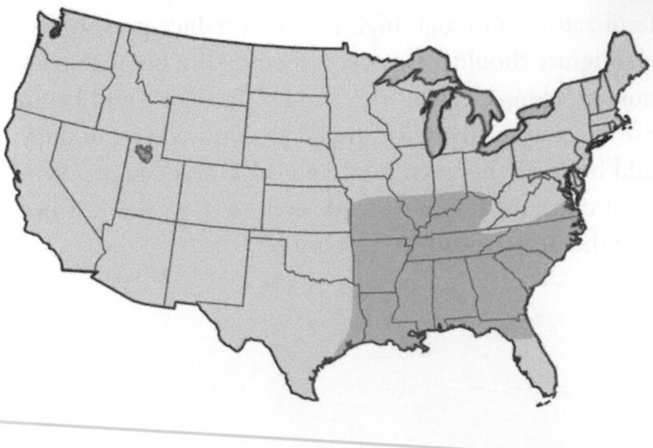

Courtesy of Lynn Sollenberger, University of Florida

PERENNIAL PEANUT

Scientific name: *Arachis glabrata* Benth.
Family: Fabaceae
Origin: South America

Adaptation and Use. Perennial peanut, a warm-season perennial legume, is well adapted to coastal plains in the southern USA. It is adapted to well-drained, sandy soils and tolerates low fertility and moderately acid soils. Once established, it will tolerate drought but does not tolerate freezing, flooding, or poor drainage. It is used mostly for hay, pasture, and conservation.

Plant Characteristics. Perennial peanut has a creeping growth habit. It spreads by means of rhizomes that form a dense mat near the soil surface. The compound leaves have four smooth leaflets that vary from long and narrow to egg shaped. The stems are mostly prostrate. Flowers are axillary and range from yellow to orange. The root system is extensive and consists of some taproots and many finely branched nodal roots on rhizomes.

Management. Perennial peanut is established by planting vegetative cuttings called sprigs using specialized planting equipment. Inoculation is not usually necessary because the sprigs are naturally inoculated. Planting is normally done in winter or early spring when the plants are not actively growing. Sprigs should be planted at 80–120 bushels/acre at a depth of 1–2 in. Perennial peanut is slow to establish, and new plantings do not compete well with grasses or weeds. Grazing should begin at a height of 6–8 in., and the minimum grazing height is 2 in. Perennial peanut tolerates close grazing but not continuous stocking. It produces two to three hay cuttings per year harvested at 5- to 8-week intervals.

PERENNIAL RYEGRASS

Scientific name: *Lolium perenne* L.
Family: Poaceae
Tribe: Poeae
Origin: Eurasia

Adaptation and Use. Perennial ryegrass is adapted to temperate climates with mild winters and cool summers. It is an important forage species in Western Europe and New Zealand. In the USA, its use as a forage has been limited to coastal areas of the Pacific Northwest and the Northeast. However, plant breeders are developing cultivars that are more winter hardy. It is adapted to soils with fine to medium texture. It prefers well-drained soils, but tolerates wet soils with good surface drainage. It does not tolerate low fertility, drought, heat stress, or severe winters. Perennial ryegrass produces very high-quality forage. It is often used for dairy pasture in mixtures with alfalfa or white clover, but is also used for hay, silage, and conservation.

Plant Characteristics. Plants have a bunch growth habit and produce most growth in spring and fall. Leaves are folded in the whorl. Leaf blades are flat, smooth, glossy, and keeled underneath with slightly rough edges. The leaf sheath is flattened to round, and is red to purple near the base. The leaf collar is narrow. The auricles are small and claw-like, and the ligule is membrane-like, rounded, and toothed. The stems grow upright and are smooth. The inflorescence is a long, narrow spike with flattened spikelets each containing 3–10 florets. The root system is fibrous.

Management. Perennial ryegrass should be planted into a firm, well-prepared seedbed in spring or late summer. The seeding rate is 15–20 lb/acre when seeded alone, and 5–8 lb/acre when seeded with a legume. The seeding depth is 0.25–0.50 in. Perennial ryegrass persists and produces better under rotational stocking. Grazing should begin at a height of 6–8 in., and livestock should be removed at 3 in. The rest period should be 2–4 weeks during periods of active growth. Hay should be cut at the early head stage. Perennial ryegrass should not be cut or grazed 4–6 weeks prior to frost, and at least a 4-in. stubble should be left for winter. Some cultivars contain a fungus that produces alkaloids which can cause acute health problems in livestock.

Courtesy of Michael Collins, University of Missouri

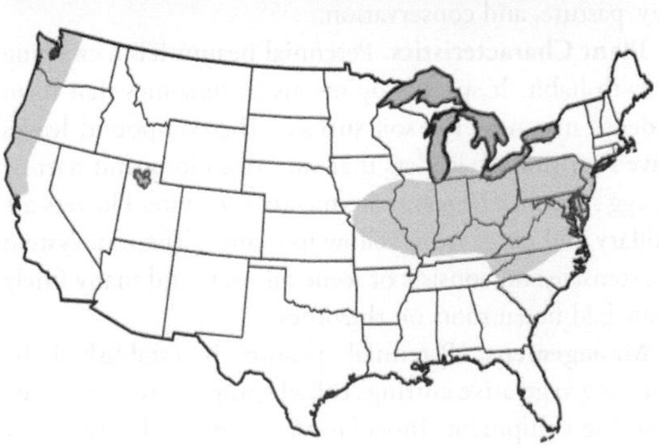

Courtesy of Gregory Bishop-Hurley and Robert Kallenbach, University of Missouri

PRAIRIEGRASS

Aliases: matua brome, rescuegrass
Scientific name: *Bromus catharticus* Vahl var. *catharticus*
Family: Poaceae
Tribe: Poeae
Origin: South America

Adaptation and Use. This cool-season perennial grass is adapted to temperate areas with mild winters. The range of adaptation is limited by poor winter hardiness and susceptibility to foliar diseases under warm, humid conditions. Prairiegrass is adapted to fertile soils with medium to coarse texture. It tolerates heat and drought, but does not tolerate flooding or poor drainage. Although used mostly as a cool-season pasture, it can also be used as a hay crop.

Plant Characteristics. Prairiegrass has a bunch growth habit and produces most growth in spring, fall, and winter in milder climates. The leaf blades are flat, with rough surfaces and edges, and rolled in the whorl. The leaf sheath is round to partially flattened and is often hairy, particularly at the base. The leaf collar is broad and divided. The leaves have a long membrane-like ligule that is sometimes notched; auricles are absent. The stems grow upright and are smooth. The panicle is open, with flattened spikelets containing several florets. The root system is fibrous.

Management. Prairiegrass should be planted into a firm, well-prepared seedbed in spring or late summer at a rate of 25–40 lb seed/acre and at a depth of 0.25–0.50 in. It does not compete well with weeds. It does not tolerate close grazing or continuous stocking. Grazing should begin at a height of 8–10 in. and livestock should be rotated at 3 in. The rest period should be 4–6 weeks. Hay should be cut at boot stage. Regrowth is relatively slow, limiting the number of cuttings. Cutting and grazing during stem elongation will reduce regrowth. Prairiegrass should not be cut or grazed 4–6 weeks prior to frost, and a 5-in. stubble should be left for winter.

Courtesy of Michael Collins, University of Missouri

RED CLOVER

Scientific name: *Trifolium pratense* L.
Family: Fabaceae
Origin: Eurasia

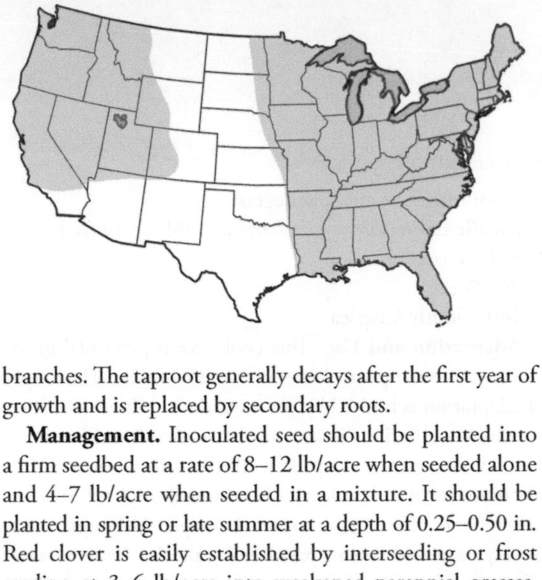

Adaptation and Use. This short-lived perennial legume is widely adapted throughout the humid, temperate regions of the USA. Red clover is grown as a winter annual in the South. Two red clover types are grown in the USA: medium types flower earlier and produce two or more harvests, whereas mammoth types flower much later and typically produce a single harvest. Adapted to a wide range of soils, red clover grows best on well-drained soils with a fine to medium texture. It tolerates low fertility and moderately acid soils, but has low drought tolerance. It tolerates cold temperatures, but in warm temperatures it is strongly affected by root and crown diseases. Red clover tolerates shade better than most other legumes. It is used for pasture, hay, silage, and conservation, and is often grown in mixtures with cool-season grasses.

Plant Characteristics. Red clover has a mostly upright growth habit. The primary stem does not elongate, but it produces many axillary branches at the lower nodes. Leaves are palmately trifoliolate, with oblong to wedge-shaped leaflets that are usually variegated with a white "watermark." Stipules are fused to the petiole. The stems are hollow and hairy. The terminal head has rose to magenta flowers. Plants produce a short taproot with many secondary branches. The taproot generally decays after the first year of growth and is replaced by secondary roots.

Management. Inoculated seed should be planted into a firm seedbed at a rate of 8–12 lb/acre when seeded alone and 4–7 lb/acre when seeded in a mixture. It should be planted in spring or late summer at a depth of 0.25–0.50 in. Red clover is easily established by interseeding or frost seeding at 3–6 lb/acre into weakened perennial grasses. Grazing should begin at early to mid-bud stage. Hay should be cut at early bloom; the minimum harvest height is 2 in. When grazed in pure stands, red clover has high bloat potential and can cause reproductive problems in livestock. Hay, especially if not properly dried, may contain a fungal toxin that causes excessive slobbering and, in severe cases, spontaneous abortion.

REED CANARYGRASS

Scientific name: *Phalaris arundinacea* L.
Family: Poaceae
Tribe: Aveneae
Origin: Europe, Asia, North America

Adaptation and Use. This cool-season perennial grass is adapted to humid and sub-humid temperate areas. Reed canarygrass is grown throughout the northern half of the USA, but is most common in the Northeast and North Central regions. It is adapted to poorly drained soils with a fine to medium texture. It is moderately tolerant of acid and alkaline soils, and tolerates flooding and moderate drought. It is very winter hardy. Although used mostly for pasture and hay on soils that are too wet for other cool-season grasses, reed canarygrass can also be used for silage. An important conservation species, it is commonly used for filter strips, grass waterways, streambank stabilization, and phytoremediation, and it provides excellent wildlife habitat.

Plant Characteristics. It forms an extensive sod by means of rhizomes, and produces most growth in spring and fall. Leaf blades are smooth near the base but become rough near the tip. Blade margins are rough, and a leaf constriction is often present. Leaves are rolled in the whorl. The leaf sheath is round, smooth, and split with overlapping margins. The collar varies from narrow to broad and may be divided. The ligule is membrane-like and round to pointed. Auricles are absent. The stems grow mostly erect and are round and coarse. The narrow to open panicle is densely populated with spikelets. The fibrous root system is extensive and deep.

Management. Reed canarygrass is difficult to establish due to slow emergence. However, once established, its growth is vigorous and it may become invasive on wet sites. It should be planted into a firm, well-prepared seedbed in spring or late summer at a rate of 8–12 lb seed/acre when seeded alone or 5–8 lb seed/acre when seeded with a legume. The seeding depth should be 0.25–0.50 in. It responds well to N fertilization. Reed canarygrass tolerates continuous stocking, but greater productivity is achieved when it is stocked rotationally. Grazing should begin at 6–12 in. The minimum grazing height is 3–4 in. Hay should be made between the boot and early head stages. Subsequent cuttings should be made every 4–6 weeks except during the 4- to 6-week period prior to frost. Some cultivars contain high levels of tryptamine alkaloids that reduce palatability and may cause chronic health problems in some classes of livestock.

Courtesy of Michael Collins, University of Missouri

SAINFOIN

Courtesy of Surya Acharya, Agriculture and Agri-Food Canada

Aliases: holy clover
Scientific name: *Onobrychis viciifolia* Scop.
Family: Fabaceae
Origin: Eurasia

Adaptation and Use. This herbaceous perennial legume is primarily grown in in the northern region of the Intermountain West and Northern Great Plains. It can be grown at elevations of up to 6000 ft, and is best adapted to semiarid and sub-humid environments on well-drained soil with a pH of 6–8. It does well on calcareous soils, but has low salt tolerance and does not tolerate wet or acid soils. Sainfoin persists well under dry conditions, but is short-lived under wet or irrigated conditions. It tolerates drought and cold and is winter hardy. It is used in rangeland improvement in drier areas, and for hay, pasture, and conservation in areas with higher rainfall. Sainfoin flowers produce large amounts of nectar that produce high-quality honey.

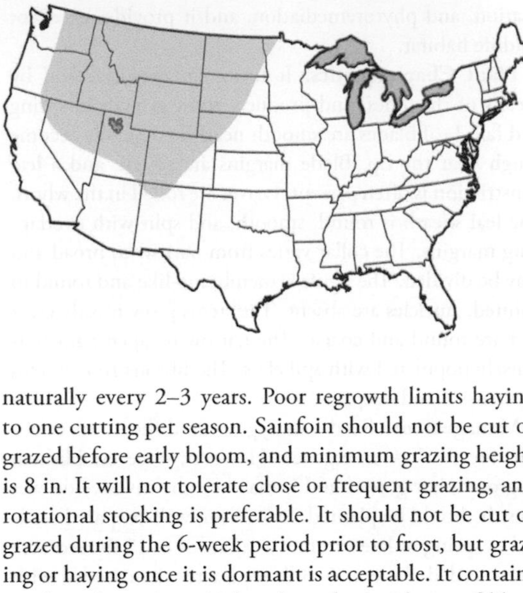

Plant Characteristics. Sainfoin has an upright growth habit with many tall hollow stems arising from a crown. The stem has a reddish-brown coloration and hairs. Leaves are pinnately compound with 5–10 pairs of opposite leaflets occurring along the rachis, with a single terminal leaflet. The inflorescence is a compact raceme with 20–50 white, pink, or purple flowers. The plant develops a large, deep taproot system.

Management. Sainfoin can be planted in spring or late summer. The seeding rate is 34 lb/acre for pasture or 2–5 lb/acre in rangeland mixture. It is most compatible in a mixture when seeded with bunchgrasses such as crested wheatgrass. Plant inoculated seed at a depth of 0.25–0.75 in. Sainfoin is susceptible to crown rot and has limited persistence (3–6 years) in infected areas. Stand longevity can be increased by allowing it to reseed naturally every 2–3 years. Poor regrowth limits haying to one cutting per season. Sainfoin should not be cut or grazed before early bloom, and minimum grazing height is 8 in. It will not tolerate close or frequent grazing, and rotational stocking is preferable. It should not be cut or grazed during the 6-week period prior to frost, but grazing or haying once it is dormant is acceptable. It contains condensed tannins which reduce the incidence of bloat in grazing livestock.

SERICEA LESPEDEZA

Alias: Chinese lespedeza
Scientific name: *Lespedeza cuneata* (Dum. Cours.) G. Don
Family: Fabaceae
Origin: Asia

Adaptation and Use. This warm-season perennial legume is adapted to the humid region of the USA, but its northern acclimation is limited by low winter hardiness. Adapted to a wide range of soils, it tolerates low fertility and acid soils. It is relatively tolerant of drought and poor drainage. It is used for summer pasture, hay, and conservation.

Plant Characteristics. Sericea lespedeza has an upright growth habit and produces most growth in summer. The leaves are pinnately trifoliolate, with wedge-shaped leaflets that are sharply pointed at the tip and are smooth on the upper surface but covered with fine hairs below. New stems are fine and become coarse with age. Two types of flowers occur in leaf axils along the stems: those with cream to purple petals, and those without petals. The plant has a deep, branched taproot system.

Management. Inoculated, preferably scarified, seed should be planted in early spring at 10–15 lb/acre and 0.25–0.50 in. deep. Seedlings establish slowly, but established plants are very competitive. Begin grazing at 8–12 in., and remove livestock at 4 in. Hay should be cut when the crop is about 12 in., leaving a 3-in. stubble. It should not be cut or grazed 6 weeks prior to frost, and a 3- to 4-in. stubble should be left for winter. Sericea lespedeza has low bloat potential. High tannin levels may reduce palatability.

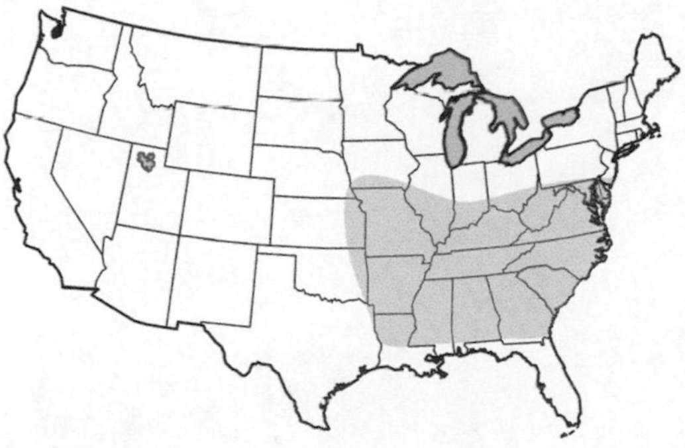

Courtesy of Michael Collins, University of Missouri

SIDEOATS GRAMA

Scientific name: *Bouteloua curtipendula* (Michx.) Torr.
Family: Poaceae
Tribe: Chlorideae
Origin: North America

Adaptation and Use. This native warm-season perennial grass is adapted throughout the natural grassland regions of the USA, but is most common in the Great Plains. Although adapted to a wide range of soils, it occurs most frequently on fine-textured soils. It tolerates drought and moderate alkalinity. An important range grass, it is used primarily for grazing, but is sometimes harvested for hay. It is also used for conservation and wildlife.

Plant Characteristics. Sideoats grama has a mostly upright bunch growth habit and produces most growth in the summer. Some types spread by means of short rhizomes. The foliage has a bluish-green color. Leaf blades are coarse, straight, and mostly basal, with sparse hairs along the margins. Sheaths are round, smooth below but with long, smooth hairs above. The ligule is truncated, with a fringe of hairs, and auricles are absent. The stems are erect, smooth, and purple near nodes. The inflorescence is a panicle with several spike-like branches and several spikelets per branch. The plant produces a deep, fibrous root system.

Management. Sideoats grama should be planted in late spring or early summer. The best results are obtained by seeding locally adapted ecotypes at a rate of 2–6 lb/acre into a firm seedbed at a depth of 0.25–0.50 in. Sideoats grama remains green longer in the fall than other grama grasses, and stockpiles well. It does not tolerate overgrazing. Grazing should begin at 12 in. and livestock should be removed at 6 in.

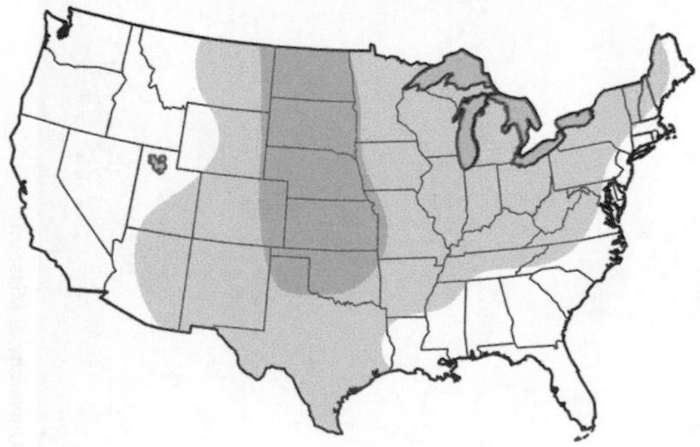

Courtesy of Vivien Allen, Texas Tech University, and Tim Phillips, University of Kentucky

SMALL GRAINS

OAT

Scientific name: *Avena sativa* L.

RYE

Scientific name: *Secale cereale* L.

TRITICALE

Scientific name: X *Triticosecale* Wittm.

WHEAT

Scientific name: *Triticum aestivum* L.
Family: Poaceae
Tribe: Aveneae (oat), Triticeae (rye, triticale, wheat)
Origin: Europe (oat, rye), Asia (wheat). Triticale is an interspecific hybrid between wheat and rye.

Adaptation and Use. Although grown primarily as grain crops, these annual grasses are also used as annual forages. They are adapted throughout the USA and southern Canada. Oat is the least winter hardy of the small grains; it is usually grown as a summer annual. Rye is the most winter hardy of the small grains; it is often grown as a winter cover crop. Both species are also grown as winter annuals in areas with mild winters. Small grains are adapted to a wide range of soils. In general, they require moderate fertility and moisture availability and do not tolerate drought or flooding. Rye is more tolerant of adverse soil conditions than are other small grains. It tolerates drought, low pH, and low fertility, but responds well to fertilization. Oat is used primarily as a silage or hay crop but is sometimes used as pasture. It is often used in companion seedings with legumes, especially alfalfa. Wheat, rye, and triticale are used primarily for winter pasture but also can be used as silage or hay crops. The straw from small-grain crops is often used as roughage.

Plant Characteristics. Small grains have a bunch growth habit with many upright tillers. Leaves are rolled in the whorl. Oat leaf blades are smooth at the base, become rough near the tip, and have rough or hairy margins. The leaf sheath is round, smooth or hairy, and split with overlapping margins. The leaf collar is broad and divided, and auricles are absent. The membrane-like ligule is rounded and finely toothed. Rye leaf blades are rough above and smooth below, and have rough margins. Blades are covered with a waxy coating that gives them a bluish appearance. The leaf sheath is round, smooth, and split with overlapping margins. The leaf collar is mostly broad, and the auricles are small. The membrane-like ligule is rounded, appears frayed, and may have a fringe of hairs. Wheat leaf blades are smooth near the base and rough near the tip on the upper side. The lower side is entirely smooth and keeled near the base. The leaf sheath is round and split

Courtesy of Vivien Allen, Texas Tech University

OAT

Courtesy of Vivien Allen, Texas Tech University

RYE

with overlapping margins. It may be smooth or hairy. The leaf collar is broad, and the auricles are small and hairy. The membrane-like ligule is rounded, appears frayed, and may have a fringe of hairs. The leaf morphology of triticale is more similar to that of wheat than of rye. Small-grain stems grow upright, have hollow internodes, and are smooth. The oat inflorescence is an open panicle with many branches each ending with a single spikelet containing several florets. Rye has a spike inflorescence with one awned spikelet per node containing two fertile florets. Triticale has a spike with one spikelet per node, each containing several florets. The wheat spike has one spikelet per node containing one to several florets, and some cultivars have awns. The root systems of small grains are fibrous.

Management. Small grains are relatively easy to establish. They should be planted into a firm, well-prepared seedbed at a depth of 1.0–1.5 in. Planting should be done during late summer or fall when grown as a winter annual, and in early spring when grown as a summer annual. Seeding rates are 60–100 lb/acre for oat and 90–120 lb/acre for rye, triticale, and wheat. Higher seeding rates (1.3–2 times higher) are sometimes recommended for hay and silage. The seeding rate of oat should be reduced by 25–50% when planted as a companion crop. Grazing should begin at a height of 9–10 in. for oat, and 6–10 in. for rye, triticale, and wheat. Grazing during stem elongation will remove the apex and prevent grain development. Small grains should not be grazed while dormant, and a 3-in. stubble should be left if grazed in the fall. Oat hay and silage are usually harvested after some grain development has occurred (for higher milk yield), but should be harvested earlier when used as a companion crop to minimize competition. Hay or silage of triticale and wheat should be harvested at boot to early head stage. Rye may cause an off flavor in milk.

Courtesy of Michael Collins, University of Missouri

TRITICALE

Courtesy of Vivien Allen, Texas Tech University

WHEAT

Courtesy of Michael Collins, University of Missouri

SMOOTH BROMEGRASS

Aliases: Austrian brome, Hungarian brome, Russian brome
Scientific name: *Bromus inermis* Leyss.
Family: Poaceae
Tribe: Poeae
Origin: Eurasia

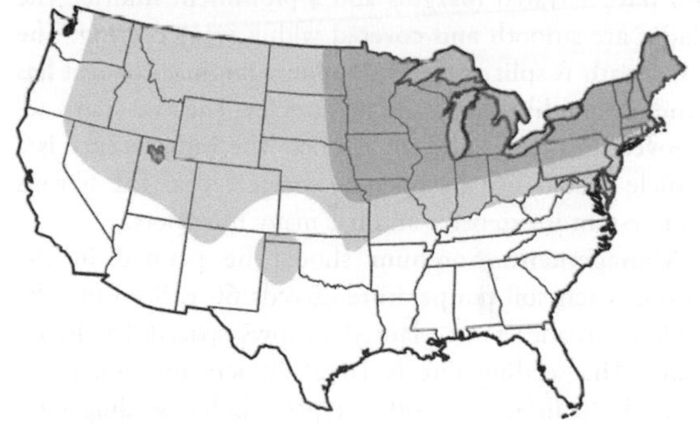

Adaptation and Use. This cool-season perennial grass is adapted to humid and sub-humid, temperate areas. Smooth bromegrass is adapted to well-drained soils with fine to medium texture. It resists heat and drought by becoming dormant. It does not tolerate flooding very well, but is very winter hardy. It is used mostly for pasture, hay, and conservation, and is particularly compatible in mixtures with alfalfa and red clover.

Plant Characteristics. Smooth bromegrass forms an extensive sod by means of rhizomes, and produces most growth in spring and fall. Leaf blades are flat and smooth and have rough margins. A W-shaped leaf constriction is usually present on the blade. Leaves are rolled in the whorl. The leaf sheath is round, smooth, and closed. Lower sheaths may be hairy. The collar is mostly broad and divided, and auricles are absent. The membrane-like ligule is short and truncate to rounded. Stems grow mostly erect and are round and smooth. The narrow to open panicle has numerous spikelets. The fibrous root system is extensive and deep.

Management. Smooth bromegrass should be planted into a firm, well-prepared seedbed in spring or late summer at a rate of 8–12 lb seed/acre when seeded alone or 6–8 lb/acre when seeded with a legume. Seeding depth should be 0.25–0.50 in. It responds well to N fertilization. It persists best when stocked rotationally. Grazing should begin at 8–10 in. and end at 4 in. Hay should be made between late boot and early head stages. Grazing or cutting during elongation will result in slow regrowth and may weaken the stand. Subsequent cuttings should be made every 4–6 weeks, but not during a 4- to 6-week period prior to frost. A 6-in. stubble should be left for winter.

SORGHUMS

Scientific name: *Sorghum bicolor* (L.) Moench
Family: Poaceae
Tribe: Andropogoneae
Origin: Africa

Adaptation and Use. Sorghum is an important cereal crop in semiarid regions of the world, including the Great Plains of the USA. It is most often grown in areas that are too dry for corn. Several types are commonly grown for forage, including forage sorghum, sudangrass, and sorghum × sudangrass hybrids. Adapted to a wide range of soils, sorghums prefer well-drained soils but tolerate poorly drained soils with adequate surface drainage. They tolerate moderately acid and low-fertility soils, but respond well to fertilization. Sorghums have low tolerance of flooding or salinity. They avoid drought by becoming semi-dormant. Although used primarily for pasture and silage, they are also used for hay and green chop.

Plant Characteristics. Sorghums have a bunch growth habit with few to many upright tillers. Forage and grain sorghums tiller less than sudangrass and sorghum × sudangrass hybrids. Sudangrass has narrower leaves and finer stems than other sorghums. The leaf blades are long and have serrated margins and a prominent midrib. The blades are smooth and covered with a waxy coating. The leaf sheath is split with overlapping margins. The leaf has a membrane-like ligule and no auricles. The oval stems are grooved and have solid internodes. The inflorescence is a panicle that is usually open in forage types. The fibrous root system is extensive and has many fine roots.

Management. Sorghum should be planted in the spring when soil temperature exceeds 60°F. Seed may be drilled, broadcast, or planted in rows spaced 15–40 in. apart. The seeding rate is 10–25 lb/acre for sudangrass and 15–30 lb/acre for other types. Higher seeding rates are recommended for narrower row spacing. Depending on soil moisture levels, seeding depth is in the range of 0.5–2.0 in. Grazing should begin at a height of 18–24 in. for sudangrass and 24–28 in. for other types. The minimum grazing height is 4–6 in. Silage should be harvested after grain has reached the dough stage. Hay should be cut at the boot to early head stage, leaving a 4- to 6-in. stubble to ensure adequate regrowth. Sorghum hay is difficult to dry, and should be conditioned at harvest to improve drying. Sorghums contain prussic acid, which when present at high levels can cause cyanide poisoning in livestock. Prussic acid is most concentrated in young, actively growing tissues. The problem can be avoided by not grazing until plants have reached a height of 18–24 in. Cutting for hay or silage generally alleviates the problem because the toxin is released and volatilizes during handling. Sorghum can also accumulate nitrate to toxic levels. This usually occurs under drought and freezing stress, which also tend to increase prussic acid levels. Therefore grazing of stressed sorghum should be avoided.

Courtesy of Vivien Allen, Texas Tech University

Courtesy of Mike Collins, University of Missouri

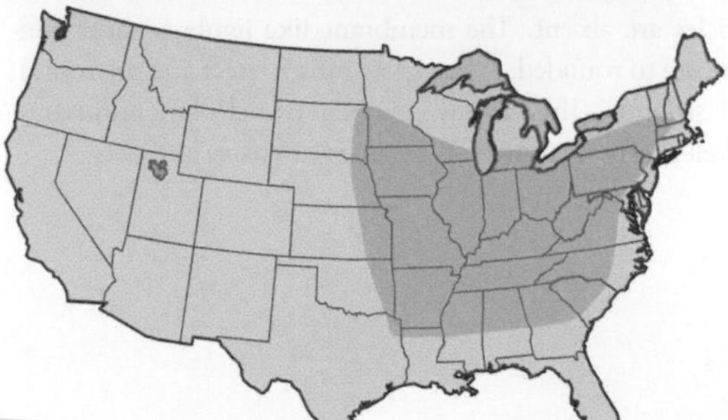

SUBTERRANEAN CLOVER

Courtesy of Michael Collins, University of Missouri

Aliases: sub clover
Scientific name: *Trifolium subterraneum* L.
Family: Fabaceae
Origin: Mediterranean

Adaptation and Use: This self-seeding winter annual legume is well adapted to Mediterranean climates. It is an important species worldwide, particularly in Australia. It is grown as a rangeland species west of the Cascades in Oregon and California, and for pasture in the southeast USA. It can be grown throughout the continental USA with the exception of the extreme North Central and Great Plains region. Subterranean clover tolerates acid to slightly alkaline soils (pH 4.0–7.5). It grows on well- to poorly drained soils, but does not tolerate flooding. It is moderately tolerant of drought and poor soil fertility. It is used primarily as a pasture species in mixtures on the west coast and southeast coastal plains, but is used as a cover and manure crop in other areas.

Plant Characteristics. Subterranean clover has a prostrate growth habit with many horizontal stems arising from the crown. Leaves are palmately trifoliolate and borne on long, hairy petioles. Leaflets are triangular to heart-shaped, hairy and variegated with a light green "watermark." Stipules are large and tapered. Stems are hairy and although decumbent do not produce nodal roots. The inflorescence is inconspicuous with 3–7 white to pinkish florets. Seed are enclosed by a bur that is placed near the ground by elongation of the peduncle. The bur has stiff bristles that facilitate partial burying of the seed in soil. The plant produces a branched taproot.

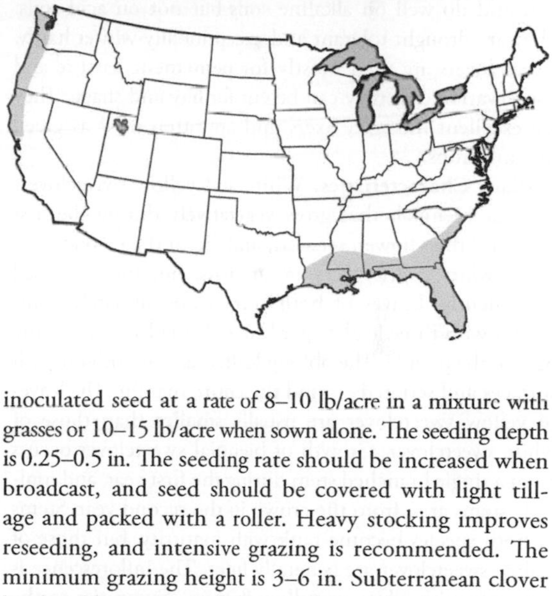

Management. Subterranean clover is usually grown in a mixture with grasses and other clovers as pasture for spring and summer grazing. Once established it will naturally reseed if residual forage is removed in the fall. It should be established in late summer or early fall in a prepared seedbed, and can be drilled or broadcast seeded. It is compatible with tall fescue, orchardgrass, and ryegrass, with the latter being preferred. Plant inoculated seed at a rate of 8–10 lb/acre in a mixture with grasses or 10–15 lb/acre when sown alone. The seeding depth is 0.25–0.5 in. The seeding rate should be increased when broadcast, and seed should be covered with light tillage and packed with a roller. Heavy stocking improves reseeding, and intensive grazing is recommended. The minimum grazing height is 3–6 in. Subterranean clover has low bloat potential, but produces phytoestrogens in concentrations high enough to sometimes cause reproductive problems in livestock.

SWEETCLOVERS

WHITE SWEETCLOVER

Scientific name: *Melilotus albus* Medik.

YELLOW SWEETCLOVER

Scientific name: *Melilotus officinalis* (L.) Lam.
Family: Fabaceae
Origin: Eurasia

Courtesy of Vivien Allen, Texas Tech University

Adaptation and Use. White sweetclover and yellow sweetclover are closely related species. Both are biennial legumes that are well adapted to humid and sub-humid, temperate climates. Although adapted throughout the USA, they are most common in the eastern Great Plains and north central states, where they are naturalized. Sweetclovers are adapted to soils with moderate to excellent drainage, and do well on alkaline soils but not on acid soils. They are drought tolerant and exceptionally winter hardy. Sweetclovers are used mostly for permanent pasture and conservation, but they can be cut for hay and silage. They are excellent nitrogen fixers and are often used as green manure crops.

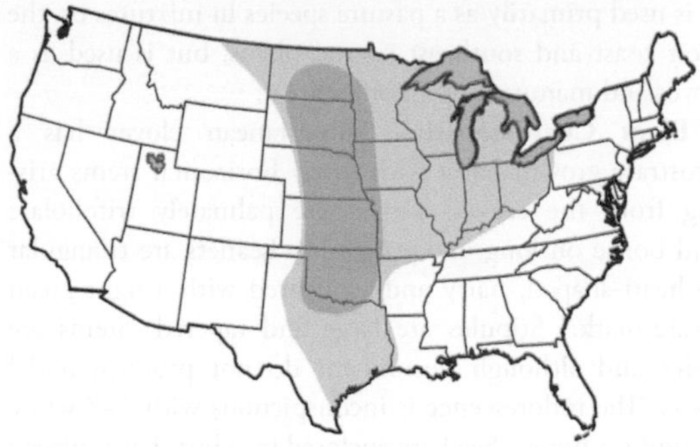

Plant Characteristics. White and yellow sweetclovers are true biennials that grow vegetatively during the first year, and then flower, set seed, and die in the second year. Some white sweetclovers are annuals, but they are used infrequently. Leaves of both species are pinnately compound with three leaflets per leaf and small stipules at the base of the petiole. The oblong leaflets are smooth on both surfaces and serrated around the entire margin. The leaves of yellow sweetclover are usually smaller than those of white sweetclover. Growth of biennial sweetclover occurs from a single branched stem during the first year, and multiple stems arise from the crown in the second year. Stems of both species become rank with maturity, but those of yellow sweetclover are generally finer. The inflorescence is a raceme with white or yellow flowers, giving rise to the common names of the two species. Biennial white sweetclover flowers 1–3 weeks later than yellow sweetclover. Both species develop a deep taproot system.

Management. Sweetclover occurs and reseeds naturally in association with other species in pastures and rangelands. It is used in mixtures with other species for pasture, but is sown in pure stands for hay or as a manure crop. Inoculated seed should be planted 0.25–0.50 in. deep in spring at 10–16 lb/acre alone or 3–5 lb/acre in a mixture. Sweetclover does not tolerate close grazing or mowing during the first year of growth because regrowth occurs from buds on the stem rather than on the crown. A stubble of at least 8–12 in. should be left to ensure adequate regrowth. Grazing can begin in the second year at 8–10 in. Hay should be cut at the bud stage before plants become too rank. Sweetclover can cause bloat, and it contains high levels of coumarin, which gives the foliage a sweet odor but tastes bitter and reduces palatability to livestock. When sweetclover is harvested for hay or silage, coumarin can be converted by storage fungi to dicoumarol, which causes hemorrhaging in livestock. Moldy sweetclover hay or silage should never be fed to livestock. Low-coumarin cultivars are available and should be used for hay and silage.

SWITCHGRASS

Courtesy of Vivien Allen, Texas Tech University

Scientific name: *Panicum virgatum* L.
Family: Poaceae
Tribe: Paniceae
Origin: North America

Adaptation and Use. This perennial warm-season grass is native to North America. It occurred naturally throughout the USA east of the Rocky Mountains, but was most common in the tall-grass prairie region. Although adapted throughout the USA, it is mainly grown in sub-humid and humid areas where introduced warm-season grasses are not adapted. Upland ecotypes are better adapted to moderately to well-drained soils, whereas lowland ecotypes are better adapted to wetter soils and can tolerate flooding. Switchgrass tolerates low fertility and acid and moderately alkaline soils. It has good drought and heat tolerance, and has good winter hardiness when managed properly. Although grown primarily as a summer forage for pasture and hay, it is also used extensively in conservation and wildlife plantings. It has high potential for use as a herbaceous biomass energy crop, and cultivars are being developed and released for this purpose.

Plant Characteristics. Switchgrass forms a sod by means of rhizomes, but young stands often appear bunch-like. It produces most growth in summer. Leaf blades are flat and have rough margins and a dense patch of hair located on the surface near the collar. The leaves are rolled in the whorl. The leaf sheath is round, smooth, and split with hairy margins. The collar is broad and often has marginal hairs. The ligule is a fringe of hairs, and auricles are absent. Stems are round and smooth and grow mostly erect. The inflorescence is an open panicle with spikelets near the end of branches. The fibrous root system is extensive and deep.

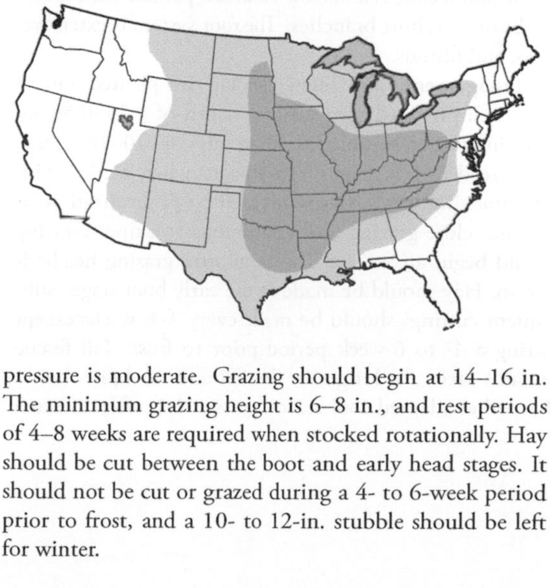

Management. Switchgrass is daylength sensitive, and care should be taken to plant cultivars that are adapted to the area where they are to be grown. Seedlings are slow to establish, and because the adventitious root system forms near the soil surface they require adequate surface soil moisture. Seed can be planted into a prepared seedbed or no-tilled into herbicide-treated sod or crop residues. Planting should be done in late spring or early summer at a rate of 3–6 lb seed/acre at a depth of 0.25–0.50 in. Weed control during establishment is critical because seedlings are not very competitive. Switchgrass tolerates continuous stocking during periods of active growth if grazing pressure is moderate. Grazing should begin at 14–16 in. The minimum grazing height is 6–8 in., and rest periods of 4–8 weeks are required when stocked rotationally. Hay should be cut between the boot and early head stages. It should not be cut or grazed during a 4- to 6-week period prior to frost, and a 10- to 12-in. stubble should be left for winter.

TALL FESCUE

Scientific name: *Festuca arundinacea* Schreb.
Family: Poaceae
Tribe: Poeae
Origin: Europe, Africa

Courtesy of Michael Collins, University of Missouri

Adaptation and Use. This cool-season perennial grass is adapted to humid, temperate areas. It is the predominant forage species throughout the humid transition zone that lies east of the Great Plains between 32° and 40° north latitude. Adapted to soils with fine to medium texture, it tolerates low-fertility, acid, alkaline, and poorly drained soils. It has good drought and heat tolerance when infected with a fungal endophyte. It also tolerates flooding and is winter hardy. Tall fescue is used mostly for pasture, hay, and conservation. It is compatible with alfalfa, red clover, and other legumes.

Plant Characteristics. Tall fescue has a bunch growth habit, but forms a weak sod by means of rhizomes when mowed or grazed frequently. It produces most growth in spring and fall. Leaf blades are flat and have rough margins and a prominent midrib. Blades are dull and ribbed on the upper surface and glossy below. Leaves are rolled in the whorl. The leaf sheath is round, smooth, and split with overlapping margins. The collar is broad and divided, often with marginal hairs. Auricles are blunt and hairy, and the membrane-like ligule is truncate. Flowering stems are round and smooth and grow mostly erect. The inflorescence is a narrow to broad panicle with spikelets borne on short branches. The root system is extensive, deep, and fibrous.

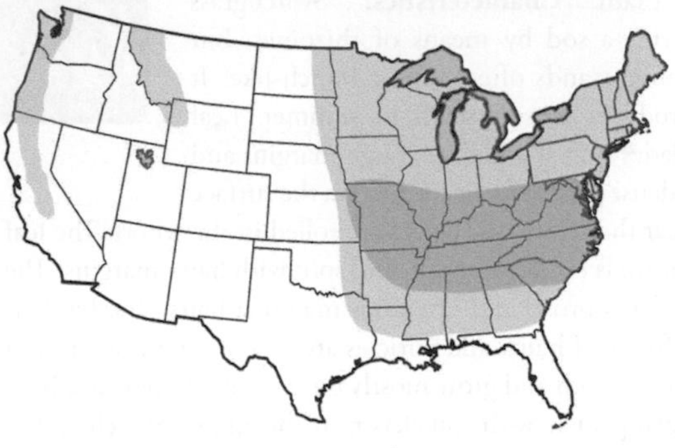

Management. Tall fescue should be planted into a firm, well-prepared seedbed at a depth of 0.25–0.50 in. in spring or late summer. Seeding rates are 10–18 lb/acre when seeded alone, and 5–9 lb/acre when seeded with a legume. Tall fescue responds well to N fertilization. It tolerates close grazing and continuous stocking. Grazing should begin at 6–8 in. The minimum grazing height is 2–4 in. Hay should be made at the early boot stage. Subsequent cuttings should be made every 4–6 weeks except during a 4- to 6-week period prior to frost. Tall fescue remains green for longer in the fall and stockpiles better than other adapted cool-season grasses. It is often infected with an endophytic fungus that improves plant persistence but produces alkaloids that are toxic to livestock. The problem is most severe when livestock are under heat and humidity stress. Planting in mixtures with legumes may offset some toxicity. Endophyte-free cultivars are available, but they are less adapted to extreme environments than are infected cultivars.

TALL WHEATGRASS

Courtesy of Kenneth Moore, Iowa State University

Scientific name: *Thinopyrum ponticum* (Podp.) Barkworth & D.R. Dewey
Family: Poaceae
Tribe: Triticeae
Origin: Eurasia

Adaptation and Use. This cool-season perennial grass is well adapted to semi-arid and sub-humid areas of the Northern Great Plains and Intermountain Region. It is adapted to a wide range of soils, but is most often used on alkaline and saline soils where other wheatgrasses will not grow. It is very winter hardy and tolerates drought, salinity, alkalinity, and poor drainage. Tall wheatgrass is used primarily for pasture, hay, and conservation.

Plant Characteristics. Tall wheatgrass has a bunch growth habit and produces best in spring and early summer. It matures later and stays green longer than other similarly adapted wheatgrasses. Leaf blades are rolled inward and have prominent veins and a blue-green color. They are smooth below and vary from rough to smooth above and along the margins. The leaf sheath is split with overlapping margins. The auricles are pointed and clasping, and the ligule is membrane-like. The stems are erect and smooth. The inflorescence is a terminal spike with one spikelet per node. The root system is deep and fibrous.

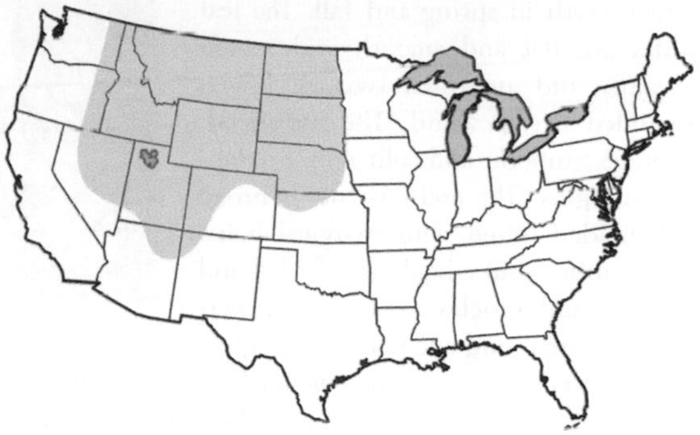

Management. Tall wheatgrass should be planted into a firm, well-prepared seedbed in spring. Seeding rates are 8–12 lb per acre and the seeding depth is 0.5–1.0 in. This grass responds well to N fertilization. It does not tolerate close, heavy grazing. Begin grazing at a height of 16–18 in., and remove livestock at 8 in. Hay should be cut at the early head stage. Tall wheatgrass should not be cut or grazed 4–6 weeks prior to frost, and an 8-in. stubble should be left for winter.

TIMOTHY

Scientific name: *Phleum pratense* L.
Family: Poaceae
Tribe: Aveneae
Origin: Europe

Courtesy of Michael Collins, University of Missouri

Adaptation and Use. This cool-season perennial grass is well adapted to humid, temperate areas with severe winters. Timothy is adapted to soils with fine to medium texture that are well to somewhat poorly drained, and it tolerates acid to moderately alkaline conditions. It requires at least moderate fertility, and tolerates flooding. It has low heat and drought tolerance but is very winter hardy. Although used mostly for hay and silage, it commonly occurs as a minor component of permanent pastures. Timothy hay is often favored for horses. It is very compatible with red clover.

Plant Characteristics. Timothy has a bunch growth habit and produces most growth in spring and fall. The leaf blades are flat and smooth with rough margins, and are often twisted. Leaves are rolled in the whorl. The leaf sheath is round, smooth, and split with overlapping margins. The collar is mostly broad and divided, often with marginal hairs. The membrane-like ligule is rounded and notched, and auricles are absent. Flowering stems are round and smooth and grow mostly erect. The lower internodes become swollen to form a storage structure called a haplocorm. The dense, spike-like panicle has spikelets borne on short branches. The root system is shallow and fibrous.

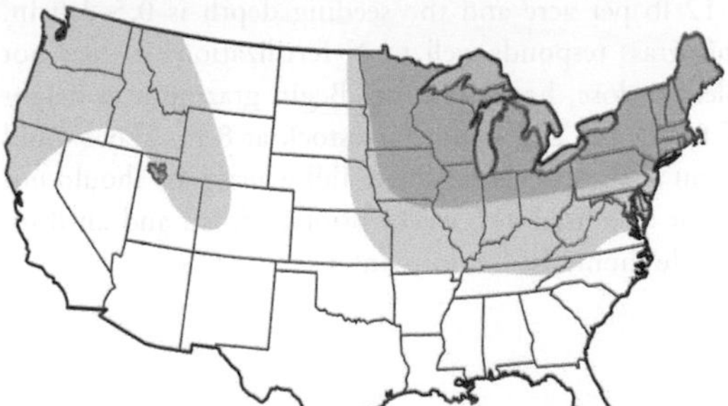

Management. Timothy should be planted into a firm, well-prepared seedbed in spring or late summer at 8–10 lb seed/acre when seeded alone or 4–6 lb/acre when seeded with a legume. The seeding depth should be 0.25–0.50 in. Timothy responds well to N fertilization. It does not tolerate close grazing or continuous stocking so it should be rotationally stocked. Grazing should begin at 6–8 in. and livestock should be removed at 3–4 in. Hay should be made between the early and mid-head stages. It is important to avoid grazing or cutting during stem elongation, as this results in slow regrowth and may weaken the stand. Regrowth should be cut before jointing when new tillers appear from the crown, except during the 4- to 6-week period prior to frost. A 6-in. stubble should be left for winter.

WESTERN WHEATGRASS

Alias: bluejoint grass
Scientific name: *Pascopyrum smithii* (Rydb.) Barkworth & D. R. Dewey
Family: Poaceae
Tribe: Triticeae
Origin: North America

Adaptation and Use. This cool-season perennial grass is native to the North American short-grass prairie. Although adapted to a wide range of soils, it occurs more frequently on heavy, lowland soils. It tolerates alkaline and saline soils that are subject to long periods of flooding. It is also very drought tolerant. Western wheatgrass is an important range grass, and is used primarily for grazing and soil conservation in the semiarid region of the Great Plains.

Plant Characteristics. This sod-forming grass spreads by means of rhizomes and forms a loose but uniform sod. Most of its growth occurs in the spring and fall. The foliage has a waxy coating that imparts a distinctive blue-green color. Leaf blades are flat with prominent veins, and the upper surface and margins are rough. The leaf sheath is round and open, and may be smooth or rough. Purple, claw-like auricles clasp the stem. The ligule is membrane-like and has irregularly shaped edges. The flowering stems grow upright and are smooth. The spike is narrow, with 1–2 spikelets per node containing 5–12 florets. The glumes and lemmas sometimes have awns. The fibrous root system is extensive and vigorous.

Management. Western wheatgrass should be planted in early spring or late fall. Late summer seedings are not recommended unless there is irrigation. The normal seeding rate is 6–12 lb/acre, but the rate should be increased to 18–23 lb/acre under poor soil conditions. The seeding depth is 0.25–0.50 in. Western wheatgrass can also be established from sod pieces. Grazing should begin at 6 in. or later, and should not be lower than 3–4 in. Rotational stocking is recommended. Western wheatgrass stockpiles well for fall and early winter grazing.

Courtesy of Michael Collins, University of Missouri

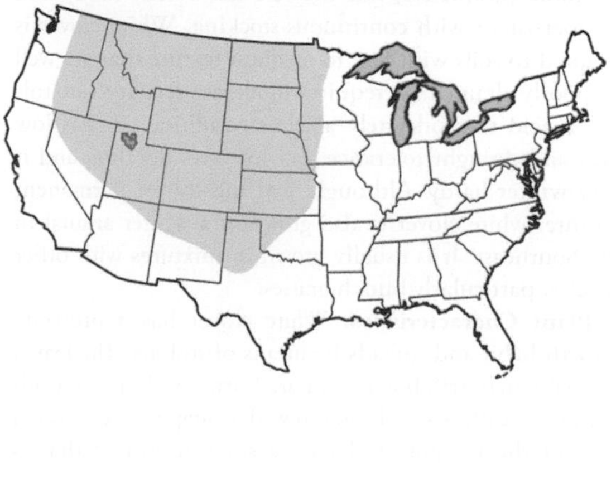

Courtesy of Michael Collins, University of Missouri

WHITE CLOVER

Aliases: ladino clover, Dutch clover
Scientific name: *Trifolium repens* L.
Family: Fabaceae
Origin: Eurasia

Adaptation and Use. This perennial legume is well adapted to humid, temperate climates. White clover is grown throughout the humid region of the USA and under irrigation in the West. It is naturalized throughout its area of adaptation. Small (wild) types have a very short, prostrate growth habit and are not very productive. Medium types, which include Dutch clover, are somewhat larger and reseed naturally. Large types, including ladino clover, have larger leaves and longer petioles and are more productive, but they are less winter hardy and less persistent with continuous stocking. White clover is adapted to soils with fine to medium texture that are well to poorly drained. It requires moderate fertility but tolerates acid to moderately alkaline conditions. It has low heat and drought tolerance, but tolerates flooding and is very winter hardy. Although used mostly for permanent pasture, white clover is also grown as a winter annual in the Southeast. It is usually grown in mixtures with other species, particularly bunch grasses.

Plant Characteristics. White clover has a prostrate growth habit and spreads by means of stolons. The leaves are palmately trifoliolate and are borne on long, smooth petioles. Leaflets are oblong to wedge-shaped, are serrated around the margin, and have a smooth surface that is

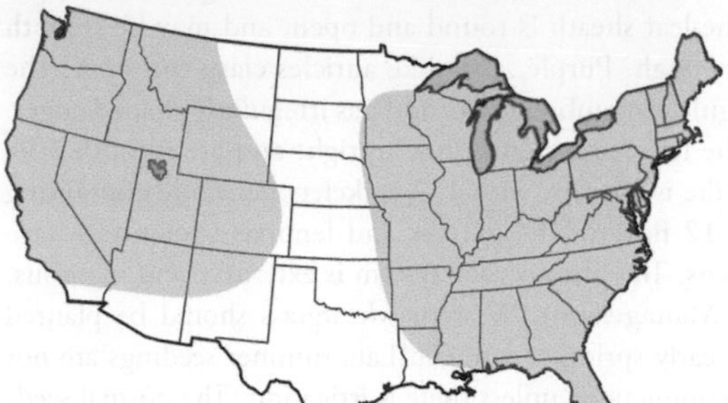

usually variegated with a white or red "watermark." The stems are smooth and grow horizontally near the soil surface. The round head has white to pink flowers on long peduncles. The short taproot decays after the first year and is replaced with an adventitious root system arising from the nodes of stolons.

Management. White clover is well adapted morphologically for grazing, but has high bloat potential, so it is nearly always planted with or interseeded in grasses. Inoculated seed should be planted or interseeded 0.25–0.50 in. deep in early spring or late summer at 1–2 lb/acre in a mixture. Grazing management is generally determined by the companion grass. The minimum grazing height is 2 in. White clover spreads vigorously by means of stolons, and reseeds naturally if allowed to mature. Bloat precautions should be taken when grazing pastures with over 25–35% white clover.

References

Barnes, RF, CJ Nelson, M Collins, and K J Moore. 2003. Forages: An Introduction to Grassland Agriculture, 6th ed., Volume I. Iowa State Press, Ames, IA.

Barnes, RF, CJ Nelson, KJ Moore, and M Collins. 2007. Forages: The Science of Grassland Agriculture, 6th ed., Volume II. Blackwell Publishing, Ames, IA.

Barnhart, SK, DG Morrical, JR Russell, KJ Moore, PA Miller, and CE Brummer. 1998. Pasture Management Guide for Livestock Producers. Iowa State University Extension, Ames, IA.

Bishop-Hurley, GJ, SA Hamilton, and R Kallenbach. 2000. Missouri Dairy Grazing Manual. University of Missouri Extension, Columbia, MO.

Gould, FW, and RB Shaw. 1983. Grass Systematics, 2nd ed. Texas A&M University Press, College Station, TX.

Leithead, HL, LL Yarlett, and TN Shiflet. 1971. 100 Native Forage Grasses in 11 Southern States. Agriculture Handbook No. 389. Soil Conservation Service, US Department of Agriculture, Washington, DC.

Moser, LE, DR Buxton, and MD Casler. 1996. Cool-Season Forage Grasses. Agronomy Monograph No. 34. American Society of Agronomy, Madison, WI.

Moser, LE, BL Burson, and LE Sollenberger. 2004. Warm-season (C_4) Forage Grasses. Agronomy Monograph No. 45. American Society of Agronomy, Madison, WI.

Phillips, CE. 1962. Some Grasses of the Northeast: A Key to Their Identification by Vegetative Characteristics. Field Manual No. 2. University of Delaware Agricultural Experiment Station, Newark, DE.

Sheaffer, CC, NJ Ehlke, KA Albrecht, and PR Peterson. 2003. Forage Legumes: Clovers, Birdsfoot Trefoil, Cicer Milkvetch, Crownvetch, Sainfoin and Alfalfa, 2nd ed. Minn. Agric. Exp. Sta. Bull., 608-2003. Minnesota Agricultural Experiment Station, St. Paul, MN.

Sotomayor-Ríos, A, and WD Pitman. 2001. Tropical Forage Plants: Development and Use. CRC Press, Boca Raton, FL.

Stubbendieck, J, SL Hatch, and CH Butterfield. 1997. North American Range Plants, 5th ed. University of Nebraska Press, Lincoln, NE.

Taylor, NL (ed.) 1985. Clover Science and Technology. Agronomy Monograph No. 25. American Society of Agronomy, Madison, WI.

US Department of Agriculture. 1948. Grass: The Yearbook of Agriculture 1948. US Government Printing Office, Washington, DC.

US Department of Agriculture, Natural Resource Conservation Service, and Grazing Lands Technology Institute. 2003. National Range and Pasture Handbook. US Department of Agriculture, Washington, DC.

Websites

Arizona Range Grasses (http://cals.arizona.edu/pubs/natresources/az1272/). Cooperative Extension, College of Agriculture, University of Arizona, Tucson, AZ.

Country Pasture Forage Profiles (www.fao.org/ag/agp/AGPC/ doc/Counprof/regions/index_all.htm). United Nations Food and Agriculture Organization, Rome, Italy.

Germplasm Resources Information Network (GRIN) Taxonomy for Plants. (www.ars-grin.gov/cgi-bin/npgs/html/ index.pl). National Plant Germplasm System, US Department of Agriculture Agricultural Research Service, Beltsville, MD.

Grass Genera of the World: descriptions, illustrations, identification, and information retrieval; including synonyms, morphology, anatomy, physiology, phytochemistry, cytology, classification, pathogens, world and local distribution, and references. Version: 5th (http://delta-intkey.com/grass/).

GrassBase – The Online World Grass Flora Descriptions (www.kew.org/data/grasses-db.html). Royal Botanic Gardens, Kew, UK.

NewCROP (New Crop Resource Online Program) (www.hort. purdue.edu/newcrop/). Center for New Crops and Plant Products, Purdue University, West Lafayette, IN.

PLANTS Database (http://plants.usda.gov/java/). National Plant Data Team, Greensboro, NC.

Plants of Texas Rangelands – Grasses (http://essmextension.tamu. edu/plants/?collection=grasses). Ecosystem Science and Management, Texas A&M University, College Station, TX.

The Identification of Certain Native and Naturalized Hay and Pasture Grasses by Their Vegetative Characters. MacDonald College, McGill University, Technical Bulletin No. 16 Revised 1938) (http://anr.ext.wvu.edu/r/download/195827). West Virginia University Extension Service, Morgantown, WV.

Weeds of the North Central States. (www.aces.uiuc.edu/vista/html_pubs/WEEDS/list.html). North Central Regional Research Publication No. 281. University of Illinois Agricultural Experiment Station Bulletin 772. University of Illinois, Urbana-Champaign, IL.

PART II

FORAGE MANAGEMENT

CHAPTER 11

Forage Establishment

Marvin H. Hall and Michael Collins

Producing the highest forage yields possible under any given set of soil and environmental conditions requires a dense stand of adapted forage species. Planning ahead with regard to **soil pH** and fertility, obtaining quality seed of the "best" species and cultivar, and undertaking weed control, as well as understanding the factors that affect forage stand establishment (e.g., seedbed preparation, time and depth of planting, seeding rate, and harvest management), are vital for optimizing stand establishment.

Species and Cultivar Selection

Forage crops, whether for livestock feeding, soil conservation, or wildlife habitat, are grown on soils that vary widely with regard to drainage characteristics, slope, fertility, and other conditions. The intended use of the forage after establishment must be considered along with these and other soil factors when selecting forage species and **cultivars**. For example, reed canarygrass tolerates poorly drained soils better than most other grasses (see Chapter 6), and birdsfoot trefoil and red clover tolerate wet conditions and low pH better than alfalfa (see Chapter 8). Conversely, because of its deep root system, alfalfa is a good choice for sandy soils with low water-holding capacity, compared with shallow-rooted species such as timothy or birdsfoot trefoil (Hall et al., 2004).

Lime and Fertility Requirements

Once the forage species or mixture has been selected, attention must be given to the soil environment. Satisfactory stands and yields of forages, like those of other crops, are obtained only when the soil is adequately limed and fertilized based on soil test information.

Low soil pH is a common limitation to stand establishment in the eastern USA. Increasing the pH to a value near 7.0 increases the availability of most plant nutrients and promotes the growth of desirable microorganisms. The main purpose of liming is to raise soil pH. However, lime also supplies calcium (Ca) and, in the case of dolomitic lime, magnesium (Mg) needed for plant growth. Ideally, lime should be applied 6–12 months prior to seeding into a tilled seedbed, and 1–2 years ahead of no-till seeding to allow time to neutralize the acidity in the rooting zone.

Phosphorus (P) is essential for the development of healthy seedlings, so it is important to have soil P levels in the optimum range prior to seeding.

The application of approximately 40 lb/acre (45 kg/ha) of fertilizer or manure N just prior to seeding forage grasses enhances early seedling growth and development under most conditions. Similar rates of N may improve establishment of **legumes** when soils are low in N or the organic matter content is less than 1.5%. Much higher rates may be needed for **annual** grasses, especially warm-season species. For example, silage corn frequently receives fertilizer N rates of 150 lb/acre (168 kg/ha) or more.

When soil nitrate levels are adequate and conditions are favorable for early nodulation, the addition of N is not recommended in legume establishment. High rates of N are generally detrimental to rhizobial infection and **nitrogen fixation**. Unlike P and N, K is seldom a limiting factor in seedling establishment. However, soil testing is recommended because forage crops typically remove large amounts of K from the soil.

Time of Planting

Forage seedings are usually timed to coincide with periods of adequate rainfall and favorable temperatures to help to ensure successful stand establishment.

Forages: An Introduction to Grassland Agriculture, Seventh Edition. Edited by Michael Collins, C. Jerry Nelson, Kenneth J. Moore and Robert F Barnes.

Spring Planting

Spring planting is typically done as early as possible, beginning in February in the southern USA and in May in northern states. Delaying spring planting to avoid frost damage reduces the likelihood that there will be adequate rainfall for rapid seedling development. The risk of frost damage is typically offset by the advantages of early seeding dates, but the possibility of frost damage must also be considered when choosing the best planting date.

The advantages of spring planting include:

- a greater likelihood of adequate rainfall
- near optimum temperatures for **germination** and early seedling growth
- the ability to harvest the crop in the seeding year.

The disadvantages of spring planting include:

- potential for heavy weed pressure compared with late-summer seedings
- the possibility that prolonged rainy periods will delay planting beyond the optimum date
- the possibility that cold soil temperatures may delay seedling emergence in northern areas
- the risk of a dry period soon after germination killing some seedlings
- the increased risk that late frosts will damage young seedlings.

Frost is less of a concern with grasses than with legumes due to differences in emergence patterns. Grasses exhibit **hypogeal germination**. If exposed leaves of grass seedlings are damaged by frost, they can regrow from **meristematic** tissues located beneath the soil surface. Generally, the danger of frost has passed by the time the growing points of grass seedlings emerge.

Legumes, on the other hand, usually exhibit **epigeal germination**. Late frosts can kill these seedlings because their growing points emerge early in the process. The degree of injury depends on both the minimum temperature and the duration of the cold temperature. Alfalfa seedlings tolerate air temperatures as low as 28°F for 4 hours before being killed. Red clover seedlings tolerate frost better than alfalfa seedlings. Companion or **nurse crops** slow down heat loss at night and help to avoid frost injury to legume seedlings.

Late-Summer Seedings

Late-summer seedings take place from early August in the northern USA to early November in southern areas. Weed problems are usually less with late-summer seedings than with seedings done in spring. In addition, summer annual weeds that do germinate are typically killed by frost. Late-summer seedings also provide the opportunity to harvest an annual grain crop before establishing the forage crop.

The advantages of late-summer seedings include:

- soil moisture levels generally being more suitable for tillage
- weed problems usually being less
- farm workload usually being less hectic than in spring.

The disadvantages of late-summer seedings include:

- the increased risk that moisture stress will limit establishment
- the typical loss of a seeding-year forage harvest due to the short growing period before winter
- the risk that winter injury may occur if plants do not become well established before a killing frost
- the possibility that winter annual weeds can become a serious problem if forage seedlings emerge late and provide little or no competition.

The risk of dry weather is minimized by delaying summer seedings until the likelihood of rainfall is higher and evaporation is reduced. However, seeding should be done at least 6 weeks before the average date of the first killing frost, in order to minimize the risk of winter injury. This normally gives the seedlings sufficient time to develop an adequate shoot and root system, to form a **crown**, and to store carbohydrate reserves for overwintering.

Nurse crops are not recommended for late-summer seedings because they compete for moisture, and weed pressure is low. In northern areas, fall forage harvest after a late-summer seeding is not recommended because this increases the risk of winter injury.

The incidence of **damping-off** of seedlings is usually less for late-summer than for spring plantings, but late-summer legume seedings are more susceptible to *Sclerotinia* crown and stem rot (*Sclerotinia trifoliorum* Eriks). When conditions are favorable, new seedings may be completely destroyed by this disease. As young seedlings develop and cell walls lignify, they become resistant to this disease. Therefore late-summer seedings should be made at the earliest possible date so that seedlings are well established by the time infection occurs. *Sclerotinia* has little effect on spring seedings.

Winter or Frost Seeding

Frost seeding takes place in northern regions of the USA from late February through March. Frost seeding is most commonly used to thicken a stand or introduce new species into an existing stand. It is done by distributing seed directly onto the frozen soil surface. Expansion and contraction associated with the night–day freeze–thaw cycles incorporate the seed into the surface layer. Frost seeding is inexpensive and can be completed with hand seeders, cyclone seeders mounted on tractors or four-wheelers, or grain drills.

Table 11.1. Success of cool-season perennial grasses and legumes planted by frost seeding

	Seeding rate (lb/acre)			
Species	4	6	8	10
Grasses		Plants/ft^2		
Perennial ryegrass	3.2			10.8
Orchardgrass	5.0			6.9
Smooth bromegrass	0.8			1.3
Reed canarygrass	0.4			0.4
Timothy	3.0			–
Legumes				
Alfalfa		–	3.1	
Birdsfoot trefoil		4.7	–	
Red clover		11.6	–	

Source: grass data from Undersander et al., 2001; legume data from Cosgrove, 2001.

Species vary widely in their ability to establish by frost seeding. Among common legume species, red clover establishes particularly well using this method, but alfalfa establishes poorly (Table 11.1). Grasses also show mixed results when frost seeded. The most successful are orchardgrass, perennial ryegrass, and timothy, and the least successful are reed canarygrass and smooth bromegrass (Table 11.1).

Seedbed Preparation

Seeds of most forage species are small compared with seeds of grain crops. This makes thorough seedbed preparation critically important for forage plantings. The goal of seedbed preparation is to ensure that the seedbed allows the seeding device to place the seeds at the proper depth with good seed-to-soil contact. Equipment availability, time availability, field slopes, soil type, and likely weed pressure should be considered when selecting a seedbed preparation method.

Tillage

Tillage practices vary, but typically consist of primary tillage with a moldboard or chisel plow, followed by secondary tillage using a disc or field cultivator. Tillage provides a smooth surface, kills existing weeds, buries weed seeds below the soil surface, and allows incorporation of lime and fertilizers. Smoothing of fields previously in row crops aids hay and silage harvesting following establishment (Fig. 11.1).

Excessive tillage creates overly loose seedbeds that lose moisture more rapidly. Such seedbeds also increase the potential for surface crusting following rainfall. Small clods or soil aggregates can help to prevent soil crusting. Tillage also increases the potential for wind and water erosion until the crop becomes established. Insufficiently tilled, cloddy conditions or seedbeds with excessive plant residues make proper seed placement difficult and reduce seed-to-soil contact.

Fig. 11.1. Tillage to level and smooth a field prior to forage seeding.

The ideal tilled **seedbed** should be smooth, firm, and free of clods (Fig. 11.2). A firm seedbed is essential for proper seed placement because this allows seeding equipment to run at a more uniform depth. A general rule of thumb is that if the heel of your shoe sinks into a seedbed by more than 0.5 in. (1.3 cm), the seedbed is not firm enough (Fig. 11.3). Placing seed in a loose or cloddy seedbed will often reduce emergence because of incorrect seeding depth and poor seed-to-soil contact.

If erosion is not a concern, primary tillage can be done in the fall. Freeze–thaw cycles in northern areas and winter rains in other areas will help to break up clods to

Fig. 11.2. Ideal seed bed for forage planting, with no soil clods and excellent potential for good seed-to-soil contact.

Fig. 11.3. Imprint of shoe heel indicating that the seedbed is firm enough to achieve an accurate seeding depth.

produce a finer seedbed, resulting in fewer field operations in spring. However, on sloping soils, primary tillage should be done just prior to planting, in order to minimize soil erosion. Secondary tillage is usually needed to break up the remaining clods and smooth the seedbed. Excessive secondary tillage leads to overly fine seedbeds that increase the possibility that hard crusts will form when rain falls soon after planting, making seedling emergence more difficult.

Grain drills, cultipacker seeders, or no-till drills can be used for successful seed placement in tilled seedbeds (Table 11.2).

Reduced Tillage

Reduced tillage, as the name implies, creates less soil disturbance than tillage seedings. It involves sufficient tillage to provide adequate seed-to-soil contact, but leaves significant amounts of plant residue on the surface to impede run-off and control erosion. This method is recommended on sloping soils where erosion is a concern.

Chisel plowing followed by a soil-finishing step is a common reduced tillage sequence. Disking or field cultivation might be adequate in low-residue situations. Plant residues remaining after tillage should cover approximately 35% of the soil surface.

Table 11.2. Effects of tillage methods and forage seeding implements on alfalfa seedling density (plants/ft^2)[a]

Tillage method	Residue on soil	Grain drill	Cultipacker seeder	No-till drill
	%	Plants/ft^2		
Moldboard plow	11	28	31	38
Chisel plow	49	31	19	23
Disc harrow	43	28	24	32
No tillage	63	–	–	21

Source: Cosgrove et al., 1991.

[a]Seeded in spring. Stand density measurements were made at the end of the establishment year.

Grain drills, cultipacker seeders, or no-till drills can be used for successful seed placement in reduced tillage systems (Table 11.2).

No-Till

No-till seeders can place forage seeds at the correct depth with good seed-to-soil contact without any primary tillage. These specialized drills are designed to cut through plant residues or sod and place the seed into the soil. Only no-till drills should be used for planting in no-till systems (Table 11.3). It is important that the drills have sufficient weight to ensure penetration of the coulters that precede the seed openers to cut through residues or sod, which is particularly difficult under dry conditions. The best designs have springs or other mechanisms on individual seeding units (Fig. 11.4) so they allow seeding to continue even if some units are raised out of the ground by rocks or other impediments. Most no-till drills have depth/press wheels to help to control seeding depth. Placing seed too deep into the soil is the most common problem encountered with no-till drills. Soil type and moisture affect planting depth, so frequent monitoring of seed depth is recommended with no-till drills.

Table 11.3. Planting rates of coated and uncoated alfalfa cultivars using the same drill and drill setting

Cultivar	Without clay coating	With clay coating
	Planting rate (lb/acre)	
A	17.0	18.3
B	15.2	19.5
C	13.8	–
D	17.5	–
E	18.0	–

Source: Vondrachek, 1991.

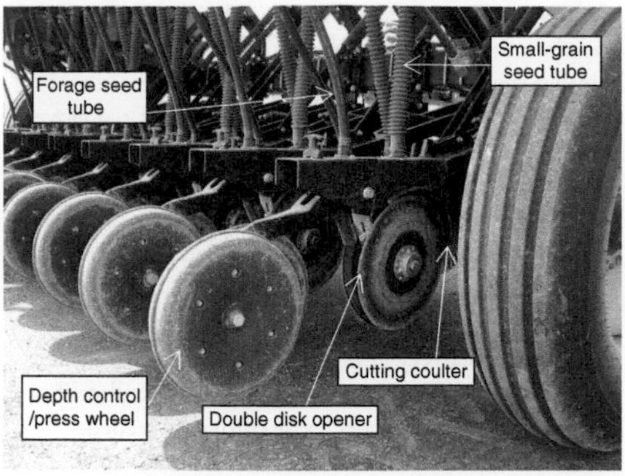

FIG. 11.4. Components of a no-till drill seeding unit. Forage seed tubes drop seed behind the double disk openers as the soil is falling back into the disk opening. Consequently, planting speed is an important factor in achieving uniform seeding depth. (Photo courtesy of Marvin Hall.)

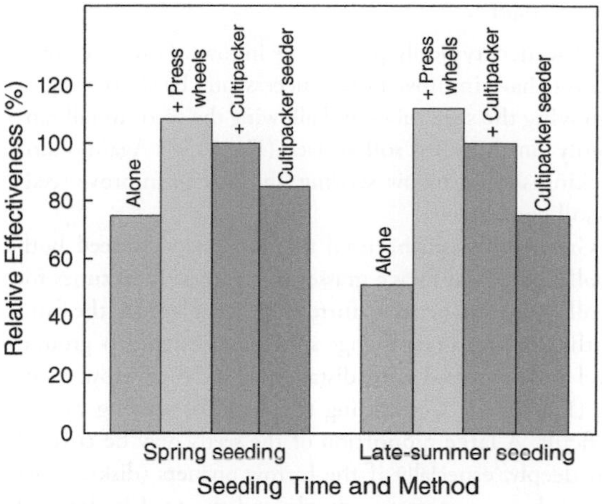

FIG. 11.5. Influence of seeder type on alfalfa stand establishment. (From Tesar and Marble, 1988. Reproduced with permission of the American Society of Agronomy.)

No-till methods usually depend on herbicides to limit growth of the existing vegetation during establishment. Fall applications of glyphosate are often used to control **perennial** weeds (e.g., quackgrass), and to limit growth of the existing sod prior to spring pasture renovation. Since incorporation is not possible, lime and P should be top-dressed 1 year or more before no-till seeding.

Seeding Implements

Forage seeding implements fall into two basic categories depending on how they arrange the seed in the soil (i.e., whether they broadcast it or sow it in rows). Accurate metering of seed, placement of seed at the correct depth, and good seed-to-soil contact (Fig. 11.5) are key characteristics of seeding implements.

Broadcast Seeder

These seeders place the seed uniformly on the soil surface but in no particular pattern.

Cultipacker Seeder

These seeders are most successful when some soil tillage precedes planting. Typical cultipacker seeders have two corrugated rollers with the seed-metering boxes mounted in between them. The first roller firms the soil and leaves shallow trenches into which the seeds drop. The corrugations on the second roller are offset from those on the first roller. This arrangement covers the seed at a consistent and shallow depth, and firms the soil. These seeders provide optimum seed placement and good seed-to-soil contact on properly prepared seedbeds.

Cultipacker seeders should not be used on moist, heavy soils because this can contribute to crusting. These seeders may not provide adequate soil coverage on seedbeds with heavy crop residues, or when seeding lightweight grass seeds such as smooth bromegrass and wheatgrasses.

Similar results can be obtained by using a single corrugated roller to firm the tilled soil before distributing seed using a spinner seeder, grain drill, sprayer, or aerial broadcast. Distributing forage seed in liquid or dry fertilizer sprayers or spreaders is a relatively new technique, often referred to as fluid or suspension seeding. This is an effective method for rapidly and uniformly broadcasting seed over large areas. This operation is usually done by custom applicators. A final pass with a corrugated roller, cultipacker, or harrow is needed to cover the seed and improve seed-to-soil contact.

Spinner or Cyclone Seeder

These seeders have a spinning plate below the seed metering box that slings the seed out onto the soil surface. Depending on the design, the seed can be distributed in a band up to about 30 ft (9.1 m) wide.

Spinner or cyclone seeders are most commonly associated with frost seeding. However, if they are used in tilled fields in the spring or late summer, a cultipacker should be used after seeding to improve seed-to-soil contact. Planting multiple species with this system generally requires separate passes because differences in seed weight, shape, and density often allow segregation in the seed box, leading to uneven seed distribution.

Drill Seeder

Drill seeders typically place seeds in rows. However, some growers have improved their success rates by shortening or removing the seed tubes and allowing the seeds to fall randomly on the tilled soil surface (Fig. 11.4). Again, cultipacking should follow seeding, in order to improve seed-to-soil contact.

Grain drills can be used to simultaneously seed both small grains and forage grasses or legumes. Seed tubes for small grains run between furrow openers placed at the front of the seeding unit. Forage seed tubes extend to ground level and drop seed some distance behind the furrow openers (Fig. 11.4). Controlling the depth of seeding can be difficult. A large proportion of the seeds may be covered too deeply, especially if the furrow openers (disks, shovels, or hoes) are set too deeply and the seed is dropped before or near the openers. A general rule of thumb is that about 10% of the seeds should be visible on the soil surface when drilling seed, to help to ensure that seed are not being placed too deep.

The addition of press wheels to drill seeders significantly improves seed-to-soil contact. Drills with press wheels generally result in better stands than cultipacker seeders, particularly on fields with crop residue (Fig. 11.5 and Table 11.2). Drills without press wheels should be followed with a cultipacker to provide optimal seed-to-soil contact.

Seeding Rates

Seeding-rate recommendations are usually given in pounds (lb) of **pure live seed** per acre. Seeding rates vary widely because the size and weight of individual seeds of forage species vary. Rate recommendations are also influenced by local environmental and soil conditions.

Specific seeding rates depend on many factors, including the following:

1. *Soil type.* Lower rates can be used on light, sandy soils because seedling emergence is more complete than on heavy soils.
2. *Seed quality.* Seeding rates must be increased for seed lots with lower germination or more inert material. To take these factors into account, seeding rates should be based on pure live seed (PLS). Pure live seed can be calculated by multiplying viable seed percentage (% germination on the seed tag) by seed purity, with both terms expressed in decimal form. Thus a seed lot with 95% germination and 97% purity would have 0.95 × 0.97 = 0.92, or 92% PLS. Planting rate can then be calculated by dividing the recommended seeding rate by the PLS value. To obtain the desired seeding rate of 12 lb/acre (13.6 kg/ha) for the seed lot described above, a planting rate of 12/0.92 = 13 lb/acre (14.6 kg/ha) would be needed.
3. *Seed size.* The weight of seed needed to obtain similar plant numbers varies widely between forage species. Similar stand densities should result from planting 4 lb (1.8 kg) of white clover or 12 lb (5.4 kg) of red clover, because of the difference in their seed sizes. Different cultivars of the same species may also differ in seed size and in flow characteristics. For example, the same drill setting delivers 13.8 lb (6.3 kg) of seed for one alfalfa cultivar, and 18 lb (8.2 kg) for another (Table 11.3). Clay seed coating increases weight but not necessarily the PLS delivery at a given drill setting by 8–28%. These differences emphasize the need for drill calibration each time a new species or cultivar is seeded.

Even when soil conditions, seedbed preparation, and seeding depth are optimum, only a small percentage of the seeds that are planted develop into established plants. Most recommended seeding rates are three to five times greater than would be necessary if every seed developed into a healthy plant. Seeding more than the recommended rates does not overcome problems with seedbed preparation, competition, fertility, or other factors.

Seeding Depth

Forage seed are usually small, with limited food reserves, and must be planted at shallow depths compared with typical annual grain crops. Most forage seedlings are not able to reach the surface if they are planted more than 1 in. deep. However, some seed coverage is needed to provide moisture for germination and seed-to-soil contact so that the emerging **radicle** can penetrate the soil. A good rule of thumb is to plant seed no deeper than seven times the seed diameter.

Optimum planting depth is also influenced by soil type (Table 11.4). Seedlings can emerge from greater depths in coarser-textured, sandy soils than in fine-textured soils. Optimum seeding depths are 0.25–0.5 in. on clay and loam soils, and 0.5–1 in. on sandy soils. Greater planting depths may be needed on dry soils or in late spring to help to ensure that seed have adequate moisture.

Table 11.4. Established alfalfa plant numbers on three types of soil at four depths of seed placement

	Depth (in.)			
Soil	0.5	1	1.5	2
	Number of plants produced from 100 seed			
Sand	71	73	55	40
Loam	59	55	32	16
Clay	52	48	28	13

Source: Sund et al., 1966.

Seeder Calibration

Planter calibration—that is, setting the seed output per unit area or distance traveled—is essential for obtaining accurate seeding rates. Calibration can often be accomplished by turning the drive wheel to simulate travel over a predetermined distance.

Typical steps in planter calibration include the following:

- Step 1. Measure the circumference of the drive wheel (the wheel that turns to operate the seeding units). On culti-packer seeders, the drive wheel is the roller.
- Step 2. Lift the machine so that the drive wheel is off the ground.
- Step 3. Attach calibration pans or bags to collect the seed as it falls out of the metering units or seed tubes.
- Step 4. Turn the drive wheel a predetermined number of times.
- Step 5. Weigh the seed collected.
- Step 6. Calculate the amount of seed delivered per acre.

Typical calculations include the following steps:

- Step 1. Determine the size of the "test" area.

  Test area = length of test run × planter seeding width.

  The length is determined by multiplying the drive-wheel circumference by the number of turns used.
- Step 2. Calculate the acreage by dividing the test area in square feet (ft^2) by 43,560.
- Step 3. Divide the seed weight (in pounds) by the area (in acres). The resulting value is the delivery rate in pounds of seed per acre.

Example calculations

Assume that a drill has 10 seeding units, 7 in. (17.9 cm) apart from each other, and that the drive-wheel circumference is 8.5 ft. (2.6 m). The drive wheel is turned 10 times and 2.0 oz (57 g) of seed are collected.

Test area length = 10 turns × 8.5 ft/turn = 85 ft (26 m)

Test area width = 10 seeding units × 7 in. = 70 in. (5.8 ft, or 1.8 m)

Test area = 85 ft × 5.8 ft = 495 ft^2

Test area in acres = 495/43,560 = 0.0114 acres

Weight of collected seed = 2.0 oz

2 oz/16 oz per lb = 0.125 lb

Seeding rate = 0.125/0.0114 = 11.0 lb/acre

Species Mixtures

Mixed stands that have at least one grass and one legume are generally used to provide benefits such as **bloat** control, greater weed suppression, and improved seasonal distribution of production (Sanderson et al., 2012). Thus establishment may involve planting two or more species simultaneously or in succession. The seed can sometimes be mixed together for seeding if they flow similarly and do not segregate in the seed box. If segregation is minor, seed can periodically be remixed in the seed box. However, in many cases individual components must be seeded from separate seed boxes or in separate operations. Calibration is also a challenge when seeding mixtures (see "Planter Calibration"). When seeding more than one species in the same seed box, a known proportion of each must be added, and the planter must then be calibrated to deliver the desired amount of the seed mixture.

Attention must also be given to the planting depth of mixtures. When different species have different optimum depth requirements, they should be planted from different boxes or seeded separately. For instance, seed of an oat nurse crop need deeper placement than the forage legume seed with which they are often planted.

Seed Inoculation

A unique association between many legume species and certain bacteria results in the conversion (commonly referred to as fixation) of atmospheric nitrogen (N_2) into forms that plants can use to meet their N needs. Nitrogen fixation results from a **symbiotic** (mutually beneficial) relationship with soil bacteria of the **genera** *Rhizobium* and *Bradyrhizobium*. The symbiotic relationship between the bacterial species and the legume species is highly specific. Therefore selection of the particular *Rhizobium* or *Bradyrhizobium* species that will be effective on the legume species that is being established is critically important.

Inoculation is the term used to describe the introduction of the appropriate bacteria to the seed prior to planting. Once the N-fixing bacteria have been introduced, they persist in the soil for a few years. However, to avoid the risk of poor nodulation, inoculation of legume seed before planting should be a standard practice.

Legume seed can be pre-inoculated during seed processing or inoculated at the time of planting. Much of the legume seed currently available in the USA is pre-inoculated by the seed processor. This method is effective, but re-inoculation is recommended if the inoculation date has expired or if seed have not been stored under the cool, dry conditions needed to maintain inoculum viability.

Inoculation at planting time is done by adding a peat-based inoculum as a sticky powder to seeds, and then mixing thoroughly. These bacteria are very sensitive to light and heat, so the inoculum must be refrigerated or stored under cool, dry conditions prior to use. Seed inoculated in this way must be planted within a few hours to ensure that the bacteria remain viable.

Other Seed Treatments

Seed of some forage species may require treatment to overcome dormancy, to protect seedlings from diseases, or to improve germination.

Scarification and Stratification

Dormant seeds have their germination inhibited by internal rather than external factors. The production of hard seed is a common form of dormancy in forage legumes. Hard seed have a waxy layer in the seed coat that is impermeable to water. As a result, they fail to germinate because they are unable to imbibe water. Radicle emergence may also be inhibited in hard seed. Dormancy caused by hard seed may be overcome by **scarification**, which is the physical or chemical treatment of the seed to create physical openings in the impermeable layer. This allows moisture uptake and/or radicle emergence, but it also shortens the viable life of the seed.

Seed of some species require exposure to cold temperatures before they will germinate. In nature, this may aid the survival of a species by ensuring that seed overwinter before they germinate. Exposing seeds to moisture and cold temperatures in preparation for planting is known as **stratification.** Eastern gamagrass is one species that requires stratification before planting. In this case, treatment involves soaking the seed in water and maintaining it at cool temperatures for several weeks prior to planting. Eastern gamagrass seed reverts to the dormant state if it dries out prior to planting. Other species that benefit from stratification include switchgrass, big bluestem, and indiangrass.

Seed Protectants

Most legume seeds of named cultivars are treated with a fungicide during seed processing, to help to protect the young seedling from common seedling diseases caused by species of *Pythium* and *Phytophthora.*

Seed Coating

Some forage seed may be coated with lime or clay. These materials are added in order to alter the pH of the seed microenvironment after planting (in the case of lime coatings) or to attract moisture to the seed (in the case of clay coatings). The results of experiments conducted to assess the usefulness of these treatments have been inconsistent (Cosgrove et al., 1991). These coatings can comprise up to 35% of the weight of the final product, and they affect seed flow, so seeder calibration is necessary whenever the cultivar or seed coating changes (Table 11.3).

Weed Control during Establishment

Weed control is critical to the development of long-lived and productive forage stands, especially during the first 6 weeks after seeding. Small-seeded species are generally not strong competitors with weeds, and cultivation is seldom an option. Cultural practices are important in weed control during forage establishment. Correct seeding rates, adequate fertility, and optimum pH help to ensure vigorous, competitive forage plants and help to control weeds. Planting should be timed to give forages the greatest possible advantage over weeds. The use of a nurse crop is a cultural practice for weed control. Weed populations can be minimized prior to forage seeding by using a crop rotation that includes 1 or 2 years of corn or soybean that provide other herbicide options to reduce the populations of many annual and perennial weeds.

The seed of many weeds are similar in size, shape, and weight to forage seed, and their removal during seed processing is difficult. For this reason, the use of high-quality seed is important to ensure that the seed has been produced and labeled according to standard procedures, and that it contains no noxious weed seed and only limited amounts of other weed seed.

Seeding with a Nurse Crop

Nurse crops are crops such as oat, barley, triticale or, in some locations, winter wheat that are sometimes planted along with forage crops to provide quick ground cover that can reduce soil erosion and weed invasion while the forage seedlings become established. The best nurse crops are annual species that germinate and emerge quickly. These same characteristics typically increase forage yields during the seeding year compared with seeding without a nurse crop. Small-grain nurse crops are often allowed to mature fully for grain harvest. The straw produced in this system is useful as animal bedding or as a cash crop.

Erosion control and seeding-year forage yield benefits of nurse crops must be balanced against the understanding that they also compete with forage seedlings for light, moisture, and nutrients. Thus, while they often increase total dry matter yields, seeding-year yields of the seeded forage are usually less with nurse crops than without a nurse crop. Final plant size and stand densities are usually similar with both methods. Forage establishment can be reduced if wind and/or heavy rainfall causes the nurse crop plants to lodge (break or fall over). The dense, compressed cover created by lodging shades young seedlings and creates conditions favorable for disease organisms. To reduce **competition** and **lodging** potential, it is recommended that small-grain seeding rates be reduced to 75% of the rates recommended for small grain alone. Normal seeding rates can be used if the nurse crop is to be harvested earlier (e.g., at the boot stage) as forage. Selection of early maturing, short-strawed small-grain cultivars also helps to minimize competition and lodging.

Small-grain nurse crop forage is of lower quality than pure legume forage, and is often used for beef cattle or non-lactating dairy animals with lower nutritional requirements.

In a modification of the nurse crop system, forages can be seeded into small-grain stubble after harvest. This avoids direct competition between forage seedlings and nurse crop plants, and allows small-grain management for optimum grain and straw yields. Establishment success in this system is highly dependent on receiving late-summer rainfall to replenish the soil moisture removed by the small-grain crop.

Another option in legume nurse crop management is to use a grass **herbicide** to kill the vegetative small-grain crop at a shoot height of 4–10 in. (10–25 cm). The dead plants still provide early erosion and weed control benefits, but competition and lodging problems are reduced. Using this approach, seeding-year forage yields are comparable to yields without a nurse crop (Fig. 11.6).

Herbicides

In this seeding system, the forage crops are planted without a nurse crop. When establishing legumes, herbicides can provide larger yields of high-quality forage during the seeding year compared with the nurse crop system. The use of herbicides rather than a nurse crop has increased since the development of selective herbicides. This method eliminates competition from the nurse crop, and often allows up to three forage harvests during the seeding year. The method is not recommended on sites with significant slopes, due to the increased erosion potential.

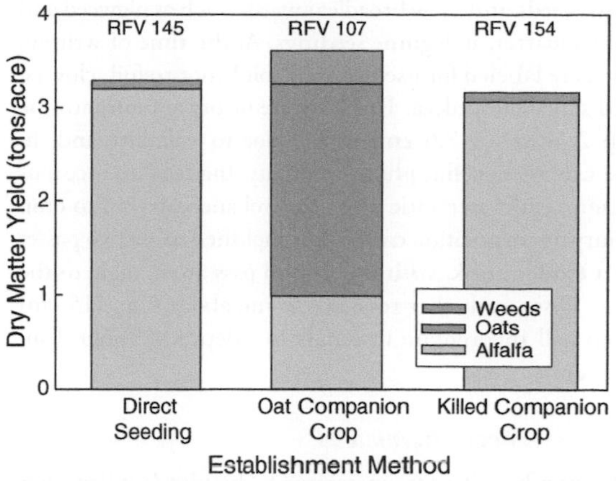

FIG. 11.6. Total dry matter (DM) yields for three methods of establishing alfalfa. The establishment-year DM yields were comparable for all three methods evaluated in this study, but the alfalfa percentage was much lower for the companion crop system. The lower quality of the companion crop forage is reflected in the weighted relative feed value (RFV) for each system. (From Becker et al., 1998. Reproduced with permission of the *Journal of Production Agriculture*.)

The number of herbicides that are labeled specifically for forage establishment is limited. Fall-applied, preplant incorporated herbicides are applied and worked into the soil prior to planting. Post-emergence herbicides are applied after the emergence of the crop (and weeds).

Herbicide Tolerance

At the time of writing, alfalfa is the only perennial forage species with glyphosate-tolerant cultivars. This tolerance allows glyphosate to be applied post emergence to tolerant cultivars for broad-spectrum weed control without affecting the alfalfa.

Preplant

As far as possible, perennial weed problems should be controlled prior to establishing forage crops, as many more options are available before than after the forage crop has been seeded. This is especially true for grass–legume mixtures. Fall applications of non-selective herbicides such as glyphosate can be used to control quackgrass, thistles, and other troublesome perennial weeds.

Preplant Incorporated Herbicides

These herbicides need to be thoroughly incorporated into the seedbed prior to forage seeding. Benefin and EPTC (S-ethyl-dipropylthiocarbamate) herbicides control annual grass weeds and a few broadleaf weeds, such as pigweed and lambsquarter, in legume seedings. At the time of writing, both are labeled for use in alfalfa, birdsfoot trefoil, clovers, and sericea lespedeza. Both require incorporation into the top 2–3 in. (5–7.6 cm) of soil due to volatility and, in the case of benefin, photosensitivity. Inadequate incorporation results in erratic weed control and can lead to crop injury. Incorporation can be accomplished using two passes of a tandem disk, with the second pass at an angle to the first. Discs and other tools set to run about 6 in. (15 cm) deep will incorporate materials to a depth of about 3 in. (7.6 cm).

Post-Emergence Herbicides

The number of post-emergence herbicides (applied after the forage plants have emerged from the soil) available to control weeds in both legume and grass forage crops is greater than for pre-emergence herbicides. In addition, their labels (the legal documents that specify where, when, and how a herbicide can be used) change periodically. Therefore it is important to seek advice from your Cooperative Extension office on which herbicides to use, and you should always read the label before using any herbicide. Many herbicides have harvest or grazing restrictions (a minimum number of days between application and harvesting or grazing) that will also limit when they can be used. For example, alfalfa seedlings should be at least at the fourth trifoliate leaf stage before being treated with 2,4-DB, to avoid crop injury, but the treated forage may not be harvested for 60 days after application. The harvest restriction for bromoxynil is shorter (30 days), but temperatures above 70°F (21°C) for 3 days after application may result in crop injury by this herbicide.

Clipping

Competition from weeds in new seedings can be reduced by timely clipping. The growing points of weeds should be above cutting height to achieve effective control. Clipping too soon (before the growing points are above cutting height) may actually stimulate weeds to produce new branches, and thus lead to even greater competition. Clipping too frequently can reduce seedling development as well as forage yields the following year. Most extension specialists also recommend that clipping of new legume seedings should end 4–6 weeks before a killing frost, to allow a build-up of carbohydrate reserves for winter.

Pasture Renovation

In pasture renovation, inputs such as sod disturbance, fertilizer and lime, weed control, and seeding of superior species and cultivars are used to enhance the productivity of an existing sod (Hall and Vough, 2007). Renovation frequently involves the introduction of legumes into perennial grass sods. Renovation with legumes often improves annual dry matter production and may also improve forage quality.

Preparation for spring pasture renovation should begin in the previous fall with close grazing or clipping to control existing vegetation. This also reduces surface residues that would shade small seedlings. Some leaf area should be allowed to accumulate prior to spring application of a contact herbicide to suppress the existing vegetation. The herbicide is generally broadcast sprayed (i.e., applied uniformly over the entire area).

Interseeding is the introduction of annual or perennial forage species into existing pastures without herbicide treatment. This method is used in the southern USA to introduce cool-season grasses and legumes into dormant bermudagrass or other warm-season perennial grasses. In northern regions, cool-season grasses or legumes are often interseeded via frost or no-till seeding into thinning pastures to increase production.

Natural Reseeding

Natural reseeding plays a critical role in maintaining plant populations in some more extensive forage systems, especially those used for pasture. **Indeterminate** species such as white clover and birdsfoot trefoil are able to produce substantial amounts of viable seed even during grazing by livestock, provided that grazing pressure is not excessive. In the transition zone and upper South, individual birdsfoot trefoil plants generally live only 2 or 3 years, but the stands can persist indefinitely as a result of this natural reseeding.

A more consistent approach is to allow birdsfoot trefoil to grow unharvested for approximately 70 days during midseason. Flowering and seed production occur during this period and, due to the indeterminate growth habit, the quality of the available forage remains high. Substantial populations of some annual species, such as korean lespedeza, may continue for several years through annual reseeding.

Sprigging

Most bermudagrass cultivars do not produce viable seed, so this species is usually established by planting vegetative plant materials; this technique is called sprigging. Sprig diggers are used to harvest the mixture of rhizomes, stolons, plant crowns, and shoot material that serves as the source of sprigs. Alternatively, hay equipment can be used to harvest shoot material for sprigging, although viability is quickly lost if the material dries before planting. Sprigging can be done from February to July in the deep South, and in June and July in the transition zone.

Plantings made later in the season in the transition zone may suffer more winterkill than stands established earlier in the season. Sprigs should always be planted as soon as possible after digging, in order to minimize drying or heating which can reduce viability. Typical planting rates for

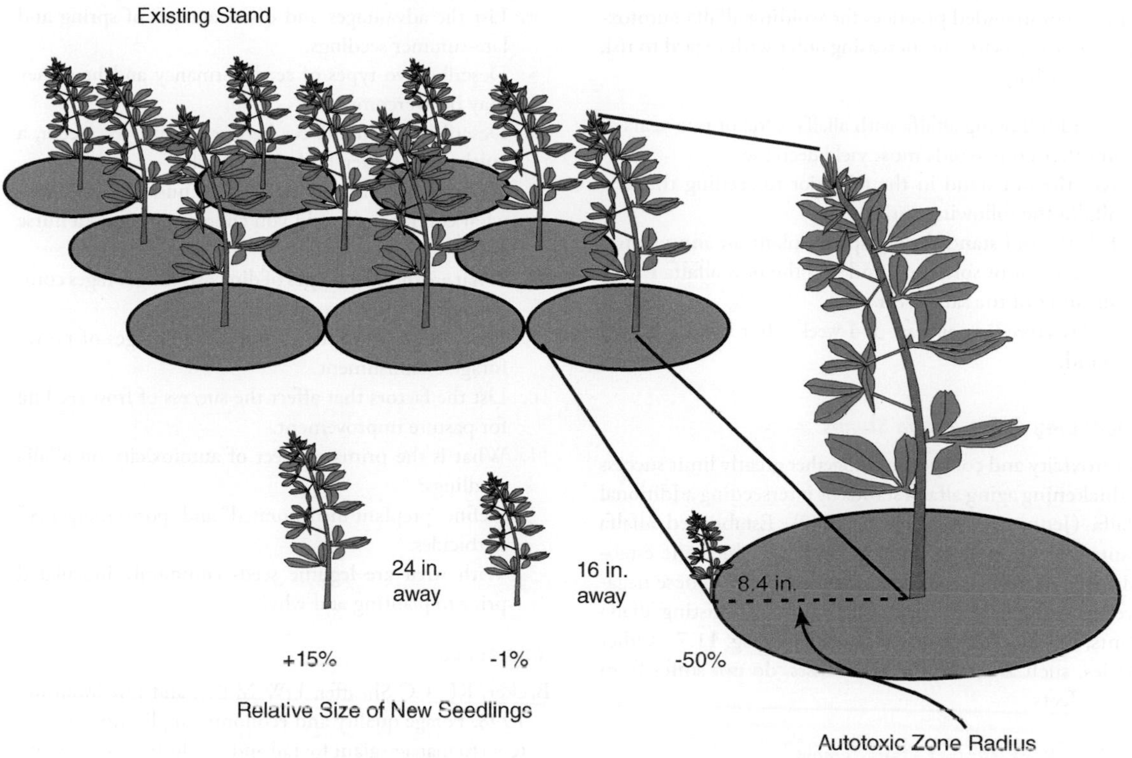

FIG. 11.7. Substantial negative effects are exerted by established alfalfa plants on growth of seedlings 8 in. away, but no such effects are observed for seedlings at a distance of 16 in. or more. (Based on data from Jennings and Nelson, 2002.)

bermudagrass are 30–40 bushels of dug sprigs/acre (2.6–3.5 m^3/ha), or about 1500 lb of tops/acre (1680 kg/ha). Sprigs can be planted using commercial sprig planters that bury pieces 2–3 in. (5–7.6 cm) deep, or they can be broadcast over the surface and disked in as soon as possible. Firming the soil with a cultipacker or similar implement after sprigging is beneficial. Limpograss, perennial peanut, and stargrass are also established vegetatively.

Autotoxicity

Autotoxicity, which is the production of chemical compounds by a plant that inhibits development of other plants of the same species, can occur when alfalfa is reseeded in fields where the same crop has been grown recently. The negative effects are manifested primarily as inhibition of root development of new seedlings, but some effects can be permanent.

The specific compounds responsible for autotoxicity have not been positively identified. Studies have shown that autotoxic chemicals are found in higher concentrations in alfalfa shoots than in roots, and are water soluble, so they are readily washed from plants by rainfall. In addition, when plants are plowed down or killed using herbicides, the compounds are released into the soil.

The duration of autotoxicity effects is shorter for sandy soils because the compounds move downward more quickly, but the effects may be amplified during this shorter period. On heavier-textured soils, the duration of the effect is longer but its intensity is lessened by adsorption on soil colloids. The autotoxicity response is less for thin existing stands and when more time has elapsed between killing of the old stand and reseeding (Seguin et al., 2002).

The effects of autotoxicity dissipate with time, but the impact on affected plants persists throughout the life of the stand. Affected plants remain stunted, have shallower root systems, and yield less long after the toxin itself has disappeared. This long-term effect has been referred to as *autoconditioning*, because young plants seem to be conditioned for lower yield by events that occurred during the seeding year.

Avoiding Alfalfa Autotoxicity

Establishment problems may persist for only a few weeks, but may be present for alfalfa planted up to a year later.

Some recommended practices for avoiding alfalfa autotoxicity are listed below in increasing order with regard to risk for autotoxicity.

1. Avoid following alfalfa with alfalfa. One or two years of another crop avoids most yield decreases.
2. Kill the old stand in the fall prior to seeding the new alfalfa the following spring.
3. Kill the old stand during spring, plant an annual crop such as oat or sorghum, and seed the new alfalfa in late summer of the same year.
4. Delay reseeding at least 3–4 weeks after killing the old stand.

Thickening Aging Alfalfa Stands

Autotoxicity and competition together greatly limit success in thickening aging alfalfa stands by interseeding additional alfalfa (Jennings and Nelson, 2002). Established alfalfa plants have a "zone of influence" within which the establishment of new alfalfa seedlings is restricted. These negative effects are large for seedlings 8 in. from existing alfalfa plants, but absent at a distance of 16 in. (Fig. 11.7). Other species, such as red clover and grasses, do not suffer from these effects.

Seeding-Year Harvest Management

Stage of growth dictates harvest time in the seeding year just as it does in subsequent years. For instance, it is recommended that alfalfa should be harvested in the bud stage to optimize yield and quality. This stage could be reached in as little as 60 days after emergence, depending on weather conditions. Additional cuttings may be taken at the same stage without affecting stand density.

New grass seedings can be harvested at the boot to early head stage. Most perennial cool-season grasses require a cold period to induce flowering, and will not show reproductive development in the seeding year. These grasses should be harvested when the leaves are fully elongated, but before senescence begins.

Summary

Stand establishment is an important step in developing productive forage systems. Optimum methods for stand establishment vary widely among forage species within a region and between regions. Reduced-tillage or no-tillage methods are especially important on sloping land, in order to maintain or enhance soil fertility and productivity.

Questions

1. List the advantages and disadvantages of seeding mixtures compared with pure stands.
2. What is "pure live seed" and how is it calculated?
3. Why are seedbed preparation and shallow planting important for forage stand establishment?
4. List the advantages and disadvantages of spring and late-summer seedings.
5. Describe two types of seed dormancy and how they may be overcome.
6. Describe the differences between a spinner seeder, a cultipacker seeder, and a grain drill.
7. What are the reasons for using a nurse crop? Under what conditions would you choose not to use a nurse crop?
8. What are the advantages of direct seeding forages compared with nurse crop seeding?
9. Describe the advantages and disadvantages of no-till forage establishment.
10. List the factors that affect the success of frost seeding for pasture improvement.
11. What is the primary effect of autotoxicity on alfalfa seedlings?
12. Define "preplant incorporated" and "post-emergence" herbicides.
13. With what are legume seeds commonly inoculated prior to planting and why?

References

Becker, RL, CC Sheaffer, DW Miller, and DR Swanson. 1998. Forage quality and economic implications of systems to manage giant foxtail and oat during alfalfa establishment. J. Prod. Agric. 11:300–308.

Cosgrove, DR. 2001. Increasing pasture quality and sward density. In Proceedings of the Upper Midwest Grazing Conference, July 17–19, Dubuque, IA.

Cosgrove, DR, PE Daley, LG Koenig, and TJ Ritten. 1991. Effects of coated seed on alfalfa stand density and yield in reduced tillage systems. In Proceedings of the American Forage and Grassland Council, pp. 166–170. American Forage and Grassland Council, Berea, KY.

Hall, MH, and LR Vough. 2007. Forage establishment and renovation. In RF Barnes, CJ Nelson, KJ Moore, and M Collins (eds.), Forages: The Science of Grassland Agriculture, 6th ed., Volume II, pp. 343–354. Blackwell Publishing, Ames, IA.

Hall, MH, JA Jennings, and GE Shewmaker. 2004. Alfalfa establishment guide. Forage Grazinglands doi:10.1094/FG-2004-0723-01-MG.

Jennings, J. and CJ Nelson. 2002. Zone of autotoxic influence around established alfalfa plants. Agron. J. 94:1104–1111.

Sanderson, MA, G Brink, L Ruth, and R Stout. 2012. Grass–legume mixtures suppress weeds during establishment better than monocultures. Agron. J. 104:36–42.

Seguin, P, CC Sheaffer, MA Schmitt, MP Russelle, GW Randall, PR Peterson, TR Hoverstad, SR Quiring, and DR Swanson. 2002. Alfalfa autotoxicity: effects of reseeding delay, original stand age, and cultivar. Agron. J. 94:775–781.

Sund, JM, GP Barrington, and JM Scholl. 1966. Methods and depths of sowing forage grasses and legumes. In Proceedings of the 10th International Grassland Congress, pp. 319–323. Finnish Grassland Association, Helsinki, Finland.

Tesar, MB, and VL Marble. 1988. Alfalfa establishment. In AA Hanson, DK Barnes, and RR Hill Jr. (eds.), Alfalfa and Alfalfa Improvement, pp. 303–322. American Society of Agronomy, Madison, WI.

Undersander, DJ, DC West, and MD Casler. 2001. Frost seeding into aging alfalfa stands: sward dynamics and pasture productivity. Agron J. 93:609–619.

Vondrachek, G. 1991. Seeding rate variety trial. In Proceedings of the Wisconsin Forage Council Forage and Use Symposium, January 22–23, Wisconsin Dells, WI.

CHAPTER 12

Forage Fertilization and Nutrient Management

David J. Barker and Michael Collins

Soil fertility is one of the most important determinants of forage production. Soils with high fertility can support up to five times more forage production than soils with low fertility. Manipulation of soil fertility through fertilization can have more effect on increasing forage production than almost any other management option for forage improvement.

Fertilization is the most practical management tool available to managers for alleviating nutrient deficiencies. **Fertilizer** and **lime** costs may account for 10–30% of total forage production costs. Expenditure on fertilizers is only justifiable when there is an economic return from increased forage or livestock production or improved forage quality.

Excessive application of nutrients can have negative impacts on the environment. Most commonly, excess nutrients in soil can leach into surface water, affecting its quality for other uses. In addition to being costly to farmers, these impacts are unacceptable to the wider community. Forage crops have an excellent ability to absorb nutrients from the soil, and can be managed for multiple ecosystem services, including both food production and nutrient management.

This chapter will focus on **nutrient balance** rather than taking the nutrient cycle approach. The net nutrient balance for a system, which is simply the difference between nutrient imports and exports, can be calculated on different scales according to the setting (e.g., a field, a farm, or a watershed). Where the balance is positive, the nutrient status of that particular setting will increase, and where the balance is negative, the nutrient status will decrease. Nutrient imports come not only from fertilizers, but also rainfall and, where livestock are involved, manure and imported animal feed. Nutrient exports include leaching, erosion, and sold feed and animal products.

What Is Soil Fertility?

Fertility is the capacity of the soil to provide the nutrients required for plant growth. It is a qualitative rather than quantitative concept, as there are no direct measures of soil fertility. Soil fertility is determined primarily by soil chemical composition (nutrient status), but is also affected by soil physical structure (including soil texture, soil **organic matter** content, gaseous movement, root growth, and water supply) and soil biological activity (including nutrient cycling, **rhizobia**, and **mycorrhiza**).

Nutrients are released from particles of the soil **parent material** by the physical and chemical processes associated with weathering. This capacity to supply soil nutrients depends on the following:

- *Chemical composition of the parent material.* Clay minerals such as basalt, vermiculite, and apatite are rich in nutrients, whereas others, such as rhyolite and silica, have a low nutrient content.
- *Soil texture.* Clay, silt, and sand have a similar chemical composition, but clay particles have a higher surface area to volume ratio and thus a higher potential weathering rate and chemical reactivity.
- *Climate.* Moderate temperatures and non-leaching rainfall favor nutrient accumulation.

Forages: An Introduction to Grassland Agriculture, Seventh Edition. Edited by Michael Collins, C. Jerry Nelson, Kenneth J. Moore and Robert F Barnes.

Released cations are attracted to soil particles by weak electrostatic charges. A quantitative measure of this charge is **cation-exchange capacity (CEC)**, which is the negative charge at the surface of soil particles that results from dissociation of H^+ ions. Cations weakly attracted to the CEC are available for release to the **soil solution**, a layer of water that covers all the soil particles, and from which nutrients are taken up by plants. Surplus H^+ ions resulting from soil acidity can displace cations and reduce the percentage of the CEC that is occupied by cations. A poor soil has a CEC of less than 12 meq/100 g soil and a fertile soil has a CEC of more than 12 meq/100 g soil.

An excess of any one cation can displace other cations from the CEC. In saline soils, Na^+ can displace other cations, or with excessive liming, Ca^{2+} can displace Mg^{2+}. Deficiencies may result for nutrients that are present in the soil but are not present in adequate levels on the cation-exchange sites.

Soil organic matter is typically 2–6% of soil weight, and is another indicator of soil fertility. The remnant carbon skeletons from organic matter decomposition (**humic acid** or **humus**) bind to the surfaces of soil particles and exert an influence considerably greater than their proportion of soil mass. Effects of organic matter include increasing CEC, releasing nutrients as the organic tissues break down to humus, binding soil particles to form aggregates that improve soil structure, and increasing soil water-holding capacity.

Pasture species vary in their response to soil fertility. Species such as kentucky bluegrass and velvetgrass are tolerant of low fertility, whereas other species, such as orchardgrass, need moderate fertility. Species such as perennial ryegrass, timothy, and most legumes need still higher soil fertility for optimum production and persistence. Soil fertility influences botanical composition by its differential impact on forage and weed species. Pasture improvement programs that change pasture species but do not improve fertility will soon revert to the original species composition.

Soil Quality

Soil quality is defined as "the capacity of a specific kind of soil to function, within natural or managed ecosystem boundaries, to sustain plant and animal productivity, maintain or enhance water and air quality, and support human health and habitation" (Karlen et al., 1997). The concept takes a holistic approach in considering the soil as a living part of the agricultural ecosystem. It does this by considering biological as well as chemical and physical properties in relation to understanding and sustaining soil productivity. Although the technical definitions differ slightly, in practice the terms "soil health" and "soil quality" are used synonymously. Indicators of soil quality depend on the soil use that is being evaluated, but generally include physical characteristics (e.g., soil aggregate stability, texture, and bulk density) and biological characteristics (e.g., soil microbial activity, plant residue decomposition rates, and earthworm activity) in addition to typical chemical indicators (e.g., pH, salinity, organic matter levels, and nutrient concentrations).

Plant and Animal Nutrient Requirements

Macronutrients are those essential minerals that typically exceed 0.1% of dry matter. They include nitrogen (N, 1–5%), potassium (K, 2–4%), calcium (Ca, 0.4–1%), magnesium (Mg, 0.25%), phosphorus (P, 0.25%), and sulfur (S, 0.2%). These macronutrients are components of specialized molecules such as amino acids, proteins, enzymes, nucleic acids, and chlorophyll (Tables 12.1 and 12.2).

Micronutrients (also known as **trace elements**) are those essential minerals that are typically less than 0.05% (500 parts per million or ppm) of dry matter. In order of decreasing abundance, they include silicon (Si), boron (B), iron (Fe), manganese (Mn), copper (Cu), zinc (Zn), chlorine (Cl), molybdenum (Mo), and nickel (Ni).

A few elements are not required by plants but are essential for animals that consume forages, namely selenium (Se), chromium (Cr), cobalt (Co), sodium (Na), and iodine (I). These nutrients are usually present in plants, having been absorbed from the soil by passive uptake. Inadequate levels can result in deficiencies for livestock even though the forage plants are healthy. The discovery of Co deficiency as the cause of "bush-sickness", a chronic wasting disease in sheep and cattle in central New Zealand (during 1930–1940), and the dramatic improvement in animal performance that resulted from direct supplementation with Co is an excellent illustration of this situation (Grace, 1994).

Plants and animals exhibit symptoms when there are deficiencies in any of the essential nutrients. In many cases, plant nutrient deficiency symptoms include a pale and stunted appearance. However, in more dramatic cases the symptoms are more diagnostic (Fig. 12.1, Fig. 12.2, and Table 12.3). Soil nutrient levels are not easily estimated, and should be measured to support the best management decisions (Table 12.4).

Nitrogen

Nitrogen is generally the most limiting nutrient in forage agriculture. There are several reasons for this, including (1) extensive use of N during plant growth (it represent up to 5% of dry matter), (2) the high solubility of plant-absorbable forms of N (NH_4^+ and NO_3^-), making it susceptible to leaching, (3) the numerous pathways for N loss from pasture compared with other nutrients, and (4) the fact that N is provided only indirectly by the soil, with the predominant source being the atmosphere. Nitrogen inputs to soil are largely dependent on prior fixation by a legume.

Table 12.1. Mineral nutrients (excluding the macro-elements C, O, and H) and their role in plants and animals

Element	Symbol	Role in plants	Role in animals
Macronutrients			
Nitrogen	N	Amino acids, protein synthesis, nucleic acids	Protein synthesis
Potassium	K (K_2O)	Enzyme activation, winter hardiness, water relations, N uptake and protein synthesis, disease resistance, translocation, starch synthesis	Maintains acid–base balance, enzyme reactions, carbohydrate metabolism
Calcium	Ca	Calcium pectate and membrane function, cell regulation	Component of structural skeleton, blood coagulation, cell regulation
Phosphorus	P (P_2O_5)	Utilization of energy from food reserves, used early in life cycle, root formation, nucleic acids	Component of structural skeleton, energy metabolism, nucleic acids
Magnesium	Mg	Component of chlorophyll, cofactor in ATP (energy) metabolism	Skeleton development, phosphorylation, enzyme activation
Sulfur	S	Sulfhydryl groups, amino acids	Present in amino acids, acid–base balance, intracellular constituent, carbohydrate metabolism
Micronutrients essential for plants and animals			
Iron	Fe	Component of chlorophyll, cytochromes, and enzymes	Component of hemoglobin (for oxygen transport in blood and muscles), component of cytochromes of the electron transport chain
Manganese	Mn	Formation of amino acids, chloroplast membrane, enzyme systems	Needed for bone matrix formation
Chlorine	Cl	Photosynthetic phosphorylation, charge balance, osmotic pressure	Regulation of extracellular osmotic pressure, maintaining acid–base balance
Boron[a]	B	Amino acids, protein synthesis, nodule formation	Possibly not required
Copper	Cu	Nitrate reduction, photosynthetic electron transfer	Components of enzymes and Fe metabolism, immune system
Zinc	Zn	Enzymatic activities	Activates enzymes, component of metalloenzymes
Molybdenum	Mo	Component of nitrate reductase, N fixation	Component of metalloenzymes
Nickel*	Ni	Component of urease, N fixation in legumes	Has been shown to be essential in rats, and is assumed to be essential in livestock
Micronutrients not essential for plants, but essential for animals			
Silicon	Si	Possible minor role in drought resistance, mechanical strength, presence in sharp leaf margins a grazing deterrent	Mineralization of bones
Sodium	Na	Not essential, but when present acts with K to regulate osmotic pressure and charge balance	Acts with K and Cl to maintain extracellular fluid balance, maintains osmotic pressure, heart and nerve function
Iodine	I	Possible minor role in tissue culture	Thyroid gland function
Chromium	Cr	Possibly not required	Glucose tolerance factor—insulin regulation
Cobalt	Co	N fixation in alfalfa	Component of vitamin B_{12}
Selenium	Se	Possibly not required	Component of glutathione peroxidase, cell membranes, immune system function

[a]Requirement by animals is so small that supplementation is never required.

Sources: Adapted from the following: Miller and Heichel, 1995; Tisdale et al., 1993.

Table 12.2. Average concentrations of chemical elements in forage foliage, and the levels at which deficiency and toxicity symptoms appear

Element	Absorption form(s)	"Normal"	Deficiency for plants	Deficiency for livestock[a]	Toxicity for plants
Macro-elements		(%)	(%)	(%)	(%)
Carbon	CO_2	45	–	–	–
Oxygen	O_2	42	–	–	–
Hydrogen	H_2O	6	–	–	–
Macronutrients		(%)	(%)	(%)	(%)
Nitrogen[b]	NO_3^-, NH_4^+	1–4	<1, <2[j]	<1 (<2.5)	Very low to none[e]
Potassium	K^+	2–4	<1	<0.9	Very low to none[e]
Calcium	Ca^{2+}	0.5–2.0	<0.0002	<0.4 (<0.6)[i]	Very low to none[e]
Magnesium	Mg^{2+}	0.2–0.8	<0.05	<0.1 (0.3)[l]	Very low to none[e]
Phosphorus	$H_2PO_4^-$, HPO_4^-	0.25–0.5	<0.2	<0.1[i]	Very low to none[e]
Sulfur	SO_4^{2-}	0.2–0.3	<0.15[d]	<0.2[d]	Very low to none[e]
Micronutrients essential for plants and animals		(ppm)	(ppm)	(ppm)	(ppm)
Iron	Fe^{2+}, Fe^{3+}	50–1000	<35	<0.1 (<1.0)	Very low to none
Manganese	Mn^{2+}	30–300	<20	<15	>500
Boron	H_3BO_3	10–50	<10	Very low to none[m]	>75
Copper	Cu^+, Cu^{2+}	5–15	<5	<0.6[f]	>20
Zinc	Zn^{2+}	10–100	<10	<4 (<6)	>200
Molybdenum	MoO_4^-	1–100	<0.2	<1[g]	>2000
Nickel	Ni^{2+}	0.2–2.0	<0.1	Very low to none[m]	>30
Micronutrients not essential for plants, but essential for animals		(ppm)	(ppm)	(ppm)	(ppm)
Silicon	$Si(OH)_4$	400–10,000	Not required	Not known[h]	
Chlorine	Cl^-	500–10,000	Not required	<2000	>20,000
Sodium	Na^+	100–200	Not required	<1000 (<2000)	
Iodine	I^-	3	Not required	<0.3 (<0.5)	
Chromium	Cr^{3+}, Cr^{6+}	0.2	Not required	Very low to none[m]	
Cobalt	Co^{2+}	0.05–2.0	<0.02[c]	<0.11	Very low to none
Selenium	SeO_3	0.15	Not required	<0.3[k]	

[a]Assuming that there is no mineral supplementation and no interference from other nutrients, values in parentheses are for lactating livestock.

[b]Percentage N × 6.25 = percentage protein.

[c]Required for rhizobia in legumes.

[d]N:S ratio should not exceed 10:1.

[e]There is no practical toxic limit. However, competitive exclusion of other nutrients is certain at high nutrient levels.

[f]Cu absorption is subject to interference from Zn, Mo, S, and Fe.

[g]Mo values above 5–10 ppm can impair Cu uptake in livestock.

[h]Si values in excess of 0.2% of diet will depress intake.

[i]Desirable Ca:P ratio is between 2:1 and 1:1.

[j]Less than 1% for grasses, less than 2% for legumes.

[k]Se values above 8.5 ppm may be toxic to livestock.

[l]Mg uptake can also be depressed by excessive K and Ca.

[m]Lower limit for requirement is not known.

Sources: Adapted from the following: Follett and Wilkinson, 1995; National Research Council, 2001; Tisdale et al., 1993; Spears, 1994.

FIG. 12.1. Effect of phosphorus and potassium on alfalfa top growth. A deficiency of any essential element will restrict forage plant growth even if other elements are present in sufficient quantities to meet plant needs. (Miller, 1984. Reproduced with permission of Miller.)

Components of the Nitrogen Balance

The nitrogen balance is complex, and soil N status is extremely dynamic. Furthermore, there is no simple *way to* measure plant-available soil N. The amounts of plant-absorbable forms (NH_4^+ and NO_3^-) are often not correlated with yield, which is less dependent on the small amounts of mineral N present at any given time, which

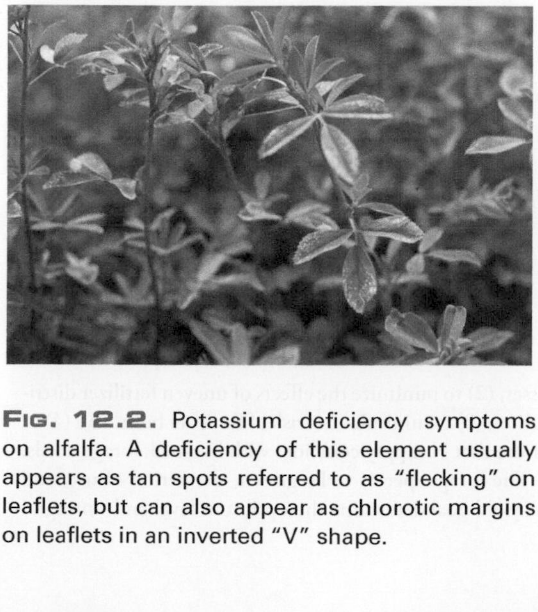

FIG. 12.2. Potassium deficiency symptoms on alfalfa. A deficiency of this element usually appears as tan spots referred to as "flecking" on leaflets, but can also appear as chlorotic margins on leaflets in an inverted "V" shape.

are in the range of 40–100 lb N/acre (45–112 kg N/ha), than on N release from the much larger soil organic N pool, which can be in the range of 1–5 tons N/acre (0.9–4.5 mt N/ha). Despite these difficulties, there are a number of important principles that emerge from the balance equation:

Soil N status = (N fixation + imported N + soil release
+ atmospheric deposition)
− (atmospheric losses + leaching losses
+ removal in plant and animal products).

Nitrogen Fixation

A major factor affecting **nitrogen fixation** (in addition to legume species) is the concentration of inorganic N (NH_4^+ and NO_3^-) in the soil. Legumes will absorb and utilize inorganic N in preference to fixing N, and fertilization in excess of 50 lb N/acre (56 kg N/ha) can suppress N fixation. As a rule of thumb, each 2 lb N/acre (2.2 kg N/ha) of fertilizer N will suppress fixation by 1 lb/acre (1.1 kg/ha).

Nitrogen fixed by forage legumes benefits associated grasses through at least seven pathways. The major pathways include the following:

- *Urinary N from grazed legumes.* This route accounts for 70–75% of the consumed N, but 50–80% may volatilize and be lost as ammonia.
- *Excretion of legume N as dung.* Dung contains 2–3% N in various organic forms. Its availability to grasses is dependent on decomposition and incorporation into the soil organic N pool.
- *Decay of nodules and legume root tissue.* Estimates are in the range of 0–100 lb N/acre/year (0–112 kg/ha/year).
- *Death and decay of leaves and* ***stolons****.* Stolons are frequently below grazing height, and depending on grazing management, up to 30% of the leaves may not be grazed.

The minor pathways include the following:

- *Leaching of N from living vegetation by rain or irrigation.*
- *Exudation of nitrogenous compounds by the legume roots or nodules directly into the soil.*
- *Direct legume-to-grass transfer.*

Fertilizer Nitrogen

The main source of imported N is fertilizer. The percentage of N in common fertilizers is as follows: anhydrous ammonia, 82%; urea, 46%; NH_4NO_3, 33.5%; $(NH_4)_2SO_4$, 21%; nitrate solutions, 28% to 32%. Ammonia-N forms increase soil acidity and may require adjustment with lime. Legume crops contribute residual N to subsequent crops. Since this N is in organic forms, a portion is carried over into the second year and beyond.

Forage dry-matter responses to N fertilizer generally fall within the range of 10–50 lb dry matter/lb N. In temperate

Table 12.3. Symptoms of mineral deficiencies in forage plants and in herbivorous animals

Element	Deficiency symptoms in forage plants	Deficiency symptoms in herbivorous animals
Macronutrients		
Nitrogen	Chlorosis/yellowing of leaves, stunted growth	Low growth rates, poor production
Potassium	Spotted leaf tips in alfalfa, reduced cold and disease resistance	Rare, but can occur in productive and lactating cattle: reduced intake, weight loss, hair loss, weakness, production loss
Calcium	Slowed development and eventually death of apical buds of shoots and roots	Impaired bone growth, resulting in slow growth and osteoporosis
Phosphorus	Purplish or reddish coloration of leaves and sometimes stems, stunted growth	Unthrifty, poor growth, poor milk production
Magnesium	Most common on sandy soils, leaf chlorosis especially of the interveinal areas, while the veins remain green	Hypomagnesemia
Sulfur	Leaf chlorosis, low cysteine, cystine, and methionine levels, low vitamin and chlorophyll synthesis, possible nitrate accumulation	Reduced protein synthesis, resulting in slow growth and poor production
Micronutrients essential for plants and animals		
Iron	Interveinal chlorosis in younger leaves	Rare, but can occur in calves: anemia and high mortality due to impaired immune response
Manganese		Impaired growth, skeletal abnormalities, poor reproduction, abnormal births
Chlorine	Leaf curling, chlorosis, abnormal root growth	Anorexia, lethargy, eye defects, reduced respiration, blood in feces
Boron	Stunted growth, yellowing (especially in young tissue)	Not required
Copper		Loss of hair pigmentation, scours, anemia, impaired immune function
Zinc	White or stripped leaves	Reduced feed intake and growth, parakeratosis of skin on head, legs, and neck
Molybdenum	Impaired N uptake and metabolism, resulting in N-deficiency symptoms	Deficiency is rare, no consistent symptoms have been found
Nickel		Reduced growth, low rumen urease activity in lambs
Micronutrients not essential for plants, but essential for animals		
Silicon	None	Si is so abundant that no deficiency symptoms have been observed
Sodium	None	Licking and chewing, salt craving, drinking urine, unthrifty, rough hair coat
Iodine	None	Enlarged thyroid (goiter), especially in calves
Chromium	None	None
Cobalt	Impaired rhizobia nodulation	Vitamin B_{12}levels <3 µg/L, chronic wasting disease, with unthriftiness and weight loss
Selenium	None	White muscle disease

Source: National Research Council, 2001.

regions, grass yields will continue to increase up to N rates of 330–400 lb N/acre (370–448 kg N/ha) annually. Most N fertilizers are topdressed in split applications during the growing season. The reasons for dividing the N applications in this way are (1) to avoid leaching and volatilization losses, (2) to minimize the effects of uneven fertilizer distribution, (3) to minimize the risk of fertilizer burn, and (4) to ensure that N supply coincides with livestock forage needs.

The forage species to be grown, the period of use, and the yield goal determine the optimum level and timing of

Table 12.4. Average, minimum, and maximum concentrations of nutrients in soil

Element	Absorption form(s)	Average	Deficient levels for plants[a]	Adequate levels for plants[b]
Macronutrients		(ppm)	(ppm)	(ppm)
Nitrogen	NO_3^-, NH_4^+	1–50[e]	NA[d]	NA[d]
Potassium	K^+	50–200	<50	100
Calcium	Ca^{2+}	500–8000	<200	500
Phosphorus[c]	$H_2PO_4^-$, HPO_4^-	2–100	<10	>50
Magnesium	Mg^{2+}	0–1000	<50	200
Sulfur	SO_4^{2-}	10	<7	>12
Micronutrients essential for plants and animals		(ppm)	(ppm)	(ppm)
Iron	Fe^{2+}, Fe^{3+}			
Manganese	Mn^{2+}	20–50	<10	
Boron	H_3BO_3	0.5–1.0	<0.1	>0.25
Copper	Cu^+, Cu^{2+}	10–30		
Zinc	Zn^{2+}	50–150	<1	>1.5
Molybdenum	MoO_4^-	50–150		
Nickel	Ni^{2+}	10–30		
Micronutrients not essential for plants, but essential for animals		(ppm)		
Silicon	$Si(OH)_4$		Not required	
Chlorine	Cl^-		Not required	150
Sodium	Na^+		Not required	
Iodine	I^-		Not required	
Chromium	Cr^{3+}, Cr^{6+}	15–25	Not required	
Cobalt	Co^{2+}		Not required	
Selenium	SeO_3	0.1–0.5	Not required	

[a]Fertilization is recommended.
[b]No fertilization response is likely to occur.
[c]Bray test.
[d]NA = no available standardized test; the soluble forms NO_3^- and NH_4^+ are a small fraction of total soil N.
[e]Nitrate only.
Sources: Follett and Wilkinson, 1995; Allaway, 1968; Watson, 1995.

N fertilization. Lower rates of N are recommended during periods when drought may limit production and therefore reduce the effectiveness of the fertilizer. Timothy recovers slowly following grazing or hay harvest, and has a relatively shallow root system, so an early-spring N application is recommended, in order to coincide with adequate soil moisture and rapid growth. Orchardgrass, smooth bromegrass, tall fescue, and reed canarygrass are more drought tolerant than timothy and kentucky bluegrass, and use higher N rates more efficiently.

Legume–grass mixtures consisting of more than 30% legume should not receive N fertilization. With this proportion of legume, sufficient N should be fixed to maintain optimum productivity of the mixture. Nitrogen tends to shift the competitive advantage to the grasses, which would otherwise be restricted by the limited N supply. Where the objective is to maintain the legume, the emphasis should be on adequate P and K fertilization, rather than on N. If there is less than 30% legume in a mixture, N fertilization may be needed to maintain optimum grass productivity. A suggested rate of N application is about 50 lb N/acre (56 kg/ha) when the legume represents 20–30% of the mixture, and 112 lb N/acre (125 kg/ha) or more when the legume is less than 20% of the mixture.

Another source of imported N is manure. Cattle manure typically contains 0.5–0.6% N, equivalent to 10–12 lb N/ton (5–6 kg N/mt), and poultry manure can contain as much as 20 lb N/ton (10 kg N/mt) (Table 12.5).

Table 12.5. Average composition of various manures in dry and liquid form

	Dry form					
	Nitrogen (N)		Phosphorus (P_2O_5)		Potassium (K_2O)	
	lb/ton	%	lb/ton	%	lb/ton	%
Dairy	11	0.5	5	0.2	11	5
Beef	14	0.6	9	0.4	11	5
Hogs	10	0.4	7	0.3	8	3
Chickens	20	0.8	16	0.7	8	3
	Liquid form (per 1000 gal)					
	lb	%	lb	%	lb	%
Dairy	26	0.3	11	0.1	23	0.2
Beef	21	0.2	7	0.1	18	0.2
Hogs	56	0.6	30	0.3	22	0.2
Chickens	74	0.7	68	0.7	27	0.3

Source: Adapted from Miller, 1984.

Regardless of the source, only about 50% of the manure N will be available to the crop during the first year. Applying sufficient livestock wastes to meet plant N needs may result in excess applications of P and K.

Release from the Organic Pool

In addition to inorganic forms (NH_4^+ and NO_3^-), N is also present in a number of organic forms in soil. Such organic N is generally not directly available to plants. Release of this N into plant-available forms (**mineralization**), largely driven by soil microorganisms, can be as much as 600 lb N/acre/year (672 kg N/ha/year). This might represent only 5–10% of the total soil N.

Atmospheric Deposition

Atmospheric deposition of N occurs when lightning converts N_2 into water-soluble forms of N, and also from nitrogenous air pollutants associated with industrial regions. Deposition rates are low in the Midwest, but can be 20 lb N/acre/year (22 kg N/ha/year) in industrialized regions of the USA, and are reported to be 50 lb N/acre/year (56 kg N/ha/year) in Europe.

Volatilization

Some fertilizer N is converted to ammonia (NH_3) and lost to the atmosphere. This is environmentally undesirable, it increases the cost of meeting plant N needs, and it reduces the benefits of fertilization. In extreme cases, all of the applied product can be lost by volatilization. Nitrate-containing N fertilizers are less subject to **volatilization** losses than are urea or ammonium-N sources, and may be preferred in warmer climates or seasons.

Nitrification and Denitrification

Nitrification and **denitrification** are microbial processes that can account for 30% of soil N losses. Nitrification is the oxidation of NH_4^+ to NO_3^-, and denitrification is the reduction of NO_3^- to N_2O, NO_2, and N_2, which are gases that are released into the atmosphere. Usually the rates are fairly low, but they can be high (up to 3% of total N) in dung and urine patches. Denitrification is also high in wet soils, and can be reduced by drainage.

Leaching

Nitrate is readily soluble in water and, since it has a negative charge, it is not held by the soil cation-exchange sites. Thus the leaching of NO_3^- into streams and groundwater can be a significant hazard of excessive agricultural N use. Perennial forage crops are generally more effective than annual crops in taking up and holding N from the soil. There might be opportunities for using forages as a component of riparian management as a filter to intercept surface movement (run-off) of N from adjacent row-cropping areas (see Chapter 19). However, leaching rates can be extremely high if total N inputs (fixation, fertilizer, and manure) exceed 250 lb N/acre/year (280 kg N/ha/year).

Hay, Silage, or Animal Product Nitrogen

Most of the N removed from soil is in the harvested forages and livestock. An alfalfa hay crop yielding 6 tons/acre/year (13 mt/ha/year) could easily contain 360 lb N/acre/year (6 tons/acre/year × 3% N) (6.7 mt/ha/year). The N removed in hay, silage, or animal products must be replaced from the soil or from fertilizer.

Phosphorus

Phosphorus is usually the second most limiting nutrient after N in forage agriculture. In the USA, soils in the South are commonly deficient in P, whereas those in the Midwest and the West vary (Tisdale et al., 1993). A few, including those in central Kentucky, are derived from high-phosphate limestone parent material and exceed 300 ppm of P despite having no history of P fertilization.

Phosphorus is crucial in stand establishment because it is utilized for seedling root growth. This element is also of critical importance for animals, for which the most desirable ratio of Ca:P is between 2:1 and 1:1.

Phosphorus is especially important for legume growth and N fixation. Grasses have a fibrous root system with a large surface area, and are more competitive than legumes with regard to absorbing P present at low levels in the soil.

The plant-absorbable forms of P are HPO_4^{2-} and $H_2PO_4^-$. The various P sources differ in terms of their availability to plants. Phosphorus is relatively immobile in soil, and the accumulation of P is becoming a significant environmental concern.

One unique aspect of P availability is the influence of **mycorrhiza**—an association between soil fungi and plant roots. Plants provide a protected environment for the fungi, while the fungal hyphae grow out into the soil, effectively extending the soil contact area of the root system and facilitating P transport to the plant. The process usually occurs naturally, and there have been few successful attempts to make mycorrhizal fungi available commercially.

Phosphorus Status and Components of the Phosphorus Balance

Compared with N, the soil P balance is relatively simple. Measurement of plant-available soil P by the Truog and Olsen tests gives a reliable prediction of potential forage response. Since excess P is lost from the soil through crop removal or soil erosion, it is recommended that available soil P status should be built up to 30, 35, and 40 lb P/acre (34, 39, and 45 kg *P*/ha) in high-, medium-, and low-P-supplying regions, respectively. Plant growth is highly dependent on soil P status.

Soil P status = (imported P + release from soil and soil organic matter) − (leaching + removal in plant and animal products).

Fertilizer Phosphorus

The P content of commercial fertilizer is measured as P_2O_5 (see "Common Fertilizer Conversions"). The percentage of P_2O_5 in several commercial P fertilizers is as follows: superphosphate, 20%; triple superphosphate, 45%; rock phosphate, 41%. There are some additional sources of P that may also contain N. Examples of these include the following: diammonium phosphate, 53% P_2O_5 and 21% N; monoammonium phosphate, 48% P_2O_5 and 11% N; ammonium phosphate sulfate, 20% P_2O_5 and 16% N.

Common Fertilizer Conversions

The nutrient analysis of all fertilizers is expressed as the relative percentages of N, P_2O_5, and K_2O. When calculating fertilizer application rates it is frequently necessary to calculate the actual amount of the nutrient applied. In this case a conversion factor of 0.436 is required to convert percentage values from P_2O_5 to P, and a conversion factor of 0.83 is required to convert percentage values from K_2O to K.

Example

(A) A common NPK fertilizer composition is 6-15-40. What is the nutrient application rate if the fertilizer is applied at 200 lb/acre?

1. Nitrogen:

   200 lb/acre × 0.06 lb N/lb = 12 lb N/acre

2. Phosphorus:

   200 lb/acre × 0.15 lb P_2O_5/lb × 0.436 lb P/lb P_2O_5 = 13.1 lb P/acre

3. Potassium:

   200 lb/acre × 0.40 lb K_2O/lb × 0.830 lb K/lb K_2O = 66.4 lb K/acre.

(B) Will this fertilizer rate be sufficient to replace the N, P, and K removed by an alfalfa field cut for hay (assume 4 harvests each of 1 T/acre, 4% N, 0.3% P, and 3% K)?

1. Nitrogen:

   8000 lb/acre × 0.040 lb N/lb = 320 lb N/acre

2. Phosphorus:

   8000 lb/acre × 0.003 lb P/lb = 24 lb P/acre

3. Potassium:

   8000 lb/acre × 0.030 lb K/lb = 240 lb K/acre

The fertilizer regime will not replace all of the nutrients removed in the hay.

Note: This system is used throughout the USA. However, in many other countries, fertilizer analysis is expressed as the relative percentages of N, P, K, and S. Therefore in international situations it is essential to check which system is being used.

On low-fertility soils, the fastest (and often the most economical) way to achieve optimum P status is with a large application of "build-up P", in the range of 30–50 lb/acre (34–66 kg/ha) of P_2O_5, which ideally should be incorporated throughout the plow layer. Once the optimum soil P level has been reached, smaller annual applications of "**maintenance phosphorus**" can be used.

Topdressed P for perennial forage crops can be applied at any convenient time, usually after the last harvest or in early fall. Fall application ensures that plants have sufficient P going into the winter to maintain healthy root structure and early spring vigor. Most of the P in commercial fertilizers is highly water soluble. Fertilizers with water solubility higher than 75–80% will not increase dry matter yields more than will those with water solubility levels of 50–80%. Water solubility is more important when band-placing small amounts of fertilizer to stimulate early seedling development; at least 40% of the applied P should be water soluble for acid soils, and preferably 80% for calcareous soils. High water solubility is desirable for calcareous soils, especially those low in available P.

Numerous researchers have reported the effect of P on various forage crops (Ludwick and Rumberg, 1976) (Fig. 12.1). In Virginia, based on a 3-year average, alfalfa yields were found to increase from 3.2 tons/acre to 5.2 tons/acre (7.1 mt/ha to 11.6 mt/ha) when 90 lb P/acre (101 kg P/ha) were applied. At the same location, orchardgrass yields were increased with P applied at rates of 22 lb P/acre (25 kg P/ha) and 90 lb P/acre (101 kg P/ha) (Lutz, 1973). Native warm-season grasses and various other grasses also respond to P application (Taliaferro et al., 1975).

About 9 lb/acre (10 kg/ha) of P_2O_5 is needed to increase the soil P test by 1 lb/acre (1.1 kg/ha). For example, a soil with a P level of 18 lb/acre (20 kg/ha) would need 288 lb/acre (323 kg/ha) of P_2O_5 (i.e., [50–18] × 9) over the 4-year period to reach the target of 50 lb/acre (56 kg/ha) of available P. Some soils will fail to reach the desired goal in 4 years with P_2O_5 applied at the suggested rate, whereas others may exceed this goal. Therefore it is recommended that every field should be retested every 4 years.

Manure typically contains 0.2% P, equivalent to 9 lb P_2O_5/ton (4.5 kg P_2O_5/mt). Although this is not a very high concentration, the large volume of manure that can be spread on land over repeated years can result in accumulation of P toward state limits as low as 300 ppm. Since there are relatively few sources of P loss, the amount of P that is being imported to a farm can be easily calculated from the amounts of purchased livestock feed and mineral supplements. This P is deposited on the farm via manure.

Release from Soil and Soil Organic Matter

The release of P from the soil and soil organic matter is complex because of the number of different forms that soil P can have. Most P that is taken up by plants is released from clay minerals. High-P-supplying soils are usually more favorable for rooting throughout the soil profile. The P-supplying power of a soil is related more to the mineralogy than to the soil P status.

Soils may supply low amounts of P for a number of different reasons:

- Low-P parent material, previous P loss during the soil-formation process, or unavailable P due to high pH or calcareous material
- Poor internal drainage, which restricts plant growth and thus root exploration of the soil for P
- Hardpans (dense, compact layers that inhibit root penetration or branching)
- Shallow soil above the bedrock, sand, or gravel
- Drought, strong acidity, or other conditions that may restrict crop growth or reduce rooting depth.

Manure is the main source of soil organic P. Foraging by animals removes herbage in a somewhat uniform pattern, but manure and urine are not uniformly distributed within the field. The efficiency of use of this P is reduced due to this irregular distribution. Producers sometimes use a chain drag or other implement to disperse manure on pastures, but the costs of doing this probably outweigh the benefits.

Hay, Silage, and Animal Product Phosphorus

The P contained in hay, silage, meat, or milk represents the largest removal of this element from a field. For alfalfa hay that yields 6 tons of dry matter/acre/year (13.4 mt/ha) and contains 0.3% P, 36 lb P/acre/year (36 kg P/ha/year) would be removed. A grazing dairy cow that produces 12,000 lb (5448 kg) milk/year with 0.1% P from a 1.25-acre (0.5-ha) land area will export 10 lb P/acre/year (11 kg P/ha/year).

Fields with excessively high soil P concentrations should be managed to maintain a negative P balance. Removal of P in forage and animal products while P inputs are discontinued is the only practical way to lower soil P levels.

Leaching

Phosphorus is relatively insoluble and does not readily leach from the soil. Many soils fix or hold P and minimize both leaching and availability to plants. Fixation varies with the clay mineral composition of the soil (i.e., it is dependent on soil type). Since losses of P from soil are most closely associated with the physical loss of soil particles, P contamination of waterways is usually associated with erosion. The perennial nature of many forage crops ensures complete soil cover in most instances, and offers a high degree of protection of soil from erosion. There is a good case for inclusion of forages in riparian management strategies.

Sulfur

Sulfur is important for maintaining root growth, is needed for protein formation, and is required for N fixation by legumes. Soils on which forage crops respond to S fertilization are mainly confined to the southeastern USA. Increasingly strict Clean Air Acts (1963, 1970, and 1990) have reduced atmospheric S deposition from burning coal, with the result that S may need to be reassessed when making fertilizer recommendations.

Plant-available S is usually in the form of SO_4^{2-}. Most S that is used by plants comes from the air or from organic matter. Sulfur is present in many commonly used fertilizers. The SO_4^{2-} ion is readily soluble, so leaching from soil can occur, but S concentrations are usually low and not a cause for environmental concern. Deposition of excessive amounts of atmospheric S (especially SO_2) from combustion of fossil fuels can lead to acidification of rainfall in heavily industrialized regions.

Soil S status = (fertilizer S + atmospheric deposition
+ release from the soil) − (leaching
+ removal in forage and animal products)

Sulfur deficiencies are most common on highly leached or sandy soils and soils that are low in organic matter. Sulfur may also become deficient under continuous alfalfa production. For alfalfa production, S should be applied if the soil test levels fall below 7 ppm. If the soil test levels are in the range 7–12 ppm, S can be applied on a trial basis, but benefits are unlikely to be obtained when S levels are above 10–12 ppm (Table 12.2). Common soil S tests are not as reliable as tests for P and K, so plant analysis is the best approach for assessment of crop S status.

Limited data are available that show an alfalfa response to S application. In Minnesota on soils that were considered to contain sufficient S, the dry matter yield was doubled when elemental S was applied as gypsum, and yields continued to increase up to a maximum of 70 lb S/acre (78 kg S/ha) (Lanyon and Griffith, 1988). Sulfur deficiency reduced the winter hardiness of the alfalfa stand. Alfalfa yields of over 7.5 tons dry matter/acre (16.8 mt/ha) contained 0.33% S and removed 50 lb S/acre (56 kg S/ha).

The ratio of N concentration to S concentration in legume herbage is an indicator of S nutrition status, and is important in ruminant nutrition. An N:S ratio of 10:1 is considered optimal for forage utilization by animals. If the ratio exceeds 15:1, forage yield and protein production may be depressed.

Cations

Although most mineral nutrients are cations, the key cations for forages are K^+, Na^+, Mg^{2+}, and Ca^{2+}. Sodium (Na^+) is included because although it is not essential for plants, it is essential for livestock. These cations are all held in the soil at cation-exchange sites.

Cation balances are relatively simple. The main inputs are fertilizer (especially K^+), lime/dolomite (especially Ca^{2+} and Mg^{2+}), and release from the soil (especially for K^+); the main losses are in plant and animal products (especially for K^+). These nutrients are water soluble and leaching does occur, but in most cases it is of minimal agricultural or environmental significance.

Nutrient status = (fertilizer and lime + release from the soil) − (leaching + forage and animal products)

Of the cations, K^+ occurs at the highest concentration in plants, and therefore has the greatest potential removal in plant and animal products. For alfalfa that is producing 6 tons of dry matter/acre/year (13.4 mt/ha/year), and assuming 3% K, 360 lb K/acre/year (403 kg K/ha/year) will be removed.

There are three forms of K in the soil:

1. Soluble K is the smallest portion of total soil K. By supplying current plant needs, annual applications of K minimize losses of this element.
2. Exchangeable K is held on the soil colloids and is readily available to plants. This fraction also makes up a small percentage of the total K in the soil.
3. Non-exchangeable K is held within the clay fraction of the soil and is neither soluble nor available to plants. Non-exchangeable K makes up the largest percentage of total K in the soil, except in highly acidic, sandy soils, or on soils with a high organic matter content, where non-exchangeable K levels are relatively low. As soil minerals weather, non-exchangeable K gradually becomes available.

The K content of commercial fertilizer is expressed as the percentage of K_2O (see "Common Fertilizer Conversions"). Potassium is often referred to as "potash" because one early source was the ash from coal furnaces in the steel industry. Today almost all K is applied as KCl (60% K_2O). Usually up to 300 lb/acre (336 kg/ha) of K_2O can be safely broadcast in the seedbed without damaging the seedlings. Large annual applications of K are often split, with half being applied after the last harvest in fall, and half after the first harvest the following spring. Lower rates are often applied in a single application in the fall when other farm activities demand less time, and to aid in the development of cold tolerance. It takes approximately 4 lb/acre (4.5 kg/ha) of K_2O to increase soil test K by 1 lb/acre (1.1 kg/ha).

Potassium has a direct effect on winter survival of alfalfa. In Canada, research showed an increase in the number of plants that survived when K was applied at rates of 100–200 lb/acre (112–224 kg/ha) as K_2O, whether the temperature was 16°F (−20°C) or 25°F (12.6°C). In another study, the number of shoots per alfalfa plant increased in a linear manner as the K fertilization rate was increased (Table 12.6) (Blaser and Kimbrough, 1968).

The main source of Ca is limestone. Liming has dual roles in forage-livestock systems. It reduces soil acidity, while Ca adds strength to plants and is essential for the formation of the bones and teeth of animals. Calcium and

Table 12.6. Effect of K on winter survival of alfalfa and number of living stems per plant, in Ontario

K_2O rate lb/acre (kg/ha)	Winter survival (%) 25°F (−4°C)	Winter survival (%) 15°F (−9°C)	Number of living stems/plant 25°F (−4°C)	Number of living stems/plant 15°F (−9°C)
0 (0)	73	56	2.8	1.9
100 (112)	97	60	3.4	2.6
200 (224)	90	80	3.8	3.0
300 (336)	97	80	4.3	3.8

Source: Adapted from Blaser and Kimbrough, 1968.

P make up about 70% of the mineral matter in livestock, and about 90% of their skeletons. Calcium is needed in particularly large amounts by growing, pregnant, and lactating animals. Most of the Ca needed by livestock can be furnished by forages, but limestone must be added to acid soils to supply the plants with adequate levels of Ca for proper growth.

Magnesium needed in forage production may be supplied by dolomitic limestone or by fertilizers containing Mg, such as sulfate of potash-magnesia. Legume forages generally contain sufficient Mg to meet the needs of animals. Soils that are high in K and low in Mg may produce an increased incidence of hypomagnesemic or grass tetany (see Chapter 16). This disease of cattle can be controlled by feeding extra Mg or by growing legumes in the pasture mixture.

Soil Carbon

Most soil C is in organic matter that has resulted directly or indirectly from the decay and breakdown of plant tissue. Soil organic matter affects plant growth indirectly through its contribution to soil fertility and its contribution to atmospheric CO_2. Organic matter makes a direct contribution to soil fertility through the provision of cation-exchange sites, the release of nutrients, a high surface area:volume ratio that increases water-holding capacity, and improvement of soil structure by acting as a bonding agent between soil particles.

It is difficult to demonstrate a quantitative relationship between soil organic matter content and forage yield. Forest soils have an organic matter content of around 4–6%, and soils derived from natural grasslands and prairies have an organic matter content of around 2–4%. Soil C to a depth of 4 in. (10 cm) is about 18 tons/acre (40 mt/ha). Due to microbial activity on soil C, it can take many years to increase soil organic matter levels by even 1%.

Typical soils have a C:N ratio of between 10:1 and 12:1. Additions of C-rich organic material will increase this ratio and will result in N **immobilization**. Immobilization can be offset by simultaneous additions of N.

Atmospheric CO_2 levels have increased over the last 100 years from 280 ppm to 350 ppm. The cause and consequences of this increase are less certain. Data from the Woods Hole Research Center in Massachusetts suggest that emissions resulting from changed land use have contributed 25% (1.6×10^{15} g) of the increase in atmospheric carbon, with the remaining 75% (5.5×10^{15} g) coming from fossil fuel emissions. Emissions due to changed land use include both the burning of forests to clear land for forage and crop production, and the release of CO_2 from the depletion of soil organic matter that occurs with intensification of land use. A huge research effort is being directed toward investigation of the factors involved in soil C sequestration, and its potential impact on global CO_2.

Soil pH and H^+

One of the most serious limitations to forage production is soil acidity. Soil pH affects plant growth in several ways.

1. The solubility of some metals (e.g., Al and Mn) may increase to toxic levels at very low pH values.
2. The populations and activity of organisms responding to transformations involving N, S, and P may decrease as acidity increases.
3. Calcium may become deficient as the pH decreases, particularly if the soil cation-exchange capacity is extremely low.
4. Symbiotic N fixation is greatly reduced on acid soils (Fig. 12.3). The symbiotic relationship requires a very narrow range of soil pH compared with that necessary for the growth of plants that do not rely on N fixation.

FIG. 12.3. Alfalfa plants sampled from different areas of the same field (in Wooster, Ohio) that have variable soil pH resulting from uneven distribution of lime. Note the effects of soil pH on plant size, the number of stems and roots, and the crown diameter, and the absence of rhizobia nodules on the plant from the more acidic areas. (Photo courtesy of David Barker, Ohio State University.)

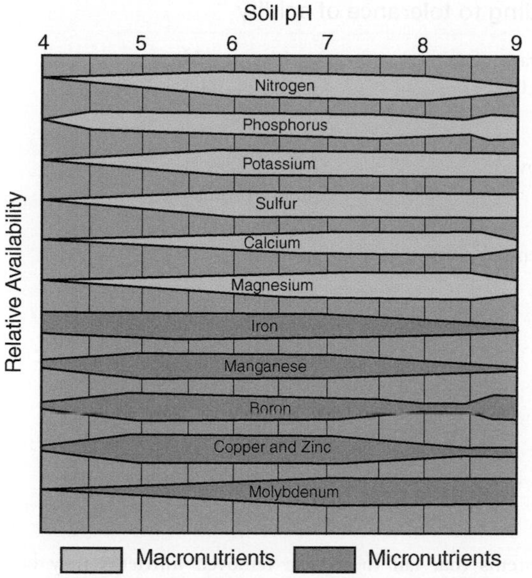

Fig. 12.4. Effect of pH on nutrient availability. Plant availability of most elements is greatest near pH 7 (i.e., neutral pH), and decreases rapidly as the pH drops below 6 for elements such as potassium, magnesium, sulfur, and molybdenum.

5. Many acid soils are poorly aggregated and have poor tilth, particularly if they are low in organic matter.
6. Acid soils have a lower cation-exchange capacity, and consequently reduced nutrient availability.
7. The plant availability of many nutrients is affected by pH (Fig. 12.4). For example, P availability is greatest when the pH is in the range 6.5–7.5. Potassium availability declines very rapidly below pH 6.0. Molybdenum availability decreases dramatically as soil pH decreases, so Mo deficiencies are usually corrected by appropriate liming. Molybdenum is important for the formation and function of N-fixing nodules on legume roots. When the soil pH exceeds 7.0, there is a decrease in the availability of Fe, Mn, B, Cu, and Zn. When the soil pH exceeds 8.5, leaching treatments with gypsum or S may help to lower the pH.
8. A pH of around 6 is ideally suited to earthworm activity. Earthworms promote nutrient cycling and maintenance of soil structure.

Low soil pH—that is, acidity—can result from a number of different factors, including N fertilizer use, high inputs of organic matter, soil weathering (depending on soil type), and N fixation. About 4 lb/acre (4.5 kg/ha) of lime are needed to neutralize the acidity resulting from 1 lb/acre (1.1 kg/ha) of N applied as ammonia or urea.

Forage crops differ in their sensitivity to pH (Table 12.7). For example, alfalfa and smooth bromegrass need pH levels near neutral (6.5–7.0) for optimum productivity, whereas alsike clover, annual lespedezas, tall fescue, and reed canarygrass may tolerate moderate acidity. A regular soil test (every 4 years or so) is the best way to keep a check on soil acidity levels.

It is preferable to apply all of the lime needed to make the required adjustment in pH at once, during establishment of the crop. Liming at this time allows the material to be worked into the soil. This is beneficial because reactivity in the soil does not extend far from the lime particle. Limestone can be applied to a prepared seedbed, disked in, the material plowed down, and the soil worked. Where necessary, lime may be left on the soil surface, but it will probably be less effective in correcting pH during the first cropping season.

Alfalfa is highly responsive to pH. In 5-year averages, more than 4.9 tons/acre (11 mt/ha) are obtained with no fertilizer and a pH of 6.5 or more, compared with 3.2 tons/acre (7.1 mt/ha) at a pH of 5.0–5.5. Adding P, K, and B fertilizer along with liming increases the forage yield to 7 tons/acre (15.7 mt/ha). If legumes are mixed with grasses, liming to the appropriate pH is advisable for the following reasons:

1. To maintain the legume in the stand
2. To meet the needs of bacteria and so maintain the optimum N-fixing ability of legumes
3. To provide optimum availability of other nutrients, which occurs when the pH is increased up to 7.0.

Nutrient Implications for Forage Production Systems

Hay and Silage

Because no nutrients are immediately recycled in urine and manure, a harvested hay or silage crop removes more soil nutrients than does grazing. Typical nutrient concentrations are listed in Table 12.2, and by multiplying by the annual yield of removed forage it is possible to calculate the actual amounts of nutrients required.

The equations in "Common Fertilizer Conversions" can be used to calculate the actual amount of fertilizer required. Where nutrients are not replaced from fertilizer, the nutrient balance will be negative, soil nutrient status will decrease, and yield potential may decrease. There are three situations where fertilization might not be required: (1) where nutrient release from the soil is adequate; (2) where the soil test values are higher than the recommended values; and (3) where animal manure can be used to supply nutrients.

Effects of Grazing

Grazing animals have a number of interactions with the nutrient dynamics of pastures. The most obvious is from

Table 12.7. Classification of forage crops according to tolerance of acidity

Tolerant of acidity (optimum pH 5.5–6.0)	Moderately tolerant of acidity (optimum pH 6.0–6.5)	Sensitive to acidity (optimum pH 6.5–7.0)
Alsike clover	Crimson clover	Alfalfa
Birdsfoot trefoil	Ladino white clover	Sweet clover
Lespedeza	Red clover	Smooth bromegrass
Vetch	Orchardgrass	Reed canarygrass
Kura clover	Annual and perennial ryegrass	Barley
Tall fescue and meadow fescue	Timothy	
Kentucky bluegrass	Dallisgrass	
Redtop and bentgrass	Chicory	
Sudangrass	Corn and wheat	
Bahiagrass		
Bermudagrass		
Oats, rye, pearlmillet, soybeans		

Source: Adapted from Miller, 1984.

manure and urine returned back to the soil. Since 80% or more of most nutrients pass through the animal, this process is a significant efficiency in favor of grazing.

Grazed nutrients removed from throughout a field are deposited onto only 5–10% of the field. This pattern increases soil variability. Forage benefits to high-nutrient areas do not compensate for reduced yields in areas that are missed. Soil samples in pastures should not include these areas of concentrated manure return, as high values from these areas could lead to overestimation of likely forage production.

Perhaps as an evolutionary adaptation to the effects of predators, animal defecation occurs less often while grazing and is more frequent when animals are ruminating or resting. Animals have a higher rate of defecation near watering and sleeping or resting areas. There is thus a concentration of nutrients in the vicinity of watering points and areas of shade, and there is a good argument for having moveable watering points. In hill land, nutrients will be moved from steeper areas to flatter (resting) areas. One of the benefits of rotational stocking is the reduction of nutrient redistribution by stock. The more frequently animals are moved, the closer nutrients (manure) will be deposited to where they were grazed. Nutrient movement is similar for sheep and cattle.

Fertilizer

Modern fertilizers are highly soluble, and the nutrients are readily available to plants. Organic fertilizers such as manure may continue to release nutrients for up to 5 years.

One recent trend is the emergence of organic farming (Fig. 12.5). Definitions of organic farming vary, but a common element is dependence on organic rather than manufactured products, including fertilizers. The principles of nutrient management are as applicable in organic agriculture as they are in conventional agriculture. Organic systems that fail to replace removed nutrients may be unsustainable. Nutrient deficiencies in organic systems can be redressed using organic fertilizers. Organic fertilizers are generally less soluble than mineral fertilizers, and will usually take several years to reach the same equilibrium in supply of nutrients.

Fertilizers for forage production are usually broadcast, often by a rotary spinner mounted on a truck or tractor, and in exceptional cases by airplane or helicopter. The global positioning system (GPS) technologies available today have led to interest in more precise spatial application of required crop nutrients (see "Fertilizer and Precision Agriculture").

Most fertilizers are granular formulations. Some are processed from mineral sources (e.g., elemental sulfur, rock phosphate, or superphosphate) and are available in a

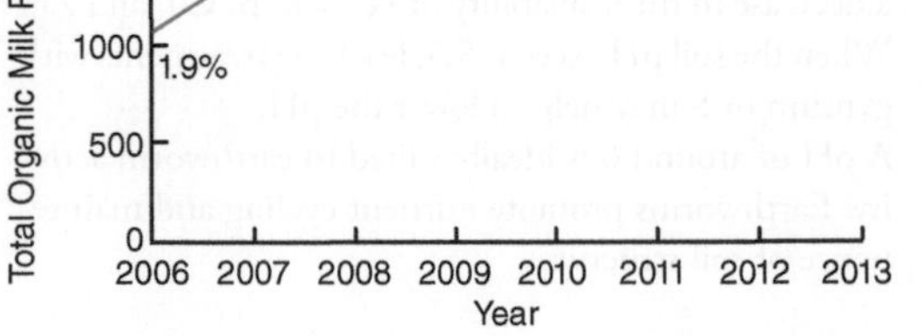

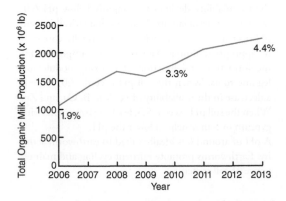

Fig. 12.5. US total organic milk production for the period 2006–2013. Data are expressed as a percentage of total US milk production. (*Source:* US Department of Agriculture, Economic Research Service.)

Fertilizer and Precision Agriculture

Reliable and affordable global positioning system (GPS) technology allows us to produce maps of nutrient distribution within fields. These maps are useful for interpreting within-field variability in crop or forage yield, as areas with higher fertility tend to have higher yield, and low-fertility areas tend to have lower yield (e.g., Fig. 12.6).

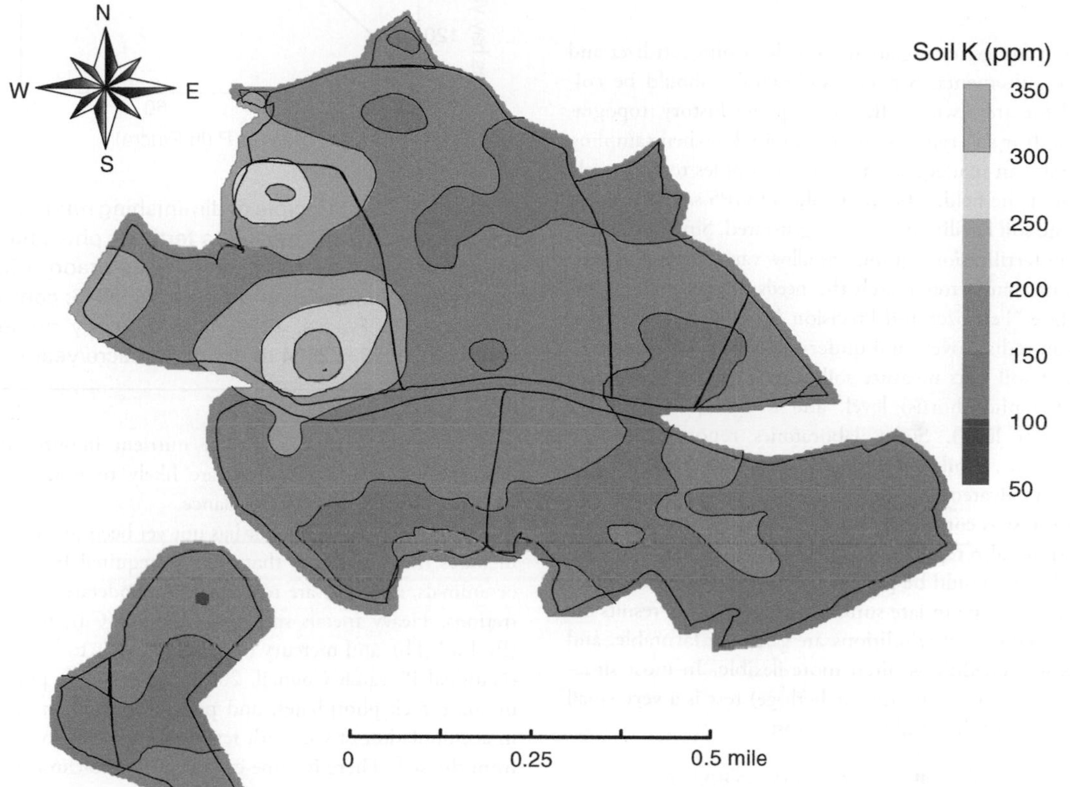

Fig. 12.6. Geographical information system (GIS) and global positioning system (GPS) technologies allow detailed mapping of soil and vegetation characteristics. This map illustrates soil-exchangeable K levels across a 400-acre pasture experiment.

Fertilizer spreaders equipped with a GPS unit can vary the fertilizer application rate. This is called "precision agriculture" because it involves the precise application of an exact rate of fertilizer based on its spatial location. The benefits of this technology are optimum production and economic responses from the fertilizer used, with minimum losses to the environment.

The fertilizer-spreading strategy depends on the objectives of the farmer.

- One strategy might be to grow a more uniform crop. This might require more fertilizer on less fertile areas, and less on fertile areas. Such a strategy will not maximize yield.
- An alternative strategy might be to put the fertilizer in areas of high yield, with the aim of getting the best production from areas with high yield potential. Such a strategy might result in excessive nutrient application if the yield resulted from high-nutrient patches.
- The most economical strategy would be to apply the fertilizer to the areas with the greatest potential to increase yield. Such a strategy requires spatial maps of nutrient distribution and crop yield, as well as knowledge of the yield response curve for the species of interest.

ground or powdered formulation. Perennial forage stands give few opportunities for incorporation, so fertilizers must usually be topdressed. This often leads to a marked decrease in fertility with depth, which can decrease deep rooting of forages.

Nutrient Testing

Soil and forage testing are used to determine fertilizer and lime requirements. Separate soil samples should be collected for areas with different cropping history, topography, and/or soil type. A common mistake when sampling is to take an inadequate number of samples to accurately represent the field. The availability of GPS satellites allow accurate soil fertility maps to be generated. Similar capabilities on fertilization equipment allow variable rates of each nutrient that better match the needs of each part of the field (see "Fertilizer and Precision Agriculture"). This system can reduce over- and under-application of nutrients.

Most soil tests measure soil acidity, the Bray P (plant-available phosphorus) level, and K (the plant-available potassium level). Some laboratories report Mehlich-3 phosphorus, a soil test that agrees well with the Bray P soil test. On calcareous or high-pH soils the Olsen phosphorus soil test is commonly used. Organic matter content is sometimes also reported in soil tests.

Soil tests should be performed every 3 to 4 years. They are usually done in late summer or fall when K results are most reliable, soil conditions are generally favorable, and the work schedule is often more flexible. In most situations, the cost of the soil (or herbage) test is a very small proportion of the total fertilizer cost.

Farm Financial Implications of Forage Fertilization

Maximum forage yield occurs at high concentrations of soil nutrients. However, maintaining these levels might not be economic, or might have undesirable environmental effects. Optimum forage yield will occur at the soil nutrient concentration that is economically attainable and, ideally, avoids environmental damage.

Optimum fertilization rates are lower than those that give maximum forage yields. This is an example of the principle of diminishing returns, in which responses to initial increments of fertilizer (soil nutrient status) are large, and responses to further applications are smaller (Fig. 12.7). The optimum soil nutrient status is usually defined as the level that gives 90% of the maximum yield.

Nutrients and the Environment

In this chapter we have raised the issue of agricultural nutrients escaping to the environment, including non-point source losses to waterways as well as gaseous losses such as NH_3, NO_2, CH_4, and CO_2. In addition to causing possible environmental damage, the potential benefits of the fertilizer to forage crops are lost. The nutrient balance approach described in this chapter provides a quantitative method of determining whether nutrient imports and exports from a forage system are likely to result in a positive or negative nutrient balance.

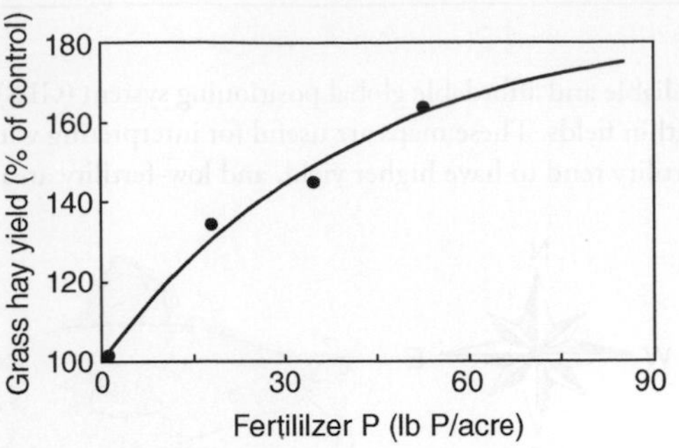

Fig. 12.7. Principle of diminishing returns for relative yield (% of control) to fertilizer phosphate, for grass hay under irrigation in Colorado. (Calculated from Ludwick and Rumberg, 1976; control yield at 0 lb P/acre, 1972 = 6870 lb dry matter/acre/year, 1973 = 2904 lb dry matter/acre/year.)

One class of elements that has not yet been mentioned includes those elements that are not required by plants or animals, and that are toxic in even moderate concentrations. Heavy metals such as cadmium (Cd), fluorine (F), lead (Pb), and mercury (Hg) can be toxic to livestock (National Research Council, 2001). Cadmium is present in some rock phosphates, and prolonged use can result in accumulation of Cd, with few options for removing it from the soil. There is some evidence that Pb can appear in animal products (meat and milk) where pastures have been exposed to exhaust fumes. However, the increased use of unleaded fuels has resulted in decreased concern about this potential hazard. The repeated application of industrial effluents to farmland has the potential to result in accumulation of heavy metals, and careful monitoring and recording are recommended.

Forages offer a number of environmental benefits compared with row-crop systems:

1. Often being perennial, they require less soil disturbance, thus greatly reducing erosion.
2. Extensively managed forages have lower nutrient inputs than cropping systems.
3. Perennials have near year-round growth and nutrient uptake.
4. Forages promote greater accumulation of organic matter and organic nutrient sources than row-cropping systems.
5. Legumes can fix N.

Despite these benefits, nutrient loss from forage systems could result from excess fertilizer application, excess manure application, or leakage from systems that are maintained with a high nutrient status. Responsible and informed nutrient management can result in sustainable forage systems that can meet the financial requirements of landowners, their families, and communities, and that can meet the demands for world food production while minimizing undesirable impacts on the environment.

Summary

Soil fertility is one of the most important determinants of forage production, and can be manipulated by producers through the addition of lime and fertilizers. Soils vary widely in their native ability to supply plant nutrients, and soil testing is essential if the optimum type and amounts of fertilizer are to be provided. Different forage species vary widely in their need for various nutrients, and these differences must be considered when making fertilization decisions. Other soil characteristics, such as organic matter and water-holding capacity, are also critically important and are major considerations in the concept of soil quality, which is the capacity of a soil to sustain plant and animal productivity, maintain water and air quality, and support human health.

Questions

1. Calculate the application rate of N, P, and K for 6-15-40 fertilizer applied at 200 lb/acre. (Hint: be sure to adjust for P_2O_5 and K_2O, to calculate actual P and K.)
2. Will the fertilizer rate in Question 1 be sufficient to replace the N, P, and K removed by an alfalfa field cut for hay? (Assume 4 harvests each of 1 ton/acre, 4% N, 0.3% P, and 3% K.)
3. From a nutrient perspective, in what ways are forage systems more sustainable than cropping systems? In what ways are they less sustainable?
4. Prepare a summary table for NO_3^-, SO_4^{2-}, and HPO_4^{2-}, showing their trend in solubility, potential for leaching, potential to accumulate in soil, and concentration in forage.
5. Why is soil nitrogen important for forages? Why is it difficult to measure?
6. What are the components of the nutrient balance for P, and what might be the consequences of this balance being positive?
7. Why might fertilization to achieve maximum yield be uneconomic?
8. What are the potential sources of N, P, and K in an organic forage system?
9. What is precision agriculture and what are its implications for fertilizer application?
10. What are the advantages and disadvantages of spreading manure onto forages?

References

Allaway, WH. 1968. Agronomic controls over the environmental cycling of trace elements. Adv. Agron. 20:235–274.

Blaser, RE, and EL Kimbrough. 1968. Potassium nutrition of forage crops with perennials. In VJ Kilmer, SE Younts, and NC Brady (eds.), The Role of Potassium in Agriculture, pp. 423–445. American Society of Agronomy, Madison, WI.

Follett, RF, and SR Wilkinson. 1995. Nutrient management of forages. In RF Barnes, DA Miller, and CJ Nelson (eds.), Forages: The Science of Grassland Agriculture, Volume II, pp. 55–82. Iowa State University Press, Ames, IA.

Grace, N. 1994. Managing Trace Element Deficiencies: The Diagnosis and Prevention of Selenium, Cobalt, Copper and Iodine Deficiencies in New Zealand Grazing Livestock. AgResearch, Palmerston North, New Zealand.

Karlen, DL, MJ Mausbach, JW Doran, RG Cline, RF Harris, and GE Schuman. 1997. Soil quality: a concept, definition, and framework for evaluation. Soil Sci. Soc. Am. J. 61:4–10.

Karlen, DL, CA Ditzler, and SS Andrews. 2003. Soil quality: why and how? Geoderma 114:145–146.

Lanyon, LE, and WK Griffith. 1988. Nutrition and fertilizer use. In AA Hanson, DK Barnes, and RR Hill, Jr (eds.), Alfalfa and Alfalfa Improvement, pp. 333–372. Agronomy Monograph 29. American Society of Agronomy, Madison, WI.

Ludwick, AE, and CB Rumberg. 1976. Grass hay production as influenced by N-P top dressing and by residual P. Agron. J. 68:933–937.

Lutz, JA, Jr. 1973. Effects of potassium fertilization on yield and K content of alfalfa and on availability of subsoil K. Commun. Soil Sci. Plant Anal. 4:57–65.

Miller, DA. 1984. Forage fertilization. In Forage Crops, pp. 121–160. McGraw-Hill, New York.

Miller, DA, and GH Heichel. 1995. In RF Barnes, DA Miller, and CJ Nelson (eds.), Forages: An Introduction to Grassland Agriculture, Volume I, pp. 45–53. Iowa State University Press, Ames, IA.

National Research Council. 2001. Nutrient Requirements of Dairy Cattle, 7th ed. National Academies Press, Washington, DC.

Spears, JW. 1994. Minerals in forages. In GC Fahey, M Collins, DM Mertens, and LE Moser (eds.), Forage Quality, Evaluation, and Utilization, pp. 281–317. American Society of Agronomy. Madison, WI.

Taliaferro, CM, FP Horn, BB Tucker, R Totusek, and RD Morrison. 1975. Performance of three warm-season perennial grasses and a native range mixture as influenced by N and P fertilization. Agron. J. 67:289–292.

Tisdale, SL, WL Nelson, and JD Beaton. 1993. Elements required in plant nutrition. In Tisdale, SL, WL Nelson, and JD Beaton (eds.), Soil Fertility and Fertilizers, 5th ed., pp. 59–94. Macmillan, New York.

Watson, ME. 1995. Research Extension Analytical Laboratory Soil Test Summary. Ohio Agricultural Research and Development Center (OARDC), Ohio State University, Wooster, OH.

Useful Websites

www.cas.psu.edu/docs/CASDEPT/AGRONOMY/EXTENSION/NMPennState/NMPennStateHome.html

www.cahe.nmsu.edu/pubs/_a/a-137.html

www.ppi-ppic.org/

CHAPTER

13

Integrated Pest Management in Forages

R. Mark Sulc, William O. Lamp, and Michael Collins

Integrated pest management (IPM) plays a key role in the practice of **grassland agriculture** because it effectively links the agronomic objectives (high yield, high **forage quality**, and long stand life) to the economic, environmental, and social goals of sustainable agriculture. Correctly applied, IPM can make forage production more profitable while responding to the environmental and health concerns of farmers and society in general.

IPM developed out of the need to suppress important pests while minimizing the use of chemical pesticides. Following World War Two, new pesticide materials offered such simple and initially highly effective methods of pest control that they were quickly adopted and widely used. As pests adapted through natural selection and became resistant to the new materials, and as scientists, farmers, and the public in general became more aware of the environmental risks and increasing economic costs associated with their use, it became clear that more effective and less hazardous options were needed. By the 1970s, major research and extension programs developed a new synthesis of old and new pest management methods that became known as IPM (Huffaker, 1980). The emphasis was on minimizing, but not necessarily eliminating, the use of pesticides. More recently, the term *ecologically based pest management* has been proposed, to reflect a move away from broad-spectrum pesticides in pest management, instead considering goals of safety, profitability, and durability (National Research Council, 1996). Alfalfa is the first **perennial** forage crop for which IPM programs were designed, and it provides an excellent model for forage pest management. The objective of this chapter is to illustrate the principles of IPM and their application to forage production systems, using alfalfa as the primary example model.

Definitions

Forage crop *pests* are biological agents that cause economic losses in three ways: (1) by reducing yield, (2) by reducing the quality of the forage, and (3) by reducing the persistence of the forage stand. As shown in Table 13.1, pests have a broad range of natural diversity. In nature, each organism has essential biological functions, and organisms are not classified as harmful or beneficial (Harlan, 1992). Classification as a pest is based on human values. A challenge for pest management is to suppress what is harmful without harming what is beneficial, and nature has not made this easy.

To *integrate* means to bring together all of the parts so that something is made whole or complete. Integrated pest management brings together several different methods in order to reduce the pest load while at the same time seeking to minimize environmental hazards and maximize the economic benefits for producers and consumers. Thus the IPM approach is more knowledge based and comprehensive than the pesticide-oriented programs that it is intended to replace. It is also an evolving concept that needs to be considered within a larger context of numerous management practices that must be integrated with the time, labor, and capital resources available to individual managers (Fick and Power, 1992). The dynamic nature of forage pest management systems is summarized in the text box "Dynamics of Forage IPM."

Forages: An Introduction to Grassland Agriculture, Seventh Edition. Edited by Michael Collins, C. Jerry Nelson, Kenneth J. Moore and Robert F Barnes.

Dynamics of Forage IPM

Integrated pest management (IPM) involves all of the decision-making and planning components of the forage crop production system. Losses associated with pests are the result of the interactions between the pest, the crop, and the environment. Integrated pest management can be illustrated as the processes associated with a series of three pest triangles (Fig. 13.1).

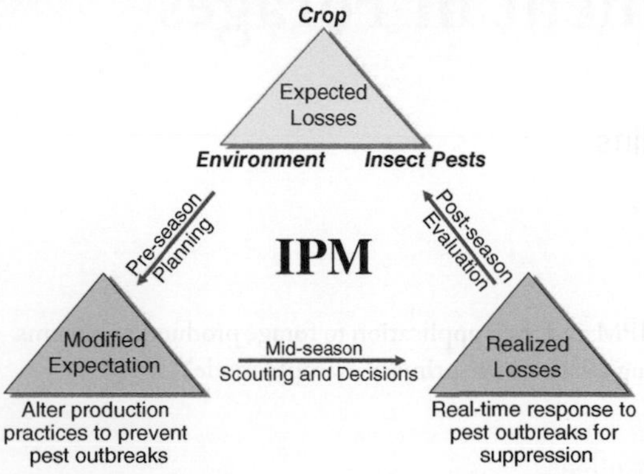

Fig. 13.1. A series of pest triangles that illustrate the concept of IPM.

Before a cropping season begins, the producer anticipates the losses from the major pests that are likely to be present (Table 13.1 and Fig. 13.1). Through pre-season planning, specific measures may be implemented to prevent outbreaks of key pests (e.g., planting a resistant variety, fertilizing to improve crop competition with weeds). Implementing these plans leads to alterations in production practices, which modify the expectations of loss (Fig. 13.1).

As the crop develops, specific environmental or crop conditions, or unusual patterns of pest populations, may result in an unexpected outbreak of a pest. Mid-season scouting of pest abundance helps the producer to keep abreast of developing pest outbreaks, and informs decision making about actions that can be taken to prevent pest populations from exceeding thresholds above which economic losses are expected. However, in IPM the use of pesticides is limited to emergency situations, and the emphasis is on prevention of significant losses from pests.

Documenting the actual losses (the "Realized Losses" triangle in Fig. 13.1) due to pest damage at the end of the season is important for post-season evaluation, which in turn affects the pre-season planning and expected losses for the next season.

Table 13.1. The various categories of agricultural pests in alfalfa

Pest category	Examples
Pathogens or diseases	
Viruses and mycoplasma-like agents	Alfalfa mosaic virus
Bacteria	Bacterial wilt of alfalfa (*Clavibacter michiganensis* subsp. *insidiosus*)
Fungi	Phytophthora root rot of alfalfa (*Phytophthora medicaginis*), spring black stem and leaf spot (*Phoma medicaginis*)
Bacteria	Bacterial wilt (*Clavibacter michiganensis* subsp. *insidiosus*)
Nematodes	Alfalfa stem nematode (*Ditylenchus dipsaci*), root-knot nematodes (*Meloidogyne* spp.), root-lesion nematodes (*Pratylenchus* spp.)
Arthropods	
Insects	Alfalfa weevil (*Hypera postica*), potato leafhopper (*Empoasca fabae*)
Mites	Two-spotted spider mite (*Tetranychus urticae*)
Mollusks	
Slugs and snails	Gray garden slug (*Deroceras reticulatum*)
Vertebrates	
Birds	Red-winged blackbird (*Agelaius phoeniceus*)
Mammals	Rodents, rabbits, wild ruminants
Weeds	
Monocots	Yellow foxtail (*Setaria glauca*), yellow nutsedge (*Cyperus esculentus*)
Dicots	Pigweed (*Amaranthus retroflexus*), dandelion (*Taraxacum officinale*)

Principles and Practices

Detailed discussion of the principles of IPM is beyond the scope of this chapter, but key concepts related to forage pest management can be summarized as follows:

1. *Many **species** of organisms inhabit forage **ecosystems**, but only a few are pests.* Only specific combinations of host crops, environments, and potential pests can cause economic losses (Fig. 13.1, "Expected Losses" triangle). Since most species are actually beneficial in some way, pest management methods should target real pests and conserve other organisms in the forage environment.
2. *Populations of all pests are naturally suppressed, and management practices should be adopted to foster natural suppression.* **Biotic** factors limit population growth of all species. Crop management practices that encourage natural enemies and competitors will aid the suppression of pest populations.
3. *Crop ecosystems can be managed to enhance pest suppression.* The condition of a crop and its physical environment influence pest populations, and crop management practices often affect losses induced by pests. Knowledge of pest life cycles and ecology is therefore critical for predicting how changes in crop management will alter pest problems.
4. *Each pest population can be managed by various methods, and generally the optimal approach is to integrate two or more methods.* Because of the complexity and unpredictability of forage systems, a single approach is rarely sufficient for permanent suppression of key pests below economic levels. Instead, IPM strategies combine various methods, including both preventive and responsive practices (Table 13.2).
5. *When a pest population density warrants additional suppression, responsive measures may be applied to limit crop damage.* The density of a pest that justifies the cost of a rescue treatment is called the **economic injury level (EIL)** (Pedigo et al., 1986). When densities are expected to exceed the EIL, action may be taken to prevent any additional damage. The pest density at which a responsive measure should be implemented is called the **economic threshold** or **action threshold**. Development of EILs and thresholds has become a standard part of IPM, thus emphasizing the prevention of economic losses caused by pests. However, the final decision to use pesticides should be based on health and long-term environmental factors as well as economics.
6. *Responsive treatments may produce unexpected and undesirable results.* Examples include non-effective control practices, development of pest **resistance**, breakdown of defense mechanisms of crop cultivars (i.e., host resistance), resurgence and outbreaks of minor pests following pesticide applications, and harm to non-target organisms, including humans.
7. *Pesticide usage results in pest evolution, so pesticides become ineffective.* Every major category of pests has now developed genetic resistance to pesticides that were once effective (Jutsum et al., 1998). Effective pesticides are important in pest outbreak emergencies, but their use increases selection pressure and speeds up the development of resistance against them. This is especially true with repeated and widespread use of the same pesticide. Therefore pesticide usage should be limited in order to retain the pesticide option for future emergencies.

Table 13.2. Methods and examples of pest control

Method	Example
Natural (none)	No human-directed changes; reliance is on natural processes that maintain pest populations below economically damaging levels
Biological control	Changing the pest's biological environment so that the natural predators of the pest are increased (sometimes by raising and releasing natural predators)
Host resistance	Changing the pest's host crop by means of plant breeding so that it is less affected by the pest (usually with improved forage cultivars)
Cultural control	Changing the pest's physical environment so that the pest is less successful (e.g., crop rotations, species mixtures, fertilization, timely harvesting, soil drainage, control of compaction by traffic)
Direct control	Changing the pest so that it dies quickly or dies without completing its life cycle, either by using chemicals (e.g., pesticides, hormones) or by physical means (e.g., trapping, sterilization, cultivation)

These principles of IPM show that specific pest management practices may or may not alleviate the negative impact of pests on crop yield, quality, and persistence. Knowledge of the effect of key pests on forage crops, critical periods leading to losses, and factors leading to pest outbreaks is needed to design optimal IPM programs for protecting forage crops. Crop management practices, such as species and cultivar selection, fertilization, harvesting schedule, and crop rotation, may also be manipulated to significantly reduce pest losses.

Steps in an IPM Program

Based on the principles listed above, the following steps are necessary components of IPM programs:

1. *Recognition of the diversity of organisms within the forage agroecosystem.* The attitude that all insects, weeds, and fungi are pests to be eliminated must be avoided, and instead the focus should be on improving understanding and education. Most non-crop organisms in an agroecosystem help to maintain a favorable balance and are beneficial in terms of human goals. Steps can be taken to protect and foster this diversity in forage systems.
2. *Identification of pest species.* Pest management is necessary because a few organisms can become very serious pests. We must be able to identify those organisms and assess their potential to cause economic injury in individual crops and locations.
3. *Understanding pest and crop biology.* Effective pest management with minimum impact on non-target organisms requires knowledge of pest biology and of the related effects of the environment and crop development. Certain life stages of a pest are more amenable to effective suppression than others. Certain environments and crop stages promote or retard development of a pest population. The more we know about a pest, the more we know about how to manage it.
4. *Selection and use of available preventive suppression strategies.* Many preventive pest suppression methods are available, including appropriate site selection for the crop, sowing pest-resistant species or cultivars, rotating crops to interrupt pest life cycles, fertilizing with the required nutrients to promote crop vigor, preventing pest movement from infested to non-infested fields via **hay**, manure, and infested machinery, and limiting pesticides to allow build-up of natural biological suppressive agents. Using IPM concepts involves anticipating problems and correctly implementing preventive suppression methods so that pest-related emergencies are usually avoided.
5. *Monitoring pest populations and implementing responsive rescue practices.* Scouting or sampling for pest abundance, distribution, and developmental stage allows growers to anticipate problems before EILs are reached. Many IPM programs involve actual sample collection and analysis procedures, while others are based on weather data and computer models. In either case, when action is needed, options for forages may include early harvesting or pesticide applications. An increasing number of growers use professional crop management consultants to provide systematic monitoring of pests and interpretation of scouting results. Such professional services often provide additional information relevant to soil, crop, and livestock management.
6. *Evaluation and refinement.* The results of a pest management program need to be evaluated with regard to the producer's objectives. Successes and failures should be reviewed and adjustments made that correct past deficiencies and incorporate new developments. "Check plots" can help to demonstrate what might have happened with some other management approach, and good field records can pinpoint problems in space and time.

Examples of Implementation of IPM Steps

There are many examples of how to effectively apply the IPM steps outlined above in forage crops. The following account focuses on specific examples of IPM applied to alfalfa, but many of these and other examples apply to other forage crops.

Promote Diversity of Organisms

Alfalfa provides a favorable habitat for many organisms, including natural predators of potentially serious pests. Farm habitats can be managed to encourage survival of those natural enemies—for example, by providing forested areas near fields, where naturally occurring predators can survive and spread into adjoining crop fields (Landis et al., 2000).

Crop rotations can be used to improve biological diversity on the farm and at the field level, which helps to limit populations of many pests, particularly by interrupting weed, insect, and pathogen life cycles. For example, corn or another monocot species is an excellent rotational crop for alfalfa because it will tend to harbor different pests to alfalfa, and can utilize the nitrogen fixed by the alfalfa. The land manager should be observant of fields where specific pest problems tend to occur less frequently for a given crop, so that if possible those fields can be prioritized for those crops within the crop rotation strategy on the farm.

Planting different forage species in a mixture is a good method for increasing diversity in the cropping environment. Many forage growers actually prefer to grow mixtures of **grasses** with alfalfa for various reasons, but in relation to IPM this practice helps to suppress weeds and tends to reduce insect pest populations in the forage crop (DeGooyer et al., 1999).

Identify Pest Species

A critical step in an effective IPM program is to identify the pests that are most likely to cause damage to the crop in that particular region. Resources to help to identify pests and their injury to the crop can be found online at state extension service websites or in several publications (e.g., Lamp et al., 2007; Samac et al., 2015). Identification of pests enables anticipation of potential problems, so the crop manager can implement preventive measures and be prepared to apply the most effective rescue treatments when pest outbreaks do occur.

In terms of insect pests, the potato leafhopper is the most consistently damaging insect pest of alfalfa in the eastern USA, but in the western and northwestern USA the alfalfa

Biological Control of Alfalfa Weevil

The alfalfa weevil, *Hypera postica*, was first introduced to the USA in 1904, and became a key pest of alfalfa throughout most of the country during the twentieth century (Fig. 13.2). Relatively few native natural enemies fed on the life stages of the weevil. Important natural enemies include ground beetles (belonging to the family Carabidae) and ladybird beetles (belonging to the family Coccinellidae) (Fig. 13.3). A fungal pathogen, *Zoophthora phytonomi*, was believed to have been accidentally introduced from Europe, and has become a major mortality factor for the larvae (Fig. 13.4). Efforts to help to manage the pest by classical biological control resulted in the release and subsequent spread of over 10 parasitoid species (Radcliffe and Flanders, 1998). Major successes include the wasps *Bathyplectes curculionis* and *Bathyplectes anurus*, which attack the larvae, and *Microctonus aethiopoides*, which attacks the adults. The combination of parasitoid species, the fungal pathogen, and native predators has been effective in suppressing alfalfa weevil abundance throughout much of its range. This pest now rarely reaches population levels that require insecticide rescue treatments in the eastern USA.

Fig. 13.2. Alfalfa weevil eggs (top left), larva (top right), pupae (bottom left), and adult (bottom right). (Reproduced with permission of Scott Bundy.)

Another example of effective biological control occurs in response to increasing pea aphid populations that commonly occur in the spring in alfalfa. Serious pea aphid outbreaks are prevented by increasing activity of natural predators such as ladybird beetles, damsel bugs, parasitic wasps, and fungal pathogens. Avoiding the use of insecticides when they are not economically warranted protects the activity of these natural predators.

Fig. 13.3. Seven-spotted ladybird beetle, one of several common predators of alfalfa weevil. (Reproduced with permission of William Lamp.)

Fig. 13.4. Alfalfa weevil larva infected with the soil-dwelling fungal pathogen *Zoophthora phytonomi*. (Reproduced with permission from University of California Statewide IPM Program.)

weevil is a greater threat in the spring, followed by a complex of caterpillar larvae (alfalfa caterpillar and armyworms) during the warmer months, whereas aphids can be troublesome in the west throughout the year.

A number of diseases and nematodes can cause damage to alfalfa, the relative severity of which varies by region and soil type. Root and **crown** diseases are especially problematic in soils that lack good drainage. Identifying the diseases, insects, and nematodes that are most likely to be problematic on the farm provides the opportunity to select plant varieties with resistance to those specific pests.

Correct identification of weed species is essential in order to select the most effective response strategy. If herbicide applications are necessary for managing the infestation, correct identification of the weed species ensures selection of the most effective herbicides.

Understand Pest Biology

Weed management in alfalfa provides a good illustration of the importance of understanding pest biology. Understanding whether a weed infestation is composed primarily of perennial, biennial, summer **annual**, or winter annual species is critical to choosing effective suppression strategies. For example, if winter annual weeds are a consistent problem on the farm, one of the best suppression tactics is to ensure that the alfalfa **canopy** completely shades the soil surface during the autumn, when winter annual weeds germinate (for further details, see "Preventative Suppression Strategies").

Many perennial weeds are more effectively suppressed by herbicide applications made on warm days in the autumn when the weeds are at an advanced stage of growth but still actively growing. Under those conditions the herbicide will be translocated to the overwintering plant structures, providing more control. The weedy plants that do survive the autumn herbicide application under those conditions will be weakened, making them more susceptible to winterkill. Winter annual weeds are also more effectively controlled by herbicides applied in the autumn than by waiting to apply herbicides in late winter or early spring. In contrast, herbicide applications in late summer or early fall are wasteful and ineffective on summer annual weeds because at that time those weeds are at the end of their life cycle. By that stage, harvesting or clipping to prevent weed **seed** production would be more useful.

Understanding insect pest biology is also critical for developing effective suppression strategies and avoiding unnecessary insecticide applications that kill natural predators. Consider the alfalfa weevil, a snout beetle that normally produces only one generation per year. In northern

Importance of Pest Species Identification: An Example Involving Aphids

Aphids are small, soft-bodied insects that reproduce rapidly and sometimes become significant pests on forage crops. They injure plants both by feeding on the plant sap and by injecting toxic saliva into the plant. Four species are often pests on alfalfa, and identification of the species is important for determining thresholds and selecting resistant varieties. Two species, the pea aphid and the blue alfalfa aphid, are very similar and require a hand lens for identification. Both appear green and are similar sized, but they differ in their antennae. The pea aphid (*Acyrthosiphon pisum*) has brown antennae with dark rings on every segment (Fig. 13.5A), whereas the blue alfalfa aphid (*Acyrthosiphon kondoi*) has antennae that are uniformly brown from the third antennal segment (from the base) to the tip (Fig. 13.5B). In comparison with pea aphids, blue alfalfa aphids can tolerate cooler conditions, feed more commonly on stems rather than leaves, and tend to cause more damage to and stunting of plants. Spotted alfalfa aphid (*Therioaphis maculate*) (Fig. 13.5C) and cowpea aphid (*Aphis craccivora*) (Fig. 13.5D) are more distinctive, but each species is associated with different biological characteristics and different thresholds for decision making. Thus identification is important for pest management.

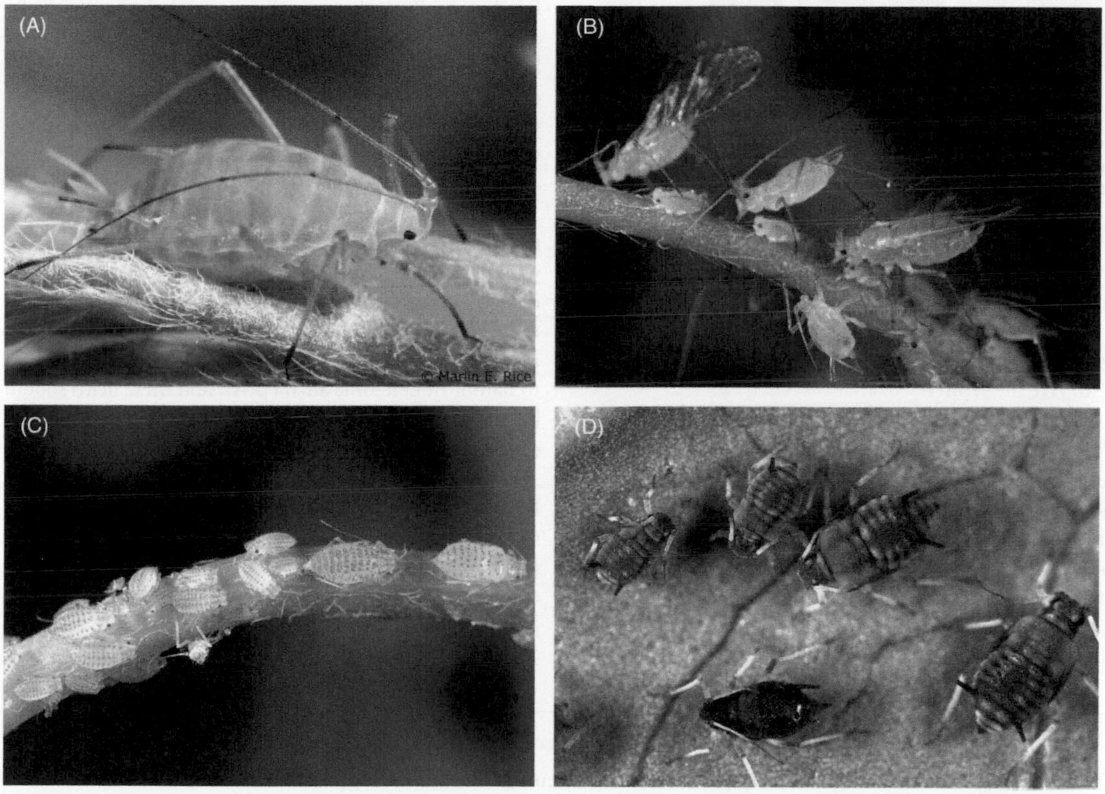

Fig. 13.5. Four species of aphid pests found in alfalfa: (A) pea aphid; (B) blue alfalfa aphid; (C) spotted alfalfa aphid; (D) cowpea aphid. (Reproduced with permission from University of California Statewide IPM Program.)

states (e.g., Minnesota, Wisconsin, New York) the weevil survives the winter only as an adult, which lays eggs in alfalfa stems during spring. The larvae that hatch from those eggs consume increasing amounts of foliage as they grow (Fig. 13.6). Peak damage usually occurs when the alfalfa is in bud stage.

Harvesting at bud stage just before severe defoliation is often the optimal action to suppress alfalfa weevil populations. In more southern states (e.g., Maryland, Oklahoma), eggs deposited in alfalfa stubble in autumn and winter may survive until spring. The larvae that hatch from those overwintering eggs emerge when alfalfa plants are less than 8 in.

Fig. 13.6. Foliar feeding damage caused by alfalfa weevil larvae. (Reproduced with permission of Scott Bundy.)

(20 cm) tall, and can cause serious defoliation while the crop is still **vegetative** (Kuhar et al., 2000). Therefore rescue treatment with insecticide is more likely to be necessary in southern regions than in the north where eggs do not survive the winter.

Select Preventative Suppression Strategies

After identifying the most likely pest problems for the region and gaining an understanding of their biology, the alfalfa manager is well prepared to select effective preventative suppression strategies that limit damage by key pests on the farm. Common preventative strategies in alfalfa include planting varieties that are resistant to common pests in the region, selecting fields with good drainage, rotating crops, using management practices that promote vigorous crop growth, and using sanitary harvesting procedures.

Plant breeders now routinely develop alfalfa varieties with multiple pest resistance. Such varieties provide an excellent front line of defense against potential problems, while at the same time reducing the use of pesticides that can harm natural predators of those pests (i.e., they promote diversity of organisms, which is Step 1 of IPM). Some forage grasses, such as tall fescue, may be infected with fungal endophytes that improve resistance to certain insects. The fungal endophyte commonly found in tall fescue produces alkaloids that are toxic to some sap-feeding insects and reduce damage by stem miners. Unfortunately, the alkaloids produced are also harmful to livestock to varying degrees. Recently it was discovered that other naturally occurring endophytes do not produce the alkaloids that are harmful to livestock. Tall fescue varieties are being introduced that contain these "novel" or "friendly" endophytes, which provide the benefits of improved plant performance without having harmful effects on livestock. Further details can be found in Chapter 16.

Proper field selection is critical for establishing and maintaining vigorous and productive stands of alfalfa, which improves crop resilience against pests. Planting disease-resistant alfalfa varieties in well-drained soils greatly reduces the risk of many devastating diseases. As mentioned earlier, some fields on the farm may be less likely to harbor troublesome alfalfa pests, and these fields should be prioritized for alfalfa if possible within the crop rotation on the farm.

Good crop management practices will improve the ability of the crop to tolerate or compete with pests. For example, timely spring planting enables alfalfa seedlings to establish and shade the soil to reduce emergence of many summer annual weeds that germinate in late spring and early summer. Timely summer plantings allow seedlings

Biology of a New Invasive Pest: The Bermudagrass Stem Maggot

The bermudagrass stem maggot (*Atherigona reversura* Villeneuve) is a new invasive pest that was first discovered in the continental USA in 2009 in the Los Angeles area, and was then found in the summer of 2010 in Georgia. It spread rapidly throughout the Southeast region of the USA, and infests bermudagrass hayfields, **pastures**, and turf. The life cycle of this pest had to be understood in order to allow effective measures to be developed to manage the damage (Baxter et al., 2014).

Field observations suggest that the life cycle of the bermudagrass stem maggot is similar to that of the sorghum shoot fly (*Atherigona soccata*), which is being used as a working model for this new pest. The life cycle appears to be about 3 weeks, with multiple generations being produced each summer. The adult fly (Fig. 13.7) lays eggs on bermudagrass, and 1–3 days later the small maggots (larvae) hatch and move towards the uppermost plant node, where they burrow into the central **shoot** to feed, causing the uppermost leaves of the plant to die and the damaged stems to stop elongating (Fig. 13.8). The damage becomes visible within 1–3 days after larval feeding begins, giving the field a "frosted" appearance. After feeding, the larvae leave the stem and move to the soil for the pupation stage. The adult flies emerge 7–10 days later.

Once the stem has been damaged by this pest, little or no further yield is added by the damaged plant. Grazing, clipping, or harvesting of the crop is recommended because the larvae do not remain in cut stems, but exit within hours of clipping and move to the soil. The larger larvae will pupate, but the younger larvae presumably die. Even if damage occurs very early in the grass regrowth cycle, cutting or grazing is preferable to subjecting the crop to further damage and allowing the fly population to build.

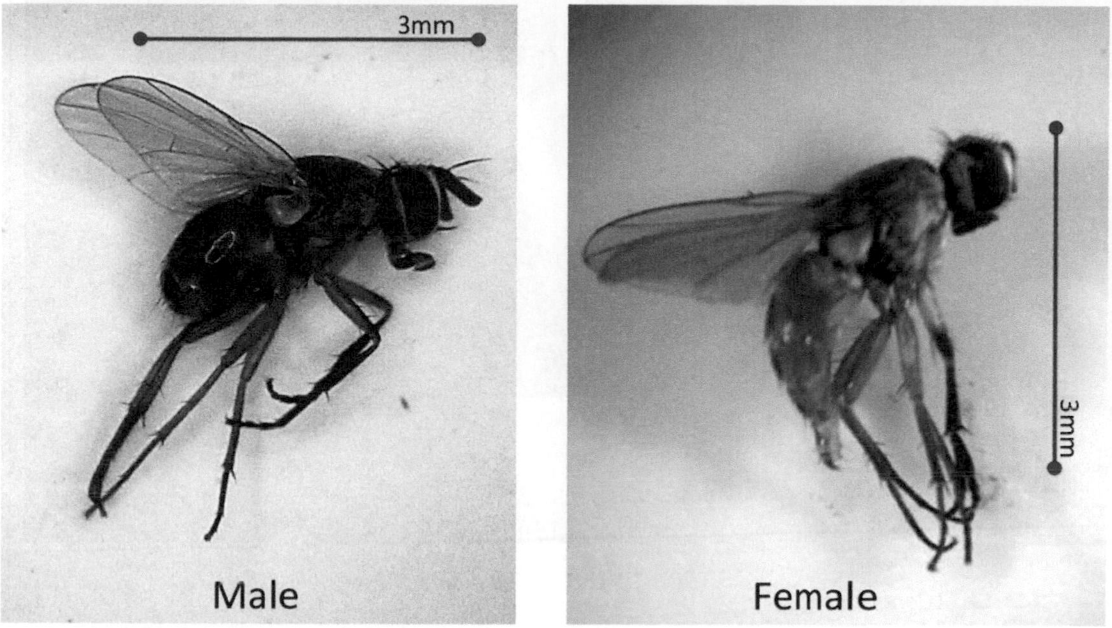

FIG. 13.7. The adult bermudagrass stem maggot is very distinctively colored, with its grey thorax and bright yellow abdomen. The male is shorter than the female due to the ovipositor on the end of the female's abdomen. (Source: Baxter et al., 2014. Reproduced with permission of Plant Management Network International.)

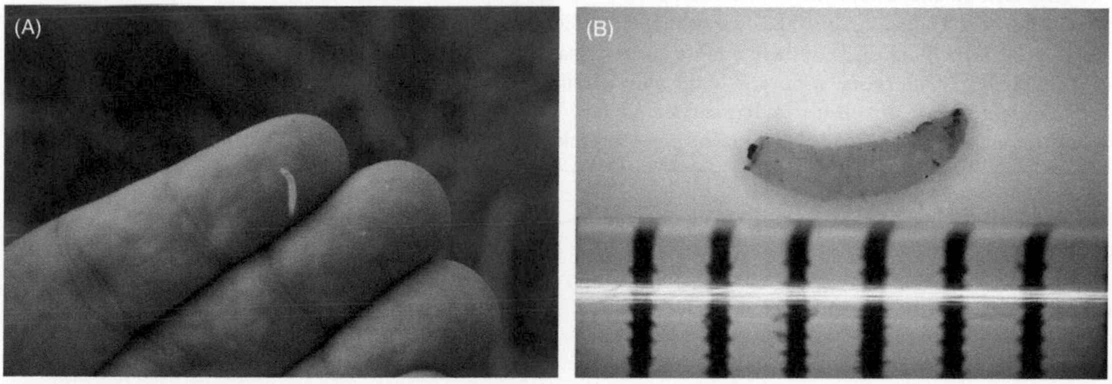

FIG. 13.8. (A) The immature bermudagrass stem maggot larva is somewhat difficult to find in the field. (B) This is partly due to the small size of the larva. (C) In addition, the larva only spends a short amount of time in the stem, and once it has matured it moves to the soil for pupation, leaving a visible hole in the bermudagrass tiller. (Source: Baxter et al., 2014. Reproduced with permission of Plant Management Network International.)

(C)

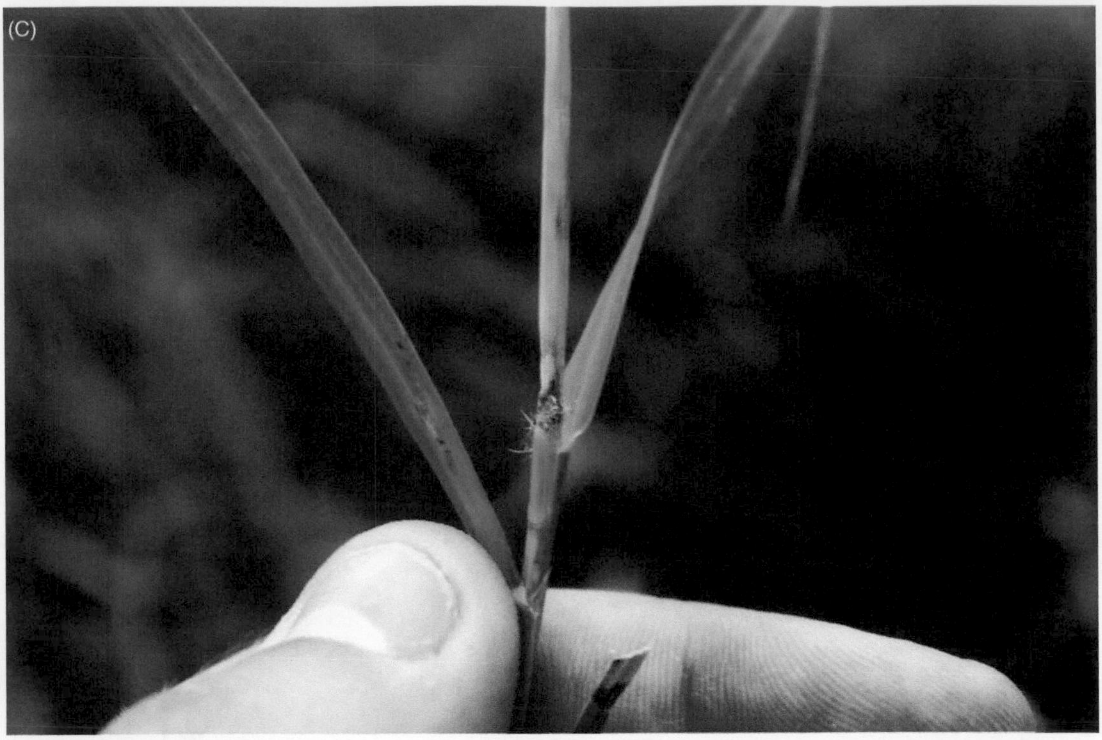

Fig. 13.8. (*Continued*)

Insecticides can be used to suppress the bermudagrass stem maggot, if there is a known history of damage by this pest and the expense of the chemical application(s) is justified by the forage yield saved. Unfortunately, economic action thresholds in response to pest abundance have yet to be established. Chemical suppression is usually not justified for the first harvest of bermudagrass. If insecticides are used, current recommendations are to make applications to regrowth 7–10 days after an affected harvest. This ensures good penetration of the chemical into the canopy (adults tend to stay in the canopy unless they are disturbed), and it suppresses flies that emerged or returned to the field as regrowth began.

Field observations indicate that there is more severe damage in cultivars with fine stems and higher **tiller** numbers. Selecting cultivars with coarse stems and lower tiller density seems to be an effective method for reducing damage. More research is needed to better understand the biology of this pest and the potential for host resistance, and to establish economic action thresholds for managing outbreaks.

to establish and shade the soil before germination of winter annual weed species begins. In established stands, harvest schedules should be adjusted so that there is sufficient regrowth to completely shade the soil during the period when winter annual weeds begin to germinate. This requires a knowledge of the germination requirements and timing of the most common winter annual weeds in the region.

Many weed species are "passive opportunists" in that they invade empty spaces created by poor stand establishment or death of the forage plants in established stands. Maintaining vigorous and thick forage stands is critical for prevention of weed infestations (Fig. 13.9). Including an aggressive perennial grass such as orchardgrass in the forage mixture serves to fill open spaces and inhibits dandelion invasion in alfalfa (Spandl et al., 1999). Correct **soil pH** and adequate fertility with good harvest management go a long way towards reducing weed problems by maintaining vigorous forage stands that are able to tolerate and recover from any insect or disease outbreaks that may occur.

Minimizing traffic of heavy equipment over the field and avoiding such traffic when soils are wet or soft will reduce damage to alfalfa crowns (Fig. 13.10), thus lowering the risk of infection by pathogens. Diseases frequently

Fig. 13.9. Weed invasion in an alfalfa cultivar with poor stand density (center) after 4 years. Weed invasion is minimal in cultivars with excellent stand density (left and right). (Photo courtesy of R. Mark Sulc.)

develop in heavily compacted areas. Pathogens may also spread by movement of hay, manure, and infested machinery from field to field. Harvesting the newest and healthiest stands before moving to older stands, and cleaning equipment thoroughly after its use in older stands where diseases are present, can help to prevent the spread of pathogens to healthy stands.

Timely harvesting will help to contain the build-up of diseases and insect pests in the crop canopy. Timely harvesting reduces leaf loss (i.e., preserves better forage quality), and is especially helpful in preventing the build-up and spread of disease-causing inoculum.

In southern states, it is sometimes possible to decrease or delay damage caused by alfalfa weevil larvae in the early spring by decreasing the amount of stubble left the previous

Fig. 13.10. Wheel-track compaction damage in alfalfa, caused by traffic of heavy equipment when soils were wet and soft. Plants in the wheel tracks were either killed or severely stunted. (Photo courtesy of R. Mark Sulc.)

Preventative Strategies for Potato Leafhopper

The potato leafhopper (Fig. 13.11) does not survive winter in the north, but annual migrations from the southern USA in upper air currents ensure that this species is present every year as a primary pest of alfalfa from June to August in the north central and northeastern USA and adjacent areas of Canada. Several preventative strategies can be used to suppress the populations of this important pest.

1. Planting alfalfa varieties with resistance to potato leafhopper (Fig. 13.12) reduces pest populations and yield losses (Sulc et al., 2014), so insecticide rescue treatments are less likely to be needed.
2. Planting nurse crops or forage grasses with alfalfa tends to reduce potato leafhopper populations and damage in the forage stand (Lamp, 1991; Roda et al., 1997).
3. Some locations on a farm tend to have higher leafhopper populations (e.g., the tops of hills). By being observant and keeping records over time, the farm manager can select fields for alfalfa that have a history of lower pest abundance.
4. Harvesting the alfalfa crop in the bud stage and cutting entire alfalfa fields with a short stubble height will reduce build-up of the leafhopper population by removing its food supply.

Fig. 13.11. Potato leafhopper nymph (left) and adult (right). (Reproduced with permission of William Lamp.)

Fig. 13.12. A potato leafhopper-resistant cultivar (left) and a susceptible cultivar (right) during a severe leafhopper infestation of alfalfa. Notice the lack of hopperburn symptoms in the resistant cultivar, compared with the severe yellowing and stunted growth in the susceptible cultivar. Further back in the susceptible cultivar are sections that were sprayed after the action threshold had been reached (lighter green color in near background) or when the action threshold was reached (dark green color in far background). (Photo courtesy of R. Mark Sulc.)

fall when eggs are being laid. This can also be achieved by selecting fall-dormant alfalfa cultivars that have a slower shoot growth rate in the fall, or by harvesting or grazing during the fall (Buntin and Bouton, 1996).

Pest Monitoring

Understanding pest biology and how it is affected by weather conditions is the foundation of a strategic pest-monitoring program. For example, the time for initiating field scouting of the alfalfa weevil in many regions is based on the accumulated growing degree days needed for it to reach the larval stage of development, which is around mid-May in the northern tier of states and earlier towards the south (Undersander et al., 2011). If alfalfa weevil populations developed to an appreciable level in that first crop, the stubble should be checked for surviving larvae that could limit regrowth after the first harvest, especially if early harvesting was used as a suppression measure, rather than an insecticide application. Avoiding insecticide use when alfalfa weevil populations are below the action threshold is an important practice to maintain the biological suppression of the alfalfa weevil. The alfalfa weevil can be commonly found along the margins of fields, but edge-of-field appearance of damage can be deceiving to many growers. Scouting to document pest abundance across the entire field may reveal that the infestation is not serious enough to warrant insecticide application, which would eliminate the suppression by natural predators occurring across the field.

Scouting weed presence is critical during the establishment phase of forages. Weeds that emerge with the establishing forage seedlings are generally most damaging because they can prevent successful establishment of plants in the new stand. New plantings should be monitored carefully during the first 60 days so that timely action can be taken if necessary to manage weed competition with developing seedlings. In the case of spring plantings, summer annual weeds can be especially damaging to stand establishment during the first summer. In late summer and fall plantings, winter annual weeds can be especially damaging the following spring, so they should be controlled during the first autumn. Actions to suppress weed competition may involve clipping, grazing, or herbicide use. A number of herbicides are labeled for use in pure stands of alfalfa, including glyphosate on "Roundup Ready" alfalfa. Herbicide options are much more limited for other species, especially when mixtures of grasses and legumes are planted. Weeds that emerge 60 days or more after the emergence of the forage crop will generally not influence forage yield that first year, but they may reduce forage nutritive value.

Evaluation and Refinement

Experiences over the years, and also during any given year, help to inform future IPM practices for the producer's individual conditions. By evaluating what has happened in the past, a producer can learn what problems are likely to exist in the future, which aids planning for the next year. Specific fields may have unusual problems because of soil conditions that can be rectified. In addition, new pests are introduced, new crop varieties become available, and new equipment provides new ways to manage crops. Changing conditions may directly or indirectly affect future pest problems. Thus careful consideration of history can aid the development of improved IPM practices and reduced pest impacts in future years.

Summary

Integrated pest management is an approach to managing pest problems that utilizes multiple methods and encompasses all categories of pests. It is based on a knowledge of ecological principles and an understanding of the biology of key organisms in the ecosystems that are being managed. It has the goals of minimizing immediate and long-term risks to the environment while maximizing economic returns. For IPM practices to be adopted successfully, they must be compatible with the total management system used on farms and ranches, and must be manageable for farm decision makers. Education programs on the principles of IPM and how to apply them play an important role.

Cultural practices that accomplish several management goals simultaneously are particularly appropriate—for example, early harvesting to maximize forage quality and to minimize pest damage. Selection of resistant cultivars is effective and can be easily utilized with forages that are regularly resown. However, many forage species are not resown frequently or managed intensively enough to warrant intervention. Natural biological controls help to reduce pest loads in all forage systems, and enhanced biological control through deliberate release of natural predators has been very effective for suppression of some introduced pests. Pesticides in the context of IPM are recommended only in response to emergency situations. It is now clear that insects, plant pathogens, and weeds develop resistance to pesticides and herbicides, so routine use not only causes environmental risks but also reduces the long-term options in pest management emergencies. Use of genetically modified crops through **genetic engineering** of pest resistance offers some promise, but except for alfalfa, that development will be slower than is the case for more valuable field crops.

Questions

1. Why are pesticides initially so appealing to crop producers, and what problems can develop from their overuse?
2. Why are pesticides sometimes used in integrated pest management?
3. Define and explain economic injury level and economic threshold or action threshold.
4. Is integrated pest management possible without pesticides? Why is this?

5. What is the role and importance of pest identification in developing strategies to manage pests in forage crops?
6. Give an example of a beneficial organism in each of the biological categories listed in Table 13.1.
7. Define and give examples of biological pest control, cultural pest control, and resistance breeding.
8. Why is early harvesting at bud stage preferable to insecticide treatment in the control of several insect pests of alfalfa?
9. Many diseases of perennial forages shorten stand life. Why does stand life have such a strong influence on the economics of forage production?

References

Baxter, LL, DW Hancock, and WG Hudson. 2014. The bermudagrass stem maggot (*Atherigona reversura* Villeneuve): a review of current knowledge. Forage Grassl. 12:1–8.

Buntin, GD, and JH Bouton. 1996. Alfalfa weevil (Coleoptera: Curculionidae) management in alfalfa by spring grazing with cattle. J. Econ. Entomol. 89:1631–1637.

DeGooyer, TA, LP Pedigo, and ME Rice. 1999. Effect of alfalfa–grass intercrops on insect populations. Environ. Entomol. 28:703–710.

Fick, GW, and AG Power. 1992. Pests and integrated control. In CJ Pearson (ed.), Ecosystems of the World, Volume 18: Field Crop Ecosystems, pp. 59–83. Elsevier, Amsterdam.

Harlan, JR. 1992. What is a weed? In Crops and Man, 2nd ed., pp. 83–99. American Society of Agronomy, Madison, WI.

Huffaker, CB (ed.). 1980. New Technology of Pest Control. John Wiley & Sons, New York.

Jutsum, AR, SP Heaney, BM Perrin, and PJ Wege. 1998. Pesticide resistance: assessment of risk and the development and implementation of effective management strategies. Pesticide Sci. 54:435–446.

Kuhar, TP, RR Youngman, and CA Laub. 2000. Alfalfa weevil (Coleoptera: Curculionidae) population dynamics and mortality factors in Virginia. Environ. Entomol. 29:1295–1304.

Lamp, WO. 1991. Reduced *Empoasca fabae* (Homoptera: Cicadellidae) density in oat–alfalfa intercrop systems. Environ. Entomol. 20:118–126.

Lamp, WO, R Berberet, L Higley, and C Baird (eds.). 2007. Handbook of Forage and Rangeland Insects. Entomological Society of America, Lanham, MD.

Landis, D, SD Wratten, and G Gurr. 2000. Habitat manipulation to conserve natural enemies of arthropod pests in agriculture. Annu. Rev. Entomol. 45:472–483.

National Research Council. 1996. Ecologically Based Pest Management: New Solutions for a New Century. National Academies Press, Washington, DC.

Pedigo, LP, SH Hutchins, and LG Higley. 1986. Economic injury levels in theory and practice. Annu. Rev. Entomol. 31:341–368.

Radcliffe, EB, and KL Flanders. 1998. Biological control of alfalfa weevil in North America. Integr. Pest Manag. Rev. 3:225–242.

Roda, AL, DA Landis, and ML Coggins. 1997. Forage grasses elicit emigration of adult potato leafhopper (Homoptera; Cicadellidae) from alfalfa–grass mixtures. Environ. Entomol. 26:745–753.

Samac, DA, LH Rhodes, and WO Lamp. (eds.). 2015. Compendium of Alfalfa Diseases and Pests, 3rd ed. American Phytopathological Society Press, St. Paul, MN.

Spandl, E, JJ Kells, and OB Hesterman. 1999. Weed invasion in new stands of alfalfa seeded with perennial forage grasses and oat companion crop. Crop Sci. 39:1120–1124.

Sulc, RM, JS McCormick, RB Hammond, and DJ Miller. 2014. Population responses of potato leafhopper (Hemiptera: Cicadellidae) to insecticide in glandular-haired and non-glandular-haired alfalfa cultivars. J. Econ. Entomol. 107:2077–2087.

Undersander, D, D Cosgrove, E Cullen, CR Grau, ME Rice, M Renz, C Sheaffer, G Shewmaker, and M Sulc. 2011. Alfalfa Management Guide. American Society of Agronomy, Crop Science Society of America, and Soil Science Society of America, Madison, WI.

PART III

FORAGE UTILIZATION

CHAPTER

14

Forage Quality

Michael Collins and Yoana C. Newman

Ruminants and other non-ruminant **herbivores** such as horses can use high-**fiber** feedstuffs because their digestive tracts are inhabited by microorganisms that utilize plant fiber by breaking down structural carbohydrates. In this way, **cellulose** and **hemicellulose** that cannot be used directly by humans are converted into high-quality human foods and produce other benefits when fed to herbivores. The majority of the total nutrients in ruminant and other herbivore diets come from **forages**.

Forage Quality and Animal Productivity

Forage quality is defined as the potential of a forage to produce the desired animal response (Fig. 14.1). **Forage nutritive value** and **forage quality** are sometimes considered to be synonymous, but in this discussion the term *forage nutritive value* will be used to refer to the concentration of available energy (total digestible nutrients or TDN) and the concentrations of other nutrients such as crude protein.

Forage quality is a broader term that includes not only nutritive value but also **voluntary intake** and the effects of any **anti-quality factors** (Fig. 14.1). Voluntary intake refers to the quantity of forage dry matter (DM) that animals will consume when they have an unrestricted supply. Research suggests that nutritive value accounts for about 30% of forage quality variation, and that voluntary intake is responsible for the remainder.

The complexity of forage utilization systems is increased by the wide variation in composition of forages compared with common concentrate feeds. Forages differ widely in fiber levels and in the susceptibility of fiber to microbial digestion, in crude protein levels, and in other important ways. This chapter addresses important factors that affect forage quality, including plant species, maturity stage, and morphology, among others.

The potential of forages to meet the energy requirements of animals frequently limits herbivore productivity. Energy value is determined primarily by the proportion of total DM or energy digested as the forage passes through the gastrointestinal tract. Forages of higher digestibility supply more energy to the animal per unit of DM consumed compared with less digestible forages. Some, especially **perennial** warm-season grasses, may require additional crude protein to meet animal needs in some cases. **Supplementation** of minerals such as P, Mg, Na, Cu, and Se is often necessary.

The quantity of DM consumed is the major determinant of digestible energy intake. Under unrestricted feeding conditions, voluntary intake is determined largely by the behavior of forage in the digestive tract. Many factors that lead to increased digestibility are also associated with increased voluntary intake, so even modest improvements in digestibility can result in large improvements in animal performance.

Forage Composition and Voluntary Intake

Forages can be divided chemically and anatomically into two main fractions: (1) cell contents and (2) **cell walls**. The cell contents include the most readily and highly digestible forage components, such as organic acids, proteins, **lipids**, **starch**, and sugars. Cell contents are digested rapidly and almost completely (90–100%) by ruminants and non-ruminant herbivores, and even by monogastric animals. The cell wall is the **fibrous** portion of the forage, and is composed of **structural carbohydrates** (cellulose and hemicellulose), **lignin**, other phenolic compounds, **cutin**, and silica. The amounts, digestibility, and digestion rates of these cell wall constituents are major determinants of animal productivity on forage diets.

Forages: An Introduction to Grassland Agriculture, Seventh Edition. Edited by Michael Collins, C. Jerry Nelson, Kenneth J. Moore and Robert F Barnes.

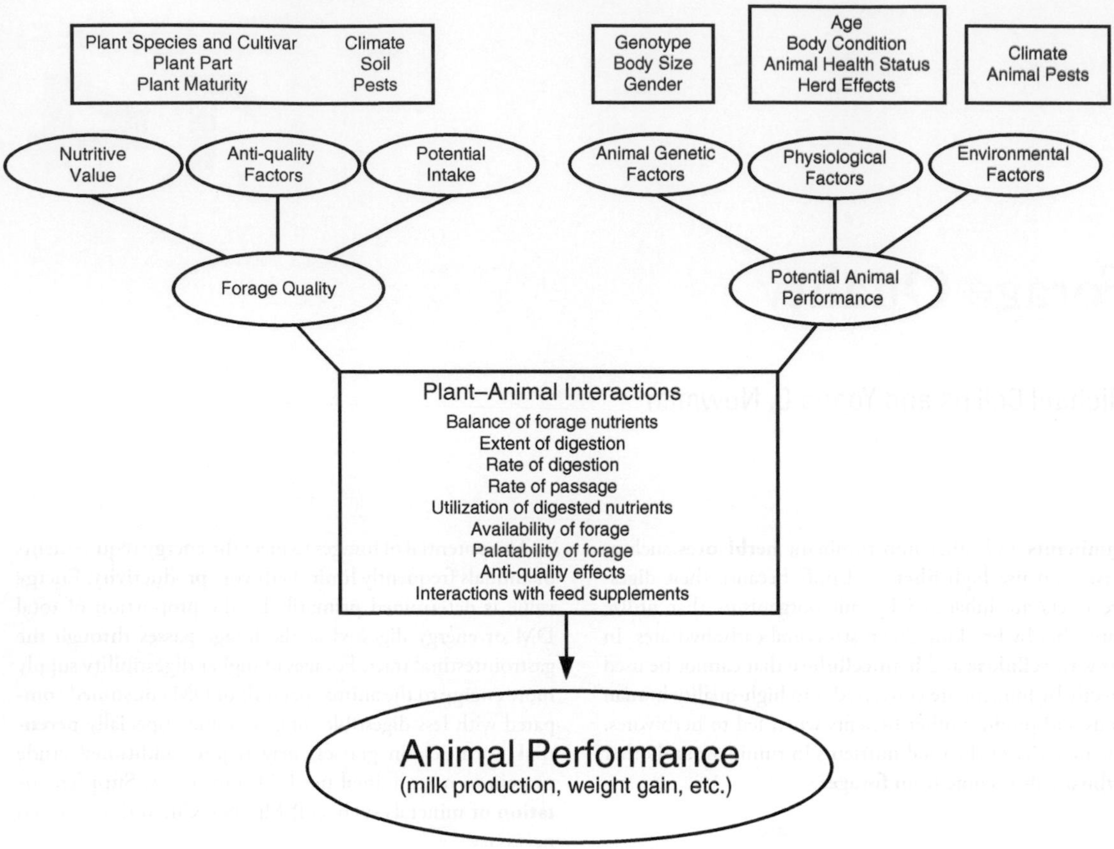

Fig. 14.1. Plant and animal factors that affect animal performance on forages. Numerous animal and plant factors interact to determine actual animal performance. (Source: Marten et al., 1988. Reproduced with permission of the American Society of Agronomy.)

With few exceptions, highly digestible forages are consumed in greater amounts than forages of lower digestibility. The relationship between DM digestibility and cell wall content is usually negative. The intake of more fibrous forages is reduced by the greater time required to digest fiber, to reduce particle size, and to move undigested residues through the digestive tract. Thus animals that are consuming high-fiber forages may be prevented by these fill effects from meeting their requirements for energy and other nutrients (Fig. 14.2A). The restrictions mentioned above are partially alleviated and voluntary intake increases when fiber concentrations decline within the fill-control region.

When fiber levels are very low, animals may meet their energy requirements and voluntary intake may actually decline as fiber concentrations decrease further, although this effect is not common in practice (Fig. 14.2B). Within the physiological control region, energy requirements of the animal are the major determinant of intake. Voluntary DM intake is adjusted upward or downward as necessary to maintain the needed energy intake.

The cell wall concentration at which fill becomes limiting in intake is not fixed, but depends on other factors, such as the physiological status of the animal (Fig. 14.2B). For example, non-lactating cows may achieve the intake necessary to meet their energy needs on diets containing as much as 65% cell walls. For these animals, fill control would be operative only when the diet exceeded this fiber level. For animals with a somewhat higher energy demand, such as moderately productive lactating cows, fill may limit intake whenever cell walls exceed 40% of diet DM. Sustained high energy demand, such as that experienced by high-yielding dairy cows, leads to stretching and increased **rumen** capacity, but this does not compensate completely for the negative effects of fiber on forage intake.

Many factors other than cell wall content influence voluntary intake. Important examples are the rate at which fiber is digested in the rumen, and the rate of passage of

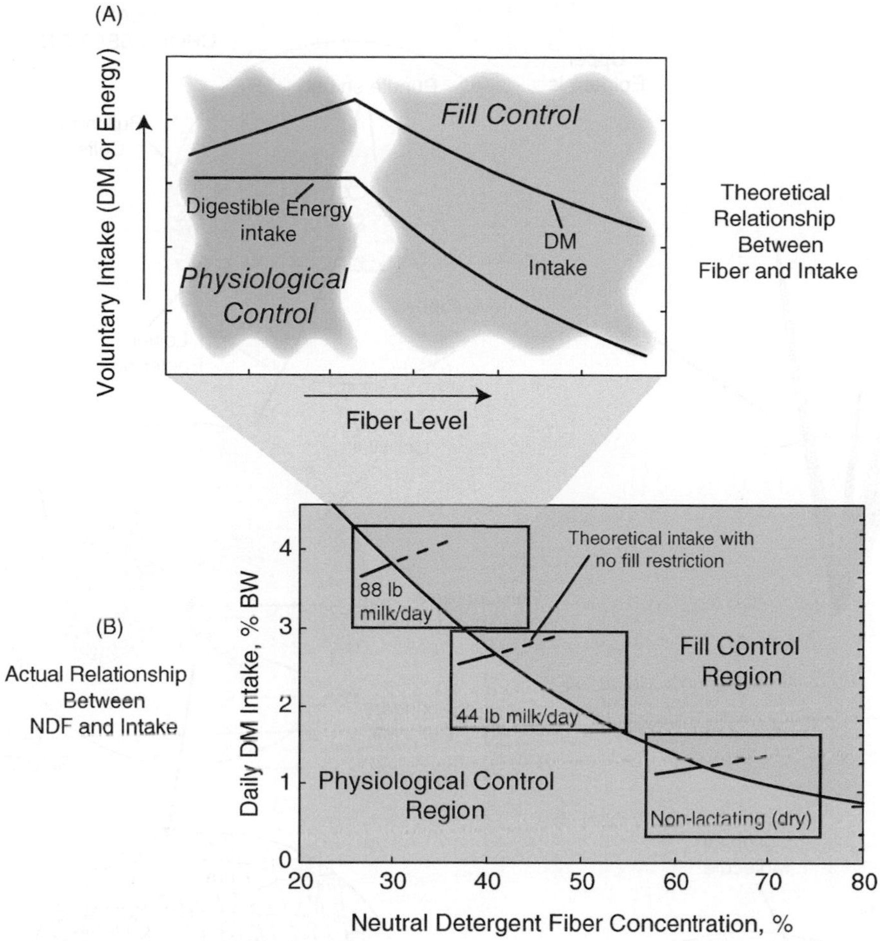

Fig. 14.2. The relationship between fiber level and voluntary forage intake by ruminants. In most feeding situations, voluntary intake declines as cell wall concentrations increase. Major factors are the fill or distention caused by forage in the rumen, and factors that affect animal physiological demand for nutrients, especially energy. DM, dry matter; BW, body weight; NDF, neutral detergent fiber. (Source: Mertens, 1994. Reproduced with permission of American Society of Agronomy.)

the undigested residues through the gastrointestinal tract. Other important factors include animal age, breed, physiological condition, and health.

Forage Composition

Although their specific characteristics and the concentrations of their constituents vary widely, all forages are composed of the same basic array of chemical compounds. The major ones include cell wall carbohydrates, lignin, crude protein, **non-structural carbohydrates**, and minerals.

Cell Wall Carbohydrates

Plant cell walls are composed mainly of structural carbohydrates together with variable amounts of lignin and small amounts of protein and minerals. Cellulose, hemicellulose, and pectins are structural carbohydrates. Unlike nonstructural carbohydrates such as starch, these structural carbohydrates are not remobilized by the plant for reuse as energy or to provide carbon for other metabolic processes.

Cellulose is the major cell wall carbohydrate in many forages, but hemicellulose concentrations are sometimes comparable with those of cellulose. Cellulose molecules are composed of long chains of glucose molecules joined together by beta 1,4-glycosidic bonds (Fig. 14.3). Cellulose molecules are linked by hydrogen (H) bonds to form larger units called **microfibrils** (Hatfield, 1993). Regions within the cellulose structure in which microfibrils are more closely packed are said to have greater crystallinity,

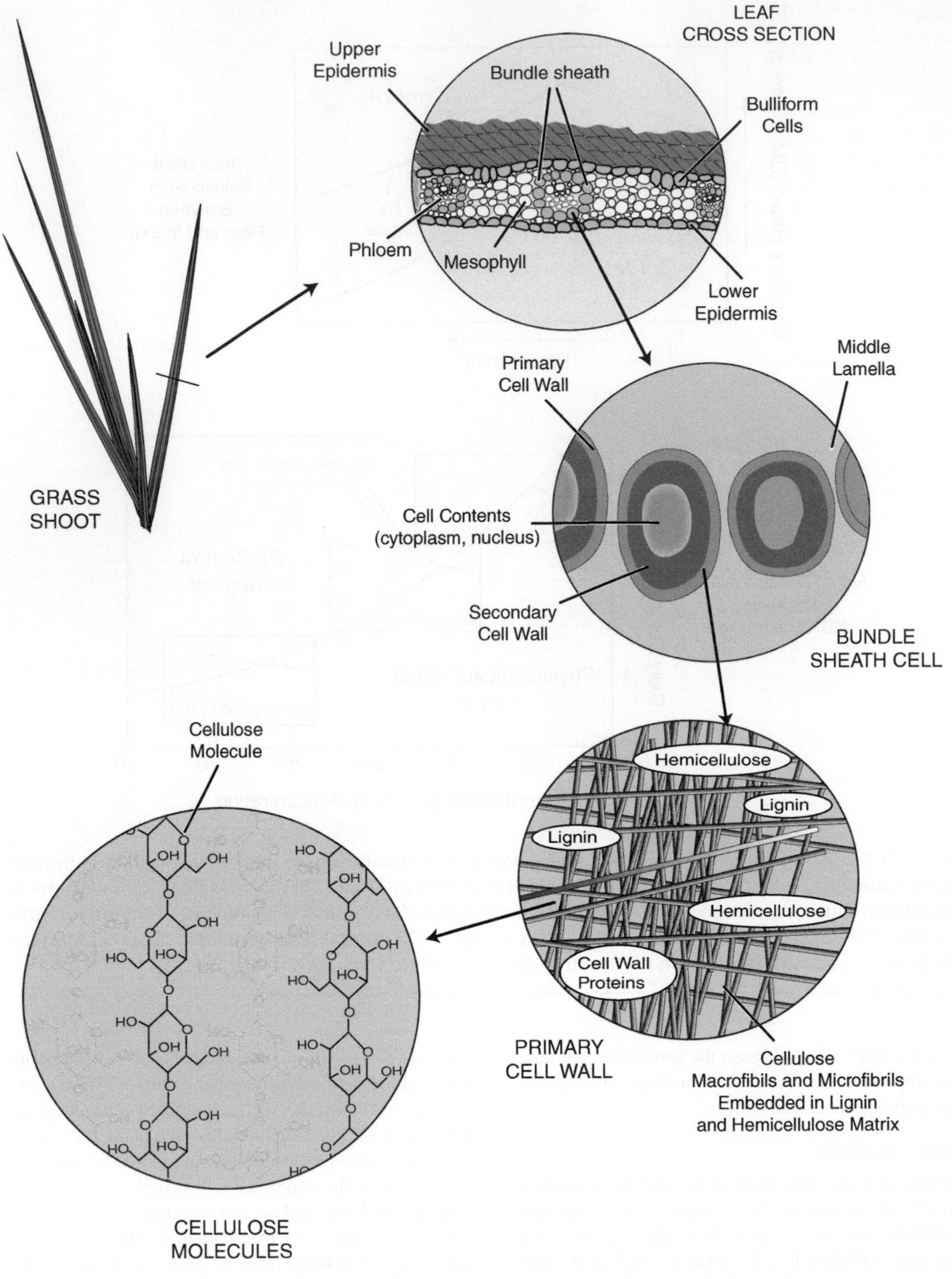

Fig. 14.3. Diagramatic representation of a cross section of a leaf from a typical warm- season grass showing the different cell types. A well-developed parenchyma bundle sheath is characteristic of warm-season grasses. Variation in the proportions of vascular tissue, epidermis, and other tissues between the major forage types contribute to the variation in their forage quality.

and this is thought to be a factor in greater resistance to microbial degradation. The subunits of hemicellulose are xylose, arabinose, mannose, galactose, glucose, and uronic acids. The cell walls of grasses typically contain three to four times as much hemicellulose as **legume** cell walls.

Pectins are found mainly in the middle lamella and primary cell wall, where they function to hold cells together. Legume cell walls usually have much higher levels of pectin than grass cell walls. Vegetative tissues have higher concentrations of pectic compounds than more mature tissues. In contrast to other cell wall carbohydrates, pectic substances are usually highly and rapidly digestible (Titgemeyer et al., 1992).

Lignin

Forage cell walls can be envisioned as an open network of cellulose microfibrils within which the hemicellulose and lignin are interspersed. It is the physical proximity of indigestible lignin to the cell wall carbohydrates that largely accounts for lignin's suppressive effects on their digestion. Chemical bonding also occurs between lignin and hemicellulose, and has additional inhibitory effects on structural carbohydrate digestion.

Lignin is a hydrophobic polymer that is highly resistant to degradation and is composed of phenylpropanoid units derived mainly from three cinnamyl alcohols (*p*-coumaryl alcohol, coniferyl alcohol, and sinapyl alcohol), known as monolignols (Hatfield and Fukushima, 2005). Lignin has the important function of adding rigidity to plant structures. Typical forages contain 3–15% lignin, with legumes generally having higher levels than grasses.

Crude Protein

Forage N is contained mainly in the protein component, but forages also contain **non-protein nitrogen**. In typical forages, protein accounts for 60–80% of the total N. **Nitrate** and free amino acids are important components of the non-protein N fraction. Proteolysis breaks down variable proportions of the protein in ensiled forages, which can contain up to 80% non-protein N. Both protein and non-protein types of N are utilized effectively by rumen microbes to meet their N needs, but excessive levels of nitrate can be toxic to animals. Forage legumes usually contain 15–20% crude protein, whereas **tropical** forage grasses contain about half as much, and cool-season grasses have levels intermediate between these extremes. Cool-season grasses sometimes have crude protein levels comparable to those of legumes, especially at high rates of N fertilization.

Forage crude protein is made up of three fractions—non-protein N, digestible protein N, and indigestible N. Non-protein N, although readily available to rumen microbes, represents a very small proportion of the total in most forages. Digestible protein N is utilized either in the rumen to support rumen microorganisms, or further down the tract in the **duodenum**. When N sources are degraded in the rumen faster than microbial capacity can incorporate N into microbial protein, excess ammonia is absorbed through the rumen wall. Some of this N is eventually excreted in the urine as urea, and represents a loss of N that could be used by the animal. Minimizing nutrient excretion also reduces the potential for environmental problems associated with livestock wastes.

The overall utilization of N in grazed alfalfa and other forages that contain rapidly digestible protein can be enhanced by increasing the amount that passes undegraded through the rumen. This fraction can be digested and its amino acids absorbed in the small intestine along with amino acids from microbial protein. The moderate levels of tannins found in birdsfoot trefoil and some other legumes reduce protein solubility and slow down rumen microbial activity, and can increase the amount of undegraded protein that reaches the small intestine.

Non-Structural Carbohydrates

The carbohydrates that serve as storage and energy reserves in plants are designated as total non-structural carbohydrate (NSC) (Smith, 1973). In forages, this fraction includes the water-soluble sugars or hydrolyzable sugars, which represent about one-third of the NSC, and also includes the fermentable carbohydrates, fructans, and starch, which represent two-thirds of the NSC. This is a very important fraction because of the digestive implications for the different herbivores. For example, in ruminants both fractions are rapidly fermented in the rumen, but this is not the case in horses.

Minerals

Forages also supply the mineral elements that are essential for animal maintenance and growth. Tables 12.1, 12.2, and 12.3 in Chapter 12 list the essential elements required by plants and animals, and describe the common deficiency symptoms in both groups. The quantities of P and Mg needed for optimum plant growth are lower than those needed by most herbivores, so supplemental feeding of these elements is usually required. In addition, animals require Na, Se, Si, Cl, I, Cr, and Co, none of which appear to be required for plant growth.

Forage Analysis

The ultimate measure of forage quality is obtained by feeding and measuring animal responses of interest, such as milk production or weight gain. Such feeding trials account for all of the important factors, including unknown or unexpected factors. Voluntary intake is measured by feeding an excess of the forage and recording DM consumption. The resulting intake data are usually expressed as a percentage of body weight (BW). The value $BW^{0.75}$, which is termed **metabolic body weight**, sometimes replaces BW because it reflects the importance of body surface area in determining animal energy requirements compared with body weight.

In practice, laboratory analyses provide useful and relatively inexpensive information about potential animal response to forages, and are usually substituted for actual feeding data. Routine forage analyses include moisture, crude protein, **cell wall constituents**, minerals, and often DM digestibility or some related estimate of energy availability.

Knowledge of forage moisture content is important for determining the amount of wet forage that should be fed to obtain the required quantity of DM. When forages are analyzed, values are often reported both on a DM basis and on an "as fed" or wet basis. Analytical values that are reported on a DM basis are calculated as if the forage was completely dry, with no moisture content at all. The DM-based values are most useful for comparing different forages, while the "as fed" values are used in feeding.

Crude protein concentration in forages is determined by measuring N and multiplying this value by 6.25. This conversion factor is used because N represents about 16% of typical proteins in vegetative plant material (100/16 = 6.25). It is important to recognize that not all proteins have this same N concentration.

Early forage analysis procedures, some of which were developed before 1900, typically included values for crude protein, **ether extract** (EE), **nitrogen-free extract** (NFE), **crude fiber** (CF), and ash. These components were determined as part of what was called the **proximate analysis** system. This system was later replaced by the detergent analysis system.

The Detergent Analysis System

During the 1960s, Peter J. Van Soest developed an alternative analysis system based on the extraction of samples with different detergent solutions (Goering and Van Soest, 1970). This chemical method, which is now the most widely used system for analyzing forages, recognizes the distinction between cell walls and cell contents (Fig. 14.4).

The procedure involves extraction of a forage sample with neutral detergent solution. This is done by boiling the forage sample in the solution, followed by filtration to isolate the **neutral detergent fiber (NDF)** fraction. The NDF is primarily the total fiber or cell wall fraction of the forage. Neutral detergent-soluble materials are primarily the cell contents but do also include pectic substances. This failure to isolate pectins with the cell walls is acutally fortuitous because their digestibility and digestion rates are more comparable to those of the cell contents fraction. The NDF concentration in various forages and concentrates ranges from around 10% in corn grain to around 80% in straws and tropical grasses. Legume forages generally have lower NDF concentrations than grasses of similar digestibility.

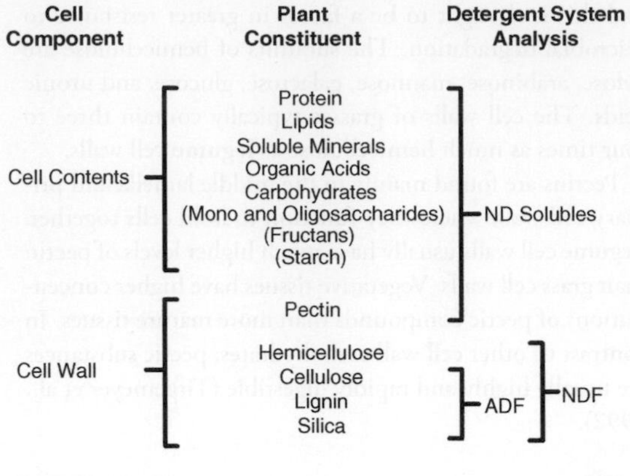

Fig. 14.4. The detergent system of forage analysis. This analytical system divides forage dry matter (DM) into cell contents and cell walls, represented by the neutral detergent fiber (NDF) fraction. The difference between NDF and acid detergent fiber (ADF) provides an estimate of the amount of hemicellulose.

The **acid detergent fiber** (ADF) procedure uses a detergent solution made using 1 N sulfuric acid to extract the sample. The ADF residue is composed mainly of cellulose and lignin, but does not include the hemicellulose, which is soluble in the acid detergent solution. Thus subtracting ADF from NDF provides a rough estimate of the hemicellulose concentration in a sample. The ADF values of typical forages and feeds range from approximately 3% in corn grain to around 40% in mature forages and around 50% in straws. Grasses and legumes of similar digestibility have very similar ADF values.

Lignin is a complex and variable compound that can be measured by one of several direct methods nondestructively using near-infrared reflectance spectroscopy (NIRS), as will be discussed later. The most popular direct methods are **acid detergent lignin (ADL)**, Klason, permanganate, and acetyl bromide-soluble lignin. These methods involve the treatment of ADF with either 72% sulfuric acid or permanganate solution. Of these four methods, the Klason method yields the highest numerical values and the ADL method gives the lowest values (Hatfield and Fukushima, 2005). Each method removes variable amounts of contaminating substances from the lignin residue, and the use of a particular method is a function of the limitations of the laboratory used.

Incorrect sample preparation can lead to the formation of artifact lignin, which consists of non-lignin compounds that show up as lignin during analysis. For example, when plant samples are dried at high temperatures (>149°F [>65°C]), sugars and amino acids may be combined in a process known as the **Maillard reaction** to form compounds that appear as lignin during detergent analysis. Artifact lignin formation is minimized by maintaining high rates of air flow through samples dried at temperatures of 140–149°F (60–65°C).

Near-Infrared Reflectance Spectroscopy

Near-infrared reflectance spectroscopy (NIRS) equipment and software systems have been developed that reduce analytical labor and costs compared with other analytical methods. Differences in reflectance of organic molecules in response to absorption of energy by H bonds with C, N, and O in the near-infrared range (1100–2500 nm) allows accurate prediction of several aspects of forage composition. Calibration equations developed from NIRS spectra and laboratory analytical data typically explain 90–99% of sample variation in CP, NDF, ADF, and *in-vitro* digestibility. Equations used by service analytical laboratories are often constructed from a broad population of samples and allow analysis of forages represented in the calibration set. Organic compounds present in significant concentrations are usually measured accurately with NIRS. Conversely, mineral concentrations and minerals added to total mixed rations (TMR) are poorly predicted using NIR.

Analysis turnaround time is shorter and analysis costs are lower using NIRS compared with standard wet chemistry procedures. NIRS analysis is non-destructive, so the sample can subsequently be used for other analytical procedures. Portable analyzers for forages are available, allowing the convenience of reliable on-farm forage analysis.

Digestibility

In-vivo digestibility is measured by feeding animals and weighing the fecal DM output. The resulting value is referred to as apparent digestibility, to take into account endogenous fecal components such as sloughed intestinal cells that are not of forage origin. The ***in-vitro* dry matter disappearance** (also called *in-vitro* dry matter digestibility) **(IVDMD)** laboratory assay uses rumen fluid to inoculate small forage samples to simulate the ruminant digestive tract. An initial 48-hour incubation under anaerobic conditions with buffer solution and rumen inoculum is followed by acidification and incubation with pepsin to digest protein. This second step simulates digestion in the lower gut. If possible, rumen fluid donors should be fed the same forage as that being tested, because diet affects rumen microbial populations. *In-vitro* digestion of structural carbohydrates and other constituents is strongly correlated with *in-vivo* measurements, and the procedure is widely used in forage research. However, the use of live inoculum makes this assay difficult to standardize across laboratories, so it has not been as widely adopted by commercial analytical laboratories. Procedures based on purified fiber-digesting **enzymes** rather than the living microbes offer some potential to standardize this method.

Relative Forage Quality and Relative Feed Value

The relative forage quality (RFQ) and the former relative feed value (RFV) are indices of forage quality. As an index, they are a single-number product of calculations that cannot be used in diet formulations, but rather as a convenient reference number for the marketing of forages or as a tool for forage quality education.

The first index, which was developed over 30 years ago, was the RFV. A decade later, the RFQ was proposed (Moore and Undersander, 2002). Both the RFQ and the RFV are indices based on negative correlations between ADF and DM digestibility and between NDF and voluntary intake. RFQ was designed to correct the variation in digestibility of the DM relative to NDF, especially between grass and legume forages. The RFQ index incorporates fiber digestibility and correlates more closely with animal performance than does the RFV. When the RFV and RFQ values are different, the RFQ is preferred, and it works acceptably with all forages except corn silage, because it fails to account for differences in starch availability (Undersander et al., 2010). Total tract NDF digestibility (TTNDFD) is a recently developed test designed to predict fiber digestibility in ration balancing for high-yielding dairy cows (Undersander, 2016). This approach adds information on rate of passage together with rate of digestion to more accurately predict animal performance. A 2- to 3-unit change in ration TTNDFD corresponds to a change in milk yield of 1 lb.

Interpreting Forage Analysis Reports

Typical forage analysis reports provide values for NDF, ADF, CP, and moisture. Concentrations of minerals may be provided, including calcium (Ca), phosphorus (P), magnesium (Mg) and potassium (K), and sometimes sulfur (S). Some values presented on the analysis report may be calculated using an equation rather than being determined directly using a specific chemical analysis. Values that are typically calculated from other measurements include **total digestible nutrients (TDN)** and **net energy of lactation (NEL)**. The report should indicate clearly whether the value has been determined analytically or calculated. Several different equations are in use for each of the constituents that are commonly calculated. Since different equations will result in different values, it is important that the report indicates which equation has been used.

Accurate Sampling for Forage Analysis

Proper sampling ensures that the analyses accurately reflect the hay, silage, or pasture which is being sampled. For hay and silage, the unit of sampling is the "lot", which is

Table 14.1. Typical steps in collection of a hay sample to ensure accurate representation of the lot of hay.

Identify the lot of hay. A "lot" is hay harvested from the same cutting, variety, field, and maturity stage, and is cut within a 48-hour period. Take multiple samples for lots of more than 200 tons.

Choose an appropriate, sharp coring device. The ideal coring device should have an inside cutting diameter of 0.38–0.63 in. (0.95–1.60 cm) with a sharp cutting edge. Dull probes push between stems without cutting. Typical coring devices have tubes that are 15–24 in. (38–61 cm) in length. Flakes or "grab" samples pulled from bales by hand do not accurately represent the hay.

Sample at random. Walk around the entire stack and sample bales at various heights. Care should be taken not to bias the sample by inadvertently avoiding bales based on their appearance.

Take an adequate number of cores. Sample 20 bales per lot (one core per bale). Up to 40 cores may be needed if the lot is unusually variable.

Use the correct coring technique. Probe the ends of bales near the center, at right angles to the bale end. Do not slant the probe.

Handling hay samples. Combine subsamples in a sealed polyethylene freezer bag. Avoid exposure of the samples to heat or direct sunlight, and send them to the lab as soon as possible.

Correct sample size. A correctly taken sample should weigh about 0.5 lb (0.2 kg). Large-diameter probes produce excessively large samples.

Splitting samples. To test the performance of a particular lab, send a fully ground and mixed sample to another lab. Do not attempt to divide the unground sample.

defined as forage from the same species and cultivar, maturity, and cutting, that is harvested within a 48-hour period.

Hay is probably the most commonly sampled forage material. Buyers, sellers, and feeders of hay all need information about forage quality. Table 14.1 shows the recommended steps when obtaining a representative sample. Subsamples are taken from 20 different bales within a lot of hay using a coring device and composited to produce a sample. Inadequate subsampling results in less accurate forage quality analysis data (see "Why Collect 20 Core Samples?"). Rectangular bales should be sampled from the end, near the center, and round bales should be sampled from the circumference, perpendicular to the bale surface (Fig. 14.5). The statistical benefits of coring multiple bales are retained even though subsamples are composited before laboratory analysis. Any weathered or spoiled hay should be represented in the sample if this portion of the hay is to be fed.

Fig. 14.5. Recommended procedures for sampling rectangular and round bales. Samples should be taken from the end of rectangular bales, near the center (top photo), and from the circumference of round bales perpendicular to the bale surface (middle photo). Samples from 20 bales should be composited to produce a representative sample for the lot (bottom photo).

Why Collect 20 Core Samples?

A small sample must accurately describe up to 200 T of hay. Bale-to-bale and analytical laboratory variation mean that the values obtained are just estimates of the "true" or actual analytical value. Adding more subsamples improves our confidence in this estimate. The *true value* is the *sample value* $\pm 1.634 * (\sigma/\sqrt{n})$ where σ is the standard deviation and n is the number of subsamples (90% probability level). The only practical way to improve accuracy is to increase n, as σ is relatively constant.

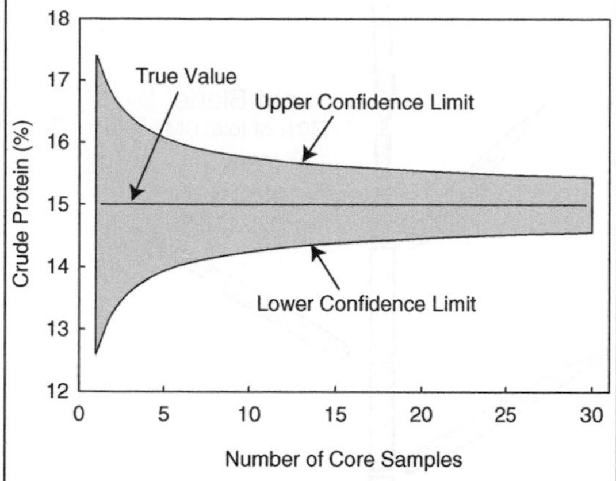

FIG. 14.6. The relationship between the number of core samples collected from a lot of hay and the size of the prediction error for crude protein concentration in the hay.

Figure 14.6 uses CP to show how increased sampling improves confidence in the laboratory value. First, the true values were determined by sampling all 40 individual bales in a small lot of hay. Based on these data, NDF = 61.2 ± 1.89, ADF = 36.5 ± 1.04, and CP = 15.0 ± 1.46. A single sample would tell us only that the true CP value was between 12.6% and 17.4% CP ($15.0 \pm 1.634 * (1.46/\sqrt{1})$, which is a very poor estimate. Increasing the number of samples to 20 gives us 90% confidence that the actual value falls between 14.5% and 15.5% ($15.0 \pm 1.634 * (1.46/\sqrt{20})$.

In terms of minimizing sampling errors, combining subsamples is equivalent to analyzing each separately and averaging the values. Furthermore, small hay lots need just as many subsamples for accurate quality analysis as large lots.

Sampling Silage for Quality Analysis

Chopped silage can be conveniently sampled during the filling process by collecting 12–20 handfuls of material from randomly selected loads, although quality changes during storage will not be reflected by such pre-storage samples. To prepare a sample, a subsample of 1–2 quarts (0.95–1.89 L) should be taken from a thoroughly mixed sample of at least 2 gallons (7.6 L) of chopped forage. The subsample should be placed into a plastic bag and refrigerated until it is transported to the laboratory. It is important to label each sample with as much information as possible relating to the field, cutting, stage of maturity, species or mixture, prominent weeds, and other factors that might aid the interpretation of the forage quality analysis.

Hay and silage composition usually changes during storage. Thus samples collected near the time of feeding provide the best representation of the forage consumed by livestock. During feeding, fermented silage can be sampled in a similar way to the procedure described above for chopped forage. Fermented silage samples should be subsampled and packaged as described previously for fresh material.

Factors That Affect Forage Quality

There are three primary factors that affect forage quality in most situations. These are forage species, maturity stage, and harvesting conditions. Secondary influences include temperature and soil moisture levels during growth, soil fertility, and cultivar. Many of these factors influence forage quality because they affect plant anatomy and morphology.

Plant Anatomy and Morphology

The shoots of forage plants consist of leaves, stems, and flowers and seed if plant maturity has advanced beyond the vegetative stage. These morphological components differ greatly with regard to chemical composition.

During the very early vegetative growth stages, stems and leaves can be very similar in composition. Occasionally very young stem tissue is more digestible than leaf tissue. In general, however, leaf material is much higher in digestibility, lower in fiber, and contains up to twice as much crude protein as stem tissue from the same plant. Figure 14.7 illustrates typical morphological component quality differences using the alfalfa and timothy from a mixture. In this case, alfalfa and timothy leaves contained two to three times as much crude protein as stems. Stems had much higher NDF and ADF levels than leaves in both species. Thus management practices or climatic conditions that affect the proportion of leaf in the total forage can have a major impact on overall forage quality.

Because there is usually such a difference in forage quality between leaf and stem, differential consumption of these components by animals can greatly affect diet quality. In addition to animal preference for leaf rather than stem, which is termed selective grazing, animals graze the upper layers of the canopy first, and this is where leaves tend to be concentrated. Animals may also avoid consuming stem tissues in some cases. Thus grazing animals commonly consume a diet that contains a higher proportion of leaf than the standing crop. (For a case that illustrates the impact that selective grazing and leaf losses during hay harvest can have on diet quality, see "Grazing Versus Hay Harvest.")

Morphological component quality is greatly affected by the types of cells from which the components are formed (Fig. 14.3). Some cell types, such as **mesophyll**, undergo rapid and almost complete degradation by microbial and mechanical action in the gastrointestinal tract.

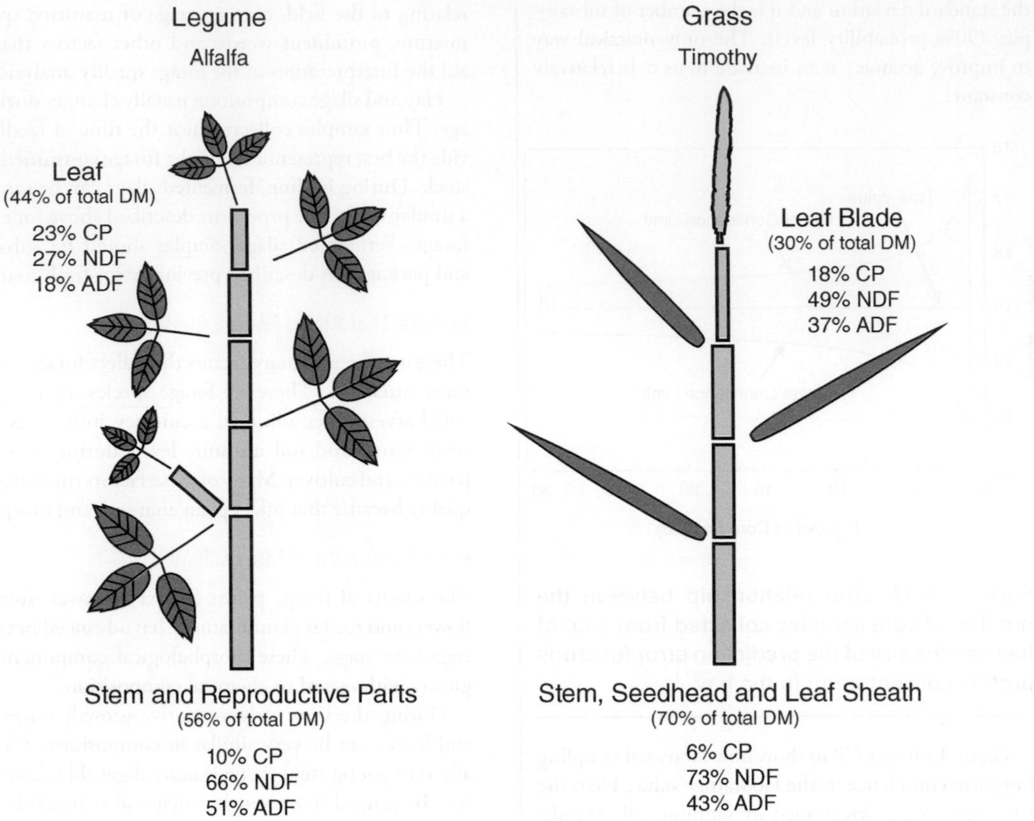

Fig. 14.7. Forage quality analysis of leaf and stem tissue from alfalfa and timothy growing together in a mixture. Leaf is usually much lower in fiber and higher in digestibility than stem. DM, dry matter; CP, crude protein; NDF, neutral detergent fiber; ADF, acid detergent fiber.

Cell wall deposition in growing cells begins with the primary cell wall. Some cells, such as those in the **xylem**, also have a secondary cell wall which is deposited inside the primary wall during shoot maturation. Cell walls from some tissue types, including the xylem, **parenchyma** bundle sheath, and epidermis, are very resistant to degradation or may pass almost intact through the entire digestive tract. **Phloem** cells are generally digestible, but the bundle sheath may provide a barrier in some forages and reduce and/or slow their digestion by rumen microbes. **Sclerenchyma** tissues, consisting of thick-walled cells that provide structural support for the plant, are sometimes present and become **lignified** as the plant matures.

Grazing Versus Hay Harvest

Many forage species are suitable for harvesting either by grazing or mechanically as hay or silage. The irregular stubble that is left after grazing gives the impression that much forage is being wasted, especially compared with the short, uniform stubbles that are left after hay harvest. However, in the case shown here, efficient grazing and hay systems would each harvest 77% of the crude protein present in the standing crop, and the grazed forage was of much better quality (Fig. 14.8). Animals graze canopies from the top down, and generally select leaf in preference to stem. Leaves are concentrated near the top of the canopy for alfalfa and many other forages, so grazing harvests the highest-quality portion of the available forage. On the other hand, around 90% of the shattering losses that occur during alfalfa hay harvest consist of leaves (Collins et al., 1987).

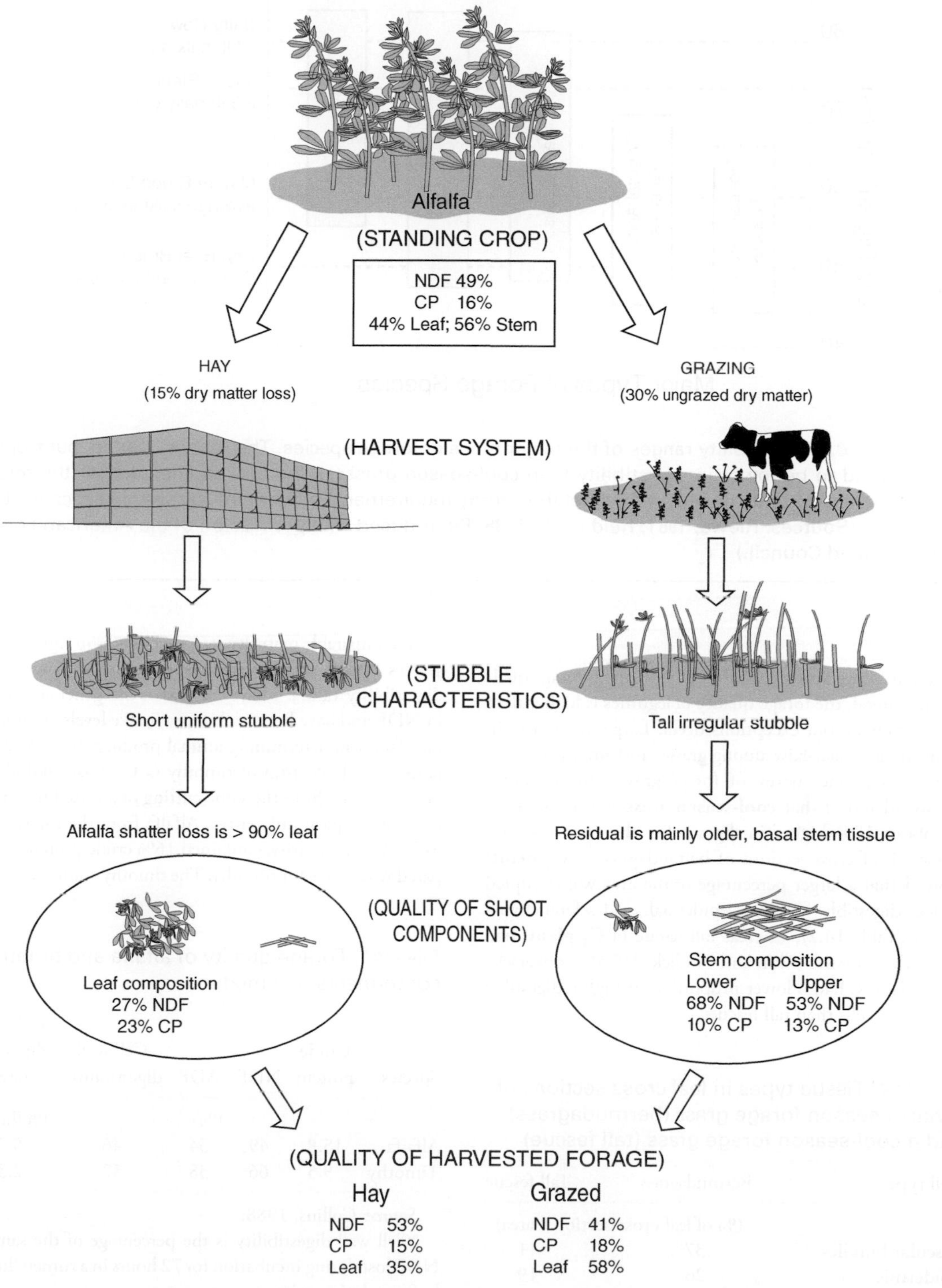

FIG. 14.8. The effect of selective grazing of alfalfa on the forage quality of the diet consumed by the animal compared with the quality of field-dried hay prepared from the same crop. NDF, neutral detergent fiber; CP, crude protein.

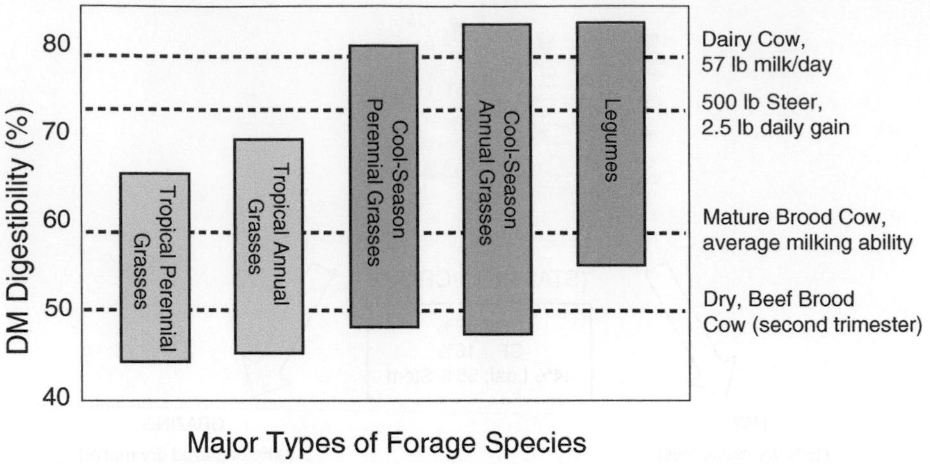

Fig. 14.9. Digestibility ranges of the major types of forage species. The ranges overlap, but tropical grasses tend to be lower in digestibility than cool-season grasses. The broken lines denote the forage digestibility levels that are needed to meet the energy requirements of different classes of beef cattle. DM, dry matter. (Sources: Riewe, 1981; Reid et al., 1988. Reproduced with permission of the American Forage and Grassland Council.)

Plant Species

Different forage species can differ markedly in forage quality. In general, the forage quality of legumes is higher than that of grasses, but exceptions occur. Large differences in forage quality also exist among grasses and among legumes (Fig. 14.9). One survey of forage grasses from around the world found that **cool-season grasses** were on average about 13% higher in digestibility than warm-season grasses. Leaf cross sections of bermudagrass (a **C_4 plant**) showed that a larger percentage of the area was occupied by less digestible **vascular**, epidermal, and sclerenchyma tissues (Table 14.2), whereas tall fescue (a **C_3 plant**) had less of these tissues (Akin and Burdick, 1975). Conversely, bermudagrass had a lower proportion of highly digestible mesophyll cells than tall fescue.

Table 14.2. Tissue types in leaf cross sections of a warm-season forage grass (bermudagrass) and a cool-season forage grass (tall fescue)

Cell type	Bermudagrass	Tall fescue
	(% of leaf cross-sectional area)	
Vascular bundles	37	11
Epidermis	26	19
Sclerenchyma	10	6
Mesophyll	27	62

Source: Akin and Burdick, 1975. Reproduced with permission of the Crop Science Society of America.

At comparable maturity stages, cool-season legumes and grasses often have similar ADF concentrations and DM digestibility values. However, grasses are generally higher in NDF and have lower voluntary intake levels, an important factor in determining animal productivity. The comparison in Table 14.3 of timothy (a C_3 grass) and alfalfa (a C_3 legume) from the same cutting of a mixed hay stand illustrates typical differences. Alfalfa from the mixture, at early bloom maturity, contained 16% crude protein, compared with 9.5% for timothy. The timothy component had

Table 14.3. Forage quality of alfalfa and timothy components of a mixture

Species	Crude protein	NDF	ADF	Cell wall digestibility[a]	Cell wall digestion rate[b]
			(%)		(%/hour)
Alfalfa	15.8	49	34	46	5.3
Timothy	9.5	66	38	57	2.3

Source: Collins, 1988.

[a]Cell wall digestibility is the percentage of the sample NDF lost during incubation for 72 hours in a rumen fluid–buffer solution mixture.

[b]Cell wall digestion rate is the percentage of the digestible cell wall material that disappears during each hour of incubation.

NDF, neutral detergent fiber; ADF, acid detergent fiber.

considerably more NDF than alfalfa. Higher NDF (total fiber) content and a slower rate of fiber (cell wall) digestion of grass forage combine to result in lower voluntary intake levels compared with typical legumes, despite the fact that the final extent of cell wall digestion in grasses is usually greater. Faster cell wall digestion and particle size reduction of legume forage allows quicker passage from the rumen and allows more forage to be consumed.

Considerable variation in forage quality also exists among the approximately 40 species that produce most of the cultivated grass forages. Forage grasses are divided into two broad classes—cool season if they are adapted to temperate regions, and warm season if they are best adapted to tropical or subtropical environments. The main cool-season grasses include orchardgrass, smooth bromegrass, kentucky bluegrass, perennial and annual ryegrass, and tall fescue. Examples of warm-season grasses include bermudagrass, bahiagrass, crabgrass, dallisgrass, big bluestem, and corn. C_3 grass forage is generally higher in quality than C_4 grass forage. Minimum crude protein levels found in warm-season grasses are lower than those in cool-season grasses. In one study, 22% of the warm-season grass samples but only 6% of the cool-season grasses were found to contain less than 6% crude protein (Reid et al., 1988).

The differences in leaf anatomy mentioned above contribute to the generally lower quality of warm-season grasses. These differences in leaf structure are related to the greater photosynthetic efficiency of warm-season grasses such as bermudagrass in converting sunlight into forage yield. However, warm-season grass leaves contain a higher proportion of highly lignified, less digestible tissues than leaves of cool-season grasses.

Environmental temperatures also contribute to differences in quality between cool- and warm-season grasses. For a given species, lower-quality forage is produced when growth occurs at high temperatures than at cooler temperatures. For example, at growth temperatures of 50–59°F (10–15°C), annual ryegrass forage contained 59% leaf, compared with only 36% leaf at temperatures of 68–77°F (20–25°C). In addition, forages grown under warmer temperatures tend to have higher lignin and fiber concentrations than those grown under cooler temperatures. Thus forage of any given species grown in hotter regions or seasons would tend to be of lower quality that produced under cooler conditions.

Maturity Stage

Forage quality nearly always declines as forages age or undergo reproductive development. In fact, maturity at harvest is usually considered to be the primary factor affecting forage quality. Young vegetative forage can approach the digestibility of concentrate feeds. For example, cool-season grasses often have DM digestibilities in excess of 80% during the first 2–3 weeks after growth begins in the spring (Stone et al., 1960). After that, forage quality usually begins to decline as the stems start to develop. Figure 14.10 illustrates a typical relationship between quality and maturity using NDF, crude protein, and digestibility of alfalfa. Digestibility and crude protein usually decline with advancing maturity, while NDF, ADF, and other fiber components increase. The digestibility of harvest-stage forages typically declines by 0.3–0.5% for each day that harvest is delayed.

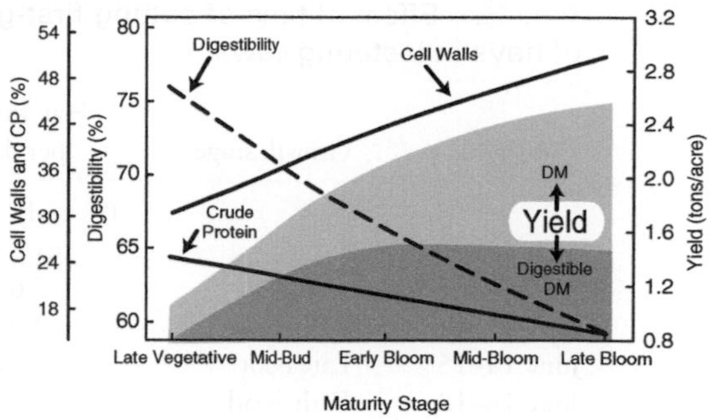

FIG. 14.10. Effects of maturity stage on alfalfa forage quality. While yield increases as shoots develop, the concentrations of crude protein and other nutrients generally decline. Concentrations of fibrous constituents such as neutral detergent fiber (NDF) and acid detergent fiber (ADF) are higher in more mature forage. CP, crude protein; DM, dry matter.

The effects of maturity stage on forage quality of corn silage, wheat, and similar forages are less because these forages usually contain significant amounts of grain. The presence of grain, which is low in fiber and highly digestible, partially or completely offsets the declining quality of the vegetative portion of the plant. Some brassica species produce forage composed almost exclusively of leaf tissue, also allowing them to maintain high quality throughout most of their productive periods.

Forage species differ in the rate at which quality declines with advancing maturity. The digestibility of spring growth forage of reed canarygrass declined more rapidly than that of tall fescue, orchardgrass, and smooth bromegrass, and the digestibility of timothy declined less rapidly than that of the latter three species. All of these grasses showed similar trends in crude protein levels, from about 25% when the shoots were 6 in. tall and vegetative, to less than 10% two months later, when the shoots had reached post-anthesis maturity (Collins and Casler, 1990).

More mature forages also generally have lower voluntary intake levels when fed to animals. Typical DM digestibility and intake values for hays harvested at different stages of maturity are shown in Table 14.4. Numerous studies have shown similar effects of maturity on forage

Table 14.4. Effect of time of cutting first-growth forage on intake and digestibility of hays in lactating cows

Cutting date	Growth stage	Hay intake per day	Hay digestibility	Relative digestible DM intake
		(% body weight per day)	(%)	
June 3–4	Vegetative	2.64	63.1	166
June 11–12	Early boot	2.36	65.7	154
June 14–15	Late boot	2.45	62.6	153
June 16–18	Early head	2.28	58.5	133
July 1	Bloom	2.30	52.7	121
July 5	Bloom	2.13	52.2	111
July 7–8	Bloom	2.05	52.2	107
July 9–10	Late bloom	1.95	51.5	100

Source: Stone et al., 1960.
DM, dry matter.

digestibility and intake of many different species. In the above example, lactating dairy cows consumed 66% more digestible DM per day on the early compared with the most mature hay. Higher fiber concentrations, lower fiber digestibility, slower DM digestion rates, and other factors account for the decrease. Fiber occupies more space in the rumen and takes longer to be digested compared with cell contents. The digestion rate of the fiber component declines as cool-season perennial grasses such as orchardgrass and timothy mature.

Leaf:Stem Ratio

Given the opportunity, herbivores will usually select leafy forages in preference to those with more stem material. Declining leaf percentage accounts for much of the loss in quality that occurs as forages mature, because leaves usually have much higher quality than stems. For example, it was found that leaf percentage in an orchardgrass crop decreased from 61% at an early vegetative stage to only 23% at late anthesis (Buxton et al., 1987). Whereas the forage quality of leaves changes little as shoots mature, maturing stems show a significant increase in NDF and decrease in digestibility. In a study covering a wide range of maturity stages (Buxton and Russell, 1988), the digestibility of cell wall material from cool-season grass stems was found to decline by 0.55% for each 0.1% increase in lignin content in the cell wall, compared with a decline of 0.34% for cool-season legumes.

Seasonal patterns of reproductive growth strongly influence the leaf to stem ratio and thus the quality of many perennial grasses. Most cool-season grasses require **vernalization** (a period of exposure to cool temperatures during fall or winter) in order for **flowering** to occur in spring. As a result, most cool-season grasses flower (i.e., produce stems) only once each year during spring. Because the regrowth forage consists mainly of leaves, its quality remains relatively stable over time. Most legumes and some grasses (e.g., bermudagrass) can flower many times during the season, and thus repeat a similar pattern of quality decline during each growth cycle.

Harvest and Storage Effects

Some loss in quality generally occurs during forage harvesting and storage, so the quality of hay and silage when fed is almost always lower than that of the original standing crop. Physical losses that have a disproportionate impact on the leaf component, especially for legumes, and **respiration** activity that utilizes non-structural carbohydrates both account for some quality loss, even under good harvest conditions. Greatly increased shattering and respiration, and leaching of soluble constituents, can cause much larger reductions in quality when rain damage occurs during harvest (Table 14.5). In this example, exposure to rain increased ADF and NDF and decreased DM digestibility of both alfalfa and red clover. Because of their higher proportions of cell contents, the quality of legume hays is reduced more by rain damage than is that of grasses. In one study, more than 60% of the total loss of DM, crude protein, ash, and digestible DM caused by rain on alfalfa hay came from the leaf component. Leaf loss, respiration, and the likelihood that rain damage will occur can all be reduced by harvesting forage as silage rather than as hay.

Crude protein levels declined only slightly in alfalfa with rain damage, and even increased slightly following rain on red clover hay. This unexpected response is thought to occur because protein molecules are large and thus less readily removed from cells by leaching compared with smaller molecules such as sugars and K, levels of which can decline sharply in rain-damaged hays.

Table 14.5. Composition of alfalfa and red clover hays harvested at late-bud maturity and either cured without rain or exposed to rain damage during field curing

Species and constituent	Well-cured hay[a]	Rain-damaged hay[a]
	(%)	
Alfalfa		
CP	26	25
NDF	32	45
ADF	28	39
DM digestibility	73	57
Red clover		
CP	24	27
NDF	30	44
ADF	25	38
DM digestibility	68	47

Source: Collins, 1983.

[a]Well-cured and rain-damaged hay for each species came from the same field and cutting.

CP, crude protein; NDF, neutral detergent fiber; ADF, acid detergent fiber; DM, dry matter.

Forage quality can also be reduced when weathering, microbial activity, and undesirable chemical reactions occur during hay and silage storage. Weathering losses can be large for hay stored outside in high-rainfall regions (Lechtenberg et al., 1974) (see also Table 17.4 in Chapter 17). Although DM digestibility of forage in the protected interior of the bales changed little during 5 months of storage, DM digestibility of the weathered portions of the bales declined sharply, especially for the mixed alfalfa–grass hay.

Grass–Legume Mixtures

Forage quality is typically higher for grass–legume mixtures than for pure stands of the same grass. Grass–legume mixtures generally have lower NDF levels and a higher crude protein content. In Georgia, mixtures of seven legumes with bermudagrass contained 11–13% crude protein, compared with 11% in pure grass receiving 90 lb N/acre annually (Burton and DeVane, 1992).

Fertilization Effects on Forage Quality

Nitrogen fertilization of grasses can substantially increase yields, and generally increases crude protein levels in the forage. For example, the application of about 70 lb N/acre (78 kg/ha) on switchgrass raised crude protein levels from 5.3% to 6.4%, and increased voluntary intake by 11%, but digestibility was unaffected (Puoli et al., 1991). The very high levels of K (more than 4% of the DM in some cases) that can be found in alfalfa forage can cause negative effects when used in dairy rations. High-K forages may have reduced availability of Mg and a higher potential for **grass tetany**. Direct supplementation of deficient minerals to the animal is generally the most economical approach to meeting animal requirements for elements such as Mg.

Environmental Effects and Diurnal Fluctuations

As discussed earlier, forage grown under higher temperature conditions is usually lower in quality than that grown under cooler conditions, in part because lignin deposition is increased at higher temperatures. Readily digestible nonstructural carbohydrates can accumulate to levels of 20% or more in cool-season grasses during fall. This response is used to advantage in the stockpiling management system described in Chapter 18, in which forage is allowed to accumulate during fall for later grazing.

Observations have shown **diurnal** changes in soluble carbohydrate levels in alfalfa due to time of day, with peak levels generally occurring late in the day. In geographical areas where rain damage during field curing is unlikely (e.g., California, Idaho), the data have shown that it is beneficial to forage quality to harvest alfalfa late in the day, to take advantage of the higher levels of soluble, highly digestible carbohydrates present at that time.

Cultivar Effects

Forage quality can be significantly improved by plant breeding. An example is Tifton-85 bermudagrass, a hybrid that has been developed which has a substantially higher yield and about 11–12% higher digestibility compared with another cultivar, "Coastal" (Clavijo et al., 2010).

Some differences in forage quality between cultivars result from changes in plant morphology, such as increased leaf content. **Genotypes** of orchardgrass have been identified that have slower rates of quality decline over time, but the DM yields of these are also reduced. Multifoliolate cultivars of alfalfa, which have five to seven leaflets per leaf rather than the usual three leaflets, are sometimes but not always higher in quality. Some multifoliolate cultivars exhibit increased leaf content and others do not, suggesting the presence of other effects that offset the higher leaflet numbers. Some trifoliolate alfalfa cultivars also have superior forage quality compared with others.

Brown Midrib Mutants

Brown midrib (bmr) mutants are certain mutants of corn, sorghum, and pearlmillet that have a distinctive reddish-brown coloration of the midribs of the leaf blades. The mutation is a simply inherited, recessive, single-gene trait. In corn, the mutation reduces the activity of O-methyl transferase, an enzyme that catalyzes the conversion of the intermediate compound 5-hydroxyferulic acid to sinapic acid. This change reduces the supply of the sinapyl alcohol precursor of lignin. In sorghum, the mutation reduces the activity of cinnamyl alcohol dehydrogenase, which catalyzes the conversion of aldehyde forms of lignin precursors into the alcohol forms prior to their incorporation into lignin. In addition to being lower in core lignin content, bmr mutants also have much lower *p*-coumaric acid contents compared with normal plants. The practical consequence of having lower concentrations of a chemically altered lignin component is a reduction in its inhibitory effect on the digestion of cell wall carbohydrates.

Low-lignin forages such as brown midrib (bmr) mutants or the more recent low-lignin alfalfas represent good examples of forage quality improvement through genetic manipulation (See "Brown Midrib Mutants"). In these cases, it appears that cell wall digestion is increased by reducing the amount of lignin rather than by any change in lignin composition (Fig. 14.11) (Grabber et al., 2009). In brown midrib mutants, mutations in lignin biosynthesis pathways reduce lignin levels by up to 50% in stem tissue and by up to 25% in leaf tissue of corn, pearlmillet, and sorghums that have the bmr characteristic (Jung and Deetz, 1993). Fiber (NDF) digestibility is increased substantially by this reduction in lignin content (Fig. 14.11); however, the rate of cell wall digestion is not affected (Fritz et al., 1990).

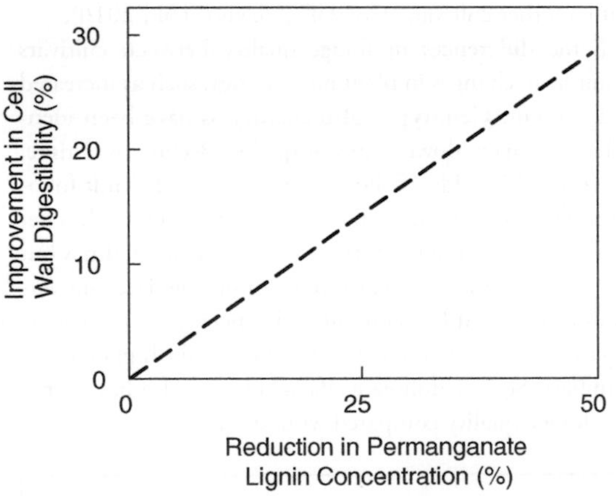

FIG. 14.11. General relationship between the improvement in cell wall digestibility in sorghum leaves and stems and the relative reduction in lignin concentration in bmr mutants. (Source: Jung and Deetz, 1993. Reproduced with permission of the American Society of Agronomy Inc.)

Summary

A number of important factors influence forage quality. Species selection, maturity stage, and the harvest/storage system used are the primary factors that must be given close attention. However, many other factors can also have an impact on forage quality. For any given plant species, maturity remains the foremost factor affecting quality (Coleman et al., 2004).

Questions

1. What is the main determinant of animal performance on forages?
2. How many core samples are needed to ensure that a lot of hay is accurately represented?
3. What are the two main constituents into which the detergent system of analysis divides a forage sample?
4. What are the three main factors that affect forage quality?
5. Name three advantages of the near-infrared reflectance spectroscopy (NIRS) forage analysis system over standard analytical methods.
6. What is relative forage quality (RFQ), and what would be an unsuitable use of RFQ?
7. How does forage quality change during maturation of the forage plant?
8. How does the bmr mutant of sorghum improve fiber digestibility?

References

Akin, DE, and D Burdick. 1975. Percentage of tissue types in tropical and temperate grass leaf blades and degradation of tissues by rumen microorganisms. Crop Sci. 15:661–668.

Burton, GW, and EH DeVane. 1992. Growing legumes with "Coastal" bermudagrass in the lower Coastal Plain. J. Prod. Agric. 5:278–281.

Buxton, DR, JR Russell, and WF Wedin. 1987. Structural neutral sugars in legume and grass stems in relation to digestibility. Crop Sci. 27:1279–1285.

Buxton, DR, and JR Russell. 1988. Lignin constituents and cell-wall digestibility of grass and legume stems. Crop Sci. 28:553–558.

Clavijo, JA, YC Newman, LE Sollenberger, C Staples, LE Ortega, and MC Christman. 2010. Managing harvest of 'Tifton 84' bermudagrass for production and nutritive value. Forage and Grazinglands doi:10.1094/FG-2010-0802-02-RS

Coleman, SW, JE Moore, and JR Wilson. 2004. Quality and utilization. In LE Moser, BL Burson, and LE Sollenberger (eds.), Warm-Season (C_4) Grasses, pp. 267–308. American Society of Agronomy, Crop Science Society of America, and Soil Science Society of America, Madison, WI.

Collins, M. 1983. Wetting and maturity effects on the yield and quality of legume hay. Agron. J. 75:523–527.

Collins, M. 1988. Composition and fibre digestion in morphological components of an alfalfa-timothy sward. Anim. Feed Sci. Tech. 19:135–143.

Collins, M, and MD Casler. 1990. Forage quality of five cool-season grasses. II. Species effects. Anim. Feed Sci. Tech. 27:209–218.

Collins, M, WH Paulson, MF Finner, NA Jorgensen, and CR Keuler. 1987. Moisture and storage effects on dry matter and quality losses of alfalfa in round bales. Trans. Amer. Soc. Agric. Eng. 30:913–917.

Fritz, JO, KJ Moore, and EH Jaster. 1990. Digestion kinetics and composition of cell walls isolated from morphological components of normal and brown midrib sorghum × sudangrass hybrids. Crop Sci. 30:213–219.

Goering, HK, and PJ Van Soest. 1970. Forage Fiber Analyses: Apparatus, Reagents, Procedures, and Some Applications. Agriculture Research Service, US Department of Agriculture. Agriculture Handbook No. 379. US Government Printing Office, Washington, DC.

Grabber, J, D Mertens, H Kim, C Funk, F Lu, and J Ralph. 2009. Cell wall fermentation kinetics are impacted more by lignin content and ferulate cross-linking than by lignin composition. J. Sci. Food Agric. 89:122–129.

Hatfield, RD. 1993. Cell wall polysaccharide interactions and degradability. In HG Jung et al. (eds.), Forage Cell Wall Structure and Digestibility, pp. 285–313. American Society of Agronomy, Madison, WI.

Hatfield, R, and RS Fukushima. 2005. Can lignin be accurately measured? Crop Sci. 45:832–839.

Jung, HG, and DA Deetz. 1993. Cell wall lignification and degradability. In HG Jung et al. (eds.), Forage Cell Wall Structure and Digestibility, pp. 315–346. American Society of Agronomy, Madison, WI.

Lechtenberg, VL, WH Smith, SD Parsons, and DC Petritz. 1974. Storage and feeding of large hay packages for beef cows. J. Anim. Sci. 39:1011–1015.

Marten, GC, DR Buxton, and RF Barnes. 1988. Feeding value (forage quality). In AA Hanson, DK Barnes, and RR Hill Jr. (eds.), Alfalfa and Alfalfa Improvement, pp. 463–491. Agronomy Monograph 29. American Society of Agronomy, Madison, WI.

Mertens, DR. 1994. Regulation of forage intake. In GC Fahey, M Collins, DM Mertens, and LE Moser (eds.), Forage Quality, Evaluation, and Utilization, pp. 450–493. American Society of Agronomy, Madison, WI.

Moore, JE, and DJ Undersander. 2002. Relative forage quality: an alternative to relative feed value and quality index. In Proceedings of the 13th Annual Florida Ruminant Nutrition Symposium, pp. 16–32. University of Florida, Gainesville, FL.

Puoli, JR, GA Jung, and RL Reid. 1991. Effects of nitrogen and sulfur fertilization on digestion and nutritive quality of warm-season grass hays for cattle and sheep. J. Anim. Sci. 69:843–852.

Reid, RL, GA Jung, and DW Allinson. 1988. Nutritive Quality of Warm Season Grasses in the Northeast. WV Univ. Agr. Exp. Stn. Bull. 699. Agricultural and Forestry Experiment Station, West Virginia University, Morgantown, WV.

Riewe, ME. 1981. Expected animal response to certain grazing strategies. In JL Wheeler and RD Mochrie (eds.), Forage Evaluation, Concepts and Techniques, pp. 341–355. CSIRO, Melbourne, Australia and American Forage and Grassland Council, Lexington, KY.

Smith, D. 1973. The non-structural carbohydrates. In GW Butler and RW Bailey (eds.), Chemistry and Biochemistry of Herbage, pp. 105–155. Academic Press, London.

Stone, JB, GW Trimberger, CR Henderson, JT Reid, KL Turk, and JK Loosli. 1960. Forage intake and efficiency of feed utilization in dairy cattle. J. Dairy Sci. 43:1275–1281.

Titgemeyer, EC, LD Bourquin, and GC Fahey Jr. 1992. Disappearance of cell wall monomeric components from fractions chemically isolated from alfalfa leaves and stems following *in-situ* ruminal digestion. J. Sci. Food Agric. 58:451–463.

Undersander, D, JE Moore, and N Schneider. 2010. Relative forage quality. Focus on Forage Vol. 12 No. 6. University of Wisconsin Extension Team, Madison, WI.

Undersander, DJ. 2016. Forage testing with greater accuracy than ever before. Prog. Forage Grower 17:14–15.

Chapter 15

Forage Utilization

Michael Collins, Kenneth J. Moore, and Craig A. Roberts

Herbivores can digest structural **carbohydrates**, such as **cellulose** and **hemicellulose**, to meet their energy needs. Due to this ability, they can have diets that are much higher in fiber than those suitable for **monogastric** animals. This chapter is concerned with the animal and plant factors that influence forage consumption and **digestion**.

Digestive Anatomy of Herbivores

In a symbiotic relationship, microorganisms in the digestive tract of herbivores are provided with a suitable environment for growth, and in turn provide energy extracted from plant cell walls for their host animals. Microflora and microfauna produce enzymes, including β-glycosidases (cellulases and hemicellulases) that degrade **cell wall constituents**, thereby gaining access to the energy contained in plant cell walls. These microbes absorb pentoses and hexoses released from the cell walls and use them for growth and development. Plant cell walls, which contain lignin in addition to cellulose and hemicellulose, are collectively referred to as fiber because of their unique nutritional characteristics. Fiber is resistant to digestion, and because it makes up a large proportion of forage diets it poses unique challenges to their utilization by livestock. Herbivores have also maintained the ability to use non-structural carbohydrates, such as sugars and starch, as energy sources.

Similar species of microflora and microfauna are found in the digestive tracts of all fiber-digesting animals. Low-oxygen conditions ensure **anaerobic respiration** (fermentation) of digestive-tract organisms. The volatile (short-chain) fatty acids (VFAs), including acetic (C2) acid, propionic (C3) acid, and butyric (C4) acid, that are excreted by these organisms are absorbed through the rumen wall into the bloodstream, and represent the primary energy source for the host herbivore.

Anatomical adaptations of simple gastrointestinal tracts to better accomplish fiber digestion have occurred over time. Domesticated mammalian species use variants of one of two fiber-digesting systems. Ruminants, including **ovine** and **bovine** species, use an enlarged multi-compartment modification of a simple stomach for fiber digestion by anaerobic fermentation, and are called foregut fiber fermenters. Horses and other **equids** have a simple stomach and employ modifications of the hindgut, namely an enlargement of the **cecum** and **colon**, for cell wall digestion by anaerobic fermentation, and are classified as hindgut fiber fermenters.

In ruminants, digestion of fiber occurs primarily in the **reticulorumen** compartment located at the beginning of the digestive tract (Fig. 15.1). The reticulorumen, which is usually simply referred to as the rumen, has a large volume because of the large amounts of forage that must be processed and the long retention times needed to complete fiber digestion to meet nutritional needs. In horses and other non-ruminant herbivores, fiber digestion occurs mainly in the cecum, which is located further down the digestive tract beyond the small intestine.

By-products of rumen fermentation include considerable volumes of gases, mainly carbon dioxide and **methane**, which must be expelled through **eructation** to relieve pressure on the rumen wall. Conditions that restrict eructation can lead to bloat (for further details, see Chapter 16). Fiber-digesting livestock contribute to global warming because methane is a potent "greenhouse gas" (see Chapter 10). Recovery of useful energy released by cell wall digestion may be increased if **methanogenesis** is suppressed by feed additives such as **monensin**.

Copious quantities of saliva help to maintain the rumen pH in the range of 6–7 on most forage diets, and low

Forages: An Introduction to Grassland Agriculture, Seventh Edition. Edited by Michael Collins, C. Jerry Nelson, Kenneth J. Moore and Robert F Barnes.

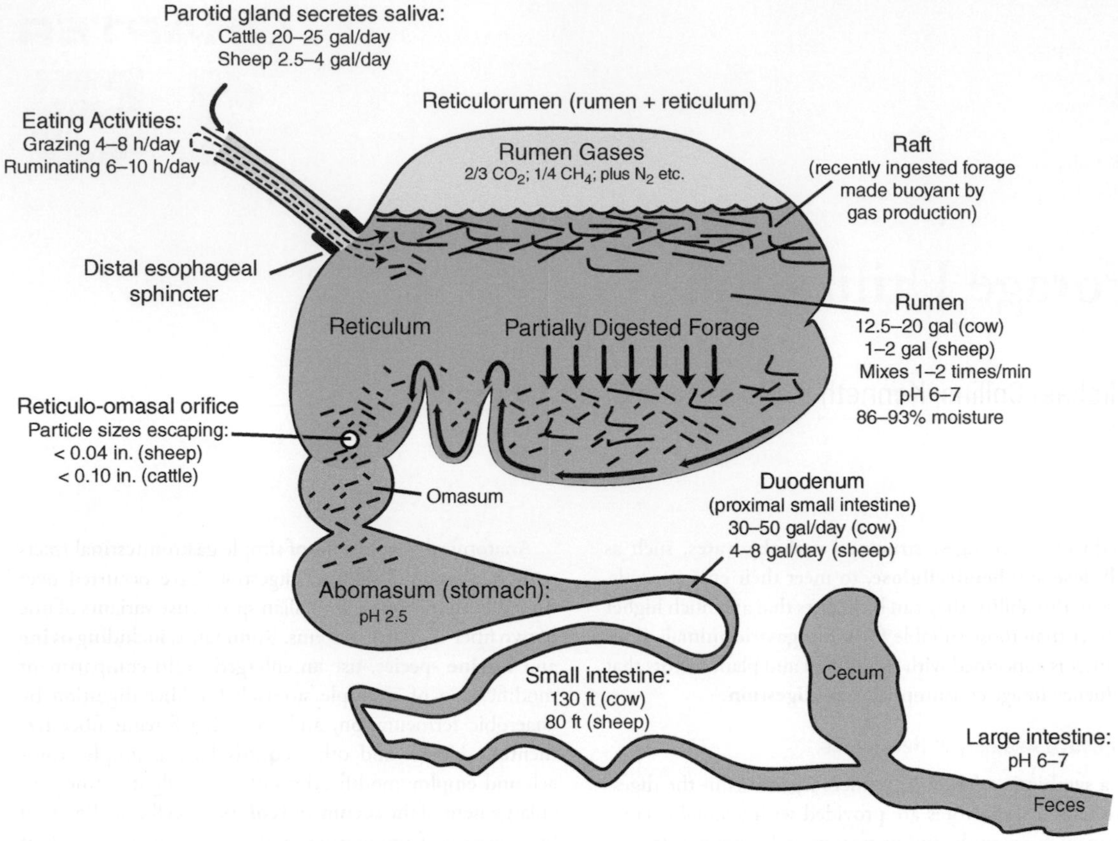

FIG. 15.1. Schematic diagram of ruminant digestive anatomy. The reticulorumen is the largest compartment, and contains a complex microbial population that is responsible for digesting structural carbohydrates, producing volatile fatty acids such as acetic acid and propionic acid that are absorbed directly through the rumen wall. (Adapted from Ellis et al., 1994.)

oxygen levels encourage rapid growth of rumen microbes. Acidity of the **abomasum** and other conditions further down the tract are more like those of monogastric animals. Microbial protein and forage protein that escape digestion in the rumen are hydrolyzed to amino acids that are subsequently absorbed from the small intestine. Forage proteins that escape digestion in the rumen are used more efficiently than those that are broken down in the rumen. Warm-season grasses usually have a lower protein concentration than cool-season grasses, but a higher amount of this escapes degradation in the rumen. This may be one of the reasons why livestock that are consuming warm-season grass forages perform better than expected based on their protein level. Legumes that contain high tannin levels also have larger amounts of intact protein bypassing the rumen. This is thought to occur because tannins form complexes with proteins that render them undegradable in the rumen.

The structures of the hindgut, namely the small intestine, cecum, colon, and rectum, play a smaller role in digestion in ruminants than they do in monogastric livestock. In ruminants, they are more concerned with nutrient and water absorption and excretion of undigested matter (Fig. 15.1).

Rumen Development in Young Animals

Newborn ruminants lack the capacity to digest fiber. Milk bypasses the immature foregut (avoiding microbial fermentation) via a transient tube-like structure (the **reticular groove**) leading directly to the **omasum**. The rumen contents of young ruminants are inoculated by fiber digesters from the environment, including their **dams**. Rumen volume and fiber digestion capability gradually increase until lambs are 2 months old and calves are 6–9 months old.

Factors That Affect Animal Performance

Animal performance is the cumulative result of numerous interactions between plant and animal factors. Potential performance depends on the physiological status of the animal, its genetic make-up, and environmental effects. For example, lactating cows have higher energy demands and thus higher forage consumption levels compared with non-lactating animals. More energy is needed to maintain body temperature when the ambient temperature falls below the **thermoneutral zone,** which in cattle is 60–75°F (16–24°C), and production is reduced. High ambient air temperatures and strong sunlight may limit daytime grazing because cattle, particularly dark-colored European breeds, may have difficulty maintaining normal core body temperature, which is 102° F (39°C).

Forage quality is a function of nutritive value, intake level, and the presence of anti-quality factors that might reduce intake or nutritive value. Plant species and maturity stage are among the factors that influence forage feeding value (see Chapter 14). Variation in forage intake accounts for around 70% of the total variation in forage feeding value, with nutritive value accounting for the remaining 30%.

Forage Digestibility

Digestibility is the percentage of dry matter (DM) or of an individual constituent such as crude protein (CP) that is digested as forage passes through the animal's digestive tract. Energy content is typically the first limiting factor in herbivore production on forages, and dry matter digestibility is a good measure of energy availability. Highly digestible forages increase animal production largely by increasing their energy intake. Sugars, organic acids, and other components of the cell contents are digested almost completely (over 90%) regardless of forage species or maturity. Cell wall constituents may be highly digestible or almost completely indigestible, depending on forage species, maturity stage, and other factors.

Tissues that become lignified with age have particularly low cell wall digestibility. **Sclerenchyma, vascular bundles**, and other **lignified** cell tissues may pass largely undigested through the digestive tract. **Rumen cellulolytic bacteria** attach themselves to fibers before initiating digestion. Lignin may chemically or physically prevent this attachment. **Mastication** and **rumination** aid digestion by increasing the surface area that is exposed for attachment by bacteria. The higher proportion of less digestible **epidermis** and vascular tissues in warm-season (C_4) grasses contributes to lower average fiber digestion compared with cool-season grasses. Forages with lower digestibility produce larger amounts of undigested material that must be passed through the gastrointestinal system to be excreted as feces, and they typically have lower intake rates.

Table 15.1. Energy partitioning of red clover forage and wheat straw; most plant materials have similar gross energy contents, but the availability of this energy to herbivores varies greatly

Crop	Gross energy	Digestible energy	Metabolizable energy	Net energy
		(Mcal/lb)[a]		
Red clover	2.04	1.09	0.91	0.45
Wheat straw	2.00	0.77	0.63	0.09

[a]1 calorie is the energy required to raise the temperature of 1 gram of water by 1°C from 14.5°C to 15.5°C. 1 Mcal = 1,000,000 calories.

Source: Adapted from Maynard and Loosli, 1969.

Energy Partitioning

Energy in ruminant diets can be partitioned into digestible and indigestible components. **Digestible energy** is the difference between feed energy and the amount of energy in the feces. This portion of the feed energy is sometimes referred to as "apparent digestible energy" because not all energy in the feces is of direct feed origin; portions are from sloughed intestinal cells, undigested microbes, and other endogenous sources.

Digestible energy may be divided into **metabolizable energy** and energy lost as methane during digestion or excreted in the urine. **Net energy** remains after subtraction of the heat lost during the processes of digestion and synthesis from metabolizable energy. A portion of the net energy goes toward maintenance requirements, and the remainder is available for productive purposes such as growth, reproduction, lactation, and the synthesis of wool. Ingested forages with low digestible energy levels may have no energy left for these productive purposes. In the example in Table 15.1, red clover had much more net energy than wheat straw.

Intake

Ad libitum intake (also referred to as **voluntary intake**) is the amount of forage consumed by animals when they have an unrestricted supply. Increasing forage intake provides more total energy and other nutrients to the animal, and decreases the proportion of dietary energy expended for maintenance purposes (see Chapter 14). Voluntary intake is not simply a characteristic of the forage, but is also affected by animal species, gender, physiological status, and health status.

Intake Regulation

The mechanisms of intake regulation are complex, and include physical limitations, physiological control, and **psychogenic** factors. Physical factors include (1) distension

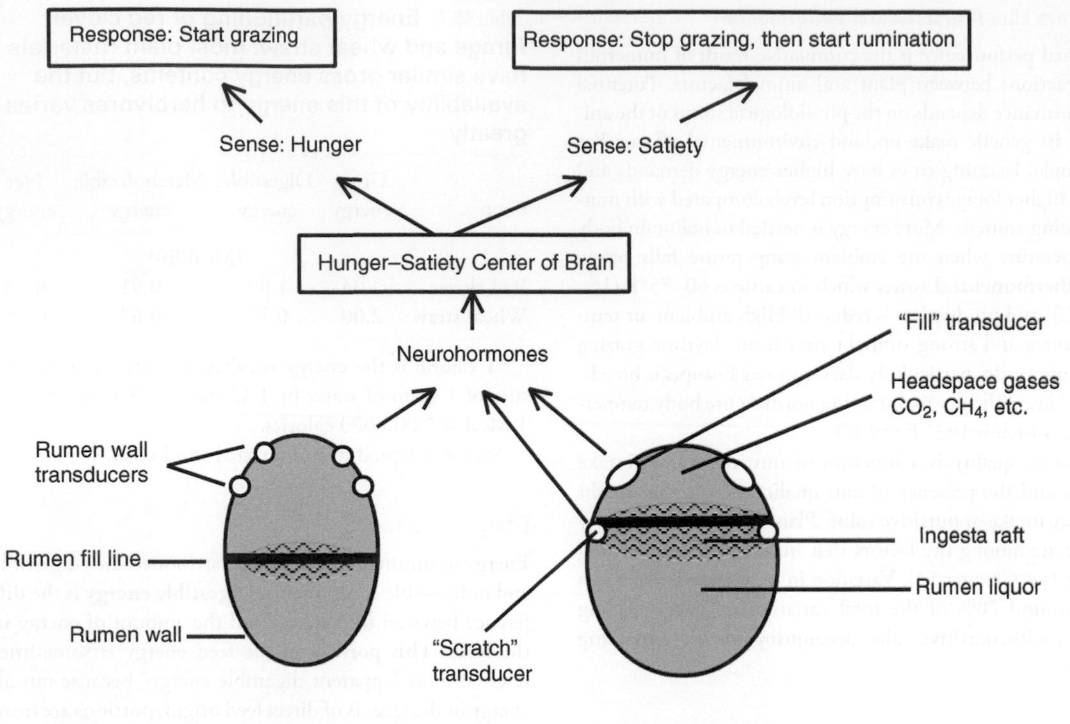

Fig. 15.2. A simple "gut fill" model for regulating the start and cessation of grazing and the onset of rumination.

that stimulates stretch sensors in the tract wall and causes cessation of feeding (Fig. 15.2), and (2) neutral detergent fiber (NDF) concentrations and composition that affect rates of digestion, particle size reduction, and **digesta** passage. Physiological factors include (1) control of hunger and satiety by the hypothalamus region of the brain, (2) cessation of feeding caused by rapid increases in **osmolarity** of the rumen/reticulum fluid, (3) a decrease in rumen fluid pH to 5.0–5.5, which decreases **rumen motility**, and (4) high concentrations of acetic acid in the rumen. Psychogenic factors include (1) herd behavior, (2) palatability, and (3) environmental and other stresses.

The general relationship between forage fiber (NDF) concentration and voluntary intake has been discussed (see Fig. 14.2 in Chapter 14). Physical limitations (**gut fill**) can cause eating to cease even though the animal has not yet fulfilled its physiological requirement for energy or some other limiting nutrient. Additional contributing factors include the facts that fiber digestion rates are usually much slower than for cell contents, and that undigested fiber particles must be reduced in size before they can exit the rumen.

These fiber relationships help to explain observations that voluntary intake of forage at any given digestibility level varies greatly among forage species (Fig. 15.3). Legumes such as alfalfa and red clover typically sustain higher intake levels than grass forages of comparable digestibility. The lower cell wall levels in legume forage help to account for these differences.

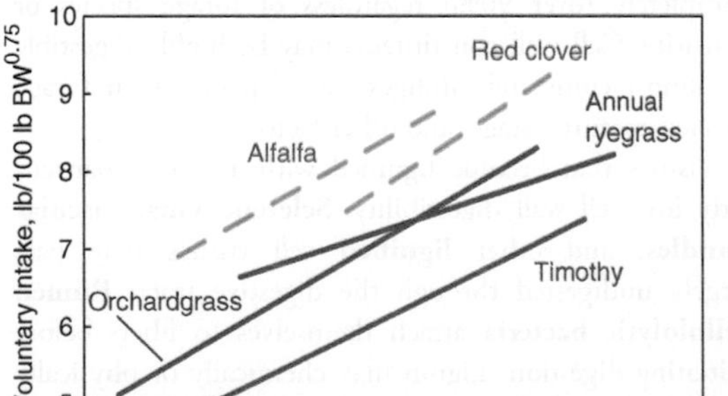

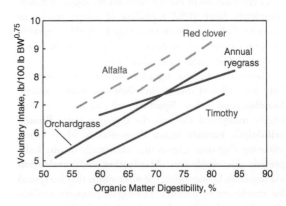

Fig. 15.3. Voluntary intake and organic matter digestibility of five different forage species. Intake varies widely at a given level of digestibility, due to differences in rate of digestion, fiber concentrations, and other factors. (Adapted from Minson, 1982.)

Voluntary intake may also be influenced by anti-quality factors or by chemical or physical factors that make the forage more or less palatable to animals. **Palatability** refers to the preference that animals have for a particular feed when given a choice. Animals fed in confinement have little opportunity to exercise complete choice with regard to diet selection, so the practical importance of palatability is difficult to assess.

Digestive Process Rates

The speed at which digestive processes occur greatly affects animal performance. Important examples are the rate of particle size reduction, the **rate of digestion**, and the rate at which forage particles move through the digestive tract (the rate of passage).

Rate of Digestion

Time is critical to the concept of forage digestion rate (Moore and Buxton, 2000). **Digestion rate** is the proportion or percentage of digestible material remaining in the rumen that gets digested per hour. Digestion rate and digestibility must be considered separately. Digestion rate (k) applies only to the behavior of the digestible portion of a constituent. It remains constant throughout the digestion process. The quantity of substrate digested per hour is greater early in the process because more substrate is present.

Cell contents are digested at a consistently high rate, but NDF (cell wall) digestion rates vary widely across forage species and tissue types. Cell wall digestion rates range from 0.02/hour to as much as 0.19/hour (Smith et al., 1972), and are generally higher for legumes than for grasses. Differences in fiber digestion rate that seem small can have a large effect on total digestion time. The time needed to complete digestion of 50% of the substrate (digestible cell wall in this case) can readily be calculated as 0.693/k. For the two forages in Fig. 15.4, it would take 30.1 hours (0.693/0.023) to digest 50% of the digestible NDF in timothy but only 13.1 hours for alfalfa. Some time elapses before cell wall digestion begins after forage is introduced

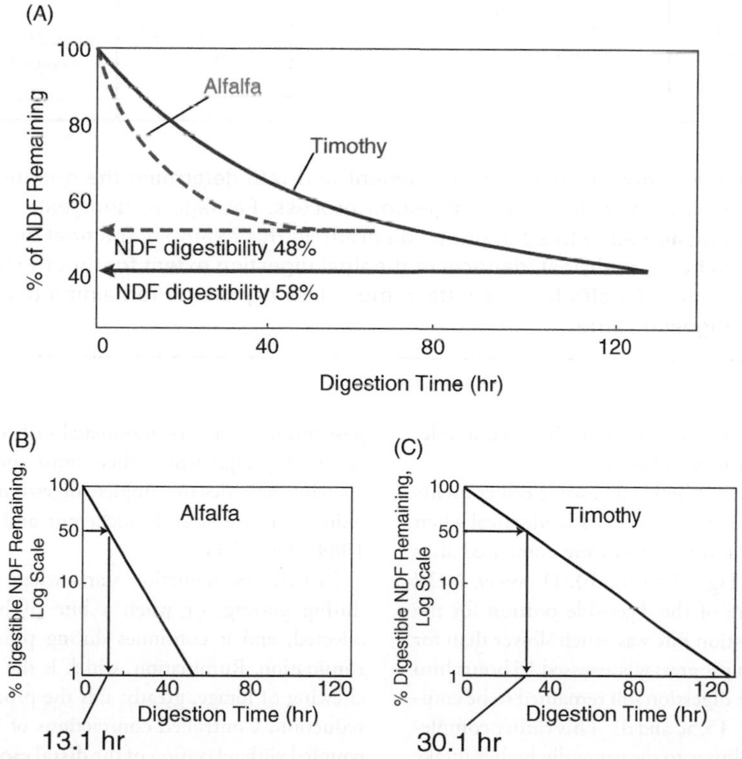

FIG. 15.4. Fiber digestion rates of grass and legume forages. (a) The grass forage had a more digestible NDF fraction, but digestion was slower (2.3%/hour) compared with the legume. For the legume forage, digestion proceeded more rapidly (5.3%/hour) even though the extent of NDF digestion was lower. (b) Viewed on a log scale, only 13.1 hours were needed to complete 50% of the NDF digestion process for the legume. (c) Viewed on a log scale, 30.1 hours were required to complete 50% of the NDF digestion process for the grass forage. (Adapted from Collins, 1988.)

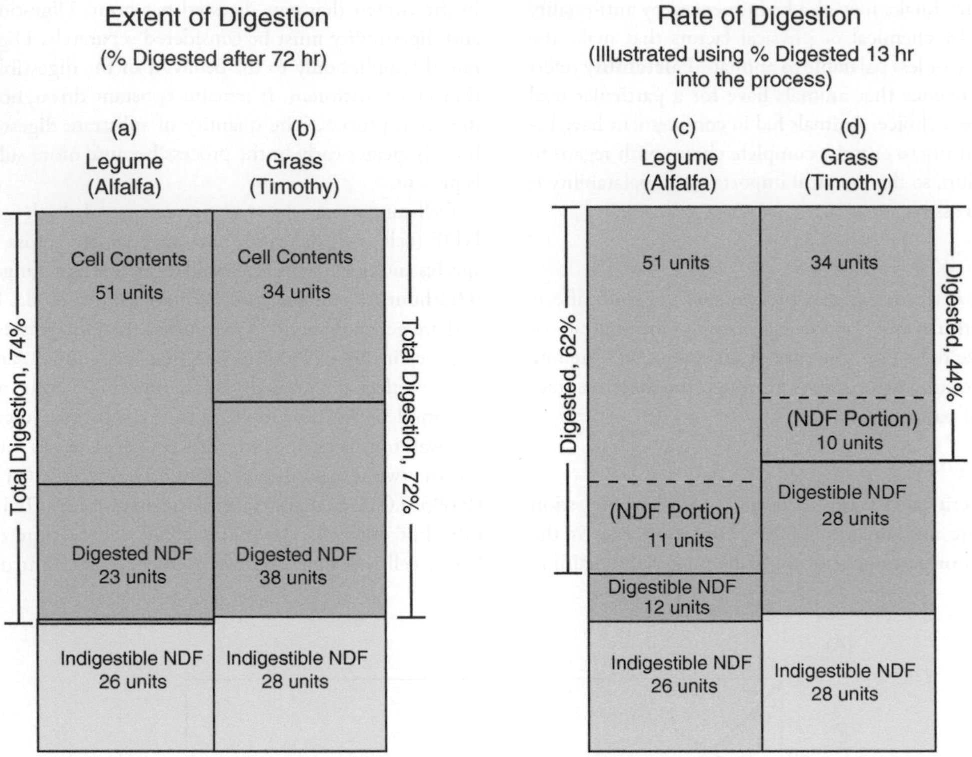

FIG. 15.5. Example showing how digestion extent and rate determine the quantity of undigested residues remaining at any point during the digestion process. Passage of undigested residues is also important, but is not considered in this schematic diagram. In this example, the final digestion extent for alfalfa (a) after 72 to 96 hours is almost identical to the final digestion extent for timothy (b). However, the faster rate of fiber digestion for alfalfa (c) results in much less dry matter remaining than for timothy (d) after a brief, 13-hour digestion time.

into the rumen. This "lag phase" was short in this example, and was omitted from the calculations.

The values of the sum of highly digestible cell contents and digestible cell wall material were nearly identical when the same alfalfa and timothy forages were compared after 72 hours of digestion (Fig. 15.5a and b). However, NDF contributed much more of the digestible portion for the grass, so its overall digestion rate was much slower than for alfalfa. As a result, when progress was assessed 13 hours into the process, much more digestion still remained to be completed for timothy (Fig. 15.5c and d). This earlier completion of digestion contributes to the generally higher intake levels observed for legume forages.

Rate of Particle Size Reduction

Concurrent with digestion, particle size reduction is needed before undigested residues can exit the rumen. Plant cell walls must be chewed, ruminated, and partially digested to reduce their particle size before they pass through the reticulo-omasal orifice. Animals that are consuming higher-fiber diets must spend more time on chewing activities to complete digestion in the rumen and reduce particle size (Beauchemin and Buchanan-Smith, 1989) (Fig. 15.1).

Particle size reduction starts when herbage is severed during grazing, or when a bite of harvested forage is selected, and it continues during primary chewing and rumination. Rumination, which is the **regurgitation** and chewing of forage, greatly aids the process of particle size reduction. Controlled contractions of the reticulum wall coupled with relaxation of the **distal esophageal sphincter** propel a wad (**bolus**) of rumen matter into the **esophagus**. **Reverse peristalsis** moves the bolus up the esophagus and into the mouth, where it may be chewed up to 40 times before it is re-swallowed.

The rate at which particles are broken down by primary (biting and chewing) and secondary (rumination) mastication, and digested to a size that can exit the rumen, is a major factor influencing voluntary intake. For example, one study found that 36% of the large particles (>0.04 in. [10 mm]) of vegetative perennial ryegrass pasture were broken down (i.e., reduced to a smaller size) during rumination, compared with 61% of the large particles of alfalfa (John et al., 1988). The grass also required more chewing (about 14,000 chews/lb [6356 chews/kg]) to achieve particle size reduction than did alfalfa (8000 chews/lb [3632 chews/kg]). Emptying of the rumen also involves the absorption of end products of digestion, such as volatile fatty acids, through the rumen wall, and the passage of soluble constituents from the rumen at a more rapid liquid flow rate.

Particle size reduction in equids is essentially limited to biting and chewing in preparation for swallowing. Consequently, grazing horses and other equids tend to favor short sward surfaces. Ingested herbage enters the **glandular stomach**, where plant cell contents are subjected to monogastric digestion. Fiber passes down the digestive tract to the cecum, where fiber digestion occurs.

Rate of Passage

With forages that have a high cell wall content or slow rates of digestion, residues spend more time in the rumen and the rate of passage is decreased. For example, it was found that NDF from leaves of rhodesgrass and pangola-grass (digitgrass) spent 32 hours in the rumen being digested, compared with 46 hours for stem material from these same species (Poppi et al., 1980). Cattle consumed 2.8% of body weight (BW) of the leaf but only 2.1% BW of the stem, even though the digestibility was almost identical for the two fractions. Faster passage, like faster digestion, makes more space available in the rumen for additional intake. Mertens and Ely (1979) calculated that each 1% increase in rate of passage would translate into a 0.9% increase in maximum digestible dry matter intake.

Non-ruminant herbivores such as horses typically have a higher rate of passage and lower levels of fiber digestion than ruminants (Cymbaluk, 1990) (Table 15.2). In this study, horses and cattle digested similar percentages of alfalfa NDF, but cattle digested a higher percentage of kentucky bluegrass NDF. There is no equivalent of a particle size gate in the equine digestive tract, and active mechanisms of particle size reduction are absent. Although the equine stomach is small, there is little evidence that it exerts more than a very transient restriction on voluntary intake, because chewing stimulates gastric emptying. In other words, the faster a horse eats, the faster its stomach empties. Intake of the grazing horse is probably regulated by hunger and satiety mechanisms via sensors in the walls of the colon and cecum.

Table 15.2. Voluntary intake and digestibility characteristics of alfalfa and kentucky bluegrass hays fed to horses and cattle; faster rates of passage in horses reduce fiber digestion compared with ruminants

Utilization characteristic	Hay type	Herbivore species	
		Cattle[a]	Horses[b]
Voluntary intake (% body weight/day)	Alfalfa	2.8	2.7
	Kentucky bluegrass	1.8	2.0
NDF digestibility (%)	Alfalfa	54	51
	Kentucky bluegrass	71	51

Source: Cymbaluk, 1990.

[a]Cattle were Hereford steers with a body weight of 500 lb.

[b]Horses were crossbred geldings with a body weight of 740 lb.

Nitrogen Nutrition

Forage utilization by grazing livestock may be limited by N intake if the ingested herbage contains less than 6–7% crude protein. Livestock that are grazing improved herbage species or pastures with at least 15% of the mass as legumes are most likely to be limited by energy intake. Energy utilization of ruminants on range or C_4 grass pastures may be limited by N because the rumen microbes that release energy from cell walls may be starved of N.

Most herbage protein (85–95%) is degraded in the rumen to ammonia by **proteases** secreted by rumen microorganisms. This portion of the dietary protein is called **rumen degradable protein (RDP)**. Ammonia released from dietary protein is absorbed by rumen microorganisms and used to synthesize new microbial protein. Excess ammonia is absorbed through the rumen wall and moves via the bloodstream to the liver, where it is converted to urea. Some urea is recycled in the saliva, but much of it is lost in the urine. Because rumen microbes have the ability to use **non-protein nitrogen (NPN)**, urea or ammonia may be added to forage to meet a portion of the animal's protein needs. In diets that contain more than 12% crude protein equivalent, the efficiency of use of this NPN is reduced.

Only small amounts of ingested dietary protein escape digestion by the rumen microflora and move intact from the reticulorumen to the abomasum via the omasum. This protein is referred to as **escape protein**, **bypass protein**, or rumen undegraded protein (RUP). If protein is broken down faster than rumen microbes can utilize it, much of it can be lost to the animal because it is excreted as urea in the urine. This explains why animals may respond positively to supplementation of RUP even when they are consuming pasture with protein levels that appear to exceed their dietary requirements.

Digestion of tract microorganisms in the **lumen** of the abomasum is initiated by **gastric lysozymes** that bind to and attack the **mucopolysaccharides** of cell walls. Protein from ruptured microorganisms is hydrolyzed by **pepsin** to amino acids, and these are absorbed from the small intestine. Gastric lysozymes secreted by the abomasum are unique because they are able to function in an acid environment and resist degradation by pepsin.

Utilization of protein in some lush legume pastures may be improved if the solubility or rate of rumen protein degradation can be reduced. Legumes that contain **tannin** have more bypass protein than legumes such as alfalfa, which lack tannin. Tannin binds proteins and reduces their solubility. Birdsfoot trefoil and sainfoin are legumes that contain some tannin, and they do have a higher proportion of escape protein than alfalfa. Red clover does not contain tannin per se, but has **low-molecular-weight phenolic compounds** that condense enzymatically via phenol oxidase to form high-molecular-weight compounds similar to tannin. Mild heating of forages during storage (see Chapter 19) reduces protein solubility, but the quality loss and the risk of spontaneous combustion prevent the practical use of this approach.

Grazing Utilization of Forages

In typical livestock systems, grazing lowers the cost per unit of production because it avoids the expenses associated with harvesting, storing, and feed forage. However, forage utilization by grazing also introduces additional factors not previously considered that may affect animal performance. These include limits on the number and size of bites, the density of forage per unit area, **canopy structure**, and the amount of time that animals can spend grazing each day.

Grazing Strategies

The gastrointestinal tracts and digestion processes of herbivores that evolved to extract energy from plant cell walls influence their grazing strategies. Foregut fiber digesters such as cattle have developed a **grazing strategy** that is compatible with their efficient system for extracting energy from fibrous materials. Evolutionary processes in ruminant **progenitors** separated eating (grazing) from particle size reduction (rumination), minimized **diet selection** activities, and established effective **rumen detoxification** processes. These features permitted high rates of intake and shorter grazing times that reduced exposure to predators on the open grasslands. These herbivores could then ruminate and rest at night in the relative safety of forest fringes.

The ancestors of horses and other equids did not separate eating from particle size reduction. Their grazing strategy involved low rates of intake, longer grazing times, higher rates of passage through the tract, and less efficient fiber digestion. Equid progenitors relied on flight as their primary defense against predators.

Energy Relationships of Grazing Livestock

Energy expended in grazing increases the net energy used for maintenance (NE_m) and decreases the net energy available for growth and reproduction (NE_p). The energy used in grazing is quite variable because of the infinite array of intake-limiting situations in natural and managed grassland habitats. Energy expended in grazing is lowest on high-quality pastures with swards that favor high rates of intake. Effort involved in grazing increases proportionally with grazing time in grassland situations that support low rates of intake. Under ideal grazing conditions, NE_m increases by about 25% over cut forage, but under suboptimal conditions (e.g., in drought or on poor rangelands), NE_m may increase manyfold, leaving no energy for production. When the amount of energy expended in grazing reaches or exceeds the amount of energy gained, the animal may cease grazing, adopt behaviors that conserve energy, or reduce NE_m by altering its metabolism.

Interactions of Grazing and Gut Fill

Grazing patterns are controlled by the hunger–satiety center in the mammalian brain. Hunger and satiety of grazing ruminants are primarily regulated by a "gut fill" mechanism (Fig. 15.2). Grazing commences because of hunger, and first slows down and then stops with the onset of satiety. In response to stimuli, transducer cells in the gut wall synthesize **neurohormones** that are transported via the vascular system of the animal to the hunger–satiety center of the brain.

Grazing Processes

In typical grasslands, herbage is sparse and of low energy density. For example, a cow with a body weight of 1000 lb (454 kg) must graze 300 lb (136 kg) of fresh herbage (with 90% water content) each day in order to attain an intake of 3% of body weight (i.e., 30 lb/day [13.6 kg/day]). In a pasture with around 500 lb/acre (560 kg/ha) of available dry matter, this would require an animal to graze the available herbage in an area of about 0.06 acre (0.02 ha) each day.

Mechanics of Grazing

Grazing herbivores, including cattle and sheep, have blunt heads that are adapted for the ingestion of large amounts of herbaceous material at high intake rates (Fig. 15.6). The shape and movements of the head and jaws, and the space between the incisors and molars permits their muscular tongue and mobile **muzzle** to sweep herbage from a greater volume of pasture than would be indicated by the width of the mouth.

Sheep and cattle use their tongues and lips to select herbage for biting, and draw it into the mouth. A wad of herbage is then clamped between the incisors and the

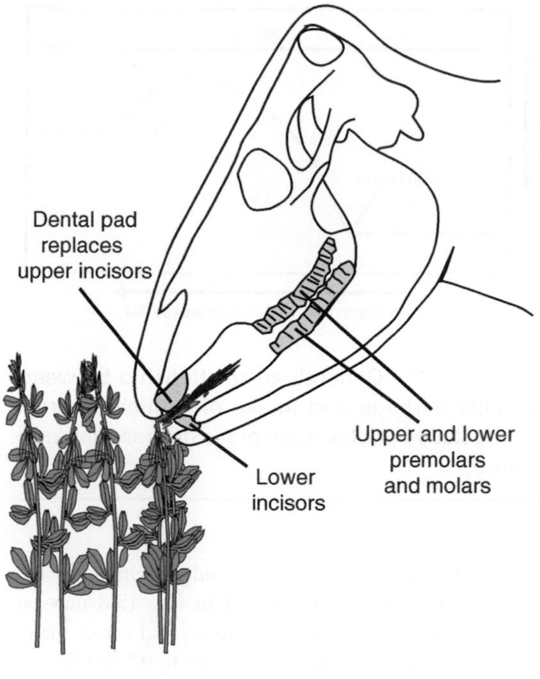

FIG. 15.6. Side view of the bovine skull. (Adapted from Hodgson, 1988.)

dental pad that substitutes for the upper incisors. The animal completes its biting action by sharply rotating the head upward about its pivot, shearing and tearing the herbage from the plant.

The condition of the teeth of grazing livestock is of great concern to managers, and is an important criterion for selecting animals for culling. Animals with excessively worn or diseased teeth may not be able to eat enough to survive.

Grazing Dynamics

Intake per day of grazing herbivores (I) is determined by the rate of intake (RI) and the time spent grazing each day (grazing time, GT).

$$I = RI \times GT$$

Rate of intake may be partly determined by animal size (mass), maturity, hunger or satiety status, herd behavior, and environment. Sward characteristics such as mass, allowance, structure, and forage quality affect intake rate. Rate of grazing is a function of intake per bite (IB) (also called **bite mass**) and rate of biting (RB).

$$RI = IB \times RB$$

To consume 3% of its body weight (13 lb [5.9 kg]) in dry matter, an animal with a body weight of 440 lb (200 kg) would have to take 6000 typical-sized bites per day. At a rate of 20 bites per minute, this would require steady grazing for 5 hours per day. However, bite size increases on denser swards. If herbage mass varied from 550 lb/acre to 2200 lb/acre (616 kg/ha to 2464 kg/ha), the number of bites needed to obtain 13 lb (5.9 kg) of dry matter could range from 12,000 to 3000 per day.

BITE MASS

Bite mass (IB) is the product of bite volume (V) and dry matter density (D) within that volume:

$$IB = V \times D$$

Because the tongue is involved in grazing, a calf with a body weight of 440 lb (200 kg) has an effective bite area of up to 12 in.^{2} (77 cm^2), more than twice the 5 in.^{2} (32 cm^2) bite area if only the incisors were used. In alfalfa with a **grazed horizon** of 5 in. (20 cm), the cylindrical volume of a bite of this area would be about 1 quart (0.95 L). An animal would need to take 6000 bites of this size to ingest 13.2 lb (6 kg) of dry matter per day, and would need to graze for about 5 hours per day.

The spatial distribution of herbage in a pasture is the single most important characteristic with regard to grazing. If, in the example presented above, the herbage density values in the 5 in. (12.5 cm) grazing horizon were 0.02, 0.04, 0.06, and 0.08 oz/quart (0.54, 1.08, 1.62, and 2.16 g/L) then that calf would need to take 12,000, 6000, 4000, and 3000 bites per day, respectively, to ingest 13.2 lb (6 kg) of dry matter per day. At 20 bites per minute this would require 10, 5, 3.3, and 2.5 hours of grazing per day, respectively.

The dry matter density of swards varies with herbage biomass, sward surface height, plant species, age of regrowth, **canopy structure**, lodging, trampling, plant disease, **flagging**, rainfall, wind, and other factors. For example, sward density may increase during the day when plants lose turgor. Herbage density is usually highest near the soil surface. In bermudagrass pastures, for instance, with a **sward surface height** of 2–3 in. (5.0–7.6 cm), there may be more than 2000 lb of dry matter per acre inch (145 kg DM per ha cm).

RATE OF BITING

Cattle are limited to about 80 jaw movements per minute, some of which involve oral manipulation of herbage, mastication, bolus formation, and swallowing. Beef cattle can sustain rates of 25–45 bites per minute for long periods on typical pastures. Sheep sustain rates of 35–50 bites per minute, and may achieve up to 134 bites per minute. Horses graze at slower rates than ruminants. Biting rate usually declines with increasing bite mass because of the additional jaw movements needed for mastication and bolus formation.

GRAZING TIME

A mature, dry, open cow in good condition will graze for about 4 hours per day on good pasture to meet her nutritional needs. However, cattle on range or pasture typically graze for 8 hours or more per day, which indicates that low allowance, low herbage quality, and low density restrict their rate of intake. Cattle seldom graze for more than 12–14 hours per day because they also need time for rumination and other individual and social activities. Because mobility is essential for grazing and livestock management, hoof health is important (see "Hoof Health").

Hoof Health

Grazing is affected by the condition of the animal's feet and legs because it must walk to feeding stations, as well as to shelter and shade, water, and supplement feeders. Herbage intake and productivity suffer when animals have misshapen, bruised, injured, or diseased hoofs. Cattle may walk 3 miles (4.8 km) per day when grazing good pastures, and much longer distances when grazing range where the density of forage per unit area is much lower. Hooves grow continuously, so hoof wear by abrasion during walking is needed for sound hoof health. Hoof diseases that affect locomotion and grazing include **laminitis** and foot abscess. Foot rot is highly contagious and potentially epidemic in sheep, and may affect productivity as well as sale and movements of livestock.

Soundness of limbs and hoofs is one criterion for culling of grazing livestock. Within some grassland farming regions, older grazing livestock with failing teeth or limbs that restrict locomotion and herbage intake may need to be moved to easier grazing conditions. For example, older ewes culled from flocks on western ranges may be shipped to eastern grasslands for one or more lamb crops before being slaughtered.

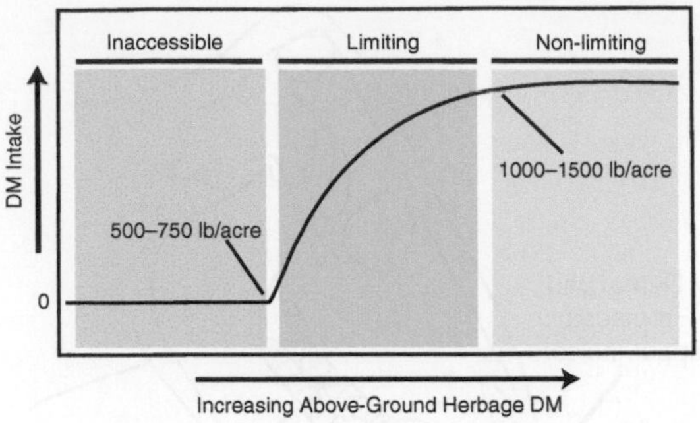

FIG. 15.7. Generalized relationship between available herbage and intake of ruminants grazing vegetative cool-season grass or grass–legume pasture.

HERBAGE BIOMASS AND INTAKE

A generalized relationship between intake and biomass as shown in Fig. 15.7 applies to many grazed temperate grass and grass–legume pastures. It also takes into account the adjustments that grazing animals make to account for low rates of intake. Herbage is considered unavailable to cattle when above-ground pasture mass falls below 500–750 lb/acre (560–840 kg/ha), because it may be too short or widely dispersed. Cool-season pastures above 1000–1500 lb DM/acre (1120–1680 kg DM/ha) do not appear to restrict intake.

Intake and Herbage Allowance

Herbage allowance describes the amount of herbage available to each livestock unit per unit of time, and is therefore related to stocking rate and carrying capacity. The generalized function depicted in Fig. 15.8 may be applicable to many grassland situations and across many time scales. Herbage intake increases up to 0.5 lb DM per 100 lb (0.23 kg/45 kg) body mass per hour of grazing as herbage allowance increases up to 1.0 lb DM per 100 lb (0.45 kg/45 kg) of body mass. The rate of intake does not increase above that allowance. Herbage intake rates decline when more than 50% of the herbage allowance is consumed. The function is displaced to the right when herbage becomes sparse, because animals spend more time traveling to new feeding stations.

Excessively high herbage allowances may lead to aging (maturing) swards, declining pasture growth rates, falling herbage and **diet quality**, **selective grazing**, increasing contamination of pasture with dung and urine, greater effort in grazing, low intake, and poorer animal production.

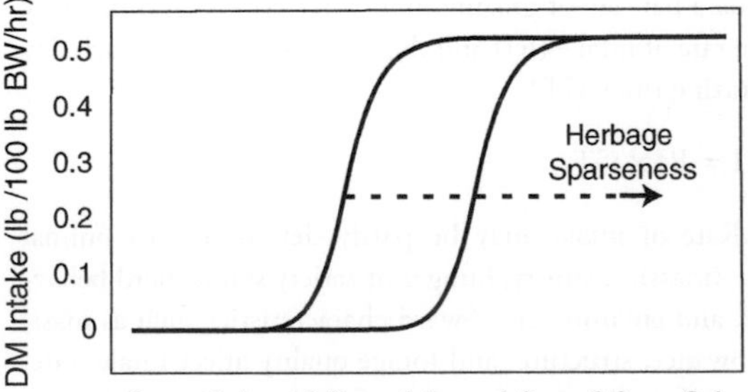

FIG. 15.8. Generalized relationship between herbage intake and herbage allowance. DM, dry matter; BW, body weight.

Feeding Deterrents

The location of the growing points of vegetative grasses at or below the soil surface protects the apical meristems from wind, fire, and grazing. Some plants also have physical deterrents to grazing, such as hairs (**trichomes**) or sharp surfaces on the edge of the leaf blade.

Chemical **feeding deterrents** may reduce herbage intake by early induction of satiety and decreasing grazing time. The **mycotoxins** of endophyte-infected tall fescue interfere with heat regulation mechanisms and alter grazing patterns (decreasing daytime grazing and increasing nighttime grazing) and herbage intake (see Chapter 16). Mycotoxins can also lower short-term intake by affecting well-being (e.g., inducing lethargy) and long-term intake by reducing energy demand (e.g., slowing growth, lowering the conception rate, reducing lactation).

Diet Learning and Selection

Young herbivores learn which plants to graze and which to avoid by mimicking their dams. **Diet learning** is particularly important for foals because horses cannot eject toxic plants by vomiting, and for young ruminants because their rumen has not yet developed the adult capacity to neutralize **toxicants**.

Herbivores use a number of cues in diet selection. Grazing ruminants usually express preferences that reflect pasture quality characteristics and diet learning (Table 15.3). Livestock use combinations of sight, taste, smell, and tactility to make diet selections.

Table 15.3. Summary of commonly observed and often highly correlated sward preferences of grazing livestock

Most preferred	Least preferred
Younger plant tissues	Older plant tissues
Greener plant tissues	Other colored plant tissues
Darker plant tissues	Lighter colored plant tissues
Leafy plant tissue	Stemmy plant tissues
Tissues with high water content	Drier plant tissues
Low-fiber plant tissues	High-fiber plant tissues
Herbage legume species	Herbage grass species
Tissues with fewer trichomes	Tissues with many long trichomes
Absence of grazing barriers	Grazing barriers (e.g., thorns, pseudostems)
High-sodium species	Low-sodium species
Sweeter tissues	Other tissues
Non-toxic plant issues	Toxic plant tissues
Clean plant tissues	Tissues contaminated with dung

Color is thought to be unimportant in diet selection by cattle, except for the fact that long (red) wavelengths are avoided, perhaps because the sight of blood may warn of predator activity. Color contrasts may help animals to select young, green tissues for ingestion. Sight is also involved in **spatial perception** and memory (e.g., the location of preferred grazing sites, water, and supplement feeders).

Taste and odor sensations are closely related, and are detected by sensors in the epithelial lining of the **oral cavity** and the tongue. These senses may not be very important in ruminants because the oral cavity is flooded with rumen fluid during rumination, and flushed with rumen gases during eructation.

Nutrition of Grazing Equids

The grazing behavior of **hindgut fiber digesters** contrasts with that of ruminants. Equids retained their upper incisors, so they bite rather than shear herbage. Horses have slow-growing, high-crowned, short-rooted teeth (**hypsodont teeth**) with extensive ridged grinding (**occlusal**) surfaces on the molars and molarized premolars, typical of fiber digesters. The equid tongue is less involved in grazing processes than that of ruminants because a lingual cartilage limits its flexibility and its ability to grasp herbage. Equids compensate for their limited tongue mobility by using their highly mobile muzzle when grazing.

Diet selection is extremely important for grazing equids because they are very susceptible to dietary poisoning. This is partly because vomiting of ingested toxic plants is restricted by the anatomy of the equine esophagus, and partly because they have no equivalent to rumen detoxification. Plant feeding deterrents that enter the equine stomach may remain unchanged or they may be activated in the acid stomach environment. Toxic substances are readily absorbed in the small intestine, and detoxification occurs by hepatic processes. In order to avoid poisoning, foals rely on their dams for diet learning by **allelomimicry**.

The nitrogen nutrition of the horse is quite different from that of ruminant livestock. Ingested protein is hydrolyzed to its constituent amino acids in the stomach, and the amino acids are then passed to the small intestine for absorption. Thus the amino acids absorbed by the equine are mainly derived from plant proteins, whereas those absorbed by ruminants are mainly derived from rumen microorganisms. Some dietary amino acids pass through to the equine cecum and colon, where ammonia is cleaved and utilized for the production of microbial protein. The cecum, colon, and rectum of equids lack lysozymes or other hydrolytic enzymes, so these microbes are excreted in the dung and little protein is recovered. However, foals may recover some protein by eating dung (**coprophagy**).

Equine dung is usually drier and has larger particles than bovine dung. Accumulations of dung in distinct piles may indicate social hierarchies within herds, but are not considered to be territorial markers. Equids are particularly averse to grazing in proximity to dung of their own species, perhaps to minimize exposure to parasites. This aversion, together with the preference for short swards, creates mosaics of shorter and taller swards. This pattern has been described as "lawns" and "roughs", and has been termed "**spot grazing.**" Because cattle prefer taller swards and are not averse to grazing near equine dung, they are compatible as co-grazers with horses in the same pasture. Sheep also prefer short swards, and are competitive with horses in mixed grazing systems.

Forage Utilization by Dairy Cattle

Lactating dairy cattle need high intakes of dry matter and energy to sustain high milk production rates and maintain animal condition. Figure 15.9 shows digestible dry matter intake of alfalfa hay cut at different maturity stages and fed in lactating cow rations containing from 20% to about 70% concentrate (Kawas et al., 1989). Cows consumed more digestible dry matter when their rations included the earlier-maturity alfalfa hays. Increasing the concentrate level in the ration could not compensate for the negative effects of late-maturity hay on digestible dry matter intake.

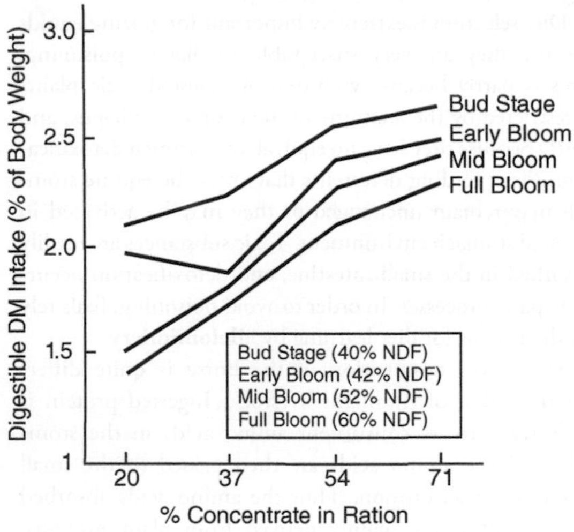

FIG. 15.9. Voluntary digestible dry matter intake (DDMI) by lactating dairy cows fed alfalfa hays at different stages of maturity. Each hay was fed at four different concentrate levels. The data show that DDMI is higher for early-maturity hay, and that increasing the concentrate feeding level does not compensate for the negative effects of advanced forage maturity. NDF, neutral detergent fiber. (Source: Kawas et al., 1989. Reproduced with permission of the College of Agricultural and Life Sciences, University of Wisconsin-Madison.)

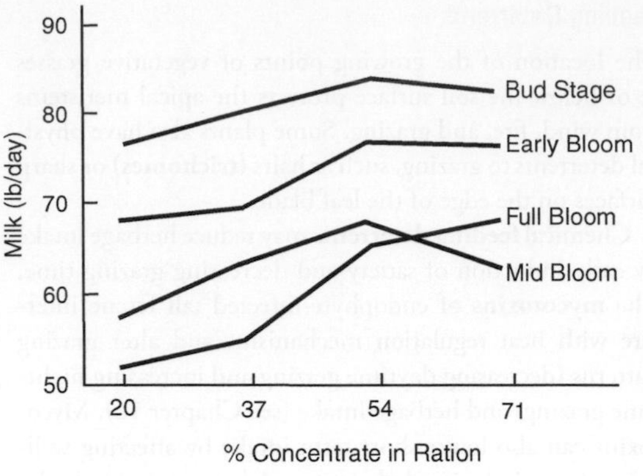

FIG. 15.10. Daily milk production levels for lactating dairy cows fed alfalfa hays at different stages of maturity. (Source, Kawas et al., 1989. Reproduced with permission of the College of Agricultural and Life Sciences, University of Wisconsin-Madison.)

Milk production reflected digestible dry matter intake, being highest for the rations with early-maturity alfalfa hay (Fig. 15.10). Cows maintained their body weight during the study period when their rations contained less than about 32% NDF or 23% ADF, so it was concluded that rations containing about 50% concentrate together with early-maturity alfalfa forage were optimal (Kawas et al., 1989). Lactating dairy cattle are also efficient users of high-quality grazed forages.

Summary

The nutritive value of forage depends not only on factors intrinsic to the plant, but also on the characteristics of the animal that is consuming it. The ability of herbivores to consume fibrous feeds reflects evolutionary changes to their digestive systems to enable them to accommodate large volumes of forage with a high fiber content. Animal performance depends on a number of interrelated factors, of which digestibility and intake are of key importance. Voluntary intake of forage is often limited by fill in ruminant livestock, due to the bulk related to its high fiber content. The rates at which forage is broken down and digested affect the amount that an animal can consume. Forages that are more rapidly broken down and digested are consumed at higher levels because their passage and absorption in the intestinal tract are faster. The intake of forage by grazing animals is often affected by their ability to consume what is available. Increasing forage allowance generally increases intake, because less time and effort need to be spent grazing.

Exploitation of grasslands and rangelands for the benefit of humans depends on the understanding and management of complex ecosystems involving herbivores that evolved gastrointestinal tracts that could extract the chemical energy stored in plant cell walls.

Questions

1. Describe the unique conditions of the rumen that allow fiber digestion in ruminants.
2. List four factors that contribute to forage quality.
3. Describe the general relationship between voluntary intake and forage cell wall concentrations. Under what conditions might this typical relationship not hold true?
4. Describe the general relationship between forage dry matter intake and pasture biomass.
5. Animals automatically adjust the amount of grazing time to compensate for differences in herbage density. Above what grazing time is such compensation no longer able to maintain intake?
6. What are typical biting rates during grazing for cattle, sheep, and horses on pasture?

References

Beauchemin, KA, and JG Buchanan-Smith. 1989. Effects of dietary neutral detergent fiber concentration and supplementary long hay on chewing activities and milk production of dairy cows. J. Dairy Sci. 72:2288–2300.

Collins, M. 1988. Composition and fibre digestion in morphological components of an alfalfa-timothy sward. Anim. Feed Sci. Tech. 19:135–243.

Cymbaluk, NF. 1990. Comparison of forage digestion by cattle and horses. Can. J. Anim. Sci. 70:601–610.

Ellis, WC, JH Matis, TM Hill, and MR Murphy. 1994. Methodology for estimating digestion and passage kinetics of forages. In GC Fahey, M Collins, DM Mertens, and LE Moser (eds.), Forage Quality, Evaluation, and Utilization, pp. 682–756. American Society of Agronomy. Madison, WI.

Hodgson, J. 1988. Grazing Management: Science into Practice. John Wiley & Sons Inc., New York.

John, A, KE Kelly, BR Sinclair, and CSW Reid. 1988. Physical breakdown of forages during rumination. Proc. NZ Soc. Anim. Prod. 48:247–248.

Kawas, JR, NA Jorgensen, AR Hardie, M Collins, and GP Barrington. 1989. Assessment of alfalfa quality with change in maturity and concentrate level. Univ. Wisconsin-Madison College of Agric. and Life Sci., Res. Rep. R3460. University of Wisconsin-Madison, Madison, WI.

Maynard, LA, and JK Loosli. 1969. Animal Nutrition, 6th ed. McGraw-Hill Book Company, New York.

Mertens, DR, and LO Ely. 1979. A dynamic model of fiber digestion and passage in the ruminant for evaluating forage quality. J. Anim. Sci. 49:1085–1095.

Minson, DJ. 1982. Effects of chemical and physical composition of herbage eaten upon intake. In JB Hacker (ed.), *Nutritional Limits to Animal Production from Pastures*, pp. 167–182. Commonwealth Agricultural Bureaux, Slough, UK.

Moore, KJ, and DR Buxton. 2000. Fiber composition and digestion of warm-season grasses. In Native Warm-Season Grasses: Research Trends and Issues, pp. 23–33. Crop Science Society of America, Madison, WI.

Poppi, DP, DJ Minson, and JH Ternouth. 1980. Studies of cattle and sheep eating leaf and stem fractions of grasses I. Voluntary intake, digestibility and retention time in the reticulo-rumen. Aust. J. Agric. Res. 32:99–108.

Smith LW, HK Goering, and CH Gordon. 1972. Relationships of forage compositions with rates of cell wall digestion and indigestibility of cell walls. J. Dairy Sci. 55:1140–1147.

CHAPTER

16

Forage-Related Animal Disorders

Michael Collins, Ali M. Missaoui, and Nicholas S. Hill

Introduction

Animals sometimes experience negative consequences from compounds contained in the forages that they consume, and managers must be aware of forage-related factors that can cause health problems. Imbalances of **nutrients** can also have negative effects on animals. The occurrence of health problems is affected by interactions between plant species, environmental conditions, and management practices.

Understanding the biochemistry involved in these disorders can lead to more appropriate management and preventative measures. Some **toxic** compounds are naturally occurring constituents of the plants themselves, while others may result from insect infestations, as in the case of blister beetle (*Epicauta* species), or from microbial activity, as in the case of botulism in hay and silage.

Why should plants contain compounds that are harmful to animals? Evidence indicates that anti-quality compounds often serve as mechanisms to prevent herbivore feeding (i.e., as anti-herbivory agents). This characteristic is beneficial to the plants because it prevents the adverse effects of overgrazing. For example, high levels of ergot alkaloids in tiller bases of endophyte-infected grasses adversely affect grazing livestock and reduce forage consumption. Such defensive roles are also clearly demonstrated in plants that have sharp thorns or spines. On the other hand, many anti-quality compounds appear simply to be metabolic intermediates or products that are not needed by the plant for growth and development. As a broad group, such compounds are called **secondary metabolites**.

Forage-related animal disorders can be categorized into three main groups:

1. Poisonous plant disorders: these are caused by plants which are toxic to animals that consume them.
2. Seasonal and conditional disorders: these occur as a result of interactions between environmental conditions, management, and animal factors.
3. Species-related disorders: these are uniquely associated with a particular forage species (Table 16.1).

Some overlap exists, as certain disorders have characteristics of more than one of these categories.

Poisonous plant disorders are distinguished from seasonal and species-related disorders by the fact that poisonous plants always or nearly always contain toxins and should be avoided. Another distinguishing characteristic is that poisonous plants are always native or invading species, not seeded species. Other disorders may occur only under certain environmental or animal conditions or, as in the case of alkaloids of tall fescue and ryegrass, may reduce animal performance on otherwise productive forage species (Table 16.2).

Poisonous Plant Disorders

Poisonous plants are undesirable native or invading plants that may be present in pasture or rangeland situations. To be considered poisonous, plants must contain some toxic compound that is harmful to livestock. The toxic effects may consist of chronic negative effects on growth rate or reproductive performance, or they can be acute, sometimes

Forages: An Introduction to Grassland Agriculture, Seventh Edition. Edited by Michael Collins, C. Jerry Nelson, Kenneth J. Moore and Robert F Barnes.

Table 16.1. Forage plant families and species that sometimes contain toxic compounds in sufficiently high concentrations to harm animals that consume them

Forage family	Common name	Anti-quality compound
Poaceae (grasses)	Forage sorghums	Cyanogens (HCN)
	Tall fescue	Ergot alkaloids (ergoline)
	Perennial ryegrass	Tremorgens (lolitrem B)
	Tropical grasses	Oxalates, saponins
Fabaceae (legumes)	Alfalfa	Saponins, phytoestrogens, bloating agents
	White clover	Cyanogens, phytoestrogens, bloating agents
	Red clover	Slaframine, phytoestrogens, bloating agents
	Alsike clover	Photosensitization agents
	Sweetclover	Coumarin (dicoumarol)
	Subterranean clover	Phytoestrogens
	Crown vetch	Glycosides
Brassicaceae (brassicas)	Turnip	Brassica anemia factor (goitrogens)
	Rape	Glucosinolates

even causing rapid death. Table 16.2 lists some common poisonous plants encountered in forage–livestock systems, and describes their effects on the animal.

It seems logical that animals should prefer to avoid consuming toxic plants. Herbivores tend to select nutritionally superior diets, especially with regard to energy and protein content, and so they have some ability to avoid toxins in the diet. Young animals in the herd learn from their dams (mothers) which plants to consume and which to avoid. Even so, livestock disorders resulting from ingestion of poisonous plants are not rare, especially under conditions of **overgrazing**, or in early spring when forage availability is limited. When animals are very hungry, the drive to feed may override their aversion to eating plants that they might otherwise avoid.

Seasonal and Conditional Disorders

Seasonal and conditional disorders are those that occur only under certain environmental conditions, at certain plant growth stages, or at certain susceptible stages for animals (Table 16.3). For example, mineral uptake patterns of forage plants are affected by environmental temperatures, and affect the occurrence of **hypomagnesemic tetany** (also called **grass tetany**). Animals are more sensitive to disorders at some ages or reproductive stages than at others. Tolerances for forage toxins are often lower during **gestation**, **lactation**, and weaning than during other life stages. Mares during gestation and lactation are particularly sensitive. Some of the most common seasonal disorders are discussed in the following sections.

Grass Tetany

Grass tetany, also called hypomagnesemic tetany or grass staggers, is a metabolic disorder characterized by low blood magnesium levels (Table 16.3). This disorder occurs throughout the USA and in other parts of the world. Clinically, grass tetany is a response to inadequate blood serum magnesium levels. Grass tetany occurs most frequently during the transition from winter to spring as temperatures rise into the range of 40–60°F (4.4–15.6 °C), encouraging a rapid flush of grass growth. It is most common on pure grass pastures, but can occur on pastures containing legumes or on grass hays.

The occurrence of tetany on grass pastures is increased when:

- The Mg concentration in the forage falls below 0.2% of dry matter
- K concentrations in the forage are high
- High rates of N fertilizer are applied
- The ratio of K/(Ca + Mg) is higher than 2.2 on an equivalent basis.

High K levels in the forage are associated with reduced availability of Mg from the ruminant digestive tract. This may reduce serum Mg levels even when forage concentrations appear to be adequate. High levels of K often occur in conjunction with elevated levels of non-protein N, fatty acids, and certain organic acids that can reduce Mg **absorption**.

Cattle with early symptoms of grass tetany exhibit nervous behavior, twitching, and may graze away from the herd. With further progression, animals collapse and have convulsions. Affected animals may arch their head back and thrash their legs in a sweeping arc referred to as "paddling" (Fig. 16.1). Older brood cows nursing calves under 2 months of age are most susceptible, because remobilization of Mg from their bone tissue is less effective than in younger animals.

Table 16.2. A few examples of plants that can be poisonous to herbivores

Common name	Scientific name	Plant identification	Toxic constituent	Affected animals	Symptoms
Common groundsel	*Senecio vulgaris*	Annual, yellow flowers with black tips in clusters	Pyrrolizidine alkaloids	All livestock	Wobbling, appetite loss, lethargy, crusty eyes and nose
Poison hemlock	*Conium maculatum*	Umbelliferae	Coniine, a pyridine alkaloid	Cattle, horses, sheep, goats	Loss of muscular strength, tremors, salivation, coma, cleft palate in newborn animals
Rattlebox	*Crotalaria sagittalis*	Annual/perennial flowers long terminal clusters	Pyrrolizidine alkaloids	Horses, cattle	Bloody diarrhea, shallow breathing, stiff gait
Tarweed, fiddleneck	*Amsinckia intermedia*	Erect, hairy, 2–3 ft tall, yellow/orange flowers	Pyrrolizidine alkaloids	Horses, swine, cattle	Diarrhea, weight loss, red urine, aimless walking
Hound's tongue	*Cynoglossum officinale*	Annual, biennial	Pyrrolizidine alkaloids	Cattle, horses	Head pressing, straight-line walking
Cherry	*Prunus* spp.	Several species (e.g., black cherry, choke cherry)	Cyanogenic glycosides	Cattle, sheep, goats, horses	Rapid or slow, difficult breathing, rapid heart rate, coma, death
Oak	*Quercus* spp.	Oak twigs, buds, acorns	Tannins	Cattle	Anorexia, nasal discharge, thirst
Nightshade, Jerusalem cherry	*Solanum* spp.	Various species, flowers similar to those of tomatoes	Alkaloids, glycoalkaloids	Cattle, horses, sheep, swine	Muscle tremors, weakness, colic, excess salivation

Sources: Burrows and Tyrl, 2001; Herron and LaBore, 1972; Kingsbury, 1964; Turner and Szczawinski, 1991.

Table 16.3. Seasonal or conditional disorders may be caused by climate, soil, plant, or animal factors, or by a combination of these; plants also produce toxic intermediary metabolites in response to climate stress or excess nutrients

Disorder	Description	Symptoms	Plants	Animals	Prevention	Treatment
Grass tetany (hypomagnesemia)	Inadequate blood serum magnesium levels	Stiff gait, staggering, twitching muscles, convulsions	Cool-season grasses in early spring	Brood cows, ewes, often heavy milkers in early lactation	Split K and N fertilizer applications	Provide Mg supplements, intravenous injections of calcium-gluconate solution, subdermal injections of magnesium
Frothy bloat (ruminal tympany)	Inability to eructate rumen gases	Distended rumen visible first on the left side	Alfalfa, red and white clover, lush grass pastures including wheat	Cattle, sheep	Use anti-bloating agents, use grass–legume mixtures, supplement ionophore monensin, strip graze	Provide hay or more mature grass forage
Nitrate poisoning	Toxic accumulation of nitrate converted to nitrites that bind hemoglobin	Labored breathing, abdominal pain	Sudangrass, oats, rape, wheat, corn, on high-N soils or under drought conditions	All livestock	Split N applications when high annual rates are used	Remove from toxic plants, dilute with other forage
Prussic acid poisoning [hydrocyanic acid (HCN) poisoning]	Young, wilted, frosted, or stunted plants; glycosides degrade to release HCN	Muscle tremors, rapid breathing, convulsions, asphyxiation	Sorghums, johnsongrass, white clover, vetch seed	Cattle	Avoid stunted, frosted plants; split N fertilizer application, test samples, prevent selective grazing	Remove from causative plants, dilute with other forage

Phytoestrogens	Plant compounds that mimic estrogens and cause reproductive problems	Poor reproductive performance; in severe cases there are visible changes	Certain species of forage legumes	Sheep are more susceptible than cattle	Replace problem species or avoid use during susceptible periods	Subclinical effects are alleviated by removal from problem forages; permanent infertility can result
Primary photosensitization	Blood compounds reacting with ultraviolet light on the skin, producing free radicals that react with dermal tissue proteins	Dermatitis affecting light skin	*Hypericum* spp., buckwheat, bishop's weed, spring parsley, lush pastures	Cattle, horses, goats, sheep	Remove from food source	Dependent upon toxicant
Facial eczema	Secondary photosensitization caused by the hepatotoxic mycotoxin *sporidesmin*	Severe dermatitis of light-skinned areas	Ryegrass pastures following warm, wet weather	Sheep, cattle	Provide zinc supplements and iron salts, manage around affected pastures	Provide shade

Sources: Cheeke, 1998; Kingsbury, 1964; Bush and Burton, 1994.

FIG. 16.1. A beef cow undergoing hypomagnesemic tetany (grass tetany) (top) followed by treatment with an intravenous injection of a Ca-Mg-gluconate solution (middle) and rapid recovery (bottom). (Photos courtesy of Vivien Allen.)

Strategies for avoiding grass tetany include direct supplementation of the animal with Mg, careful N and K fertilization management, Mg fertilization of the forage, supplementing dietary energy, and S fertilization of the forage. Direct supplementation of Mg in the diet usually ensures that animals consume adequate quantities of Mg. Supplementation is usually accomplished by adding 75–150 lb (34–68 kg) of MgO per ton of salt/mineral mix. Salt (NaCl) improves the **palatability** of MgO, which is not well liked by animals. For dairy cows, an intake of 1.1 oz Mg (2 oz MgO) [31 g Mg (57 g MgO)] per day is recommended. For lactating ewes, 0.11 oz (3.1 g) of Mg per day is typically recommended. Other Mg supplements include epsom salts ($MgSO_4 \cdot 7H_2O$), which can be added to the drinking water. Splitting fertilizer applications, especially N, can help to avoid the negative effects of high N and K levels in the forage on Mg availability to animals. Applying dolomitic limestone, which contains significant amounts of Mg, to pastures instead of regular limestone can increase plant concentrations of Mg. In most production situations, forage plants generally do not benefit much from Mg addition, so direct supplementation to the animal may be more efficient.

Feeding supplemental sources of readily available energy, such as cracked corn or molasses, enhances rumen microbial growth and N utilization and decreases the incidence of grass tetany. The more acidic rumen pH that results from rapid carbohydrate digestion appears to increase Mg absorption. Fertilizing with S if needed to maintain an N:S ratio in the diet at or below 12:1 can also improve protein utilization in the rumen.

Animals that are showing early symptoms of grass tetany (prior to coma) can be given intravenous injections of a Ca-Mg gluconate solution (Fig. 16.1). Subcutaneous injections of a saturated solution of magnesium sulfate may also be effective. Certain animals appear to be more susceptible to grass tetany, so this factor should be considered when making herd-culling decisions.

Although this approach is generally impractical, the incidence of grass tetany can be reduced by holding cattle off grass pastures in early spring until they reach a shoot height of about 6 in. (15 cm). Maintaining beef brood cows on a moderate or higher plane of nutrition and/or feeding legume hay to cattle grazing grass pastures in spring also reduces the incidence of grass tetany.

Bloat

Legume (frothy) bloat occurs when stable foam forms at the surface of the floating raft of actively digesting forage in the rumen and blocks access to the esophagus, causing gases to accumulate (Fig. 16.2). Calculated death losses in the USA and Canada due to bloat are 0.5% of the cattle population annually. Gases, especially carbon dioxide (CO_2) and methane (CH_4), are produced in large quantities during fermentation of forages in the rumen (Table 16.3 and Fig. 16.2). Ruminal gas production in animals that are grazing lush legume forages can exceed 0.5 gal (1.9 L) per minute in cattle, and about 0.2 gal (0.8 L) in sheep. These gases are produced during forage fermentation, and normally cause no problem for the animal because they escape through the distal esophageal sphincter and up the esophagus in a process called **eructation**. Bloat can also occur in cattle that are consuming high-grain rations in feedlots, but acute bloat is most common when ruminants graze succulent legume pastures. Bloat occurs rarely in animals that are consuming hay.

Bloating animals exhibit distention of the rumen on the left side (Fig. 16.2). The onset of bloat may occur very quickly after initiation of grazing, due to the large quantities of gas that can be produced rapidly in the rumen. Other symptoms include cessation of grazing, frequent urination

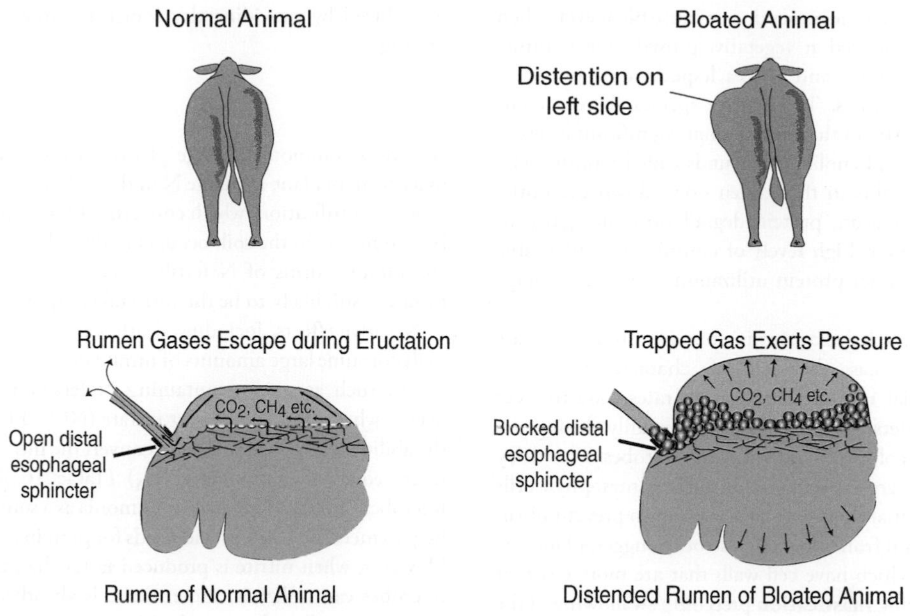

Fig. 16.2. Pasture (frothy) bloat in ruminants typically occurs when animals consume lush vegetative forage that ferments rapidly upon entering the rumen. The formation of stable foam at the surface of the rumen raft blocks access of gas to the distal esophageal sphincter and prevents eructation. Large volumes of gas can be produced under these conditions, so the onset of bloat can be apparent soon after grazing begins.

or defecation, labored breathing, and restless movements. In more severe cases the animal also becomes distended on the right side. Death may occur within minutes after the onset of acute bloat. Subacute bloat may cause a slight distention of the left side with no observable negative effects on the animal.

Attempts to identify a single causal factor in legume bloat have consistently failed. Rapid gas production, larger amounts of foam production, and greater foam stability characterize legumes that cause bloat. Pasture bloat is most common in ruminants grazing lush swards of legumes such as alfalfa, red clover, white clover, other clovers, and vetches. However, bloat can occur in stocker cattle grazing lush winter wheat pastures. Among grass forages, **vegetative** winter wheat is unusually low in fiber and high in soluble protein. It also appears that low concentrations of Ca in wheat pasturage may make bloat more likely by reducing **rumen motility**. The factor that is consistently present when pasture bloat occurs is the formation of stable foam in the rumen, which blocks eructation.

Foam production in bloating animals results from a complex interaction of animal, plant, and microbiological factors. Majak et al. (1995) reviewed Canadian research on this subject and concluded that alfalfa is most likely to cause bloat when:

- The forage is vegetative rather than more mature forage
- Grazing takes place early in the day rather than later in the day
- The forage is wet with dew
- The forage contains high levels of K and/or Ca and Mg
- The forage contains high levels of soluble protein
- Animals graze for only a few hours each day rather than for the entire 24-hour period.

The presence of high levels of soluble protein is also associated with increased occurrence of bloat. MacAdam and Whitesides (1996) implicated higher soluble protein concentrations resulting from cooler growth conditions in explaining the increased incidence of bloat observed in the cooler, high-altitude valleys of the intermountain West.

The incidence of bloat is greatly reduced by grazing grass–legume mixtures rather than pure stands of legumes. Hungry animals should not be introduced onto lush pastures of bloating legumes, especially in the early morning when dew is present on the forage.

Some "bloat-safe" legumes do not cause bloat even when pure stands are grazed at vegetative growth stages. Birdsfoot trefoil, sainfoin, and sericea lespedeza are examples of bloat-safe legumes. These three species share a common characteristic in that they contain significant levels of protein-binding phenolic compounds called tannins. The presence of tannins in the rumen slows down the initial rates of fermentation, protein degradation, and gas production. However, high levels of tannins are undesirable because they hinder protein utilization and reduce forage palatability.

Cicer milkvetch lacks tannins but is also a bloat-safe legume. In this case the bloat-safe characteristic is due to slower initial rumen fermentation rates, due to cicer milkvetch leaflets breaking down less readily during the early stages of colonization by rumen microbes. This delay in access of microbes to some of the leaf mesophyll cells slows down initial digestion rates enough to prevent bloat. Some protection from bloat has also been suggested for certain legumes which have cell walls that are more resistant to rupture during **mastication** preceding swallowing of the forage bolus. The slower release of fermentable materials and protein into the rumen reduces the initial rates of gas production.

Individual animals differ in their susceptibility to bloat. Possible explanations for these inter-animal differences include (1) differences in salivary flow and composition, (2) differences in forage intake rate, (3) differences in total forage intake, (4) differences in rumen gas production rates, and (5) rumen conditions that affect foam stability and other factors. Rapid intake rates and higher intakes would increase initial rumen gas production rates and the likelihood of bloat. Researchers found that rumen clearance rates (i.e., the rate at which liquid and forage material leave the rumen) were slower in animals that bloated compared with animals that did not bloat (Majak et al., 1995).

Bloat can be treated by introducing vegetable oil or another antifoaming agent directly into the rumen using a 0.75-in. (1.9-cm) diameter hose. The hose itself may relieve pressure by providing a route for gas to escape. Bloat can largely be controlled by feeding antifoaming agents such as poloxalene (Bloat Guard) or laureth-23. In New Zealand, where dairy cows routinely obtain most of their nutrients through grazing, the use of bloat-control chemicals is common. This approach works well for dairy cattle because milking presents an easy opportunity for dosing.

Lick blocks containing bloat-preventive materials are available for beef cattle, but it is difficult to ensure adequate daily intake for each individual animal. Supplementation of cattle diets with monensin, an ionophore, reduces bloat incidence. **Strip grazing** (see Chapter 19) reduces the incidence of bloat because animals consume a greater proportion of the available forage, including stem tissue that is lower in protein and less rapidly digested, rather than selectively grazing succulent shoot tips. Bloat can also be reduced by providing dry forage to animals prior to grazing.

Nitrate Toxicity

Nitrate is common in forage plants because it is the primary form of plant-available N in the soil under most conditions. Nitrification, which converts ammonium forms of N into nitrate in the soil, occurs rapidly. Thus, even when ammonium forms of N fertilizer such as urea are used, nitrate is still likely to be the form taken up by the crop.

Negative effects, including death, can occur when animals consume large amounts of nitrate in forage or in other forms, such as nitrate-containing N fertilizers. Toxicity occurs when ruminants ingest nitrate (NO_3^-) in excess of the ability of rumen microbes to convert the nitrite (NO_2^-) intermediate to ammonia (NH_3) (Table 16.3). Rumen microbes can effectively utilize ammonia as a source of N to help to meet their amino acid needs for protein production. However, when nitrite is produced faster than the rumen microbes can utilize it, some nitrite is absorbed through the rumen wall and enters the bloodstream. Once it is in the bloodstream, nitrite converts hemoglobin into methemoglobin, restricting its ability to transport oxygen to the body tissues.

Small amounts of nitrate are not harmful. Forage with up to 0.44% nitrate in the dry matter is considered safe to feed for all classes of cattle (Table 16.4). Forage containing more nitrate should be fed with increasing caution, especially to pregnant animals.

Symptoms of nitrate poisoning include rapid breathing, muscle tremors, incoordination, diarrhea, and frequent urination. Blood from affected animals takes on a "chocolate brown" color instead of its normal bright red color. If nitrate poisoning is suspected, the animals should be removed to a different forage source. Feeding an energy supplement such as corn can hasten microbial utilization of nitrate, thus reducing nitrite accumulation in the rumen.

Table 16.4. Recommended feeding management of forages with different levels of nitrate; these values are relatively conservative

Nitrate concentration	Recommended management
0–0.44%[a]	Safe to feed
0.44–0.88	Generally safe for non-pregnant cattle, up to 50% of the ration for pregnant animals
0.88–1.5%	Not more than 25% of the ration
> 1.5%	Do not feed

[a] If nitrate N is reported, multiply these nitrate concentration values by 0.23.

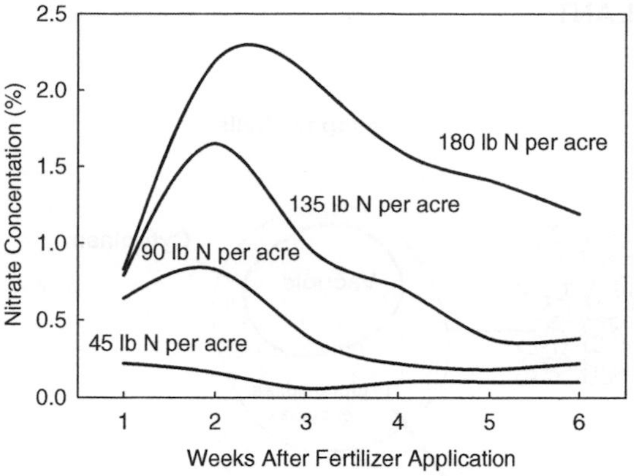

Fig. 16.3. Illustration of nitrogen fertilization effects on nitrate patterns in tall fescue forage. Nitrate accumulates following fertilization in proportion to the amount of nitrogen added. Nitrate is metabolized under suitable growing conditions, and the level in forage declines over the 2 or 3 weeks following fertilization. (Adapted from Hojatti et al., 1973.)

Treatment involves intravenous injection of a methylene blue solution.

Nitrate toxicity usually occurs with grasses, because they are most likely to receive routine N fertilization. For example, Hojatti et al. (1973) found that nitrate concentrations in tall fescue pastures reached over 2% of dry matter during the period of 2–3.5 weeks following N application (Fig. 16.3). Levels normally decline rapidly as nitrate is metabolized in the plant. Warm-season annual grasses such as corn, sorghum, and pearlmillet receive high rates of N fertilizer and may contain high concentrations of nitrate at certain growth stages.

Weather conditions that slow or prevent nitrate metabolism in the plant can cause high concentrations to be maintained indefinitely. Drought conditions, frost, plant diseases, and cloudy weather are examples of situations that contribute to the maintenance of high nitrate levels. Drought stress can prevent plants from metabolizing the nitrate taken up from the soil. A common situation in which the possibility of nitrate toxicity should be considered is when forage from drought-stressed corn, sorghum, or other grass-type grain crops not suitable for grain harvest is grazed or harvested for feeding livestock.

Among shoot morphological components, stems usually have the highest nitrate levels, especially the lower one-third of the stem. Young plants or younger tillers on older plants generally have the highest nitrate content. If this portion of the shoot is found to have low levels of nitrate, it is likely that the remainder of the plant also has low levels.

Green-chopped forage is especially prone to causing nitrate toxicity, because nitrite can be produced within the mass of chopped forage itself even before consumption by animals.

Nitrate is very stable during storage of dry hay. Microbial activity associated with silage fermentation generally reduces nitrate levels by 50% or more. The low-moisture silage (below 50%) that may result from drought conditions can lead to greatly reduced microbial activity and thus to the maintenance of high nitrate levels.

A field test for nitrate can be done on forages such as corn and sorghums by cutting a shoot near the soil surface, splitting the stem lengthwise, and dropping a solution of diphenylamine in sulfuric acid onto the exposed pith (see "Field Testing for Nitrate in Forages"). Development of a dark blue color indicates the presence of nitrate. If this qualitative field test is positive, a sample should be collected for more accurate laboratory analysis. Silage samples from a bunker silo should consist of a composite of six or more grab samples from the silage face, while hay samples should include cores from 20 randomly selected bales (see Chapter 14).

Field Testing for Nitrate in Forages

A quick field test of forage nitrate levels is useful for assessing the risk of harming livestock. A common field test kit uses 0.5 g of diphenylamine in 20 ml of distilled water, with concentrated sulfuric acid added to make a total volume of 100 ml. The solution is stored in a dark dropper bottle to exclude light, and should be stored in a dark place. To test for nitrate, stems of corn and other coarse grasses should be split, and the solution should be dropped on the inner portion near the base, where nitrate tends to accumulate. Low levels in this area generally indicate low levels throughout the shoot. Development of a blue color indicates the presence of nitrate. If a dark blue color develops within a few seconds, dangerously high levels of nitrate may be present. Test several locations within a field to take into account variability in plant composition. A positive response on this qualitative test should be followed up by laboratory testing.

Prussic Acid Poisoning

Prussic acid poisoning, also referred to as hydrocyanic acid poisoning or cyanide poisoning, occurs when livestock consume plants containing cyanogenic glycosides from which hydrogen cyanide (HCN) is released (Table 16.3) (Stanton, 1998). In the case of sorghums, epidermal cells contain a cyanogenic glucoside called dhurrin (Fig. 16.4). During ingestive mastication, cell walls and membranes are ruptured, resulting in the mixing of the cellular contents of mesophyll cells containing β-glucosidase enzymes with those of epidermal cells containing the dhurrin. These

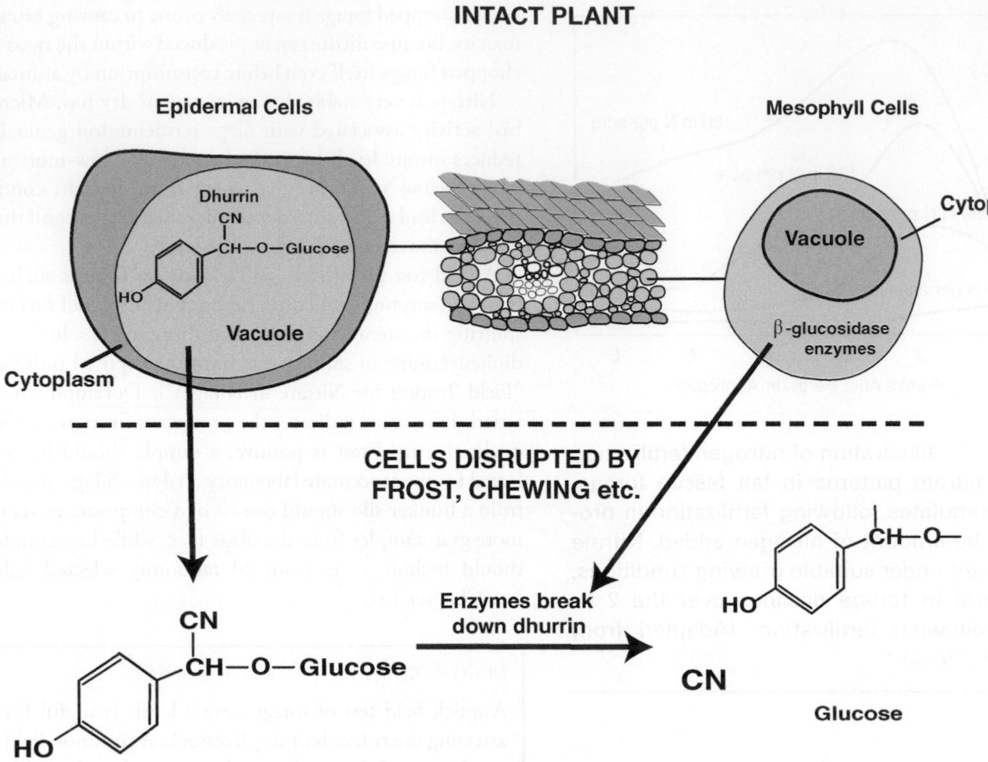

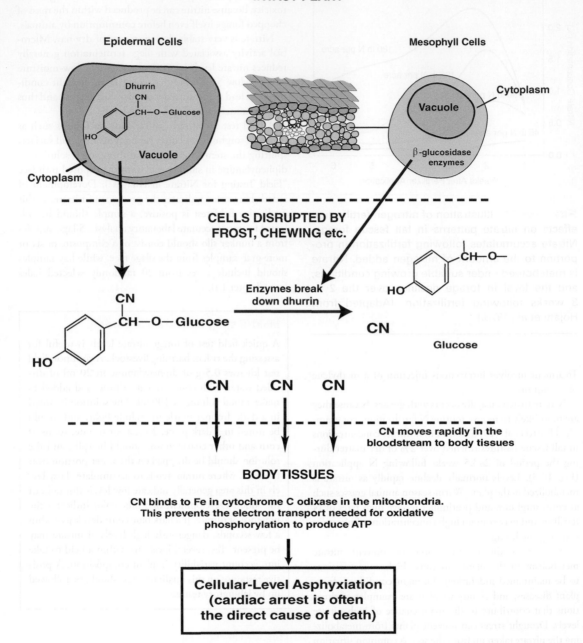

Fig. 16.4. An example of hydrogen cyanide (HCN) poisoning in forages. Dhurrin, a cyanogenic glycoside in sorghum, is broken down by plant or rumen microbial enzymes to HCN, sugar, and aglycone components. HCN is absorbed through the rumen wall and moves quickly to the body tissues. Within cells, cyanide prevents electron transfer by cytochrome C oxidase, a critical step in aerobic respiration.

enzymes cleave the sugar component of dhurrin, leaving the aglycone from which the toxic HCN is released. Cyanide is then absorbed through the rumen wall into the bloodstream, and is transported to the body tissues. Cattle, sheep, and other ruminants are more susceptible than horses because HCN release may also result from the activity of rumen microorganisms.

Other forage species contain cyanogenic glycosides other than dhurrin (Bush and Burton, 1994). In white clover, the cyanogenic compounds are lotaustralin and linamarin. In wild cherry species the cyanogenic compound is amygdalin, and poisoning is often the result of consumption of wilted leaves after wind or lightning has brought down branches. The fruit of these species also contains the cyanogenic glycoside.

Symptoms of cyanide poisoning include rapid breathing followed by labored, slow breathing, muscle spasms, and dilated pupils. Death may occur quickly, within 15 minutes to 2 hours after ingestion of a toxic amount of HCN. Cyanide poisoning results from inhibition of cytochrome oxidase, a terminal enzyme of aerobic respiration in cells (Fig. 16.4). Oxidative phosphorylation, which is required for the generation of ATP, is inhibited, leading to ATP deprivation. Affected animals effectively suffer from **asphyxiation**. In contrast to the brown blood coloration typical of nitrate toxicity, animals affected by cyanide poisoning have bright "cherry red" blood.

Prussic acid poisoning can occur on any forage or weedy species containing cyanogenic glycosides. Sorghum, sudangrass, sorghum × sudangrass hybrids, johnsongrass, and wild cherry are the plant species most commonly implicated in cyanide poisoning, but cases have occurred with white clover (Turner and Szczawinski, 1991). The problem does not occur with some other warm-season grasses, such as pearlmillet and corn, and there are large differences in cyanogenic compound levels between cultivars in sudangrass and sorghum. Iodine supplementation may be required to overcome goitrogenic effects resulting from long-term consumption of subacute levels of cyanide that inhibit iodine uptake by the thyroid gland.

Prussic acid poisoning occurs most commonly when livestock graze sorghum forage that is less than 15–18 in. (38–46 cm) tall, or when grazing occurs soon after frost. Young shoots of these species have the highest concentration of dhurrin, which declines rapidly as the plants develop. The incidence of poisoning increases after frost because freezing disrupts cell membranes and allows the mixing described above to occur prior to ingestion of forage by animals. In addition, cyanide release occurs when sorghum forage is green-chopped for feeding, because chopping also disrupts cells.

High rates of N fertilizer can increase cyanogenic glycoside levels in the plant. Management recommendations for minimizing the risk of HCN poisoning include splitting larger applications of N fertilizer. Poisoning can also occur for sorghums taller than 18 in. (46 cm) if grazing takes place after frost. In this case, young tillers at the base of older shoots may have been protected from freezing. The likelihood of poisoning is increased because these young, unfrosted shoots are preferentially grazed by livestock. Delaying grazing for 1–2 weeks after frost greatly reduces the risk of poisoning.

Sorghum hay and silage are generally safe. The crop is usually above the critical height at the time of harvest, and selective consumption of young shoots is more difficult than when frosted stands are being grazed. Ensiling greatly reduces the potential for cyanide poisoning.

Phytoestrogens

Phytoestrogens are naturally occurring plant phenolic compounds that, following further metabolism, can mimic the effects of estrogen hormones in animals, resulting in reproductive problems. Large outbreaks of infertility, originally called clover disease, have occurred in sheep grazing subterranean clover pastures in Australia. Although the reasons are unclear, sheep are more susceptible to the negative effects of phytoestrogens than are cattle. The introduction of low-phytoestrogen cultivars of subterranean clover has reduced the incidence of clover disease, but it has been suggested that more than 1 million ewes in Australia are permanently infertile due to the subclinical effects of phytoestrogens.

Subterranean clover contains particularly high levels, with up to 5% of its dry matter consisting of phytoestrogens, including formononetin (Fig. 16.5), biochanin A, and genistein. Several other forage legumes contain phytoestrogen compounds, including alfalfa, annual medics, white clover, red clover, berseem clover, soybean, and birdsfoot trefoil (Adams, 1995). In many clovers, the major estrogenic compound is formononetin, which is metabolized in the rumen to equol, which is then absorbed into the bloodstream. Alfalfa plants infected with leaf diseases such as common leaf spot [*Pseudopeziza medicaginis* (Lib.) Sacc.] contain much higher levels of the phytoestrogen coumestrol than do healthy plants. Breeding efforts to improve plant resistance to these diseases should reduce the likelihood of estrogenic activity.

Symptoms of temporary infertility in ewes include delayed rebreeding, reduced ovulation rates, and reduced twinning percentages. Castrated males may exhibit mammary development. Permanent infertility in cattle may result from cystic ovary disease, whereas in ewes the cervix redifferentiates and loses its ability to transport spermatozoa after insemination, a process called defeminization. Temporary infertility of both male and female animals is alleviated by removal from the estrogenic forage. However, if phytoestrogens persist in the diet over longer periods, infertility can become permanent. Changes during field curing and storage reduce the estrogenic activity of hay by about 70% compared with pasture. However, ensiling does not reduce estrogenic activity, and may actually increase it.

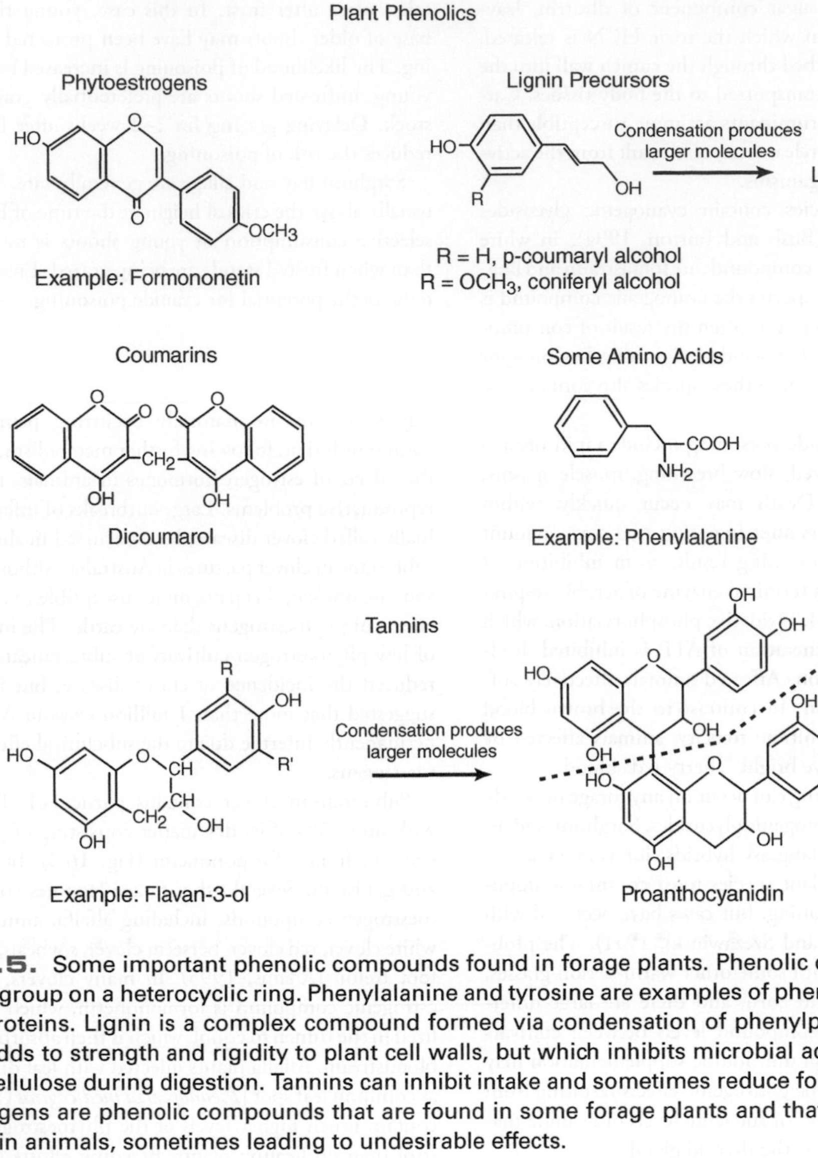

FIG. 16.5. Some important phenolic compounds found in forage plants. Phenolic compounds have a hydroxyl group on a heterocyclic ring. Phenylalanine and tyrosine are examples of phenolic amino acids found in proteins. Lignin is a complex compound formed via condensation of phenylpropanoid precursors that adds to strength and rigidity to plant cell walls, but which inhibits microbial access to cellulose and hemicellulose during digestion. Tannins can inhibit intake and sometimes reduce forage digestibility. Phytoestrogens are phenolic compounds that are found in some forage plants and that mimic estrogen hormones in animals, sometimes leading to undesirable effects.

Photosensitization

Cattle, sheep, horses, and other livestock grazing forages sometimes develop photosensitization in which unpigmented skin areas become sunburned, reddened, or blistered (Table 16.3). Primary photosensitization occurs when compounds in the plant move directly to the skin and cause the reaction. Secondary or hepatogenous photosensitization occurs when liver damage prevents normal metabolism of certain compounds, which in turn cause the reaction. An example of hepatogenous photosensitization called facial eczema occurs in the USA and in other countries on perennial ryegrass pastures infected with the fungus *Pithomyces chartarum*. Infected forages contain sporidesmin, which results in liver damage. Liver damage prevents normal metabolism of phylloerythrin, which is produced from chlorophyll. Phylloerythrin moves to the skin tissues, where it is broken down by light to the photosensitizing compounds.

Species-Related Disorders

Forage species are known to contain anti-quality compounds that limit animal productivity (Table 16.5). In some cases, levels of anti-quality compounds have been

Table 16.5. Forage species or cultivar-related livestock disorders

Disorder	Description	Symptoms	Plants	Animals	Prevention	Treatment	Future
Tall fescue toxicosis (summer slump, fat necrosis, fescue foot)	Hyperthermia (elevated body temperature), hypoprolactinemia (low serum prolactin levels)	Weight loss, dull, rough coat, excessive salivation	Tall fescue	Cattle	Check your field before grazing	Removal from contaminant	Use only fungus-free seed stock for pasture
Ryegrass staggers	Hyperthermia (elevated body temperature), hypoprolactinemia (low serum prolactin levels)	Weight loss, dull, rough coat, excessive salivation	Perennial ryegrass	Sheep, cattle, horses	Change fields	No treatment	Use endophyte-free seed
Alsike clover poisoning	Photosensitization, possible liver damage	Jaundice, head pressing, aimless walking	Alsike clover	Horses	Change fields	Removal from pastures containing alsike clover	Avoid alsike clover in horse hay or pasture
Sweet clover poisoning	Internal blood loss	Swellings under skin, pale membranes, weakness	Spoiled sweetclover hay	Cattle, horses, sheep	Avoid feeding moldy sweetclover hay	Remove from contaminated feed; give vitamin K and blood injections	
Tannins	High levels of tannins reduce intake, DM digestibility, and weight gain		Mostly legumes and woody browse species	Sheep, cattle, goats	Avoid high-tannin forage		Use low-tannin cultivars where available

Sources: Cheeke, 1998; Kingsbury, 1964; Bush and Burton, 1994.

reduced through the efforts of plant breeders. In other cases, altered forage and animal management practices can minimize their negative effects.

Endopohyte-Related Animal Disorders

Broadly speaking, endophytes are fungi that grow within plant tissues, and they can be either **antagonistic** or **mutualistic**. A special group of endophytic fungi are agronomically mutualistic, and benefit the grasses that they infect but have adverse effects on grazing livestock. These fungi are members of the Balansieae tribe of the Clavicipitaceae family of fungi. The grain disease known as "ergot" belongs to the same family. Thus the Balansieae endophytic fungi are close relatives of *Claviceps purpurea*, the organism that causes ergot.

Balansieae endophytes occur in clusters within grass genera, and genetically related grass species are infected with similar but genetically distinct endophytes. It is believed that these modern-day plants are the result of an early infection of grass plants by Balansieae endophytes, and that endophytic adaptations occurred as plants evolved into their respective species. The ecological significance of endophytes is not completely understood because most grass species infected by endophytes are not economically important. Three distinct exceptions exist in which endophytes are present in important forage species. Understanding the mutualistic role of these endophytes can contribute to management of their impact on grazing animals as well as aiding in manipulation of the plant–endophyte association in cultivar development. The mutualistic role of endophytes is best understood in tall fescue, annual and perennial ryegrasses, and drunken horse grass (*Achnatherum inebrians* [Hance] Keng). These grasses often are infected with *Neotyphodium* species endophytes which live in a symbiotic association with the grass hosts. The endophytes grow in the intercellular spaces of the leaf sheath and vegetative and reproductive meristems. They are passed from one generation to the next via embryonic infection of the seed, and thus endophyte-infected plants produce endophyte-infected seeds.

Tall Fescue

Tall fescue is well adapted to a range of climatic and geophysical conditions worldwide (see Chapter 6). The release of the cultivar "KY31" in 1943 coincided with federally funded soil erosion control measures in the Southeastern and Midwestern regions of the USA. Cost-sharing programs encouraged farmers to use "KY31" in their conservation plans, which resulted in the planting of nearly 35 million acres (14.2 million ha) of tall fescue. It is estimated that more than 25% of the current US cow-calf herd graze tall fescue pastures. Early observations indicated that livestock grazing tall fescue often performed poorly or suffered hoof loss, a gangrenous condition known as "fescue foot." A second condition known as "fescue toxicosis", or "summer slump", was often manifested as poor weight gain and compromised reproduction when grazing fescue pastures. Fescue toxicosis and fescue foot were originally thought to be independent anomalies, but are currently attributed to toxins produced by the endophytic fungus *Neotyphodium coenophialum* (formerly known as either *Epichloe typhina* or *Acremonium coenophialum*) (Hill, 2005).

The association between livestock anomalies and *N. coenophialum* was made in the late 1970s. Surveys indicate that over 95% of all tall fescue pastures had *N. coenophialum* infection levels of 65% or greater. The symbiotic nature of the plant–endophyte association was not initially understood, but became apparent with the observation that endophyte-free cultivars that were developed to avoid animal disorders lacked the persistence of infected cultivars. Early research revealed that the endophyte provided plants with increased vigor via drought and cold tolerance, insect and nematode resistance, and limited disease resistance (Fig. 16.6). The available options for producers at that time were either to plant endophyte-infected tall fescue cultivars which were toxic to livestock, or to use endophyte-free cultivars that lacked adequate plant persistence. Researchers have since identified *N. coenophialum* strains that do not produce toxins, and have incorporated these into improved tall fescue cultivars that capitalize on the endophyte-mediated agronomic properties without the toxicity effects on grazing livestock.

Ergot alkaloids (Fig. 16.7) are the cause of fescue toxicosis. The specific ergot alkaloids produced by *N. coenophialum* are simple ergot alkaloids (lysergic acid and lysergic acid amides) and ergopeptine alkaloids. The alkaloids cause vasoconstriction, which limits blood flow to the coronary band of the hooves and compromises the health of fleshy tissue above the hoof, which can result in

FIG. 16.6. Plant morphology of genetically identical tall fescue plants, one infected with endophyte (on the right) and one which is endophyte-free (on the left).

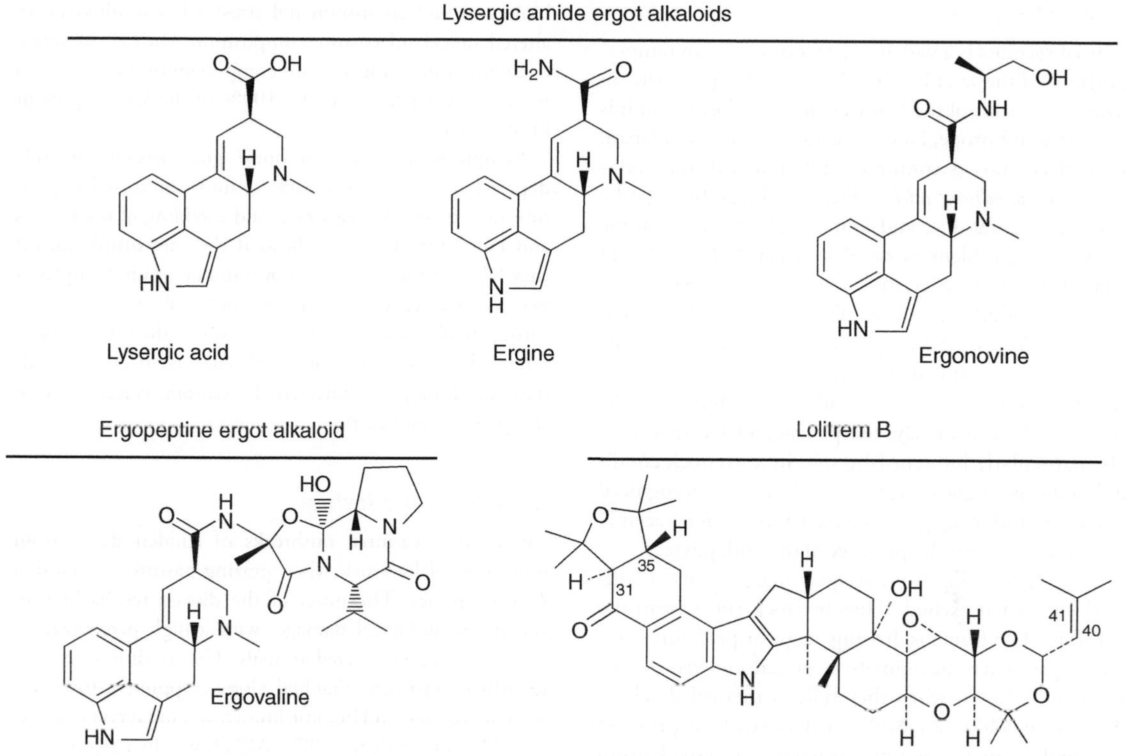

FIG. 16.7. Toxic alkaloids found in endophyte-infected forage grasses.

sloughing. Vasoconstriction also affects the ability of livestock to dissipate heat from the core of the body to the body surface, resulting in elevated body temperatures in summer and lower body temperatures in the winter months. The alkaloids also affect the endocrine and neurological systems. Collectively, the effects of the alkaloids on animal physiology result in low weight gain, poor conception rates, fetal abortion, reduced milk production, sensitivity to light and heat, and avoidance of midday grazing during the summer months (Table 16.6).

Table 16.6. Grazing behavior of steers grazing low- and high-endophyte tall fescue pastures

Grazing period	Endophyte level Low	High
	(%)	
Daylight	52	34
Dark	14	22

Source: Bond et al., 1984. Reproduced with permission of the American Society of Agronomy, Inc.

The modes of action of the ergot alkaloids depend largely on whether they are lysergic acid amides or ergopeptine forms. Lysergic acid amides are more toxic, with a lower **lethal dose (LD_{50})**, than ergopeptine alkaloids, but the latter are more effective at inducing vasoconstriction. Microbial communities in ruminant animals are capable of converting ergopeptine alkaloids into lysergic acid and lysergic acid amides. Lysergic acid is the predominant alkaloid that is excreted. Microbial metabolism in hind-gut fermenters (i.e., equine species) occurs posterior to the majority of the absorptive surfaces in the gut. Thus hind-gut fermenters do not experience bioconversion of the ergopeptine alkaloids to lysergic acid amides prior to absorption. The toxic effects of the alkaloids are different for each digestive system. The effect of endophyte toxins on horses (hind-gut fermenters) is pronounced in reproduction, with a higher incidence of abortion, dystocia (abnormal fetal presentation at birth), and agalactia (inability of the mare to produce milk). Reproductive rates in ruminants are compromised by the endophyte toxins, but not as dramatically as in horses. Lower conception rates, unthrifty calves, lower weaning weights, and poor carcass condition are the major losses in ruminant systems.

Perennial Ryegrass

Perennial ryegrass is a widely used pasture grass in temperate regions of the world. Its use in the USA is primarily on heavier moist soils of the Midwest and Northeast, and it is widely used in Europe, New Zealand, Australia, and Japan. Perennial ryegrass is commonly infected with the endophyte *Neotyphodium lolii,* which produces biologically active ergot, peramine, and lolitrem alkaloids. Argentine stem weevil is problematic on New Zealand's North Island because it feeds on vegetative meristems and prevents further growth of affected tillers. The peramine alkaloid has potent anti-insecticidal properties, and enhances the survival of infected plants in areas where the weevil is a problem, often resulting in pastures with 100% infection by the endophyte. Unfortunately, the presence of lolitrem alkaloids (particularly lolitrem B) results in a livestock condition known as ryegrass staggers, which was first recognized in New Zealand many years ago but has only more recently been associated with the presence of the endophyte.

Livestock suffering from ryegrass staggers have a stiff-legged gait when excited. Thus livestock suffer from the condition when humans, herding dogs, or predators enter the pastures, when the animals are placed into vehicles for transport, and even when they suffer a thermal shock by walking into cold water. Stiff-gaited livestock are prone to falling and are unable to right themselves. Livestock losses can be severe during shipping, when animals that have succumbed to lolitrem often suffocate from the weight of others on top of them. There have been cases where ponds have been ringed with the carcasses of sheep that have drowned because of their inability to right themselves after they have fallen into the water. Ryegrass infected with *N. lolii* also produces ergot alkaloids that cause more subtle toxicoses, similar to those found in tall fescue, such as reduced weight gain and reproductive losses.

The endophyte in ryegrass has a similar life cycle to that in the tall fescue plant, and is passed from one generation to the next via seed. Certain strains of *N. lolii* produce the anti-insecticidal alkaloid peramine, but do not produce either lolitrem or ergot alkaloids. Improved cultivars of perennial ryegrass with these endophytes maintain resistance to the Argentine stem weevil without the toxigenic effects of the ergot and lolitrem alkaloids on livestock.

Drunken Horse Grass

Drunken horse grass (*Achnatherum inebrians* [Hence] Keng) is a perennial bunchgrass that is found in the alpine and sub-alpine grasslands of Northern China. It is infected with the endophyte *Neotyphodium gansuense,* which produces lysergic acid amide and ergonovine ergot alkaloids. Endophytes benefit the infected plants by providing resistance to aphids, grasshoppers, mites, and seed harvesting by ants. The infected plants also have increased drought and cold tolerance. Endophyte-mediated tolerance of biological and environmental stresses has resulted in an altered rangeland pasture composition, with *A. inebrians* changing from being a minor component of the ecosystem structure to representing 45–100% of the species present in these native grasslands.

Symptoms of drunken horse grass toxicity include depression, teary eyes, muscle tremors, increased respiration rate, increased heart rate, and shedding of the hooves and tail. Animals may walk as if they are drunk, and if they fall over they often cannot stand up again. Symptoms persist for 6–18 hours, but livestock will resume normal eating habits within 24 hours once they have been removed from the grass, although severely affected animals may die. Endophyte-infected *A. inebrians* is also toxic to sheep, goats, and cattle.

Annual Ryegrass Toxicity

There are occasional outbreaks of sudden death from poisoning of livestock when grazing pastures containing *Lolium* species. The onset of the disease results in irreversible neurological damage, with a high percentage of mortality among affected animals. The condition was first identified in pastures that had a long cropping history with annual ryegrass, and became known as annual ryegrass toxicity (ARGT) (Edgar, 1987). ARGT was first reported in Western Australia, and was initially thought to be confined to this region, but there have been periodic outbreaks of ARGT in annual ryegrass pastures in South Africa as well. ARGT is not confined to either ryegrasses or pasture conditions. There were numerous outbreaks of ARGT associated with tall fescue as well as ryegrasses in the Willamette Valley of Oregon between 1945 and 1961. The Willamette Valley is the seed production region of Oregon.

ARGT is a result of infection of vegetative tillers by larvae of the *Anguina agrostis* nematode. The nematode larvae feed at the tiller meristem, develop galls, and complete their life cycle within the galls. The next generation of larvae remains dormant until the following year, when they emerge and repeat the infection process. The nematode larvae are surface contaminated with *Clavibacter toxicus*, a bacterium that is capable of producing a powerful class of toxins called corynetoxins. In some cases, *Clavibacter toxicus* colonizes and proliferates in the galls, resulting in the production of corynetoxins. These toxins inhibit the synthesis of glycoproteins, a class of proteins that are essential components of enzymes, hormones, cell membranes, and membrane receptors. Normal neurological cell structure and function ceases when corynetoxins are present.

In the early 1960s the life cycle of the *A. agrostis* nematode was interrupted when Oregon seed producers began burning straw residues in their seed fields. As a result, the incidence of ARGT declined. However, the burning of fields was banned in 2009 because of concerns about traffic safety and the effects of secondary smoke on human health.

Table 16.7. The impact of adding clover to "Jesup" tall fescue containing non-toxic or toxic endophytes; the results are means of an experiment conducted over 3 years at Eatonton, GA

Endophyte	Number of grazing days	Average daily gain	Gain/acre
		(lb)	
Non-toxic	102	1.83	186
Non-toxic + clover	97	2.61	252
Toxic	114	1.10	126
Toxic + clover	100	1.60	160

Source: Bouton, Andrae, and Hill, unpublished data.

This suggests that a resurgence of *A. agrostis*/*C. toxicus*-mediated toxicity could occur in the future.

Managing Endophyte-Related Disorders

The most dependable method of eliminating livestock toxicity effects caused by wild-type endophytes in pastures is to eradicate existing pastures and hayfields and replant with improved grass varieties containing non-toxic endophytes. This can be accomplished by using a herbicide in spring, followed by a summer row crop, followed by a second herbicide application prior to planting the tall fescue in the fall of the year. Care must be taken to minimize the impact of carry-over tall fescue seed in the soil.

Inclusion of other pasture species, especially high-quality legumes, is another way of reducing the impact of endophyte-derived toxins on animal performance. The presence of other pasture species offers a diverse diet for grazing livestock, and as a result of dietary selection the toxins are diluted in the digesta. However, it must be noted that dilution is not the same as elimination of the toxic effect, and animal performance is still compromised because of the toxins (Table 16.7). Toxins can also be diluted by grain supplementation or the use of high-quality non-toxic hay.

In recent years, cattle that exhibit some tolerance to the negative effects of endophyte-infected tall fescue have been identified (Campbell et al., 2014). Genetically tolerant animals have smoother hair coats, wean heavier calves, and/or gain more weight than non-tolerant animals. In the future, selection of sires or replacement females based on their expected genetic tolerance of tall fescue toxicosis may contribute significantly to the management of this disorder.

Reed Canarygrass Alkaloids

Some reed canarygrass cultivars contain the alkaloid gramine, which can reduce forage intake by animals. In reed canarygrass, the alkaloids are present at highest concentrations in leaf tissue, with lower concentrations in the leaf sheaths and stems. Gramine concentrations have been shown to be under plant genetic control, and breeding has successfully reduced alkaloid levels. "Palatin", "Venture", and "Rival" are low-alkaloid varieties of reed canarygrass. Symptoms of reed canarygrass alkaloid toxicity include muscle tremors and loss of coordination.

Sulfur-Containing Anti-Quality Constituents

Glucosinolates and S-methylcysteine sulfoxide, a non-protein amino acid, in brassica forages can have adverse effects on animals grazing these species (Cheeke, 1998). Enzymatic hydrolysis of glucosinolates produces thiocyanate, isothiocyanate, and nitrile products, all of which can be toxic. Thiocyanates are goitrogenic, and nitriles can cause liver damage. The S-methylcysteine sulfoxide (SMCO) found in some brassicas can cause a form of **anemia**. Brassica anemia does not occur in horses.

Tannins and Other Phenolic Compounds

Phenolic compounds contain a free hydroxyl group on an aromatic ring structure (Fig. 16.5). Examples of phenolic compounds that are found in plants include the amino acids tyrosine and phenylalanine, and lignin, a complex compound formed by condensation of phenolic subunits. Phytoestrogens (discussed earlier) are also examples of phenolic compounds. Dicoumarol is a phenolic compound that can be formed from coumarin during spoilage of sweetclover hay. If present in moldy sweetclover hay it can lead to internal bleeding in cattle.

Tannins are complex polyphenolic compounds found in some forages. Negative effects associated with forage tannins include reduced DM intake, reduced animal growth rates, and reduced fiber digestion. High levels of tannins (well above 15% of DM) in some cultivars of sericea lespedeza act as feeding deterrents, and can reduce digestibility. Mature forage of this species is very unpalatable to cattle. Many woody browse plants contain high levels of tannins. Leucaena, a tree legume, contains tannins in addition to the anti-quality non-protein amino acid mimosine.

At more moderate concentrations, tannins can be beneficial to animals. At low levels, the protein-binding characteristic of tannins, also used in "tanning" leather, enhances forage protein utilization by reducing initial rates of fermentation and protein breakdown in the rumen. The increase in undegraded intake protein levels (see Chapter 14) reaching the duodenum can improve crude protein utilization of species with highly soluble N fractions. The presence of tannins is the mechanism by which bloat-safe legumes avoid this problem.

Acute toxicity can be caused by the tannins found in acorns (seed) of some oaks (Table 16.1). The compounds responsible are hydrolyzable tannins that are broken down in the rumen into compounds that can cause hemorrhagic gastroenteritis and other problems (Reed, 1995). Toxicity most commonly occurs when livestock on overgrazed pastures have ready access to wooded areas.

Disorders Associated Primarily with Stored Forages

In many cases, hay curing or silage fermentation reduces the levels of toxic compounds in forages. In other cases, anti-quality factors associated mainly or exclusively with stored forages arise.

Botulism

Clostridium botulinum is a common anaerobic bacterium that sometimes proliferates in silages or in hay. Many types of *C. botulinum* exist, several of which produce toxins. Proteolytic Clostridia also break down protein, and can increase non-protein N levels to more than 50% of total forage N. In addition, these bacteria may utilize lactic acid as an energy source, thereby increasing silage pH and leading to further deterioration. Silages with low pH levels (at or below about 4.6) have a low incidence of botulism. When this pH cannot be attained through fermentation or direct acidification, wilting forage so that it contains less than 70% moisture greatly reduces the incidence of botulism by increasing the pH at which clostridial growth is inhibited (see Figure 17.11 in Chapter 17). The lower water content of wilted silage limits the growth of these undesirable organisms more than the growth of lactic acid bacteria, and thus improves silage preservation.

Death losses due to botulism are not common, but there have been cases in which large numbers of cattle have been killed. Botulism may occur in hay when dead animals are incorporated into hay bales. In one widely publicized case, 427 dairy cows from a California dairy died over a 1-week period after a badly deteriorated bale of oat hay containing a dead cat was incorporated into their mixed ration.

The bacterium that causes botulism is common in the environment, so prevention of the problem is achieved by avoiding proliferation of the organism. Thus hay bales containing dead animals should not be fed to livestock, nor should high-moisture silage in which low pH cannot be attained.

Listeriosis

Listeria monocytogenes may proliferate in aerobically deteriorated or soil-contaminated silages. Animals affected by listeriosis walk in circles, and this gave rise to the alternative name of "circling disease." Listeriosis causes an increased incidence of abortion, and death may result. Growth of *L. monocytogenes* is greatly restricted when silage pH is kept near or below 4.5, but populations of the organism can be very high in visibly spoiled silage.

Moldy Hay and Silage

Molding of hay or silage during storage leads to dry matter losses and a reduction in forage quality. Moldy hay can be detrimental to animals due simply to the presence of large numbers of mold spores, or it may contain mycotoxins. Horses are particularly sensitive to spores inhaled during feeding of moldy hay, and they can acquire chronic obstructive pulmonary disease from prolonged ingestion of moldy hay. In silage, the conditions that encourage mold development also encourage the growth of undesirable bacteria such as *L. monocytogenes*. Moist hay can support considerable growth of *Aspergillus flavus* and other fungi during storage (see Chapter 17), leading to reduced forage palatability for all livestock, and the risk of colic or respiratory problems for horses.

Blister Beetle

Blister beetles are insects that are sometimes found in alfalfa or other hays, having been killed by mechanical treatments during hay production, especially mowing. They contain cantharidin, a toxic compound to which horses are particularly sensitive, although cattle are also affected. The striped blister beetle (*Epicauta vittata*) contains relatively high concentrations of cantharidin (4.5–5 mg/beetle), and is most common in the western USA. The black blister beetle (*E. pennsylvanica*), which is more common in the eastern USA, has lower levels of cantharidin (0.08–0.40 mg/beetle). Areas within fields with high concentrations of blister beetles are associated with infestations of grasshoppers, the eggs of which provide a food supply for developing blister beetle larvae.

In the animal, cantharidin is absorbed from the gut and causes severe irritation of the urinary tract and other mucosal membranes. The digestive tract is also affected, and the occurrence of colic is increased during subacute blister beetle poisoning. Symptoms include frequent urination and elevated temperatures. In acute cases, affected animals may die within 3 days. Effective treatments are not available.

Red Clover Slobbers

Profuse salivation may occur when horses, cattle, sheep, and other animals consume red clover infected with the fungal disease *Rhizoctonia leguminicola*. Black patch, the common name for this disease, appears as irregular-shaped areas within the field where the leaves of infected plants have a dark brown or black color.

The adverse effects on animals are caused by slaframine and swainsonine, two alkaloids that are found in the forage of infected (but not of uninfected) plants. In addition to excessive salivation, other symptoms include diarrhea, stiff joints, frequent urination, and abortion. The incidence of black patch is highest during hot, humid weather, and the disease is thus most common at mid-summer harvests (the second and third cuttings in most red clover-producing areas). The disease can occur in pasture, but hay is most commonly associated with the slobbers syndrome. Taller canopies under hay management might encourage disease development, or animals consuming hay might have

Solving Leucaena Toxicity

Leucaena is a tree legume, native to Mexico, that provides quality forage for livestock in tropical regions with options. Mimosine is a non-protein amino acid (Fig. 16.8) that is present in high concentrations in leucaena forage. Actively growing shoot tips can contain 8–12% mimosine, and young leaves contain 4–6%. During chewing and when in the rumen, mimosine is converted to 3-hydroxy-4(1H)-pyridone (3,4-DHP), which acts as a goitrogenic agent in cattle. Mimosine is a depilatory agent, which causes affected animals to shed their hair. Weight gains are also low in affected animals.

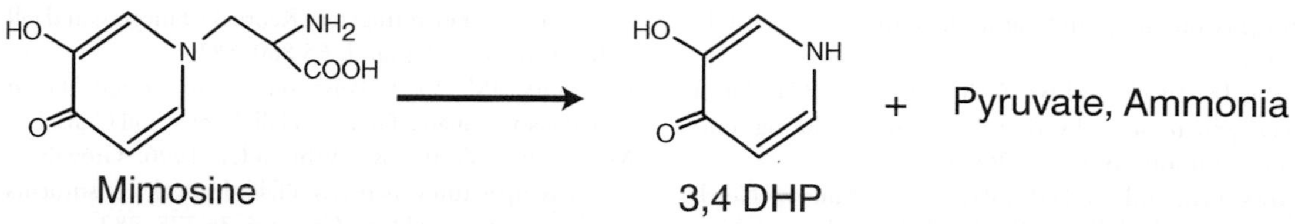

FIG. 16.8. The free amino acid mimosine and its degradation product, 3,4-DHP, have toxic effects on animals that consume leucaena forage.

Researchers noted that some livestock in regions where there had been long-term use of this species exhibited no adverse effects, even when leucaena constituted more than 50% of the diet. Australian researchers found that goats from Hawaii had a unique species of anaerobic bacterium in their rumen, *Synergistes jonesii*, that utilized the 3,4-DHP as an energy source, thereby eliminating the toxic component. Ruminal infusion of this bacterium into cattle transferred the ability to metabolize 3,4-DHP and alleviate leucaena toxicity. After an initial adjustment period, steers dosed with *S. jonesii* gained 2.27 lb/day (1 kg/day) whereas undosed steers gained only 1.15 lb/day (0.5 kg/day) (Quirk et al., 1988). *S. jonesii* is transferred among animals within the same herd, but levels decline if leucaena is absent from the diet for a few months.

less opportunity to avoid infected forage compared with grazing animals. Some other forage legumes (e.g., kudzu, lupines, korean annual lespedeza) are also susceptible to black patch and thus to the slobbers disorder.

Using Science to Reduce Forage Anti-Quality Factors

The levels of several anti-quality factors have been reduced through plant breeding. Examples include reed canarygrass cultivars with lower levels of gramine alkaloids, lupines in which toxic alkaloids are absent, and low-coumarin types of sweetclover. Tall fescue cultivars without the fungal endophyte do not contain toxic alkaloids. A good example of using research to address anti-quality problems is the discovery that leucaena toxicity can be eliminated by rumen inoculation. Inoculation transfers bacteria with the capacity to metabolize the toxic compound, thereby decreasing the levels present (see "Solving Leucaena Toxicity").

Summary

Poisonous plants are undesirable species that contain toxins; they should be eliminated from pasture and hayfields whenever possible. Forage and animal management can minimize the risks of seasonal disorders such as grass tetany, bloat, nitrate toxicity, and HCN poisoning. Successful plant breeding efforts are reducing health problems and productivity losses associated with species disorders such as tall fescue toxicosis. Livestock reproductive performance is particularly sensitive to poisonous plants and other toxins.

Questions

1. What is the active compound in blister beetle poisoning?
2. How does pasture management influence the probability of livestock consuming poisonous plants?
3. Which forage-related disorders commonly occur during spring?
4. List four management measures that are used to minimize the likelihood that bloat will occur in grazing animals.
5. List some factors that affect the occurrence of grass tetany in grazing livestock.
6. Describe the single most important silage management practice for minimizing the occurrence of botulism.
7. Explain why frost temporarily increases the likelihood of prussic acid poisoning in animals that are grazing sorghum or sudangrass during autumn.

8. Give one example of a forage anti-quality factor that has been reduced or eliminated by plant breeding.
9. Name two bloat-safe forage legume species.
10. Describe a simple field test for the presence of nitrate in forages.

References

Adams, NR. 1995. Detection of the effects of phytoestrogens on sheep and cattle. J. Anim. Sci. 73:1509–1515.

Bond, J, JB Powell, and BT Weinland. 1984. Behavior of steers grazing several varieties of tall fescue during summer conditions. Agron. J. 76:707–709.

Burrows, GE, and RJ Tyrl. 2013. Toxic Plants of North America, 2nd ed. Wiley-Blackwell, Ames, IA.

Bush, LP, and H Burton. 1994. Intrinsic chemical factors in forage quality. In GC Fahey (ed.), Forage Quality, Evaluation, and Utilization, pp. 367–405. American Society of Agronomy, Madison, WI.

Campbell, BT, CJ Kojima, TA Cooper, BC Bastin, L Wojakiewicz, RL Kallenbach, FN Schrick, and JC Waller. 2014. A single nucleotide polymorphism in the dopamine receptor D2 gene may be informative for resistance to fescue toxicosis in Angus-based cattle. Animal Biotech. 25:1–12.

Cheeke, PR. 1998. Natural Toxicants in Feeds, Forages, and Poisonous Plants, 2nd ed. Interstate Publishers, Danville, IL.

Edgar, JA. 1987. Annual ryegrass toxicity: aetiology, pathology, and related diseases. In: Australian Standard Diagnostic Techniques for Animal Diseases, pp. 1–8, CSIRO Publishing, Clayton, Australia.

Herron, JW, and DE LaBore. 1972. Some Plants of Kentucky Poisonous to Livestock. University of Kentucky College of Agriculture, Lexington, KY.

Hill, NS. 2005. Absorption of ergot alkaloids in the ruminant. In CA Roberts, CP West, and DE Spiers (eds.), *Neotyphodium* in Cool-Season Grasses, pp. 271–290. Blackwell Publishing, Ames, IA.

Hojatti, SM, WC Templeton, Jr., TH Taylor, HE McKean, and J Byars. 1973. Postfertilization changes in concentration of nitrate nitrogen in Kentucky bluegrass and tall fescue herbage. Agron. J. 65:880–883.

Kingsbury, JM. 1964. Poisonous Plants of the United States and Canada. Prentice-Hall, Englewood Cliffs, NJ.

MacAdam, JW, and RE Whitesides. 1996. Growth at low temperatures increases alfalfa leaf cell constituents related to pasture bloat. Crop Sci. 36:378–382.

Majak, W, JW Hall, and WP McCaughey. 1995. Pasture management strategies for reducing the risk of legume bloat in cattle. J. Anim. Sci. 73:1493–1498.

Quirk, MF, JJ Bushell, RJ Jones, RG Megarrity, and KL Butler. 1988. Live-weight gains on leucaena and native grass pastures after dosing cattle with rumen bacteria capable of degrading DHP, a ruminal metabolite from leucaena. J. Agric. Sci. Camb. 111:165–170.

Reed, JD. 1995. Nutritional toxicology of tannin and related polyphenols in forage legumes. J. Anim. Sci. 73: 1516–1528.

Stanton, TL. 1998. Prussic acid poisoning. Colorado State University Cooperative Extension Service Circ. 1.612. Colorado State University, Fort Collins, CO.

Turner, NJ, and AF Szczawinski. 1991. Common Poisonous Plants and Mushrooms of North America. Timber Press, Portland, OR.

CHAPTER 17

Preservation of Forage as Hay and Silage

Michael Collins and Kenneth J. Moore

Stored forages provide nutrients for livestock when pastures are inadequate, and are a consistent, reliable, and predictable feed supply for confinement feeding systems such as beef feedlots and dairies. Common forms are **hay**, usually below 20% moisture, and **silage**, preserved by anaerobic **fermentation**.

Forage Harvest and Storage Systems

Key considerations when selecting a harvesting system include the following: (1) the efficient preservation of crop nutrients; (2) suitability for the forage **species** being grown and for local growing conditions; (3) building, equipment, and labor costs associated with each system; (4) the nature of the livestock or cash forage enterprise.

Some crops are clearly suited to a particular harvest system. For example, corn is ideally suited for silage preservation because it ferments readily and would be difficult to dry for hay. Grain **crop residues** for feeding are normally harvested using hay techniques because they require no additional drying.

Preservation as silage usually minimizes dry matter (DM) losses during harvest and storage (Fig. 17.1). A large part of the DM loss in silage systems occurs during the storage phase, while the greatest losses for hay normally occur during harvest. The advantage of silage in reducing field losses is especially apparent in humid regions, because it avoids most **weather**-related harvesting losses associated with hay. Wilted silage retains the highest proportion of crop DM by avoiding excessive losses during both harvest and storage. Typical DM losses in silage production are in the range of 15–20%. Harvest and storage losses for legume crops are around 25% for dry hay, but somewhat less for grasses.

Hay is suitable for transport and marketing because it weighs less than silage and does not require **anaerobic** storage. Livestock systems that only occasionally utilize stored forage generally use hay because it is most suitable for long-term storage. Silage is most suitable for mechanized handling and feeding systems.

Hay

Hay Curing

Fresh forage often contains 75–85% moisture, so a large amount of water must be removed in order to produce hay. To produce 1 ton (0.9 mt) of hay at 20% moisture, 4–5 tons (3.6–4.5 mt) of water need to be removed. Only a few hours would be needed to dry hay if **transpiration** rates typical of living shoots could be maintained during hay curing. In actual practice, however, 3–5 days or more of field curing are typically required to dry hay down to 20% moisture. Quick hay curing minimizes **respiration** losses, as the respiration rate is nil once the moisture content of the crop falls below 40%. Rapid removal of moisture also helps both to maintain green color in the dry crop and to avoid rain damage.

Higher crop yield levels, higher fresh forage moisture levels, lower radiation levels, cooler temperatures, and higher **relative humidity** levels are associated with slower rates of moisture loss during hay curing. Conditions such as these are prevalent during spring in much of the USA. One useful approach is to produce silage from early and late

Forages: An Introduction to Grassland Agriculture, Seventh Edition. Edited by Michael Collins, C. Jerry Nelson, Kenneth J. Moore and Robert F Barnes.

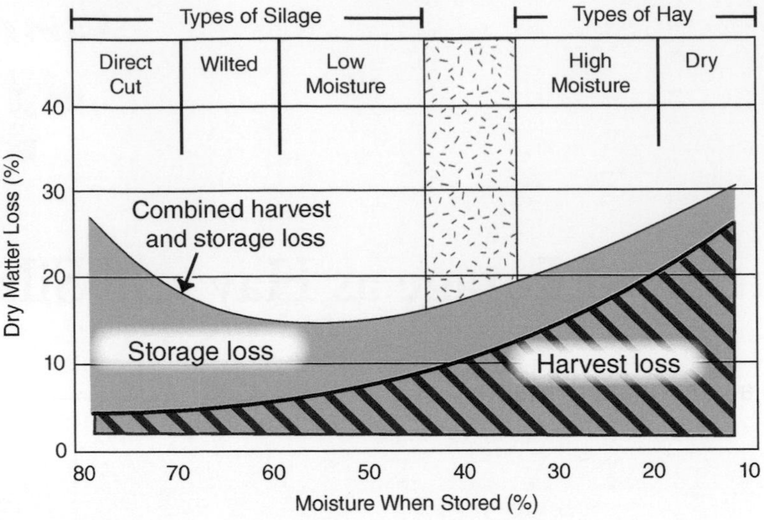

Fig. 17.1. Total, harvest, and storage DM losses for legume–grass forages harvested at different moisture levels. Silage losses occur predominantly during storage, whereas hay losses occur mainly during the harvest phase. (Adapted from Hoglund, 1964.)

cuttings when curing conditions are least favorable, and to produce hay from summer cuttings.

Weather Factors

Weather affects how readily moisture moves to the air from the drying crop. Important factors include **solar radiation intensity,** air temperature, relative humidity, wind speed, and soil moisture levels. High temperatures and low humidity are conducive to hay drying. Solar radiation, air temperature, and relative humidity are highly correlated. Higher solar radiation levels usually increase temperatures, and these in turn decrease relative humidity.

Moisture loss from hay may cease if relative humidity rises above 65–70%. At these humidity levels, especially under cool temperatures, the crop can reach **equilibrium moisture** with the atmosphere. Moisture can also move from the air to the crop when relative humidity is very high, such as when dew or rain occurs. Hay **drying rates** also improve as wind speed increases up to about 12 mph (19 km per hour).

Wet soils slow moisture loss except on sandy soils in which the surface dries quickly. In most regions, hay moisture follows a **diurnal** pattern in which drying occurs during the day, while the lost moisture is partially replaced from the air during the night as humidity levels rise or dew forms (Fig. 17.2). The cycle repeats until the hay reaches the target moisture level for baling (often 20%).

Crop Factors

Crop factors that affect drying include plant **species**, crop maturity stage, crop yield level, and initial moisture content. Species differ in initial moisture level and in the rate at which they lose moisture. Thick-stemmed crops dry slowly. Red clover usually has thick stems that are densely **pubescent**, and it dries slowly compared with alfalfa. Most

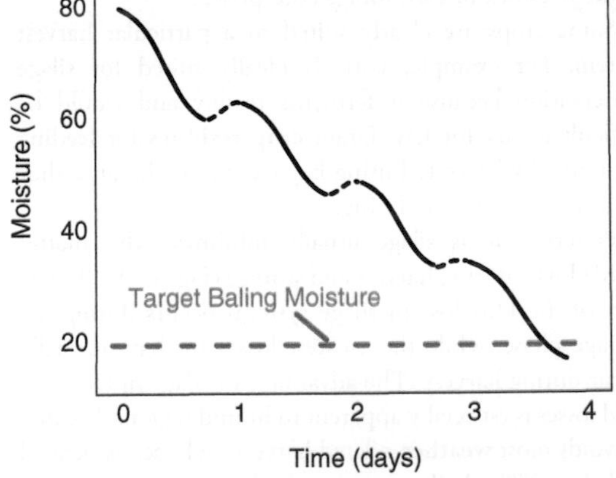

Fig. 17.2. A typical pattern of moisture loss during hay curing. Initial moisture content at cutting is usually around 80%. Moisture loss is most rapid during the day, but drying ceases at night, or moisture may be gained from the air as the humidity increases or dew falls. This pattern is repeated until the crop reaches the target moisture level for baling (often 20%).

fresh crops contain 75–85% moisture. Some, such as chicory and brassicas, can contain more than 90% moisture, which makes their use as hay crops problematical.

In general, grass hay crops dry faster than legumes. In fact, having 10–15% smooth bromegrass with alfalfa improved drying of the alfalfa component by creating a more open **swath** structure (Collins, 1985). Some tropical grasses dry slowly because they have very thick stems. Heavy (high-yield) crops produce thick swaths and **windrows** that restrict moisture loss to the atmosphere.

Management Factors

Management factors that affect hay drying include time of mowing, swath structure, windrow inversion, tedding, mechanical or chemical conditioning, and raking. These factors are discussed in the following sections.

Hay Production Practices

Typical hay production involves mowing the crop and placing it into swaths or windrows to cure. Conditioning, which is the process of crushing or crimping the stems, is usually done simultaneously with mowing by using a mower-conditioner. Tedders may be used to spread the crop over the field surface to increase interception of solar radiation and to create a thinner layer of hay. Raking is done if necessary to gather the crop into windrows before baling. In regions where drying conditions are more favorable, hay is often windrowed immediately during mowing to avoid bleaching and some of the DM losses caused by physical manipulation during raking.

Mowing

Mowing involves cutting shoots 2–4 in. (5–10 cm) above the soil surface. Taller **stubble** heights are necessary only if the species that is being used depends on axillary stem buds for regrowth. Alfalfa and many other species have regrowth almost exclusively from the **crown** near the soil surface, so close cutting is not detrimental. Reciprocating sickle mowers are the traditional method for mowing, but rotary disk mowers, which allow faster field speeds, have become common in many regions.

Conditioning

Mechanical **conditioning** is typically accomplished by moving the mowed crop between tensioned intermeshing rubber rollers before depositing it back onto the field. The bending and crushing action exerted on the crop creates physical openings in the **cuticle** layer and hastens moisture loss. Conditioning normally reduces curing time by 1 or 2 days on first cutting but somewhat less on second and subsequent cuttings. Conditioning may be ineffective when humidity levels are very high, because moisture does not move readily from crop surfaces into the atmosphere.

Plant stems usually dry less rapidly than do leaves, so mechanical and chemical conditioning should be directed mainly at hastening stem drying. Thus the stems of properly conditioned crops should show evidence of crushing or breaking (Fig. 17.3). Intermeshing roller conditioners detach some leaves and increase DM losses by 1–2%. Flail conditioners use rotating plastic or metal elements to abrade forage or bend and break stems. These

FIG. 17.3. The stem of alfalfa conditioned using an intermeshing rubber roller conditioner. Openings in the cuticle allow moisture to escape. Leaves should not show excessive bruising or detachment.

FIG. 17.4. Typical rakes used when gathering forage for baling or chopping. Parallel-bar rakes (top) and wheel rakes (middle) are the most common types, but some tedder units (bottom) can also be used as rakes. (Photos courtesy of Alan Rotz and Michael Collins.)

conditioners provide good drying benefits, but attention is needed to avoid excessive leaf loss on legume crops.

Chemical conditioning is accomplished by applying a drying agent or **desiccant** material, usually a water solution of potassium carbonate (K_2CO_3), to the crop at the time of mowing. These compounds alter the waxy materials of the cuticle layer that covers plant surfaces, and allow moisture to pass through. The uniform distribution of the desiccant over the stem surfaces that is needed to improve drying requires the use of large volumes of solution. These logistical difficulties have limited the use of chemical conditioning agents.

Tedding and Raking

When hay curing is done in swaths or windrows, much of the solar radiation is intercepted by crop stubble or by the soil. Tedding disperses the crop so that it covers the entire field. Such well-dispersed hay crops intercept most of the solar radiation and make better use of the sun's energy. In addition, tedding creates thinner crop layers that lead to more uniform moisture levels, which are useful during hay storage. It also breaks up dense clumps of forage that sometimes form during mowing. Under good drying conditions, tedding shortens curing time by about 0.5 days.

On most crops, tedding should be done within a few hours of mowing or during the early morning on successive days. At these times the leaf component is partially hydrated. Losses from tedding overly dry legume crops can exceed 10% DM, which is well above the usual loss of 1–3% DM. Tedding should be avoided on low-yielding, short crops in order to reduce losses associated with raking of short materials; moisture loss is usually adequate without tedding under these conditions.

Raking gathers the crop into windrows for baling. The timing of raking in relation to crop moisture content greatly affects DM losses associated with this step, especially for legume hay crops. When raking is done at the ideal moisture level of 35–40%, DM losses are usually less than 4%, and total curing time is not unduly extended. Dry matter losses can exceed 20% when raking is delayed until the crop is at baling moisture level. Windrowing usually slows drying compared with swaths, so raking too early (at much over 40% moisture level) extends curing time and delays baling. Grass leaves are firmly attached to stems, so grass hays resist shattering over a wide range of moisture levels.

There are several equipment options for raking (Fig. 17.4). Parallel-bar rakes have flexible tines that sweep across the field surface as the machine moves forward. Wheel rakes have tines on circular "wheel" units that catch and move the hay as they roll. Tedder rakes are combination units that can disperse the crop or, by reversing the direction of rotation, gather the crop into windrows. Raking losses are greatest for tedder rakes, then for wheel rakes, and least for bar rakes.

Windrow inverters are implements designed to turn raked hay windrows, placing them upside down in the same spot or shifted slightly to one side. Studies with windrow inverters have generally not shown any improvement in drying rate. They are useful for gently turning windrows exposed to rain.

Baling

Baling collects hay into packages for ease of handling, transport, and storage. Small rectangular bales predominate in some regions of the USA, whereas large rectangular bales are most common in cash-hay-producing regions. Round bales are the most common hay package type across much of the USA.

Small rectangular bales are readily handled and fed by hand. Collection and stacking for storage can be done mechanically or manually. Typical small rectangular bales are 14 in. (36 cm) high, 18 in. (46 cm) wide, and about 40 in. (102 cm) long, and weigh 45–60 lb (20–30 kg), depending on crop species, crop moisture level, and bale density (Table 17.1).

Larger, higher-density rectangular bales weighing 0.4–1.0 ton (0.36–0.91 mt) predominate in low-humidity cash-hay-producing regions. The combination of large size, rectangular shape, and high density makes these bales ideal for long-distance transport. Large rectangular bales are 2–4 ft (0.6–1.2 m) in width and height, and 5–8 ft (1.5–2.4 m) in length. Typical DM densities for these bales are higher than for small rectangular bales and round bales. Bales that are larger in volume or that have higher DM density experience greater heating at any given hay

Table 17.1. Typical hay package weights and sizes: dry matter densities are generally greatest for large rectangular bales; stacks are not shown, but have lower DM densities than any of the package types shown in this table

Bale shape	Diameter	Width	Length	Volume (ft^3)	Typical bale weight (lb)[a]	Dry matter density (lb/ft^3)	Safe baling moisture (%)
Rectangular	14 in.	18 in.	38 in.	5.5	60	8–11	20
Rectangular	32 in.	36 in.	7 ft	56	900	14–16	12–16
Rectangular	4 ft	4 ft	8 ft	112	1800	14–16	12–16
Round	4 ft	–	4 ft	50	500	10–13	18
Round	4 ft	–	5 ft	63	850	10–13	18
Round	5 ft	–	4 ft	79	1000	10–13	18
Round	5 ft	–	5 ft	98	1300	10–13	18
Round	6 ft	–	5 ft	141	1900	10–13	18

[a]Assumes 18% moisture concentration.

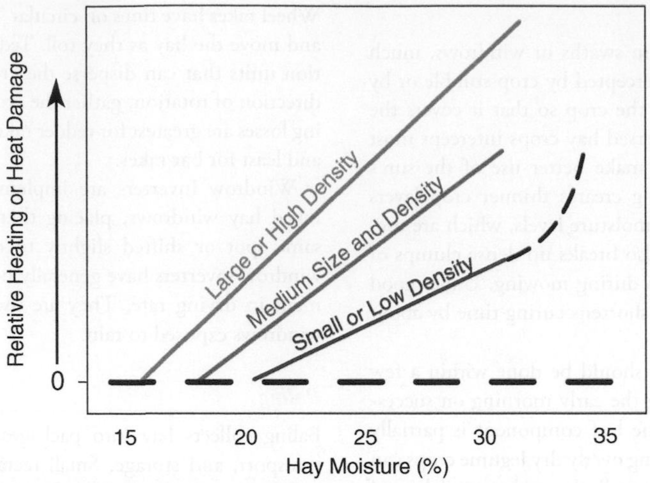

FIG. 17.5. As a general trend, heating during hay storage increases as hay moisture level increases. Using larger or denser hay packages increases heating at any given moisture level.

moisture level (Fig. 17.5). Volume and density differences affect safe baling moisture recommendations for different package types (Table 17.1).

To maintain quality and avoid excessive DM losses during storage, hay in small rectangular bales should be at or below 20% moisture, hay in round bales at 18% moisture, and hay in large rectangular bales at 16% moisture when baled (Table 17.1). In low-humidity regions, hay is often allowed to reach much lower moisture levels in the field (often under 10%), and is then baled at night when the relative humidity is higher. This approach ensures that stems are uniformly dry. Night-time baling under these unique environmental conditions allows retention of the partially rehydrated leaves while maintaining low overall crop moisture levels to avoid heating and molding during storage. Hay baled using this system often contains near 12% moisture. This approach is not feasible in eastern, southeastern, and other more humid regions of the country because relative humidity rises quickly to high levels after sunset, rewetting the forage.

Understanding and Measuring Forage Moisture

Forage dry matter (DM) and moisture are most often calculated and reported on a wet basis because the standing crop has much more water than DM. This can lead to confusion because analyses of other forage components are reported on both a dry and an as-fed basis.

A ton of silage with 40% DM on a wet basis contains 150% moisture on a dry basis:

$$\%\text{ moisture (DB)} = \frac{\%\text{ moisture (WB)}}{100 - \%\text{ moisture (WB)}} \times 100$$

$$= \frac{60}{100 - 60} \times 100 = 150\%$$

Thus silage with a dry-basis moisture concentration of 60% actually contains 1.5 times as much water as DM. For example, forage with 80% moisture actually contains four times as much water as DM.

The amount of water that forage contains must be taken into account when feeding. A silage that has 60% moisture on a wet basis and 8% crude protein (CP) concentration on a dry basis actually contains much less protein per unit of silage fed:

$$\%\text{CP (as fed)} = \frac{\%\text{CP (DB)} \times \%\text{DM}}{100} = \frac{8 \times 40}{100} = 3.2\%$$

For the same forage preserved as hay at 15% moisture (wet basis), the as-fed CP concentration would be:

$$\%CP\left(\text{as fed}\right) = \frac{8 \times 85}{100} = 6.8\%$$

Accurate estimation of moisture level at harvest is critical because the storage behavior of both hay and silage is greatly affected by moisture. Baling hay that is too wet or too dry can lead to high storage or harvesting losses, respectively (Fig. 17.6). Misjudging moisture content when making silage can lead to spoilage or excessive DM loss.

Fig. 17.6. Moisture levels in field wilting silage and curing hay can vary widely within the field, so thorough sampling is critical in order to obtain accurate estimates.

There are four common methods of testing forage moisture in the field prior to harvest—by feel, use of electronic probes, forced-air drying, or microwave drying. Estimating forage moisture by feel is generally the least reliable method, and depends heavily on moisture range and the skill of the tester. Moisture probes that estimate moisture on the basis of electroconductivity are very quick and easy to use. Forage density and other factors affect the accuracy of these meters, but they generally provide a reasonable estimate of moisture. The most accurate methods involve drying the forage using a forced-air dryer or a microwave oven. Both methods involve weighing a sample before and after drying to a constant dry weight. Both methods require electricity and are more time consuming, but they are generally quite accurate.

To yield accurate results, samples must be representative of the entire volume of forage to be harvested. It is also important to remember that, in field-drying situations, forage moisture content can change very rapidly depending on the drying conditions.

Round bales require less labor in terms of baling, handling, and feeding than small rectangular bales. Typical round bales are 4–6 ft (1.2–1.8 m) in diameter and 4–5.5 ft (1.2–1.7 m) in length, with a mass of 500–2000 lb (227–908 kg) (Table 17.1). Variable-chamber round balers use belts to apply pressure (which is usually adjustable) during the entire bale formation process, whereas fixed-chamber balers apply pressure only when bale formation is nearly complete. Bales from fixed-chamber balers have a lower-density core (the central portion of the bale) but similar or greater density near the surface compared with variable-chamber bales.

Dry-matter losses during rectangular baling are usually in the range of 2–5%. Losses are minimized by increasing ground speed during baling, and by making larger windrows. Baling DM losses for round balers are in the

Table 17.2. Dry matter losses during storage of tall fescue round bales in Kentucky (bales were 4 ft long and 5 ft in diameter)

Storage location	Binding material	Thickness of weathered layer (in.)	DM loss (%)	Total DM loss (including weathering) (%)	Proportion of bale volume weathered (%)
Outside storage	Twine	4.4	18	34	27
	Plastic netting	2.1	11	23	14
	Solid plastic	0.6	4	8	4
Hay storage shed	Twine	0	6	6	0

Source: Collins et al., 1995.

range of 8–10%, and can be higher if the moisture level falls below 15%. Baling losses are often higher for fixed-chamber than for variable-chamber balers.

Hay Storage

Storage losses are about 5% for hay harvested near 15% moisture content and stored under dry conditions. Wetter hay loses approximately 1% in additional DM for each additional 1% of moisture it must lose to reach equilibrium moisture (commonly near 15%). Equilibrium moisture is typically lower for hay baled with less than 15% moisture content. **Forage quality** is decreased during storage of hay baled above 20% moisture content unless steps are taken to avoid microbial growth and heating.

Rectangular Bale Storage

Rectangular bales are usually protected by being stored inside, especially in wet regions. Their flat surfaces do not shed water, so storage losses can be very large if they are not protected. Common storage structures include pole barns with no walls, or sheds enclosed on one to three sides. In the eastern USA, many hay sheds are completely enclosed. Good ventilation is needed during the first 1–3 weeks after baling, in order to eliminate moisture quickly as it exits the stack. In low-rainfall regions, rectangular bales may be stacked outside, but such stacks should be covered to shed water.

Round Bale Storage

Round bale storage losses range from 5% to as much as 40%, depending on climate and on the degree of protection from weathering (Collins et al., 1995). Round bales are commonly stored outside in contact with the ground and unprotected from the weather. This eliminates the need for storage structures and makes the selection of storage location very flexible. Outside storage losses are small in dry regions or in areas where winter precipitation is mainly snow. However, in regions where total annual precipitation is more than about 30 in. (76 cm), round bales can lose one-third of their initial DM during a single season of outside storage (Table 17.2).

Weather-related storage losses are deceptively large. About 40% of the loss occurs on the bottom of the bale, where it is not visible. Using crushed stone, poles, or other methods to elevate hay off the ground reduces total DM loss. In addition, even thin weathered layers over the entire bale surface affect a substantial percentage of the total bale volume (Table 17.2). The weathered layer that was 4 in. (10 cm) thick on the 4 ft × 5 ft (1.2 m × 1.5 m) bales in a study by Collins et al. (1995) represented 27% of the total volume. The percentage of bale volume affected by a given weathered depth declines as the bale diameter increases.

Weathering dramatically reduces DM **digestibility** (Table 17.3) and usually increases fiber levels. This is because the sugars, minerals, and other compounds that

Table 17.3. Effects of weathering on quality of grass round bales in Southern Indiana

Hay type	Bale fraction	DM digestibility	Crude protein
		(%)	
Grass[a]	Unweathered	59	13.5
	Weathered	43	16.4
Alfalfa–grass	Unweathered	57	14.3
	Weathered	34	16.9

Source: Lechtenberg et al., 1979.
[a]Kentucky bluegrass, tall fescue, and orchardgrass mixture.

are lost during microbial growth and leaching come almost exclusively from the cell contents fraction. Weathering effects on crude protein levels are mixed. Weathering-induced increases in crude protein levels can occur because protein molecules are large and more difficult to remove during leaching.

Hay that lies beneath the weathered material on outside-stored bales has similar forage quality to that of hay stored in a shed. For example, the weathered tall fescue hay in Table 17.2 had a DM digestibility of 28%, compared with 47% for unaffected hay from the same bales.

In humid regions, the use of hay sheds, superior binding materials, or covers to protect the hay during storage greatly reduces DM losses. The use of solid plastic binding material reduces DM losses to about 8% DM, which is similar to DM losses for hay stored in a shed (Table 17.2). Plastic sleeves or plastic covers (bonnets) for rows of bales protect round bales from weathering. Heavy tarps placed over pyramid-shaped stacks (Fig. 17.7) efficiently protect a larger number of round bales. These stacking and covering procedures restrict moisture and heat loss. Round bales with more than 18% moisture are more susceptible to quality loss and excessive DM loss during storage than are similar bales that are not stacked or covered. Increasing twine usage from the usual 6- to 8-in. (15- to 20-cm) spacing down to a 4-in. (10-cm) spacing has also been shown to reduce outside storage losses.

Other considerations for minimizing storage losses for twine-tied bales stored outside include the following: (1) selecting an open area away from trees to hasten drying following wet periods; (2) arranging bales in a single layer with 3–4 ft (0.9–1.2 m) of space between rows, to improve air circulation around the bales; bales should not be stacked if they are not to be covered, because water shed from the upper layers penetrates directly into the lower layers, where it can cause severe damage; (3) the flat ends of bales within a row can contact adjacent bales without significantly increasing storage losses.

Hay Preservatives and Barn Drying

Rain damage may occur on 30–60% of spring cuttings in some regions. Hay yield and quality are reduced when rain falls on the curing crop (Table 17.4). Rain damage leaches soluble constituents, may increase respiration rates, and increases **leaf shattering** losses, especially as additional raking or tedding steps are usually necessary to complete the curing process.

The mowing date can be adjusted slightly to avoid imminent rainfall, but predicting 4- to 5-day rain-free periods for haymaking is difficult. Long delays in harvest to avoid rain damage are ineffective because forage quality declines steadily before harvesting, due to maturation. When rain appears likely and the hay crop is nearly dry, rain damage can be avoided by baling hay at moisture levels of

Fig. 17.7. Economical storage alternatives are available for round-baled hay. This photograph illustrates the use of the "pyramid" stacking pattern for round bales. These stacks are covered by heavy-duty tarps to exclude water. Ventilation is greatly restricted in such storage systems, so hay moisture levels for round bales should be 18% or less before stacking.

Table 17.4. Comparison of forage quality of standing alfalfa forage just prior to harvest with the quality of hay produced without rain damage, and hay exposed to rain damage

Type of forage	Yield[a] (%)	Crude protein (%)	Digestibility (%)	NDF (%)
Before curing	100	23	70	43
Well-cured hay	85	20	64	46
Rain-damaged hay[b]	75	20	57	54

Source: Collins, 1990.
NDF, neutral detergent fiber.
[a]The values obtained before curing are averages of 54 hay harvests over a 4-year period.
[b]Measurable rain fell at some point during hay curing in about 50% of cases.

21–35%, well above the 20% level that is normally considered safe. Field time can be reduced by 0.5–1 day, but steps will be needed to avoid heating and molding during storage of this moist hay.

The amount of heat generated in stored hay is positively correlated with its moisture level (Fig. 17.8). Aerobic microorganisms (mainly fungi such as *Aspergillus* and *Fusarium* species) proliferate in moist hay, and generate heat by their respiration. Limited microbial growth occurs in dry hay, generates a small amount of heat, and usually raises temperatures only slightly (Fig. 17.8). Forage quality is not greatly affected, and this minor heating (often called sweating) has the beneficial effect of removing excess moisture. Once the hay equilibrates at 15% moisture or less, it becomes stable for long-term storage.

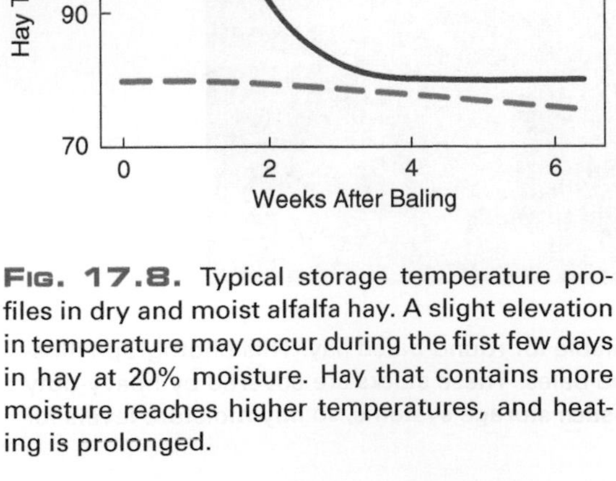

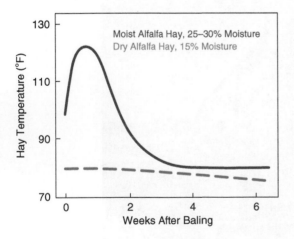

Fig. 17.8. Typical storage temperature profiles in dry and moist alfalfa hay. A slight elevation in temperature may occur during the first few days in hay at 20% moisture. Hay that contains more moisture reaches higher temperatures, and heating is prolonged.

Excessive microbial growth can raise temperatures to 130–150°F or more, increasing DM loss and encouraging Maillard reactions that reduce DM and crude protein digestibility. In extreme cases, microbial heating can lead to exothermic chemical reactions that further elevate hay temperature into the range where spontaneous combustion can occur. Dustiness caused by mold spores is increased, and **palatability** is reduced in heat-damaged hay. Effective **preservatives** to inhibit microbial growth, or artificial drying to remove excess moisture, help to avoid loss of quality and yield in moist hay.

Spontaneous Heating in Hay

Hay temperatures are commonly higher than the surrounding air temperature for 2–3 weeks after baling. They usually peak after 4–6 days, and then cool as evaporation dries the hay to below the moisture levels needed for microbial activity. Heat is generated by respiration during growth of fungi such as *Fusarium* and *Aspergillus* species. Optimum growth of these aerobic organisms occurs when the relative humidity is above 70%. Minor heating to temperatures near 130°F has little effect on forage quality.

In wetter hay, bacteria and **thermophilic** (heat-loving) actinomycetes thrive and raise temperatures up to 160°F, at which level most microbes are killed. Above 200°F the actinomycetes are inactivated, but exothermic chemical reactions may commence and increase temperatures to 500–800°F, when ignition can occur.

Prolonged heating of hay uses sugars and reduces protein digestibility. Heat-damaged hay turns brown, and the degree of brown color development is a good indicator of heat damage. Non-enzymatic browning chemical reactions convert amino acids and sugars to indigestible Maillard products that show up as acid detergent fiber (ADF) during forage analysis (Fig. 17.9). Heat damage is assessed by measuring N in the ADF fraction.

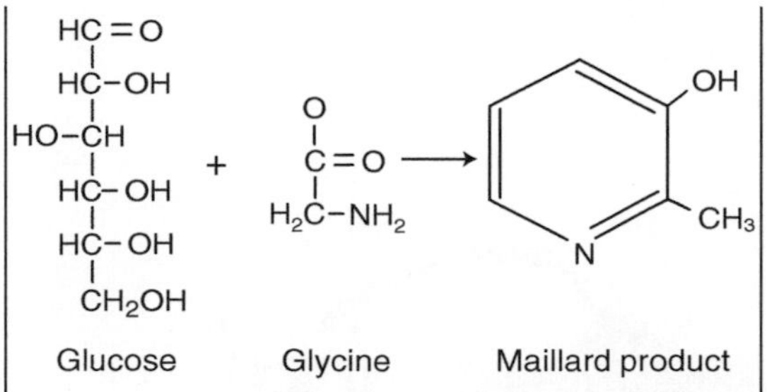

Fig. 17.9. Spontaneous heating of moist hay leads to chemical reactions between sugars and amino acids, resulting in the production of compounds known as Maillard products. These compounds reduce the digestibility of the crude protein in the hay.

Effective **hay preservatives** include organic acids, buffered acids, and ammonia sources such as anhydrous ammonia or urea. Organic acids, mainly propionic and acetic acids, inhibit mold growth and reduce heating and dustiness in hay. Propionic and acetic acids effectively control heating in moist hay, but their **corrosiveness**, high **volatility**, and handling problems have limited their use. Ammonium propionate, which is made using ammonia and propionic acid, has a pH near 6 and is as effective as propionic acid in preserving hay. The higher pH means that handling and other problems associated with the use of acids are avoided. Application rates of around 10 lb/ton (4 kg/mt) are needed for hay with 20–25% moisture, whereas rates of around 20 lb/ton (8 kg/mt) are needed for wetter hay with 25–30% moisture.

Accurate determination of hay moisture content is difficult, but recently developed preservative application equipment helps to overcome this problem by enabling automatic adjustment of preservative rates based on moisture measurements taken as the crop enters the baler. Poor material distribution can lead to well-preserved and moldy areas within the same bale. Distribution is improved by reducing windrow size, increasing spray tip pressure, diluting the preservative by 1:1 with water to increase the volume that is being applied, and increasing the number of nozzles (especially if this makes it possible to apply material to the bottom of the windrow).

Anhydrous ammonia reduces microbial growth in moist hay, and can improve fiber digestibility by acting on lignin–carbohydrate bonds in cell walls. Ammonia rates of about 1% should be applied to stacked bales that have been covered with plastic to prevent loss. Ammonia increases non-protein N levels, which can help to meet the protein needs of ruminants. Difficulties in application and safety concerns have limited the use of ammonia in hay preservation. Ammoniation of hays with a high sugar content can lead to health problems in cattle. Urea is safer to handle than anhydrous ammonia, and has similar benefits in terms of storage and fiber digestibility.

Artificial drying avoids storage problems by removing excess moisture by using heated or ambient air. Such systems are designed so that the air is forced through the bales, which are usually small rectangular bales stacked on edge. Batch systems require the hay to be moved to a permanent storage area when drying is complete. Combined drying and storage facilities utilize portable air ducts or permanent air ducts that can be closed off once drying is complete for a particular lot of hay. Uniformly dense bales are desirable, to ensure consistent drying and acceptable handling characteristics of the dried bales.

Silage

Silage is forage preserved by **anaerobic** storage, usually under conditions that encourage fermentation of sugars to organic acids such as lactic acid, acetic acid, and propionic acid. Silage systems are widely used to preserve forage in regions where climatic conditions are not conducive to hay production, and also to preserve crops such as corn that are poorly suited to hay production. The production and feeding of chopped silage is highly mechanized, so labor inputs are lower than for hay (Fig. 17.10). Equipment and structure costs are usually higher for chopped silage systems than for hay systems.

Silage avoids most of the harvesting losses encountered in hay production because silage is handled at a higher moisture level, so it is much less vulnerable to mechanical losses (Fig. 17.1). However, poorly preserved silage has undesirable microbial activity that can lead to higher DM losses, low quality, and low **palatability**. Combined harvest and storage DM losses in poorly preserved silage can equal or exceed those in hay.

Silages are classified according to their moisture levels. **High-moisture silage** contains more than 70% moisture, **wilted silage** usually contains 60–70% moisture, and **low-moisture silage** (commonly referred to as **haylage**) contains 40–60% moisture.

Silage Microorganisms

The lactic acid bacteria (LAB) are mainly responsible for the reduction in pH that occurs in properly fermented silages. Lactic acid bacteria are either *homofermenters*, which produce only lactic acid, or *heterofermenters*, which produce acetic acid, ethanol, and CO_2 in addition to lactic acid (Table 17.5). Homofermentative bacteria are preferred because lactic acid reduces pH more effectively than does acetic acid, and because the DM losses associated with gas production by heterofermentative bacteria are avoided.

Enterobacteria (also called coliform bacteria), fungi (yeasts and molds), and **clostridia** bacteria are undesirable organisms that may be present on ensiled plant material. Excessive proliferation of these organisms has negative

Fig. 17.10. Chopping alfalfa for silage production. The crop was windrowed and allowed to wilt in preparation for harvest. Chopped silage systems allow a high degree of mechanization of forage harvest, storage, and feeding.

effects on silage quality (McDonald et al., 1991). Recommended silage management practices are designed to encourage the development of desirable lactic acid bacteria and to discourage the growth of undesirable microorganisms.

Table 17.5. Species of lactic acid bacteria that are of importance in silage

Homofermentative	Heterofermentative
Lactobacillus acidophilus	*Lactobacillus brevis*
Lactobacillus casei	*Lactobacillus buchneri*
Lactobacillus coryniformis	*Lactobacillus fermentum*
Lactobacillus curvatus	*Lactobacillus viridescens*
Lactobacillus plantarum	*Leuconostoc mesenteroides*
Lactobacillus salivarius	
Pediococcus acidilactici	
Pediococcus cerevisiae	
Pediococcus pentosaceus	
Enterococcus faecalis	
Enterococcus faecium	
Lactococcus lactis	
Streptococcus bovis	

Source: McDonald et al., 1991.

Enterobacteria also produce lactic acid, but are undesirable because they compete with lactic acid bacteria for sugars, have acetic acid as a major end product, increase **buffering capacity** (which makes pH reduction more difficult) (see "Buffering Capacity"), and produce CO_2 (which wastes energy that could be utilized by animals). **Protease** enzymes break protein down to non-protein forms of N, including peptides, amino acids, amides, amines, and ammonia. Around 75% or more of the N in fresh forages is in the form of protein. In silage, especially from high-protein legumes such as alfalfa, as little as 20% of silage N may be in protein (Albrecht and Muck, 1991). The acceptability and intake of silage can be reduced by these non-protein N compounds. This breakdown of protein, called **proteolysis,** is minimized by shortening the aerobic phase, by quickly achieving a low pH, by wilting to less than 70% moisture before ensiling, and by maintaining anaerobic conditions to prevent heating. Enterobacteria are not common when silage pH is below 5.0.

Buffering Capacity

Ideal silage fermentation involves reducing the pH as quickly as possible from the initial value of about 6 in fresh forage down into the range of 4.0–4.5. The magnitude of the pH reduction that results from a given quantity of organic acid production differs widely among common silage crops. *Buffering capacity* is the term used to describe the resistance to pH change in a solution or material. More acid or base is required to alter the pH in highly buffered materials compared with less buffered materials. The data below (summarized by Pitt et al., 1985) illustrate the generally higher buffering capacities of legumes compared with grass forages. The values are equivalents of acid needed to decrease the pH value of 1 g of forage DM from 6 to 4. High buffering in legumes results from high levels of cations such as Ca, Mg, and K, and from high protein concentrations, especially if proteolysis produces non-protein N.

Buffering capacity of silage crops

Type of forage	Species	Buffering capacity (equivalents)
Legumes	Alfalfa	6.0
	Red clover	4.3
Grasses	Timothy	2.4
	Orchardgrass	2.4
	Corn	2.4

Source: adapted from Pitt et al., 1985.

Clostridia are undesirable anaerobic bacteria that ferment sugars, organic acids, and proteins (McDonald et al., 1991). Clostridial growth results in the production of butyric acid and increases DM losses. Butyric acid negatively affects silage in two ways: (1) a pungent odor is generated that makes the silage unpalatable to animals, and (2) the silage pH increases, allowing other undesirable microorganisms to proliferate. Proteolytic clostridia break down amino acids and amides, and release ammonia and CO_2. The ammonia that is produced further raises the pH. Conditions that contribute to clostridial growth include moisture levels above 70%, elevated storage temperatures, inadequate fermentable sugar levels, high buffering capacity, and an extended aerobic phase.

Phases of Silage Preservation

Silage preservation consists of an aerobic phase, a fermentation phase, a stable phase, and a feedout phase (Fig. 17.11).

Aerobic Phase

Air trapped in ensiled forage is utilized in aerobic respiration and proteolysis during the aerobic phase. Both biochemical processes are related to plant enzyme activity. With adequate packing, correct moisture content, rapid filling, and good sealing, the aerobic phase lasts for about 1 day. Aerobic respiration oxidizes plant sugars to CO_2 and water, with the release of heat, thereby wasting fermentable carbohydrates that could be used in silage fermentation. If there is significant heating during the aerobic phase, Maillard reactions occur and this can decrease crude protein digestibility, as described earlier.

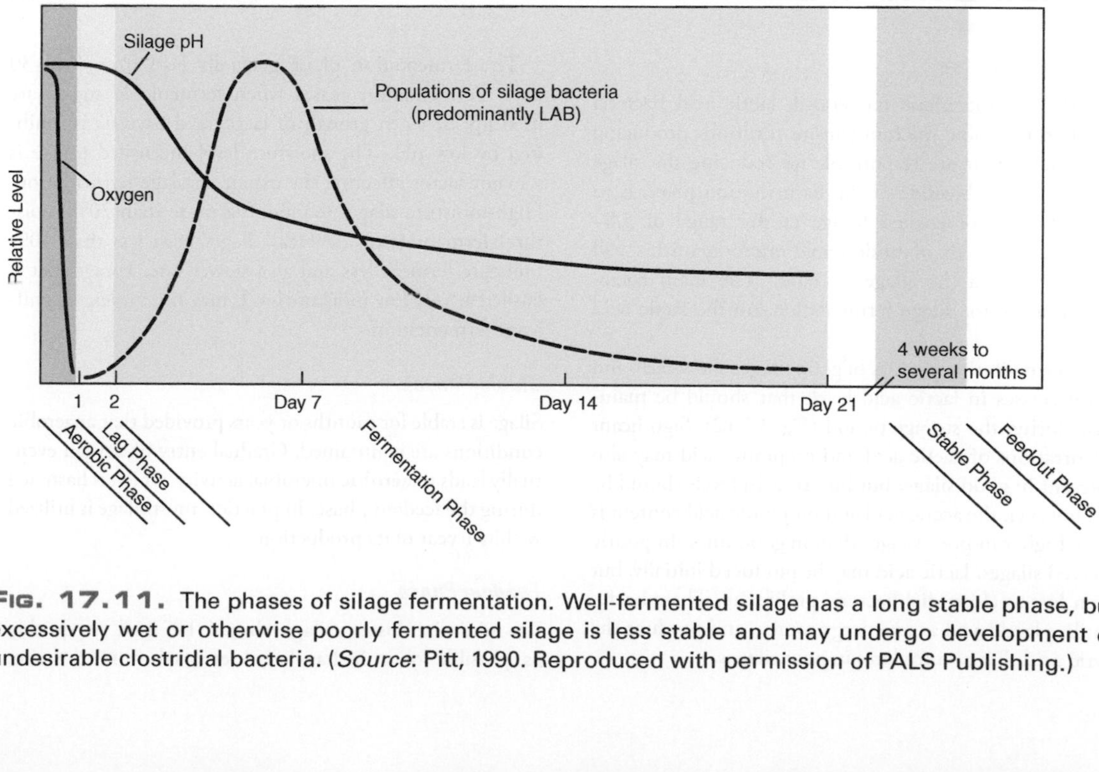

Fig. 17.11. The phases of silage fermentation. Well-fermented silage has a long stable phase, but excessively wet or otherwise poorly fermented silage is less stable and may undergo development of undesirable clostridial bacteria. (*Source*: Pitt, 1990. Reproduced with permission of PALS Publishing.)

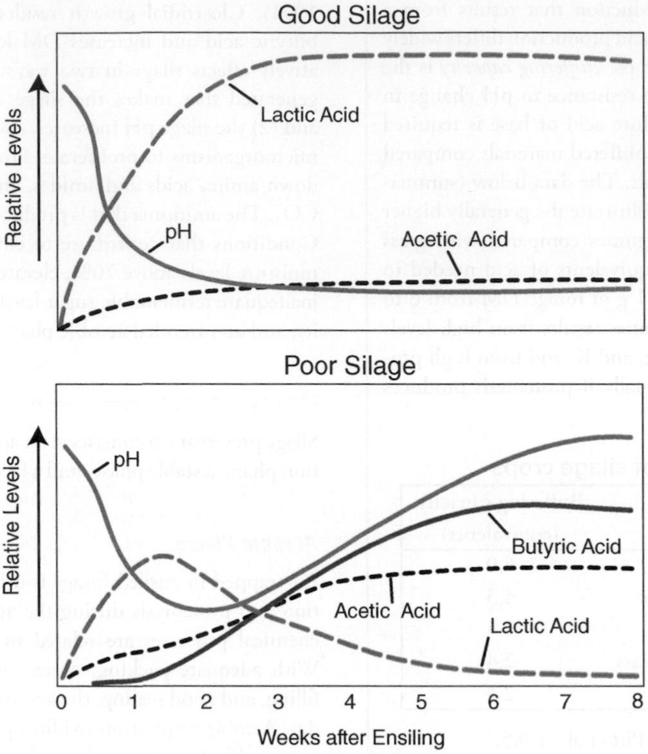

Fig. 17.12. Typical organic acid and pH profiles in good- and poor-quality silages. Good silages have a rapid drop in pH, a lower final pH, and low levels of butyric acid. Ideally, lactic acid should be the dominant organic acid present, but propionic acid and acetic acid can together exceed lactic acid levels in some well-preserved silages. (Adapted from Langston et al., 1958.)

Fermentation Phase

Once the aerobic phase has ended, lactic acid bacteria and other anaerobic microorganisms multiply, producing organic acids that are responsible for reducing the silage pH. The main objective of the fermentation phase is to reduce the pH of ensiled forage to the range of 3.8–5.0, so that growth of undesirable microorganisms will be restricted and the silage is stable. The main bacteria responsible for silage fermentation are the lactic acid bacteria.

The fermentation profiles of good silages show early and rapid increases in lactic acid levels that should be maintained during the storage period (Fig. 17.12). Significant concentrations of acetic acid and propionic acid may also be present in good silage, but butyric acid levels should be low. However, the acetic acid and propionic acid content is usually higher in poor silages than in good ones. In poorly preserved silages, lactic acid may be produced initially, but is then lost as clostridial bacteria proliferate. These bacteria utilize lactic acid as an energy source, and produce the butyric acid that accumulates in poor silages.

The fermentation phase generally lasts from 7 to 30 days. Fermentation ceases when fermentable sugars are used up, or when growth of lactic acid bacteria is inhibited by low pH. The moisture level in ensiled forage is a major factor affecting the extent of silage fermentation. High-moisture silages (containing more than 70% moisture) ferment faster, whereas silages with less than 50% moisture ferment less and at a slower rate. Forage that is ensiled at very low moisture levels may not undergo significant fermentation.

Stable Phase

Silage is stable for months or years provided that anaerobic conditions are maintained. Gradual entry of oxygen eventually leads to **aerobic** microbial activity, which is hastened during the feedout phase. In practice, most silage is utilized within 1 year of its production.

Feedout Phase

To minimize losses, silage should be used as quickly as possible once it has been removed from the silo.

Dormant aerobic microorganisms begin to grow rapidly when silage is exposed to the air, especially at the exposed surface where oxygen levels are highest. Lactic acid, acetic acid, and remaining sugars are used, producing CO_2, water, and heat (McDonald et al., 1991). Aerobic deterioration occurs sooner in silages such as corn that still have high levels of unfermented sugars, and in low-moisture silages that often have a high pH and low density. Silage palatability is reduced by the changes that occur during aerobic deterioration.

Dry matter losses and silage refusal due to aerobic deterioration can be minimized by frequent feeding of amounts of silage that can be utilized quickly, before significant deterioration occurs. Corn silage should be fed especially frequently to avoid DM losses and quality deterioration. Alfalfa silage is somewhat more stable and can be utilized over periods of 1–3 days. The longer times are suitable in cooler weather, but the shorter times should be adhered to in warm or hot weather.

Feeding rate is an important consideration when designing silage systems. Silo size and animal numbers should be matched so that at least 4–12 in. (10–30 cm) of silage is removed from the face of horizontal silos each day to minimize aerobic deterioration. For upright silos, at least 2 in. (5 cm) should be removed daily during the feedout phase.

Silage Storage

The main types of silage storage include upright silos, upright "oxygen-limiting" silos, horizontal silos, bag silos, and round bale silage. Typical upright silos are concrete structures that are generally unloaded from the top. Moisture levels of 60–65% reduce **effluent** losses and minimize unloading problems in upright silos.

Upright "oxygen-limiting" silos may be made of metal, poured concrete, or concrete staves. Many of these silos unload from the bottom, enabling second and third cuttings of perennial silage crops to be placed on top of earlier harvests. When using these silos, forages can be ensiled at somewhat lower moisture levels of 40–50%. These silos have the highest constructions costs, but DM losses are low and the silages produced are very stable.

Horizontal silos usually have concrete bottoms and side walls. The side walls should slope outward at a rate of 1 ft (0.3 m) for each 8 ft (2.4 m) of height to allow proper packing of silage along the walls. Simpler horizontal silos include open pits dug into the side of a hill, and open piles with side walls made of wood, straw bales, or hay bales, or with no side walls. Spoilage losses can exceed 50% if such silos are not covered and weighted down (Bolsen, 1997).

Bag silos are plastic tubes 8–12 ft (2.4–3.7 m) in diameter and of various lengths, into which chopped silage is packed using machines designed for this purpose. Bag silos are not well suited to long-term storage because of the risk of tearing of the plastic. To avoid the bags becoming punctured or ripped, they should be placed on a smooth hard surface such as asphalt or concrete.

In the case of round bale silage, multiple layers of UV-treated polyethylene stretch film, usually 1 mil (0.025 mm) thick, are used to cover round or rectangular bales of high-moisture forage. The film can be applied to individual bales or to rows of bales. Typical platform wrappers apply film to individual bales from a fixed roll as the bale is simultaneously revolved and rolled on the platform. In-line wrappers apply film from rolls moving on a large hoop through which bales are slowly passed to form long rows of bales. The latter system uses less film per bale by avoiding film application to the bale ends within the row.

Ensiled round bales have a large surface area to mass ratio that reduces the effectiveness of oxygen exclusion compared with some other silage storage systems. For this reason, round bale silage is best utilized within 8–12 months. When longer storage times are desired, or if moisture content is below the optimum range of 50–65%, six or more layers of film can be applied.

Silage Management

The aim of silage management is to maintain forage quality while providing adequate fermentation to ensure good preservation and minimal growth of undesirable microorganisms. Just as with grazed forage or hay, silages that support high **voluntary intake** levels usually have low fiber levels and high digestibility. To ensure high intakes, silage should also have a low pH, high lactic acid levels, low ammonia levels, and low butyric acid levels. Well-preserved silage should have a pH below 4.5, less than 10% of the total N in the form of ammonia N, high levels of lactic acid, and almost no butyric acid.

Moisture

Silage containing more than 70% moisture should usually be avoided because these conditions increase the likelihood that clostridia and other undesirable bacteria will develop. These organisms produce butyric acid and amines such as putrescine and cadaverine that appear to be responsible for the reduced intakes often seen with high-moisture silages. Butyric acid fermentation also increases DM loss. Excessively wet silage produces effluent or seepage, leading to the loss of readily **digestible** nutrients. Effluent is not usually a problem in horizontal silos or in tower silos if silage moisture content is below 70%.

Wilting reduces the amount of fermentable carbohydrate required to properly preserve the silage, and restricts the growth of undesirable microorganisms (Fig. 17.13). Wilting has the beneficial effect of increasing the minimum pH necessary to inhibit clostridial growth. Wilting may be almost essential to the production of high-quality silage from crops that have very low levels of fermentable carbohydrates, as is often the case with legumes.

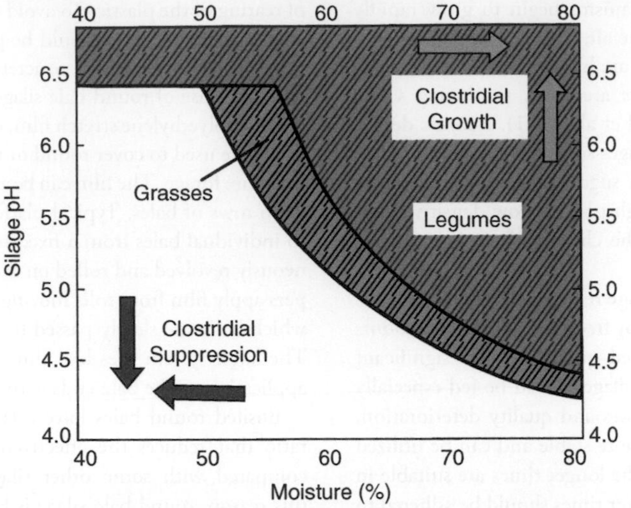

Fig. 17.13. Silage pH values necessary to avoid clostridia growth at different moisture levels. More wilting or a slightly lower pH is needed to avoid clostridial development in grass silages than in legume silages. (Adapted from Leibensperger and Pitt, 1987.)

Legume silage crops, especially alfalfa, are especially in need of wilting because they commonly lack sufficient fermentable carbohydrates to generate the acidity required to properly preserve unwilted silage (Fig. 17.14) (Leibensperger and Pitt, 1988). Grass forages typically have higher levels of fermentable carbohydrates and lower buffering (see "Buffering Capacity"), and therefore can be adequately preserved at somewhat higher moisture levels. Corn silage contains high levels of fermentable carbohydrates.

Silage moisture should match the storage system that is being used (Table 17.6). For example, wilted silage (60–70% moisture) is recommended for horizontal silos and round bale silage, whereas haylage is most suitable for sealed structures with better exclusion of oxygen.

Chopping

Finely chopped forages consolidate better and lead to increased silage density. Precision silage choppers allow accurate setting of theoretical length of cut (TLC), which is the distance that forage moves before it encounters the next knife. Most legume and grass silages should have a theoretical length of cut of 0.39 in. (1 cm), but will have some particles that are 2 in. (5 cm) or more in length. These longer particles occur because some stems are not oriented perpendicular to the shear bar as they move through the chopper. Corn silage should be chopped somewhat finer, to 0.25–0.39 in. (0.6–1 cm), in order to fracture the kernels and improve starch utilization by animals. Corn silage utilization often benefits from the use of post-chopping processing or from re-cutter screens in the chopping system to produce a more uniform particle length.

Chopping releases fermentable carbohydrates by rupturing cells. Soon after chopping, desirable silage bacteria begin to proliferate by using these carbohydrates as substrates. One limitation of typical round bale silage systems is that the forage is not chopped. Some silage balers incorporate cutting mechanisms that coarsely chop the forage to lengths of 2–6 in. (5–15 cm) during the baling process.

Oxygen Exclusion

Oxygen exclusion is needed to encourage good silage fermentation and to lengthen the stable phase. Polyethylene sheeting is often used to cover silage in horizontal silos. Uncovered silage develops a thick layer, 10 in. (25 cm) or more, of spoiled material in which the pH is 7.0 or greater, temperatures are higher, and lactic acid levels are less than 0.5%.

Holes or tears that occur in the plastic covering of round bale silage should be repaired using UV-treated repair tape. The use of oiled sisal twine should be avoided during production of round bales to be ensiled by using stretch film. Chemical reactions between the oils used during twine manufacture and the UV inhibitors in the film lead to the loss of UV protection. A visual indication of imminent film failure is the appearance of light yellow or orange streaks on the surface of the film, coinciding with the twine layers beneath. Film failure results in oxygen entry and leads

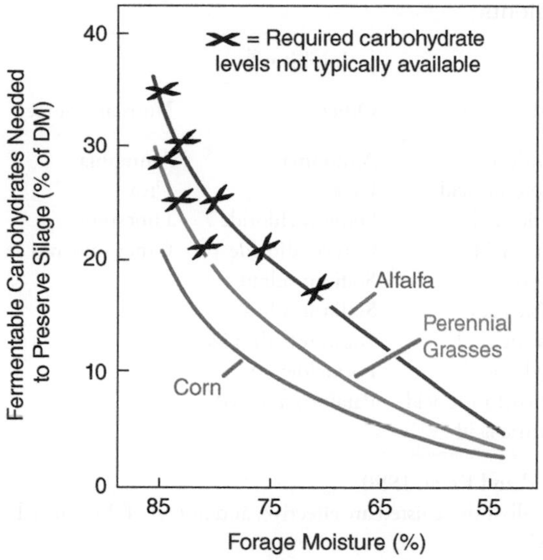

FIG. 17.14. Wilting requirement for fermentation in relation to typical fermentable carbohydrate (FC) levels of major silage types. Alfalfa should be wilted below 30% moisture before ensiling, as preservation of silage made above this moisture level would require more FC than this species typically contains (the area denoted by black crosses). Although the supply of FC would not limit silage fermentation of whole-plant corn even above 80% moisture, optimum crop management also suggests that lower moisture levels are required for this crop. Grass silages are intermediate in FC levels and in the FC requirement for proper preservation. (Data from Leibensperger and Pitt, 1988.)

to rapid silage deterioration. Unoiled sisal twine, plastic twine, and plastic netting are acceptable binding materials that do not hasten the rate of UV inhibitor loss in stretch film.

Silage Additives

Ensiled forages sometimes lack sufficient populations of bacteria to support adequate fermentation. In other cases, levels of fermentable carbohydrates may be too low, or wilting below 70% moisture may not be feasible. Silage additives can be used to remedy some of these deficiencies. Silage additives may be broadly categorized as fermentation stimulants, inhibitors, and nutrient sources (Table 17.7). Additives are applied as forages are chopped or during loading into the silo.

Stimulants have the goal of improving overall fermentation and/or increasing the production of lactic acid. These include bacterial inoculants, enzymes, and sugars. Silage inoculants are relatively inexpensive, safe, and noncorrosive, and are used at low rates. Bacterial inoculants reduce pH, shift the fermentation toward lactic acid production, and reduce ammonia production in about 60% of cases. They improve DM recovery in less than 50% of cases and bunk life in less than 30% of cases. Inoculants are more useful on grasses, alfalfa, and clovers than on corn or small-grain silages, but also tend to be less effective for crops that are very low in sugar.

McDonald et al. (1991) have suggested that an ideal silage inoculant should:

- Grow vigorously and compete with other microorganisms
- Be homofermentative
- Tolerate pH values down to at least 4.0
- Ferment glucose, fructose, sucrose, fructans, and preferably pentose sugars
- Not utilize organic acids in the silage
- Grow at temperatures of up to 120°F
- Be able to grow in low-moisture environments.

Enzyme additives contain products that break down structural and non-structural carbohydrates and release fermentable sugars. These include hemicellulases, cellulases, amylases, and pectinases. Enzymes sometimes improve forage quality by reducing the concentration of neutral detergent fiber (NDF) and acid detergent fiber (ADF). These

Table 17.6. General moisture concentration recommendations for some of the major silage crops and types of silage structures

Crop	Silo type			
	Upright	Oxygen limiting	Horizontal	Bag
	(% moisture)			
Corn and forage sorghum	63–68	50–60	65–72	60–70
Forage legumes	60–65	50–60	65–70	60–70
Cereals and perennial grasses	63–68	50–60	65–70	60–70

Source: adapted from Holland and Kezar, 1990.

Table 17.7. Types of silage additives and additive ingredients

Stimulants			Inhibitors[d]		
Bacterial inoculants[a]	Enzymes[b]	Substrate sources[c]	Acids	Others	Nutrient sources
Lactic acid bacteria	Amylases	Molasses	Formic acid	Ammonia	Ammonia
	Cellulases	Glucose	Propionic acid	Urea	Urea
	Hemicellulases	Sucrose	Acetic acid	Sodium chloride	Limestone
	Pectinases	Dextrose	Lactic acid	Carbon dioxide	Other minerals
	Proteases	Whey	Caproic acid	Sodium sulfate	
	Xylanases	Cereal grains	Sorbic acid	Sodium sulfite	
		Beet pulp	Benzoic acid	Sodium hydroxide	
		Citrus pulp	Acrylic acid	Formaldehyde	
			Hydrochloric acid	Paraformaldehyde	
			Sulfuric acid		

Sources: Muck and Kung, 1997; McDonald et al., 1991; Holland and Kezar, 1990.

Note: Not all additives or ingredients for silage use are listed, not all of those listed are effective, and not all of those listed are approved for use on ensiled material intended for livestock feed.

[a]Most contain live cultures of LAB from the genera *Lactobacillus, Pediococcus, Enterococcus,* or *Streptococcus.*

[b]Most enzymes are microbial by-products that have enzymatic activity.

[c]Most ingredients can also be listed under nutrient sources.

[d]Some inhibitors work aerobically, suppressing the growth of yeasts, molds, and aerobic bacteria; others work anaerobically, restricting undesirable bacteria (i.e., clostridia and enterobacteria), plant enzymes, and possible LAB.

products may improve silage by producing fermentable carbohydrates in crops that are very low in sugars.

Inhibitors reduce detrimental aerobic and anaerobic microbial activity during silage filling into the silo, storage, and feedout. These include formic acid and ammonia or urea. Most acids are utilized on direct-cut silage containing more than 75% moisture to produce a rapid fall in pH, to prevent clostridial spoilage, and to reduce proteolysis.

Nutrient sources that also serve as inhibitors include anhydrous ammonia, liquid ammonia, and urea. Nonprotein N increases the crude protein content of low-protein crops such as corn, sorghum, and small grains, and may improve fiber digestibility. The addition of ammonia increases the pH. Consequently, more sugars are needed to achieve the desired pH reduction in ammoniated silages.

Corn Silage

Corn is uniquely suited to preservation as silage. Whole-plant corn forage is well supplied with fermentable carbohydrates and has a low buffering capacity (see "Buffering Capacity"). The optimum harvest stage for maximum energy yields corresponds well with the ideal moisture range for silage fermentation (64–68%). These characteristics help to produce the rapid fall in pH and the low final pH (near 4.0) that is typical of well-preserved corn silage.

The maturity stage that is widely agreed to be optimal for harvesting corn silage is 2/3 milk line. The milk line can be seen on the endosperm side of corn kernels, and progresses from the tip (away from the cob) down towards the base of the kernel as the grain matures. The milk line can be viewed on the tip half of a representative ear that has been broken in half.

Summary

Stored hay and silage can provide forage of the quality needed to support many different livestock systems. Attention must be given to all of the factors that affect forage quality, including plant species, maturity stage, and leaf to stem ratio. In addition, losses in DM and quality associated with harvesting and storage, especially under wet conditions, should be kept to a minimum. Correct matching of forage moisture percentage to the harvest system that is being used, and maintaining the correct storage environment, are important for producing high-quality hay and silage.

Questions

1. What moisture level is considered safe for baling hay in small rectangular bales?
2. Name three crop factors that affect the rate of moisture loss from hay during field curing.
3. What are the advantages of the round bale package for harvesting and storing hay?
4. What are the advantages of a rotary disk mower compared with a sickle mower?
5. What are the advantages of a conditioner? Are there any disadvantages?

6. List the various types of balers and the advantages of each of them.
7. Describe the silage fermentation process.
8. Name several silage additives.
9. What are the advantages of using horizontal silage structures? Are there any disadvantages?
10. Consider a forage-harvesting system for your farm, and outline the various considerations that led to your decision.

References

Albrecht, KA, and RE Muck. 1991. Proteolysis in ensiled forage legumes that vary in tannin concentration. Crop Sci. 31:464–469.

Bolsen, KK. 1997. Issues of top spoilage losses in horizontal silos. In Silage: Field to Feedbunk, pp. 137–150. Northeast Regional Agricultural Engineering Service, PALS Publishing, Ithaca, NY.

Collins, M. 1985. Wetting effects on the yield and quality of legume and legume-grass hays. Agron. J. 77:936–941.

Collins, M. 1990. Composition of alfalfa herbage, field-cured hay and pressed forage. Agron. J. 82:91–95.

Collins, M, LD Swetnam, GM Turner, JN Hancock, and SA Shearer. 1995. Storage method effects on dry matter and quality losses of tall fescue round bales. J. Prod. Agric. 8:507–514.

Hoglund, CR. 1964. Comparative Storage Losses and Feeding Values of Alfalfa and Corn Silage Crops When Harvested at Different Moisture Levels and Stored in Gas-Tight and Conventional Tower Silos: An Appraisal of Research Results. Agricultural Economics Mimeo, Report 946. Michigan State University, East Lansing, MI.

Holland, C, and W Kezar. 1990. Pioneer Forage Manual: A Nutritional Guide. Pioneer Hi-Bred International, Des Moines, IA.

Langston, CW, H Irvin, CH Gordon, C Bouma, HG Wiseman, CG Melin, LA Moore, and JR McCalmont. 1958. Microbiology and Chemistry of Grass Silage. USDA Tech. Bull. 1187. US Department of Agriculture, Washington, DC.

Lechtenberg, VL, KS Hendrix, DC Petritz, and SD Parsons. 1979. Compositional changes and losses in large hay bales during outside storage. In Proceedings of the Purdue Cow-Calf Research Day, 5 April 1979, West Lafayette, IN, pp. 11–14. Purdue University Agricultural Experiment Station, West Lafayette, IN.

Leibensperger, RY, and RE Pitt. 1987. A model of clostridial dominance in ensilage. Grass Forage Sci. 42:297–317.

Leibensperger, RY, and RE Pitt. 1988. Modeling the effects of formic acid and molasses on ensilage. J. Dairy Sci. 71:1220–1231.

McDonald, P, AR Henderson, and SJE Heron. 1991. The Biochemistry of Silage, 2nd ed. Chalcombe Publications, Kingston, UK.

Muck, RE, and L Kung Jr. 1997. Effects of silage additives on ensiling. In Silage: Field to Feedbunk, pp. 187–199. Northeast Regional Agricultural Engineering Service, PALS Publishing, Ithaca, NY.

Pitt, RE. 1990. Silage and Hay Preservation. Northeast Regional Agricultural Engineering Service, PALS Publishing, Ithaca, NY.

Pitt, RE, RE Muck, and RY Leibensperger. 1985. A quantitative model of the ensilage process in lactate silages. Grass Forage Sci. 40:279–303.

CHAPTER

18

Grazing Management Systems

Robert L. Kallenbach and Michael Collins

Grazing Systems

As managers, we have developed **grazing systems** that integrate and balance the sometimes competing concerns with the animal, plant, soil, environmental, and managerial components of the enterprise. Individual management strategies, such as **rotational stocking**, are often mistakenly considered to be a system when they are in fact only one part of a system. Because grazing systems are composed of many parts, they are specific to the conditions for which they are designed. For instance, a grazing system designed for the Upper South, such as Kentucky, where cool-season perennials are well adapted, will be different from a system designed for the Lower South, including the Gulf States, where warm-season perennials are dominant. Even within these general regions, many differences exist.

Because grazing systems are site specific, they must be described in terms of the following:

- *Land:* soil type, fertility status, drainage, gradient, salinity, erosion potential, and cropping history,
- *Plants:* species, varieties, weeds, stage of growth, nutritive value, toxins, seasonal production, ground cover, and **forage mass**,
- *Animals:* species, number, genotype, gender, size, production status, and age.
- *Management:* intensity, **stocking method**, number and size of **paddocks**, irrigation, fertilization, pesticide use, and harvest events.
- *Human factors:* education, experience, local customs, political boundaries and regulations, markets, and values.
- *Location:* latitude, longitude, and elevation.
- *Climate:* temperature, precipitation, humidity, and season of the year (Allen et al., 2011).

When plants and animals are managed within a system, their responses and behavior may differ from that observed when they are managed alone. This is because each part of a system behaves as a consequence of its relationships with other parts of the system. When that part is managed in isolation away from the influence of the other system components, it is no longer under the same influences and may behave differently. This phenomenon can be seen in **animal performance.**

Animal performance within a system is not always predicted accurately based on performance on the individual forage species that make up that system (Table 18.1). In other words, performance on forage A and forage B is not necessarily additive. This can be due to several reasons. Forages differ in **nutritive value;** thus, if combined correctly, two forages may supply a better balance of nutrients than either forage alone, resulting in improved animal performance. In addition, these forages may grow at different times, providing a more uniform distribution of feed over a longer period of time than either forage alone. Likewise, animal performance on a given forage or at a given time is strongly influenced by previous conditions. For example, growing animals with reduced growth rates due to restricted intake or low nutritive value commonly exhibit compensatory gain (see "Compensatory Gain") in which they gain faster than expected once the restriction is removed.

Grazing lowers the cost per unit of production in typical livestock systems because expenses associated with the buildings, equipment, and labor needed for stored forages are reduced. However, designing grazing systems that can ensure an adequate daily supply of forage for each animal on a year-round basis is challenging. Forage species differ in their season of growth and use (Fig. 18.1). Livestock

Forages: An Introduction to Grassland Agriculture, Seventh Edition. Edited by Michael Collins, C. Jerry Nelson, Kenneth J. Moore and Robert F Barnes.

Table 18.1. Annual cattle gain per acre estimated from measured performance on forage components compared with actual gains when the two forages were grazed sequentially within a system

		Gain (lb/acre)		
System	Forage species	Computed[a]	Actual[b]	Difference (%)
1	Tall fescue–switchgrass	279	271	3
2	Tall fescue–caucasian bluestem	367	341	8
3	Tall fescue–pearlmillet	346	314	10
4	Orchardgrass–switchgrass	245	290	19
5	Orchardgrass–caucasian bluestem	332	306	8
6	Orchardgrass–pearlmillet	310	327	6
7	Smooth bromegrass–switchgrass	226	273	21
8	Smooth bromegrass–caucasian bluestem	313	319	2
9	Smooth bromegrass–pearlmillet	219	272	7

Source: Adapted from Matches, 1989.
[a]Computed as the average of separate component values (A + B)/2.
[b]Actual gain when the two forages were grazed sequentially within a system.

differ in their nutritional needs depending on their age, production status, and use for work. Matching the seasonal potential for quality and quantity of forage growth to the livestock's need for nutritional quality and feed quantity is one of the major challenges in grazing-system design. When this is successfully achieved, the need for harvesting or purchasing feeds is minimized, and profitability is generally improved.

Grazing systems must address the following objectives:

- Provide the appropriate quality and quantity of forage for the livestock present
- Ensure the productivity and survival of the forage stands by promoting healthy, vigorous plant growth and maintaining the desired botanical composition
- Obtain an economic return.
- Provide the desired quality of life for the manager.
- Protect the environment while conserving natural and non-renewable resources.

Increasingly, other objectives also need to be considered:

- Develop and protect wildlife habitat.
- Sequester carbon from the atmosphere.
- Improve soil health, function, and productivity.

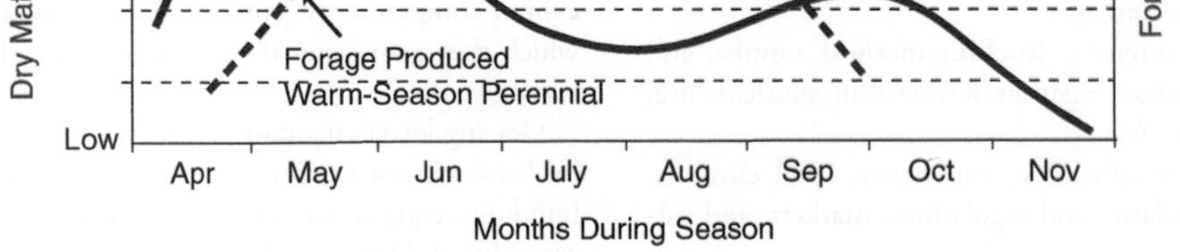

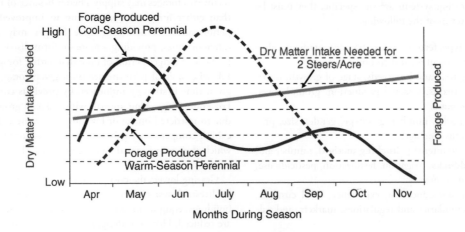

Fig. 18.1. Production rates of perennial forages generally do not match animal needs perfectly. This figure shows typical seasonal production patterns for cool-season (solid line) and warm-season (broken line) perennial forages compared with the dry matter intake needs for two growing steers per acre. Note that using some warm- and cool-season forages together in the system provides a better match with animal needs. (Adapted from Blaser et al., 1973; National Research Council, 2000.)

Compensatory Gain

Restricted nutrient intake limits average daily gain (ADG) of growing animals. However, for a time after returning to adequate nutrition these animals often gain faster than others that have been under conditions of unrestricted (ad libitum) feeding all along, This response is called **compensatory gain**, and it is illustrated in Fig. 18.2. To study the concept, one group of 540-lb (245-kg) beef steers were restricted to ADG of 0.9 lb (0.4 kg). After transfer to unrestricted feeding, these same steers gained weight at a rate of 0.8 lb/head/day (0.36 kg/head/day), more than the steers that were fed well all along (Carstens et al., 1991). The solid lines represent gains for the steers that were initially fed a restricted diet, and the broken lines represent gains for the consistently well fed steers.

Animals on restricted nutrient intakes develop reduced resting metabolic rates. Then, for 2 to 3 months after the restriction is removed, their net energy use for gain is more efficient (18% more in one study; Carstens et al., 1991), so gains increase. These animals also exhibit elevated levels of insulin-like growth factor and have a physiological drive to eat more.

Thus much of the potential gains that are not achieved because feeding is restricted during winter may be recouped when these animals go onto pastures during the following season. This practice may reduce the cost of gain because grazed DM is usually less expensive than harvested DM. Animals do not recover all of the gain lost due to previous management (see Fig. 18.2).

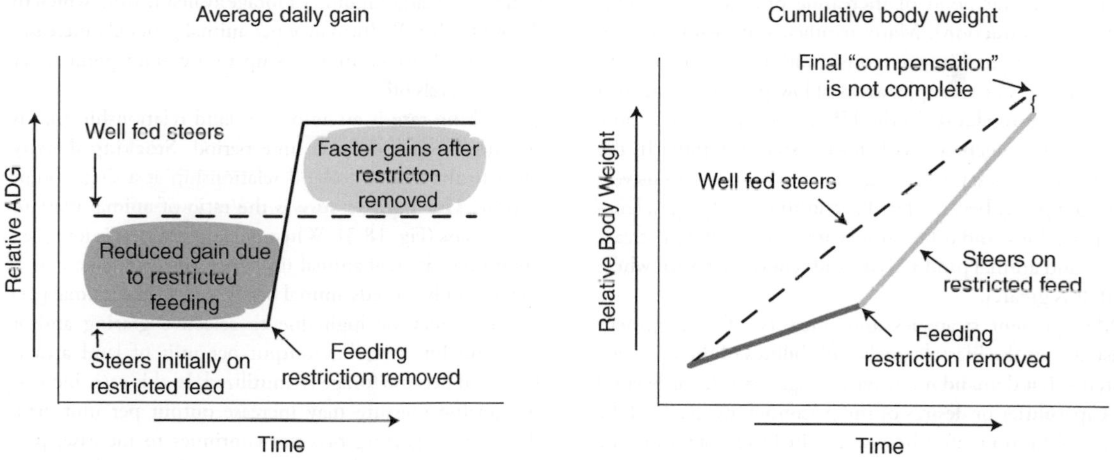

FIG. 18.2. Compensatory gain describes the increased gain that occurs for animals that had been on restricted feeding. This increase makes up some but not all of the weight gain that was lost earlier.

- Protect water quality and quantity, animal health and welfare, animal product quality, and safety.
- Protect aesthetic value and open spaces.
- Offer hunting, ecotourism, or recreational opportunities.

In the future, integrating grazing systems may offer one of the best opportunities to manage and protect our natural resources while providing an economic return and maintaining the levels of food and fiber production required to meet the needs of an increasing global population.

Grazing Management

Grazing management involves manipulation of the soil–plant–animal complex to achieve the desired results. Thus it is important to recognize that it is the system that is being managed, not just the livestock or the forages alone. The pieces must act in concert with each other.

Either **intensive grazing management** or **extensive grazing management** can be utilized. Intensive grazing management uses additional inputs of resources, labor, or capital to increase livestock production per acre or per head or to improve forage production and utilization. Extensive grazing management uses lower inputs of labor, resources, and capital. Both management styles try to achieve profitable and sustainable plant and animal production.

Grazing systems have been classified from the most extensively managed systems, exemplified by nomadic herders, to more highly managed systems located on a fixed land area. In nomadic systems, herds, flocks, and people follow seasonal rains that promote plant growth in different areas. Conversion from nomadic systems to overly intensive systems has sometimes led to **overgrazing**, and thus a deterioration in environmental quality. Whereas nomadic systems moved frequently to find adequate forage for their animals, a fixed location means that grazing may continue during periods when there is little or no forage growth, which can lead to overgrazing and hence to reduced ground cover and plant vigor.

More or less intensive systems are not inherently "good" or "bad." In fact, managers do not simply choose between extensive and intensive systems. They select from a continuum of possible inputs that might include more managerial time, more fertilizer nutrients, more fencing, or other inputs. Choices are influenced by assessment of input costs in relation to the value of the anticipated productivity increases and economic return. The inputs chosen first are often those judged to be most limiting (i.e., those with the greatest likelihood of increasing productivity or efficiency of production). Many weather, soil, plant, animal, and environmental impact considerations must enter into the decision. For example, areas of low precipitation, such as western **rangelands** in the USA, generally receive more extensive management with fewer external inputs. In the humid, temperate USA, forages tend to be more intensively managed partly because fertilizer, introduced forages, fencing, pesticides, and other inputs are more likely to increase forage and animal productivity and economic return where rainfall is greater.

Management strategies and intensity of management must also fit the lifestyle goals and abilities of the manager. Systems that demand intensive management that is beyond the capabilities or desires of the personnel involved will be unsuccessful no matter how well suited the system may be in other respects (Kallenbach, 2015).

Nomenclature of Grazing Systems

Grassland managers and scientists have developed a set of terms to facilitate communication about grazing systems. These include, for example, **grazing pressure, forage allowance, forage mass, forage yield, stocking rate, stocking density,** and **animal unit**.

Livestock differ in terms of species, breed, size, and productive status. All of these factors result in differences in the amount of forage that an animal will voluntarily consume. The term *animal unit (AU)* was developed to facilitate comparison of different kinds of livestock. An animal unit is defined as an 1100-lb (500-kg) non-lactating mature cow (*Bos taurus*) fed at maintenance or its equivalent in other kinds and classes of livestock (Allen et al., 2011). Defining the equivalent presents challenges and has been the subject of much debate. A widely used rule of thumb is that five ewes (*Ovis aries*) are equivalent to one mature cow or one horse (*Equus caballus*).

A more exact way to equate livestock is on the basis of metabolic body size, which is generally considered to be the animal's weight in pounds to the 0.75 power ($BW^{0.75}$) (National Research Council, 2000; Allen et al., 2011). Thus 1 AU = $(1100 \text{ lb } [500 \text{ kg}])^{0.75} = 233$. The metabolic body size of an 882-lb (400-kg) animal would be equal to 197, or 0.85 AU (197/233 = 0.85). This mathematical approach was developed to equate animals based on their body surface area rather than their true **body weight.** It is based on the assumptions that heat loss is more nearly related to body surface area than to body weight and that, expressed on this basis, heat loss is a constant for all species. Using this formula, comparisons have been made literally from mice to elephants.

Forage mass is the amount of forage dry matter per unit land area at a specific time. **Forage** accumulation is the increase in forage mass unit per unit land area over a specified time. **Forage allowance** is the relationship between forage mass per unit area and the number of animal units at any point in time. At high forage allowance levels, each animal has a large amount of forage available from which to choose its diet. Performance per animal generally increases as forage allowance increases up to a certain point, after which it levels off.

Stocking rate is an animal-to-land relationship that is measured over a defined time period. **Stocking density** refers to the animal-to-land relationship at a single point in time. **Grazing pressure** is the ratio of animal mass to forage mass (Fig. 18.3). When grazing pressure is low (i.e., when there are few animal units per unit of forage mass), forage supply exceeds animal needs. Individual animal performance may be high due to selective grazing and/or optimum bite size, but output per unit of land area is decreased because forage is unutilized. In this case, increasing grazing pressure may increase output per unit area. However, as grazing pressure continues to increase, performance per animal and per unit land area will begin to decline.

Stocking Methods

Stocking methods are grazing management procedures designed to achieve specific objectives (Fig. 18.4). They are used to achieve specific defoliation strategies for plants or to allocate nutrition to different classes of livestock. They may influence nutrient recycling or they may be used to discourage trailing—a social behavior in which cattle move in single file, forming paths of bare soil that can lead to soil erosion or streambank degradation. There are many different stocking methods, including **continuous stocking, rotational stocking, buffer stocking, strip stocking, creep stocking, first–last stocking, mixed stocking, sequence stocking,** and **frontal stocking,** among others. No one stocking method is best. Each is designed to accomplish specific objectives, and thus selection of the most appropriate method is critical to success. Indeed, several different stocking methods are usually included within a grazing system.

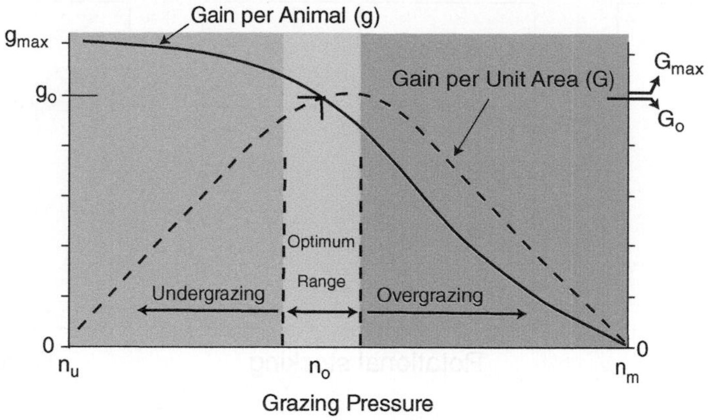

FIG. 18.3. Effects of grazing pressure (n) on gain per animal (g) and gain per unit area (G). n_u = low grazing pressure; n_o = optimum grazing pressure range; n_m = high grazing pressure. G_{max} = maximum gain per unit area; g_o = optimum gain per animal; G_o = optimum gain per unit area. (Adapted from Mott, 1973.)

Continuous and Rotational Stocking

Continuous stocking allows livestock unrestricted and uninterrupted access to a specific area for a specific time. No subdividing fences are used during this period (Fig. 18.4). Continuous stocking does not imply season-long use, and may in fact be a fairly short period, but it does require a specifically defined period. In response to changes in forage supply during the period of use, the manager can add or remove livestock, increase the total size of the area being grazed, or provide supplemental feed to maintain the desired forage-to-animal ratio.

Rotational stocking involves recurring periods of grazing among two or more paddocks, with periods of rest and regrowth for the forage between defoliation events (Fig. 18.4). Grazing periods can be shortened or lengthened to achieve optimum forage management. When forage productivity is high, rotational stocking allows the harvest of some paddocks for hay or silage before they become overly mature. When growth slows and additional forage is needed for grazing, some or all of the paddocks are reinserted into the rotation. Normally a paddock is grazed until 70–80% of the forage mass has been utilized, or until the remaining dry matter substantially limits intake (500 lb/acre [560 kg/ha]). Forage utilization is usually lower with continuous stocking, typically in the range of 50–65%.

Rotational stocking is particularly useful for forage species that benefit from **rest periods.** For instance, most alfalfa cultivars persist better and yield more if they have a rest period of about 4 weeks between **grazing cycles**, due to their cyclical use and replenishment of root and crown **non-structural carbohydrates.** Alfalfa is generally sensitive to defoliation periods that extend into this carbohydrate replenishment phase, which normally begins after shoots reach a height of about 6 in. (15 cm). If grazed continuously, alfalfa plants can be weakened, become less competitive with grasses, or even die. However, grazing-tolerant alfalfa cultivars perform better under continuous stocking than traditional alfalfa cultivars.

At other times of the year, continuous stocking may be preferable for alfalfa. If autumn temperatures are sufficiently cold to prevent alfalfa regrowth, forage that accumulated during late summer and early autumn can be stocked continuously without harming the alfalfa. Giving animals access to the entire **pasture** at this time allows selective grazing of alfalfa before freezing damage causes excessive leaf and quality losses.

Buffer Stocking

Pasture supply is not constant throughout the year, due to growth habit and environmental effects. One approach to adjusting forage supply with continuous stocking is to use temporary electric fencing to exclude livestock from part of the area, which can be harvested later if it is not needed for grazing (Fig. 18.4).

Closing off a variable area gives the flexibility to accommodate low production in the event of drought, high production, or changing animal demand. As the season progresses, the temporary fence is moved forward or backward to adjust the amount of forage available for grazing.

Strip and Mob Stocking

Strip stocking and **mob stocking** both confine animals to an area to be grazed within a relatively short time period (Fig. 18.4). Grazing pressure is set high enough to utilize the available forage quickly, usually within 0.5–7.0 days. This stocking method is particularly applicable to forage crops where no regrowth is expected during grazing.

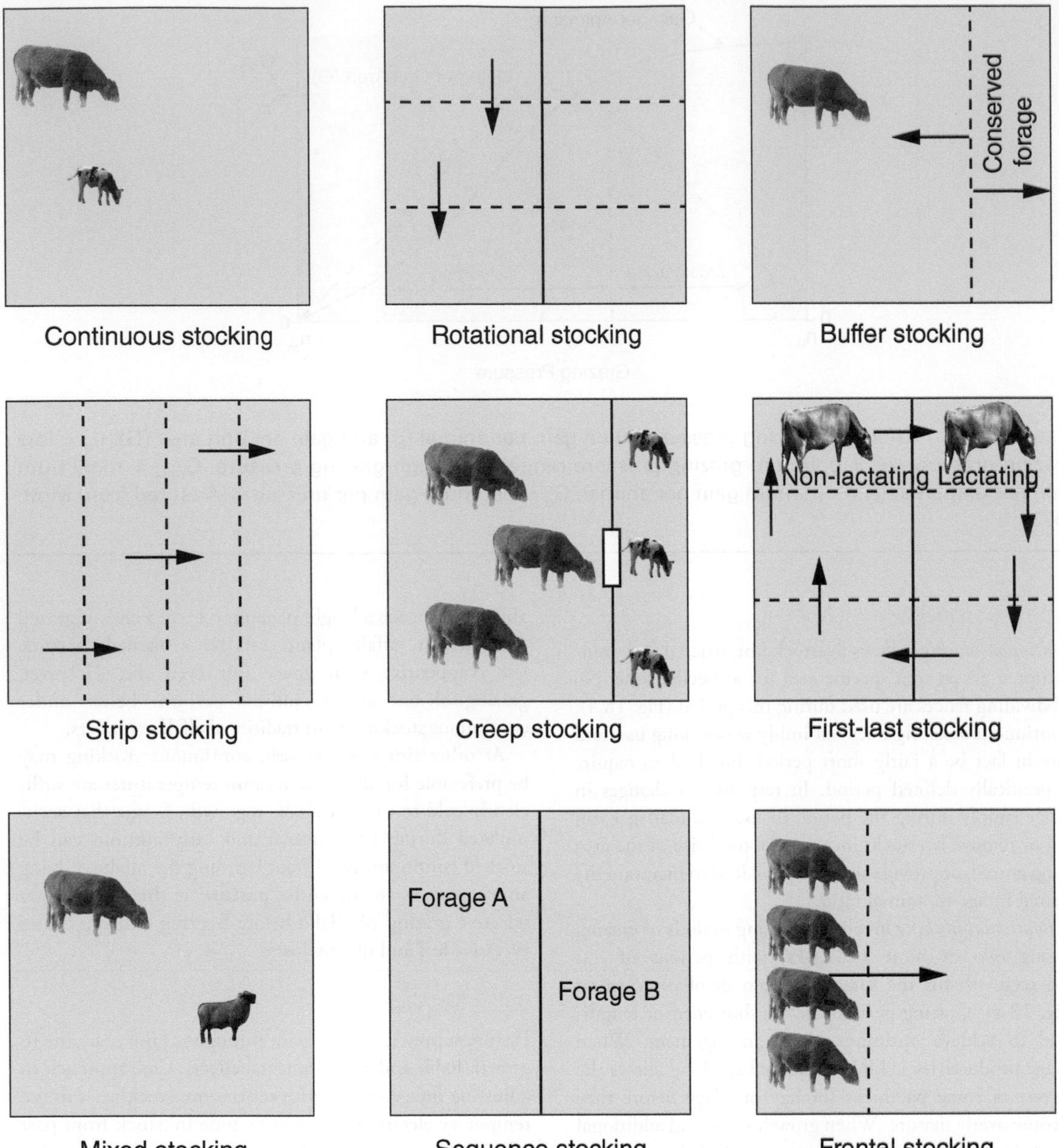

FIG. 18.4. One or more stocking methods may be used in a grazing system. Managers may employ a specific stocking method for any of the following reasons: partitioning of nutrients to different classes of livestock; matching forage produced with forage demand; allowing for harvest of excess forage; permitting rest and recovery of forages that are sensitive to prolonged defoliation; allocating predetermined amounts of forage for grazing; providing for more uniform feed production over the grazing season.

Actual utilization with these stocking methods varies substantially depending on forage quality. Often managers use these methods either because they wish to allocate or ration a limited supply of forage in order to make it last longer, or because there is an excess of overly mature forage that they wish to force animals to eat or trample into the soil. Strip and mob stocking are considered to be non-selective stocking methods that attempt to reduce selective

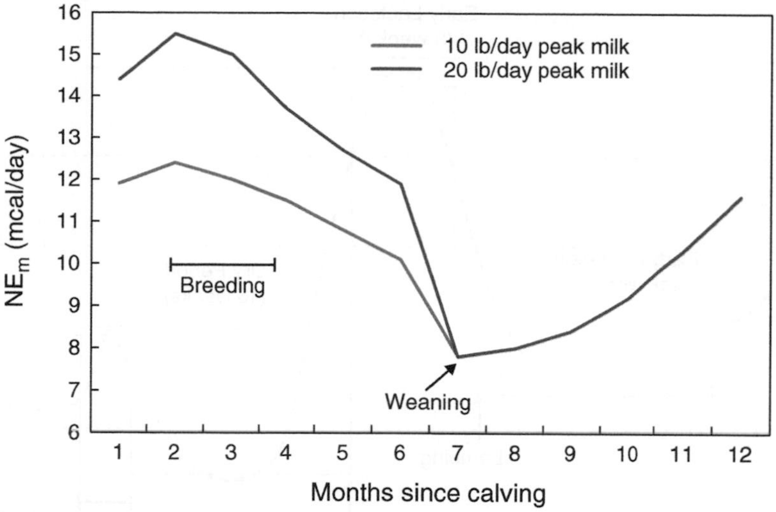

FIG. 18.5. Daily net energy (NEm) requirement for two levels of milk production from a 1000-lb mature beef cow. Energy requirements are equal for the two groups from weaning to calving. (Adapted from Hersom, 2017.)

grazing. In practice, stocking to overcome preference is rarely achieved.

Strip and mob stocking should not be confused with rotational stocking. These stocking methods explain how forage is allocated; they do not imply rest, recovery, and reuse of the area.

Creep Stocking

The nutritional needs of livestock differ according to production stages and animal objectives. Beef cows need high nutrition in late gestation, during breeding, and during the first 4 months of lactation (Fig. 18.5). In contrast, the nutritional needs of ewes increase 6–8 weeks before lambing, but peak lactation needs occur just after lambing (Fig. 18.6). The nutrition level of ewes should be increased again prior to breeding to improve conception rates.

Nutrient intake and milk production by the cow or ewe during early lactation is important to the offspring, but as milk production declines, nutrient requirements also decrease. Conversely, the nutrient requirements of the offspring are increasing during this time. Research has demonstrated that feeding the cow or ewe above nutrient requirements does not improve performance of the offspring. Milk typically provides about 50% of the nutrients needed by a 2- to 3-month-old beef calf. Research in Virginia demonstrated that daily gains of calves increased by 80% when they were creep fed in addition to having milk, compared with calves that had milk only (Fig. 18.7). Feeding cows to gain during this period did not improve calf performance.

Creep stocking partitions feed nutrients between the dam and its offspring by allowing calves to **graze** areas from which cows are excluded (Fig. 18.4) (see "Creep Feeding and Creep Stocking"). The calves can selectively graze highly **palatable** and digestible plant parts to optimize intake and increase gains. Cows graze an area that, while providing adequate quality and quantity to meet their lower nutritional needs, forces them to utilize a higher percentage of forage dry matter. Because an ample supply of high-quality creep forage is provided directly to the calf, cows can be forced to utilize their pasture more completely by consuming lower-quality plant parts.

Creep Feeding and Creep Stocking

Creep feeding and creep stocking are two management strategies that can partition feed nutrients to produce a heavier offspring, supplement deficient nutrients, and improve economics. After cows have been rebred and milk production is declining, their intake can be deliberately limited to increase efficiency of production while still meeting their nutritional needs. Providing the high nutrition needed by the calf can be accomplished by either creep feeding or creep stocking.

Creep feeding is the practice of providing suckling offspring access to feed supplements to which the dam

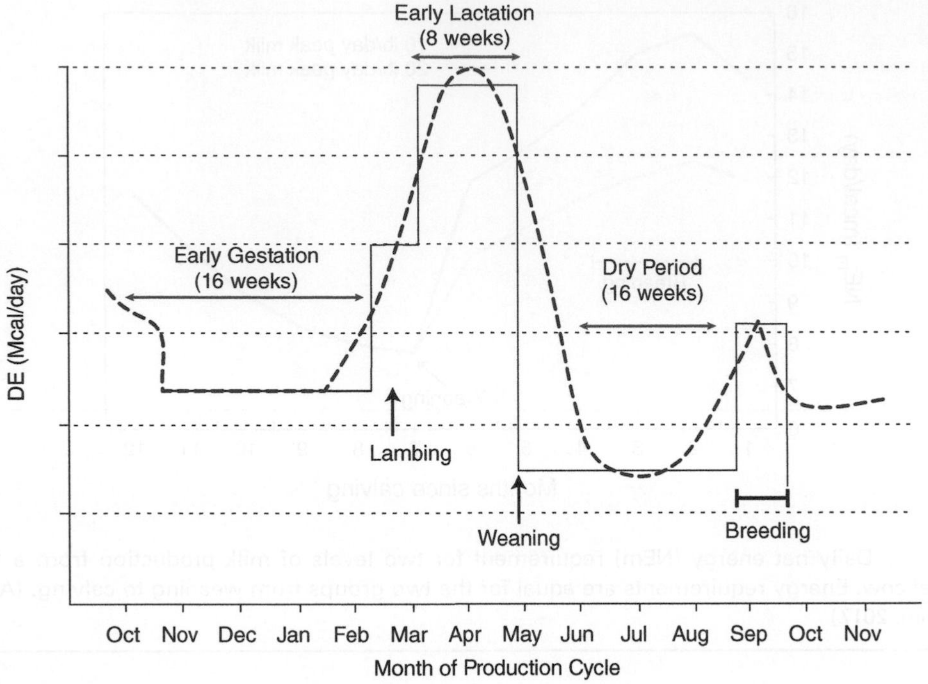

Fig. 18.6. Approximate daily digestible energy (DE) requirements of 140–155-lb (64–70- kg) breeding ewes at various production stages. (Adapted from National Research Council, 1985.)

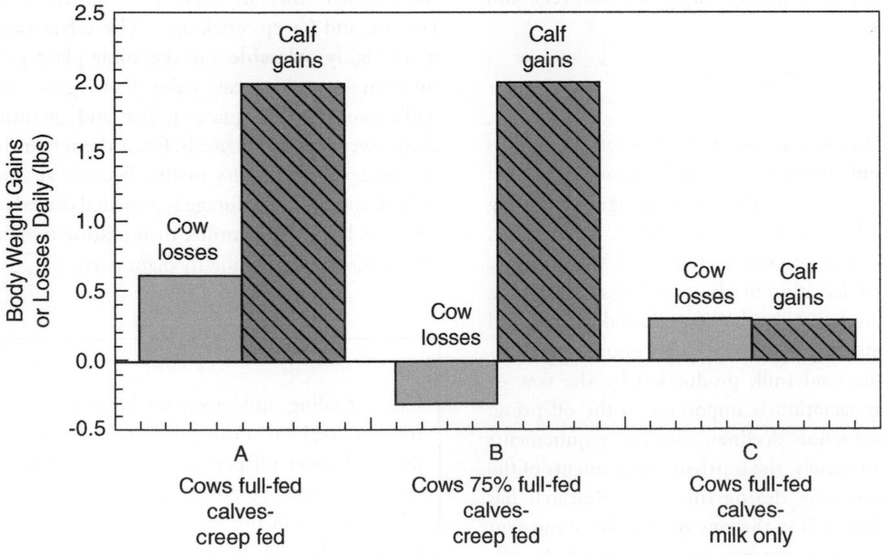

Fig. 18.7. Daily gains of creep-fed calves when cows are fed to gain or lose weight and non-creep-fed calves restricted to the dam's milk. (Adapted from Blaser et al., 1986.)

does not have access. These supplements are usually grain based to supply additional energy. They can also provide protein and minerals if forage quality is low. Creep feeding has the advantage of being easy to manage and being able to supply a precise quantity of nutrients, but it requires a structure that holds feed and excludes the dams.

Creep stocking accomplishes this by allowing the calf to graze an area away from competition with the dam. Creep stocking requires less labor and is generally more economical than creep feeding. For creep stocking to be effective, forage in the creep area must be readily accessible, must be of higher quality and quantity than the base pasture, and must be palatable to the calves.

A creep gate with openings large enough to allow offspring to pass through can be placed between the base pasture grazed by dams and the adjoining creep paddocks (Fig. 18.8). To encourage their use, creep gates should be placed in areas frequented by the livestock. The actual forage species used in the creep area is less important than the provision of a sufficient quantity of **vegetative,** leafy, high-quality forage to allow selective grazing and maximum intake.

Weaning stresses calves, foals, and lambs due to disruption of the social bond with dams and to the abrupt change in their diet. Creep stocking helps to acclimate young animals to physical separation from their dams, and hastens incorporation of significant forage intake into their diet. Weaning can be done simply by shutting the creep gate with little stress to dams or offspring.

First–Last Stocking

First–last stocking divides the forage in a paddock into higher- and lower-quality portions by sequentially grazing with two or more groups of animals that have different nutritional requirements (Fig. 18.4). First–last grazer groups can even be different livestock species.

The forage quality of most pasture **swards** is stratified, with the highest-quality forage at the top of the canopy and the lowest-quality forage at the bottom, nearest the soil surface. When introduced into a new paddock, animals also selectively graze the upper portion of the canopy first. Thus the first grazers, which are animals with high nutritional requirements, such as lactating dairy cows or finishing lambs or cattle, utilize only the best-quality, upper portion of the plant canopy. This maximizes selective grazing and bite size, ensuring these animals both the highest quality and highest intake-potential forage. This group moves forward to the next paddock before their nutrient intake becomes restricted.

The last grazers are then brought in to graze the forage left behind by the first grazers. Last grazers should be animals with lower nutritional requirements, such as dry cows or cows in late lactation. Their lower dry-matter intake and nutrient requirements allow these animals to perform adequately despite the reduced quality and quantity of the remaining forage.

Mixed Stocking

Mixed stocking combines two or more animal species in the same grazing system. The animal species may graze the area together or separately, at different times (Fig. 18.4). Mixed stocking can be useful for altering parasite cycles. Increasing the period when a particular species is off the

FIG. 18.8. A creep gate is designed to allow calves to graze areas that their dams cannot access at the same time. (Adapted from Blaser et al., 1986.)

pasture is beneficial because parasite eggs gradually die over time. Rejection of forage due to fouling with dung and urine is also reduced in mixed stocking systems, because animals are less prone to reject forage fouled by another species. Mixed stocking also takes advantage of differing animal preferences for forage species, canopy heights, and chemical composition characteristics. For example, goats prefer to **browse**, which helps to control woody and other weed species that are rejected by cattle.

Sequence Stocking

In *sequence stocking,* two or more land units with different forage species are used in succession (Fig. 18.4). Each unit can be subdivided for rotational stocking, or they can be continuously stocked. This may help with regard to matching forage quality and quantity with livestock needs. In a grazing system that is based on monocultures or a single mixture, all paddocks would have the same basic seasonal pattern of forage growth and quality and the same response to environmental stresses.

Different forage species or mixtures may have peak production at different times, or may be more heat or drought tolerant, and thus improve the seasonal distribution of production compared with a single species. Cool-season and warm-season forage species have different temperature optima for growth, and thus provide forage at different times of the year.

Frontal Stocking

Frontal stocking takes advantage of animal behavior. This method uses a sliding fence that is moved by the livestock as they push to gain access to ungrazed forage (Fig. 18.4). Grazing efficiency can be high due to uniform use and less trampling of ungrazed forage. Manure is more evenly distributed as the livestock advance across a field.

For frontal stocking to work well, a relatively flat pasture free of trees and topographical features that would impede forward movement of the fence is needed. Where appropriate, frontal stocking may increase grazing days per unit land area due to efficiency of forage use, but it does not normally increase individual animal performance compared with either continuous or rotational stocking.

Stockpiling

Stockpiling is a management technique whereby forage is allowed to accumulate for grazing at a later time. In temperate regions this technique is often used to provide forage for grazing in winter, but it could be used at any time during the year, depending on the climate and the management plan. Some forages lend themselves to stockpiling more than others.

Tall fescue is well suited to this management strategy (Curtis et al., 2008). Tall fescue fields are commonly hayed or grazed in early August, after which about 60 lb N/acre (67 N kg/ha) are applied. Livestock are excluded while the tall fescue grows during late summer and autumn. During this period, temperatures are declining but conditions are usually favorable for forage growth. Nitrogen encourages growth and leaf production, which promotes **photosynthesis.** Lower temperatures slow down **respiration** more than photosynthesis, so non-structural carbohydrates accumulate.

By early November, there is typically around 1–2 tons/acre (2.2–4.5 mt/ha) of high-quality fescue forage available for grazing. Leaves of tall fescue in the vegetative stage in fall can often be 2–3 ft (0.6–0.9 m) long. On average, 1 acre (0.4 ha) of stockpiled fescue and about 1 ton (0.9 mt) of moderate-quality hay will provide the feed nutrients needed to maintain one beef cow through the winter. Although the protein content and digestibility of stockpiled tall fescue declines over winter, forage quality may remain relatively constant for tall fescue infected with the wild-type endophyte (Kentucky-31), because toxin levels in the forage decline rapidly in winter to balance the reduction in nutritive value (Kallenbach et al., 2003). Nevertheless, it may be desirable to have some good-quality hay to supplement livestock even when forage is still available for grazing, particularly if cows are calving during late winter.

Warm-season grasses can also be stockpiled for winter grazing, and their potential should not be overlooked. Examples include bermudagrass, old world bluestem, switchgrass, and forage sorghums. Quality is usually lower than for cool-season grasses. Warm-season grasses become dormant during autumn and do not accumulate nonstructural carbohydrates. Phosphorus and sulfur are particularly likely to be deficient for animal needs when warm-season grasses are stockpiled.

Considerations When Developing Grazing Systems

The ability of a grazing animal to ingest the nutrients that it requires is influenced by animal factors, chemical and physical attributes of the plants, sward characteristics, the environment, and management. Many factors affect the nutritive value of the forage (see Chapter 14). In general, **legumes** (such as clovers and alfalfa) and cool-season annual grasses (such as annual ryegrass, wheat, oats, barley, and cereal rye) are highest in quality. Cool-season perennials (such as orchardgrass and smooth bromegrass) rank second, while warm-season perennials (including bermudagrass and old world bluestems) are lower in nutritive value. However, there are many exceptions. Growth stage, **canopy morphology,** environmental conditions, and other factors have a major influence on forage quality and animal performance.

Nutrient requirements vary among different kinds and classes of livestock. The highest-quality forages are generally most appropriate for lactating dairy cows, finishing cattle or sheep, preweaned calves, lambs, or foals, and animals used for heavy work. Breeding stock and growing animals have more moderate nutrient requirements, while mature, non-working or non-producing animals, and dry cows or ewes have the lowest nutrient requirements.

A key goal of any grazing system is to match forage quantity and quality with animal requirements. For example, seasonal dairying is based on late winter calving to match early lactation of cows with the high quality and quantity of forage normally available during spring in humid temperate regions (Fig. 18.1). By the time forage quality and growth rates decline later in the season, peak lactation has passed and nutrient demand by cows is lower. Cows are dry in winter and thus have lower nutrient requirements when forage is scarce.

Systems should be designed to minimize supplemental feed needs, but almost all systems will require some supplementation during periods of low forage growth. This could be forage harvested previously during periods when forage growth rate exceeded livestock demand. **Crop residues** such as corn or grain sorghum stalks, co-products from other industries, such as stems, leaves, and burrs removed during the ginning process for cotton "gin trash", distiller's grains, and other plant and animal industry co-products can also provide valuable supplemental feed for livestock.

Grazing system design must take into consideration fixed factors such as climate, amount and distribution of precipitation, soil types, and perhaps the capital available to invest. Other factors, such as forage species, soil fertility, and soil pH, can usually be modified. Cool-season perennial forages form the basis of grazing systems in much of the northeastern USA, whereas warm-season perennials should form the basis of systems in the southeastern USA and throughout much of the western USA. Warm-season forages may complement a cool-season base, or cool-season forages may extend grazing for a system based on warm-season perennials.

Grazing systems should be easy to manage and have sufficient flexibility to accommodate varying environmental, market, and production conditions. There are many ways to build flexibility into systems. Buying and selling animals may offer opportunities to match forage demand with forage growth. Stored or stockpiled forages buffer against times of low forage growth.

Successful systems should:

- Maximize the number of grazing days
- Minimize requirements for stored feeds and **supplements**
- Closely match the nutritional needs of livestock with potential for growth and quality of forage
- Use grazing management to minimize the need for pest control
- Conserve excess forage for hay, silage, or stockpiled grazing
- Allocate nutrition to meet the varying nutritional needs of livestock
- Provide flexibility
- Use plants and animals that are adapted to local conditions
- Recycle nutrients
- Be practical and profitable to manage.

Fencing Considerations

Widely used livestock fences include high tensile wire (electrified or non-electrified), barbed wire, electrified tape or wire (polywire), various types of woven wire, and specialized types such as board or stone fences. They differ widely in initial cost and maintenance requirements. The optimum choice depends on the economics, availability of materials, durability requirements, specific safety concerns, and the ability of the fence type to accomplish its specific purpose.

Barbed wire is generally avoided for horse pastures because the aggressive, nervous behavior of horses can result in fence-related injuries, and because some horses have high individual animal value. Electric fences are less effective for sheep, as their wool provides a natural insulation, but these fences can help to control predators such as coyotes (*Canis latrans*). Fences are expensive and should be minimized. Less fencing material is needed to surround a square or circle compared with other shapes that enclose the same unit land area. The system and the landscape should dictate the needs for fencing, and the system should be designed before the fences are built.

Nutrient Management Considerations

Grazing animals remove only a small proportion of the N and minerals in the forage that they eat (Table 18.2). For example, a 650-lb (295-kg) steer contains only about 5 lb (2.3 kg) of P. In contrast, a 5 ton/acre (11.2 mt/ha) yield of alfalfa hay that averages 20% **crude protein** and 0.3% P will remove 320 lb/acre (360 kg/ha) of N and 30 lb/acre (34 kg/ha) of P. Nutrient export may be desirable to remove excessive nutrients. Conversely, pastures require far less fertilizer input to maintain fertility than do areas where hay or silage is produced.

Most of the nutrients that are consumed by grazing animals are excreted in urine and feces. For example, more than 80% of the N in forage grazed by steers can be returned to the pasture through urine and feces. Most urinary N is readily available to plants, whereas fecal N is primarily in organic forms that must be mineralized by microorganisms before becoming available for plant uptake. Around 30–50% of the N voided by grazing animals is volatilized and lost to the atmosphere as ammonia. Phosphorus is excreted mainly in the feces. The value of waste nutrients from grazing livestock for forage growth is reduced by their typically uneven distribution over the pasture surface. Factors including landscape features, the position of and distance traveled to water, and grazing

Table 18.2. Disposition of nutrients in forages harvested by grazing animals

Nutrient	Grazing animals	
	Milking cows	Finishing sheep
	(% of total ingested)	
Nitrogen		
Retained by animal	25	4
Returned in dung and urine	75	96
Minerals[a]		
Retained by animal	10	4
Returned in dung and urine	90	96

Source: Adapted from Matches, 1989.
[a]Primarily P, K, Ca, Mg, and S.

system design can influence distribution patterns of manure and urine.

Nitrogen and P are currently the nutrients of primary concern with regard to environmental pollution. Applications of large amounts of animal waste from confinement feeding have increased concerns about excessive levels of soil P in recent years. Phosphorus is relatively immobile in soils, and accumulates to levels well above the needs of plants or animals. Excessive soil P levels contribute to pollution of surface waters primarily through direct movement with eroded soil particles.

Nitrogen follows many different pathways through the soil–plant–animal system, and is subject to losses through leaching, **volatilization, denitrification,** and erosion. Once in the soil, other forms of N can be quickly converted to nitrate, which is highly water soluble. Leaching of **nitrate** into water presents health hazards, and volatilized forms of N can contribute to air pollution and odor problems. Nitrous oxide, which is produced during denitrification, is a "greenhouse" gas and contributes to ozone depletion. Nitrate can also accumulate in plant tissues when soil levels are high. Forage nitrate levels in excess of 0.25% raise concerns about animal health, and levels above 0.5% should be considered potentially toxic (see Chapter 16). Including legumes with grasses contributes N and decreases or eliminates the need for N fertilizer, but it does not necessarily prevent nitrate leaching primarily due to recycling of N through the grazing animal.

Livestock Water Considerations

Although livestock obtain some water from snow, dew, and moisture in forages, an adequate supply of good-quality water is essential to maintain animal health, performance, and feed intake. Animals that are consuming green succulent forage have decreased water consumption, whereas dry feeds and high salt or protein intake increase water consumption. Water consumption increases as temperatures increase. For example, a 900-lb (408-kg) lactating cow consumes about 11.4 gallons (43 L) of water per day when the temperature is around 40°F (National Research Council, 2000). Daily water consumption increases to about 16.2 gallons/day (61 L/day) when temperatures rise to about 90°F (37°C). Cattle that are grazing in steep, rough terrain should not have to travel more than 0.5 mile (0.8 km) to obtain water, and those grazing on level or rolling land should not have to travel more than 1 mile (1.6 km). Giving livestock direct access to streams, ponds, and other water sources can kill **riparian** (adjacent bank and border area) vegetation, push soil into the stream, and contribute to water pollution.

Designing Forage and Livestock Systems

Land resources and climate are important considerations when designing forage and livestock systems, because these factors are least amenable to control by managers. Next we need to assess what forages are best suited to the farm, and how these forages complement each other in terms of seasonal growth patterns. It is also important to consider the nutritional needs of the livestock in the context of forage production and quality patterns. The following sections utilize some common systems used with beef cattle to illustrate these concepts.

Cow–Calf Systems: Northeastern USA

The Middleburg three-paddock system was designed for the cool, temperate environment of northern Virginia (77°43′30″ west longitude, 38°57′30″ north latitude; elevation 510 ft [156 m]). Precipitation is well distributed throughout the year, with a total of about 39 in. (99 cm) annually. Tall fescue, orchardgrass, and kentucky bluegrass are well adapted and widely used in the region. Stockpiling of tall fescue is a common practice for extending the grazing season during late autumn and early winter. Angus and other British-type breeds of cattle predominate in this

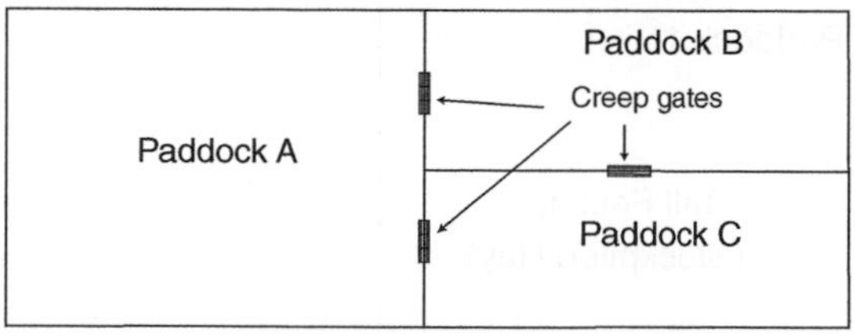

Fig. 18.9. The Middleburg three-paddock system for year-round grass–legume and cow–calf management (Blaser et al., 1977). Paddock A was used for grazing by cows during spring, summer, and early autumn. Paddocks B and C (45% of total area, divided equally) were used for hay and creep stocking by calves, and were stockpiled during autumn for grazing by cows during late autumn and winter. (Adapted from Allen et al., 1992.)

region. Cows are commonly bred to calve in late winter, with calves weaned and sold in October.

Because this climate favors cool-season perennials, a cow–calf system was designed around tall fescue that does not produce the ergot-like alkaloids from the endophyte *Neotyphodium coenophialum.* A tall fescue ladino clover mixture is continuously grazed by Angus cows during spring, summer, and early autumn in Paddock A, the largest paddock, which contains 55% of the total system (Fig. 18.9). Paddocks B and C (each representing 22.5% of the total area) consist of a tall fescue–red clover mixture that is harvested for hay, used for calf creep stocking, and stockpiled for winter grazing by the cows. Cows also graze in Paddocks B and/or C after the first hay cutting, but remain in Paddock A when stockpiling begins in early August. Calves are always given access to creep graze at least one paddock where cows are excluded.

Cow–Calf Systems: Southeastern USA

Cow–calf systems predominate in the southeastern region, which extends from eastern Texas and Oklahoma to the Atlantic Ocean, with calves generally being sold at weaning. European cattle breeds are sometimes crossed with Brahman to improve tolerance of heat and other stresses. Warm-season perennials such as bermudagrass, bahiagrass, and dallisgrass are well adapted and are widely used in pasture systems. Production on these species declines in late summer, and they become dormant during winter. Precipitation averages 40–50 in. (102–127 cm) annually, and is relatively well distributed over the year but tends to be lower in fall. Winter temperatures in the region are high enough to support the use of winter annual grasses and legumes. These annual species can be overseeded into the perennial warm-season grass, or they can be grown separately to extend the grazing season and provide high-quality forages.

Working with cow–calf systems in southern Georgia (83°13′ west longitude, 31°23′ north latitude; 380 ft [116 m] elevation), Hill et al. (1985) compared bermudagrass and bahiagrass as the base pasture with and without annual ryegrass overseeded in late October. Cows calved between January 1 and March 15. Calves were weaned in September. Cows were supplemented with hay previously cut from the system if forage was limited for grazing. Cows that did not have access to the high-protein ryegrass were supplemented daily with cottonseed meal at 11 lb/cow (5 kg/cow).

Grazing ryegrass-overseeded pastures increased calf weaning weights (Table 18.3), and the calves from the bermudagrass systems outperformed those from the

Table 18.3. Cow–calf systems for the southeastern USA based on warm-season perennials

	"Coastal" bermudagrass		"Pensacola" bahiagrass	
	Ryegrass	No ryegrass	Ryegrass	No ryegrass
Cows				
Pasture area/cow (acres)	0.8	1.0	1.1	1.1
Hay (tons/cow)	1.3	1.8	1.4	1.8
Cottonseed meal (lb/cow)	0	11	0	11
Average final weight (lb)	1027	1008	1015	1009
Cows pregnant (%)	94	93	92	87
Calves				
Birth weight (lb)	71	70	70	68
Weaning weight (lb)	515	496	480	447
Annual returns to capital, land, labor, and management				
Per cow (US$)	129.86	110.07	67.64	65.12

Source: Hill et al., 1985.

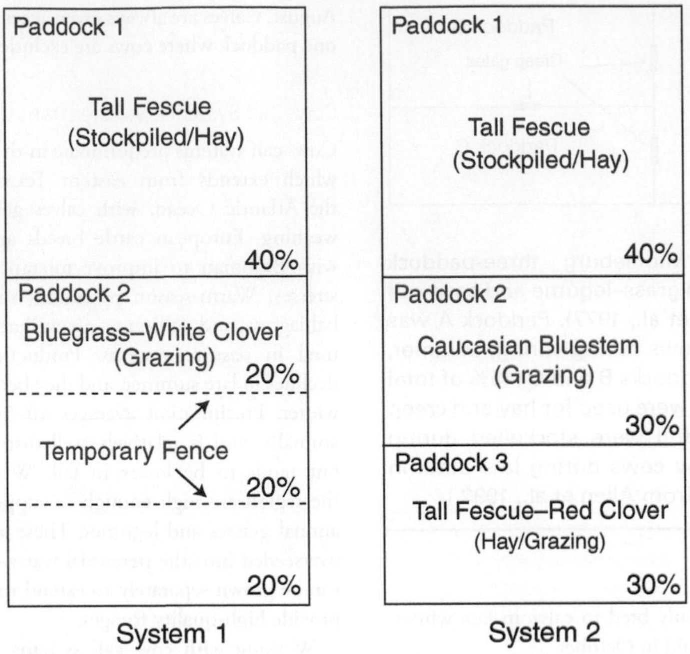

Fig. 18.10. Forage systems for production of stocker steers from November to the following October. System 1 was based on all cool-season perennials, whereas System 2 used both cool-season and warm-season perennials. (Adapted from Allen et al., 2000.)

bahiagrass systems. Cows that grazed bahiagrass alone tended to have a lower pregnancy rate than cows on other systems.

Stocker Systems

Another common decision that cow–calf producers must make is whether to retain calves after weaning to add additional gain before they are sold (Bailey and Kallenbach, 2010). Researchers compared pasture systems for these **stocker** cattle in the transition zone between cool- and warm-season grasses, near the Virginia/Tennessee border, comparing the use of caucasian bluestem as a component of the pasture (System 2) with another system that used tall fescue and a kentucky bluegrass–white clover mixture (System 1) (Fig. 18.10). In System 1, 40% of the acres were used to produce tall fescue hay and stockpile pasture for fall, winter, and early spring grazing.

Weaned Angus crossbred steers grazed the mixture of kentucky bluegrass and white clover from late April until mid-October, along with midsummer grazing of the tall fescue as needed. Rotational stocking was used for the bluegrass–white clover pastures.

System 2 uses three paddocks, with one producing tall fescue for hay production and stockpiling as in System 1 (Fig. 18.10). Paddock 2 (30% of the area) is planted with a warm-season perennial, caucasian bluestem, and grazed from mid-June to mid-September, along with some grazing of the tall fescue. Paddock 3 (tall fescue–red clover) is harvested for hay in early summer and the regrowth is stockpiled for grazing from mid-September to mid-October, when the steers are sold.

System 2, with the warm-season component, produced 77 lb (35 kg) more total gain per steer and 90 lb (41 kg) more total live weight gain per acre than the cool-season system (Table 18.4). Furthermore, System 2 produced about 30% more hay while using 40% less hay for feeding, so twice as much hay was available for sale compared with the cool-season system. Inclusion of the legume reduced the need for N fertilization.

In Oklahoma, eastern gamagrass and old world bluestems are well-adapted warm-season perennial grasses, but they need to be maintained in a vegetative state to produce high-quality forage. Two systems were evaluated for crossbred yearling steers with an initial weight of 527 lb (239 kg) in western Oklahoma, where total annual precipitation is around 24 in. (61 cm), concentrated between April and September (Table 18.5). Steers in System 1 sequence grazed **native forages** and "Iron Master" old world bluestem continuously, and steers in System 2 sequence grazed eastern gamagrass with rotational

Table 18.4. A comparison of two systems for production of stocker steers on cool- and warm-season perennial forages

	Forage system	
	TF/KBG/WC[a]	TF/CBS/RC[b]
Initial weight (lb)	372	370
Total gain (lb)	337	414
Gain/acre (lb)	405	495
Hay harvested (tons/system)	3.18	4.58
Hay fed (tons/system)	1.23	0.75
Surplus hay (tons/system)	1.94	3.83
Days/year forage was fed	61	38

Source: Adapted from Allen et al., 2000.

[a]TF, tall fescue; KBG, kentucky bluegrass; WC, white clover.

[b]TF, tall fescue; CBS, caucasian bluestem; RC, red clover.

stocking, and old world bluestem using continuous stocking (Table 18.5).

Over the entire grazing season, there was no difference between daily gains of steers for the two forage systems (Table 18.5), but total beef production per acre on the system with eastern gamagrass was more than twice that of the native grass system, due to the higher stocking rate that could be imposed. Steer gains on the native forage–old world bluestem system were higher than would be expected from a system consisting solely of native forages.

It should be kept in mind that maximum live weight production may not necessarily result in maximum net returns, because input costs must also be considered. For example, the eastern gamagrass–old world bluestem combination in this example had the highest production per acre, but also had higher input costs. These included initial establishment and N fertilization costs, and increased capital costs for the greater number of livestock.

Summary

Successful forage and livestock systems are profitable and sustainable enterprises that exist in harmony with nature. They should always include the goal of environmental stewardship of land, water, and air. Nutrient management plans must balance imports against exports of N and minerals. Protection of water quality and quantity is a responsibility of every livestock and farming operation.

Table 18.5. A comparison of two systems for production of stocker steers on native and introduced warm-season perennial forages in Oklahoma

	Grazing system	
	Eastern gamagrass/old world bluestem	Native grasses/old world bluestem
Pasture size	(acres)	
Eastern gamagrass	4.0	–
Native grass	–	16.0
Old world bluestem	4.0	4.0
Total area	8.0	20.0
Stocking rate	(acres/head)	
Eastern gamagrass	0.5	–
Native grass	–	2.0
Old world bluestem	0.5	0.5
Total	1.0	2.5
Daily gains by grazing period	(lb/head/day)	
Period 1 (eastern gamagrass or native)	2.6	2.5
Period 2 (old world bluestem)	2.7	2.4
Period 3 (eastern gamagrass or native)	1.3	2.0
Overall	2.2	2.2
Beef production	(lb/acre)	
Eastern gamagrass or native grass	199	59
Old world bluestem	260	224
Total	229	92

Source: Adapted from Gillen et al., 1999.

Forage species selection and use must consider land resources, climate, and intended use, while harvest management should be based both on the nutritional requirements of the livestock and on the architecture and regrowth characteristics of the plant. Grazing-based systems that minimize stored feed may require more management but can improve economics, enhance nutrient management, and address environmental problems.

Questions

1. How do stocking methods differ from grazing systems?
2. Why is the concept of time important when selecting the most appropriate grazing method?
3. Why is it useful to compare livestock on the basis of animal units rather than as numbers of animals per acre or live weight per acre?
4. What soil, plant, animal, and environmental characteristics should be considered when planning a forage–livestock system?
5. List at least five objectives of grazing systems.
6. Compare and contrast grazing systems for stocker cattle in eastern and western environments.
7. How does stocking rate differ from grazing pressure?
8. What nutrients are generally of greatest concern environmentally?
9. If 3.5 tons of orchardgrass–red clover hay are harvested per acre and the hay contains 13% crude protein and 0.35% P, how many pounds per acre of N and P will be removed from the field?
10. Describe the different grazing methods. List two benefits of each method.
11. What is metabolic body size? Determine the metabolic body size of a steer weighing 895 lb (406 kg).

References

Allen, VG, JP Fontenot, DR Notter, and RC Hammes Jr. 1992. Forages systems for beef production from conception to slaughter: 1. Cow-calf production. J. Anim. Sci. 70:576–587.

Allen, VG, JP Fontenot, and RA Brock. 2000. Forage systems for production of stocker steers in the upper south. J. Anim. Sci. 78: 1973–1982.

Allen VG, C Batello, EJ Berretta, J Hodgson, M Kothmann, X Li, J McIvor, J Milne, C Morris, A Peeters, and M Sanderson. 2011. An international terminology for grazing lands and grazing animals. Grass Forage Sci. 66:2–28.

Bailey, NJ, and RL Kallenbach. 2010. Comparison of three tall fescue-based stocker systems. J. Anim. Sci. 88:1880–1890.

Blaser, RE, DD Wolf, and HT Bryant. 1973. Systems of grazing management. In ME Heath, DS Metcalfe, and RF Barnes (eds.), Forages: The Science of Grassland Agriculture, 3rd ed., pp. 581–595. Iowa State University Press, Ames, IA.

Blaser, RE, RC Hammes Jr, JP Fontenot, and HT Bryant. 1977. Forage-animal systems for economic calf production. In E Wojhan and H Thons (eds.), Proceedings of the XIII International Grassland Congress, Vol. 2, p. 1541. Akademie-Verlag, Berlin, Germany.

Blaser, RE, RC Hammes Jr, JP Fontenot, HT Bryant, CE Polan, DD Wolf, RS McClaugherty, RG Kline, and JS Moore. 1986. Forage–animal management systems. Va. Agric. Exp. Stn. Bull., pp. 86–87. Virginia Polytechnic Institute and State University, Blacksburg, VA.

Carstens, GE, DE Johnson, MA Ellenberger, and JD Tatum. 1991. Physical and chemical components of the empty body during compensatory growth in beef steers. J. Anim. Sci. 69:3251–3264.

Curtis, LE, RL Kallenbach, and CA Roberts. 2008. Allocating forage to fall-calving cow-calf pairs strip-grazing stockpiled tall fescue. J. Anim. Sci. 86:780–789.

Gillen, RL, WA Berg, DL DeWald, and PL Sims. 1999. Sequence grazing systems on the southern plains. J. Range Manage. 52:583–589.

Hersom, M. 2017. Basic nutrient requirements of beef cows. University of Florida Pub. AN190. 10 pp.

Hill, GM, PR Utley, and WC McCormick. 1985. Evaluation of cow-calf systems using ryegrass sod-seeded in perennial pastures. J. Anim. Sci. 61:1088–1094.

Kallenbach, RL. 2015. Describing the dynamic: measuring and assessing the value of plants in the pasture. Crop Sci. 55:2531–2539.

Kallenbach, RL, GJ Bishop-Hurley, MD Massie, GE Rottinghaus, and CP West. 2003. Herbage mass, nutritive value, and ergovaline concentration of stockpiled tall fescue. Crop Sci. 43:1001–1005.

Matches, AG. 1989. Contributions of the systems approach to improvement of grassland management. In Proceedings of the XVI International Grassland Congress, pp. 1791–1796. Association Francaise pour la Production Feurragere, Versailles Cedex, France.

Mott, GO. 1973. Evaluating forage production. In ME Heath, DS Metcalfe, and RF Barnes (eds.), Forages: The Science of Grassland Agriculture, 3rd ed., pp. 126–135. Iowa State University Press, Ames, IA.

National Research Council. 1985. Nutrient Requirements of Sheep, 6th revised ed. National Academies Press, Washington, DC.

National Research Council. 2000. Nutrient Requirements of Beef Cattle, 7th revised ed. National Academies Press, Washington, DC.

CHAPTER

19

Managing Grassland Ecosystems

Charles P. West and C. Jerry Nelson

Grasslands constitute the world's major land form used for grazing livestock production. Sustaining grassland productivity requires an understanding of how ecological forces act on **soil health**, plant vigor, and nutrient conversion by animals. Grassland sustainability also considers aspects beyond profitable livestock production, such as conserving water flow and quality, soil and streambank stability, carbon (C) storage, habitat for wildlife including pollinators, human recreation, and preserving landscape heritage. Society values such ecosystem services, and managed grasslands are well suited to meet these expectations. This chapter reviews the tenets of grassland ecology in relation to current challenges and strategies for managing grasslands for economic and environmental sustainability.

Grasslands are plant associations that consist mainly of grasses (family Poaceae), but usually also include **legumes** (family Fabaceae), **composites** (family Asteraceae), **sedges** (family Cyperaceae), shrubs, and sparse trees. These areas dominate much of the earth's land surface, especially the more arid regions where grassland can compete effectively with trees (Fig. 19.1). Natural grasslands or **prairies** generally exist in semiarid to sub-humid regions that receive 12–40 in. (300–1020 mm) of annual precipitation. **Savannas** (grasslands mixed with trees and shrubs) form a transition zone between forests and prairies, and **steppes** (short grasses mixed with low shrubs) form a transition zone between prairies and desert (Kephart et al., 1995).

Grassland **biomes** are labeled by the dominant plant communities that existed before large-scale human habitation. As a result of human exploitative uses and infrastructures, the composition of much of today's grasslands differs greatly from their native composition. Human interventions have included tillage, crop production, irrigation, fire suppression, introduction of new species of plants and herbivores, and energy extraction. The most obvious change in North America has been the widespread replacement of prairies and savannas (Fig. 19.2) with high-input annual crops, resulting in exposed soil and a drastic decline in biodiversity. Hay meadows and **permanent pastures** were also established with mixtures of introduced perennial species, often by converting exhausted cropland on marginal sites to non-native grassland. Introduced plants are preferred in the wetter zones of grasslands because of their longer growing season and their ability to respond economically to fertilizer and weed or brush control.

Rangeland consists of extensively grazed areas in drier environments in which plant composition changes slowly, but remains dominated by native species. Rangelands typically receive no fertilizer, irrigation, or pesticide inputs. Therefore management knowledge is required to closely align grazing timing and intensity to mitigate rainfall patterns and the need for stand recovery. Chronic water deficits and rough landscapes favor diverse plant communities, but these have slow growth and poor survival if they are grazed short or frequently. Sustainability in rangelands involves a more complex consideration of the interacting factors that confer a balance between productivity and stability.

Many chemical, physical (**abiotic**), and biological (**biotic**) factors interact to determine the species composition, productivity, and dynamics of grasslands. Abiotic factors include limitation of plant productivity by low light, water (especially drought), low or high temperature, lack of **nutrients**, and frequency of fire. Biotic factors include

Forages: An Introduction to Grassland Agriculture, Seventh Edition. Edited by Michael Collins, C. Jerry Nelson, Kenneth J. Moore and Robert F Barnes.

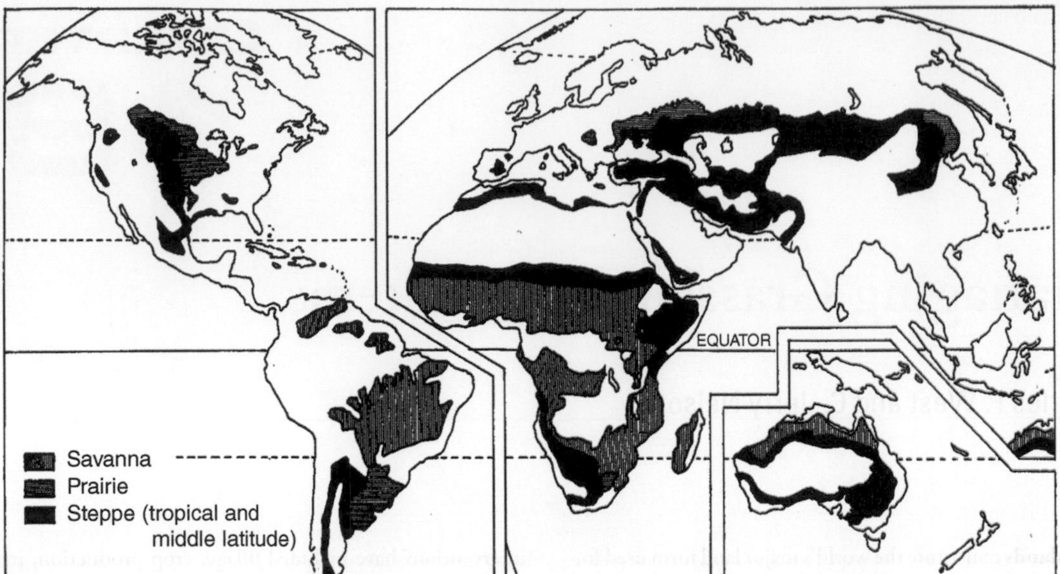

FIG. 19.1. World distribution of native grasslands. Each area that is classified as savanna, steppe, or prairie has a climate and vegetation structure similar to that of other areas with the same designation. This allows transfer of technology and prediction of germplasm usefulness.

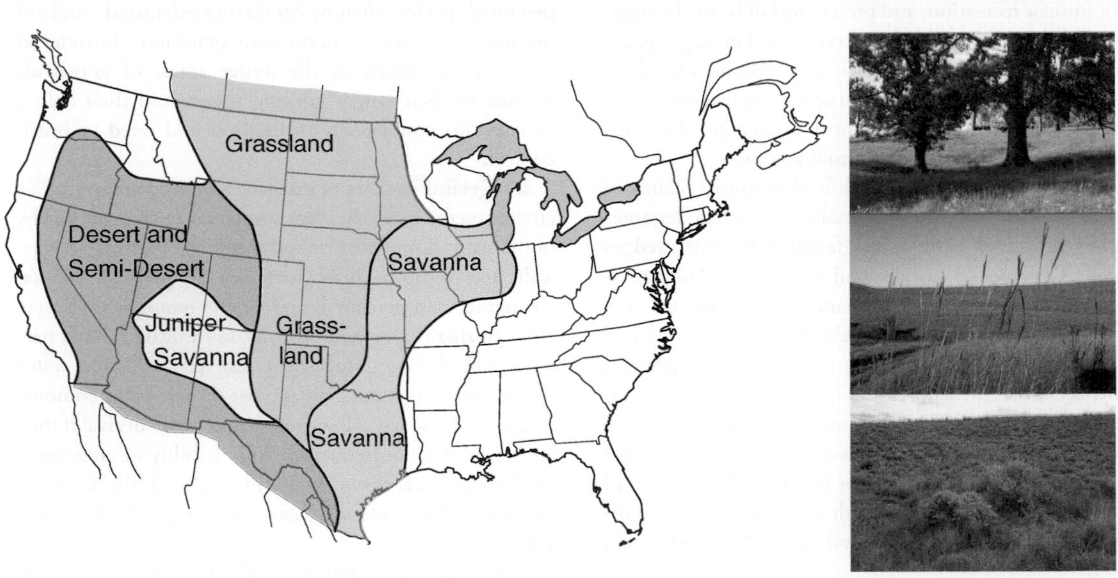

FIG. 19.2. General regions for grasslands and savannas in North America. Areas east of the tall-grass prairie region have naturalized grasslands consisting mainly of introduced species. Savannas are transitional areas between grassland and forest. Savannas are often managed as permanent pastures or meadows for harvesting forage. Top photo, a savanna site in Kentucky; middle photo, the Konza Prairie site in Kansas, which has more than 500 plant species (compared with typically around 10–20 species in seeded pastures in the eastern USA); bottom photo, arid shrubland in Arizona. (Adapted from Raven et al., 1992.)

plant and animal genetics, **herbivory** by above-ground and below-ground animals, symbionts, and decomposing organisms.

The biotic component is large and complex. Plants make up the largest portion of the **biomass**, with annual above-ground dry weight production ranging from as low as 300 lb/acre (336 kg/ha) in very dry sites to over 5000 lb/acre (5600 kg/ha) in tallgrass humid grasslands. The mass of the roots and tubers can exceed that of the above-ground biomass by more than twofold. Other biota include very diverse types of animals, insects, and microorganisms, all of which interact to process energy and nutrients.

Climate and the Distribution of Grasslands

Major land areas can be divided roughly into biomes, which comprise ecosystems of native vegetation that evolved within and are adapted to a particular climate (Raven et al., 1992). The grassland and savanna biomes of North America occupy the central portion of the continent, extending from northeast Mexico to the prairie provinces of Canada (Fig. 19.1 and Fig. 19.2). The North American Corn Belt was naturally a savanna biome dominated by tall perennial grasses such as big bluestem, switchgrass, and indiangrass, except in pockets of woodland on wetter sites. Fire and forest clearing combined with intermittent drought causes woody vegetation to be more stressed than grasses. Another approach to classifying major vegetation is based on defined ecoregions (Bailey, 1995). Major biomes are each subdivided into a number of vegetation and climatic zones to provide guidance for management of more defined ecosystems.

Water deficit is the principal climatic factor controlling both the development of grasslands and the distribution of grassland versus forest (Fig. 19.3). Actual rainfall is not as critical as available soil water, especially during those periods when temperatures are favorable for growth. High temperature and solar radiation intensity increase potential **evapotranspiration**. For example, Minneapolis in Minnesota has a mean annual precipitation of 28.5 in. (724 mm) and a mean annual temperature of 42°F (6°C), whereas the corresponding values for Manhattan, Kansas are 32.9 in. (836 mm) and 55°F (13°C), respectively. Although Manhattan receives more precipitation, there is less available soil water because it has proportionately more evapotranspiration, at 29.1 in. (739 mm) per year compared with 23.2 in (589 mm) per year in Minneapolis, due to higher temperature, higher sun angle in summer, and a longer growing season. Thus Manhattan is a tallgrass prairie site, whereas Minneapolis is a humid forest site.

Natural grasslands may exist in regions with relatively high precipitation because of unusual soil conditions. For example, tallgrass prairies once occurred as far east as Long Island, New York, and the coastal plain of the southeastern USA. These grasslands occupied sandy soils, which have low nutrient content and low water-holding capacities compared with the nearby loamy soils that supported forests. Fire helped to maintain these grassland communities by suppressing tree encroachment. In contrast, the Blackland Prairies of Alabama, Arkansas, and Mississippi developed on poorly drained clay soils that were waterlogged from fall to late spring but were dry during summer.

In the western USA, topography has a major effect on temperatures and interacts with precipitation in determining the vegetation (Fig. 19.4). Moist air from the Pacific Ocean cools as it moves up mountain slopes, and its water vapor condenses as rain or snow. The dried air flows over the mountains to the leeward side, thus creating semiarid conditions. These rain-shadow regions include valleys east of the Pacific Coast mountain ranges, the Great Basin region east of the Cascade and Sierra Nevada mountains, and the Great Plains region east of the Rocky Mountains.

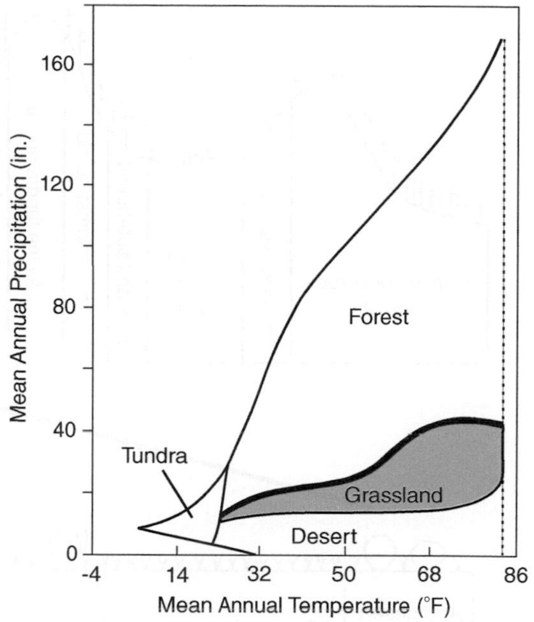

FIG. 19.3. Seasonal temperatures and precipitation are the major climatic factors that influence the class of dominant vegetational ecosystem. Tundra forms in cold, dry climates. Grasslands form in regions that are too wet for deserts but too dry for forests. Savannas are transition zones between grasslands and forests, and steppes are transition zones between grasslands and deserts. At high temperatures, the grasslands are mainly C_4 species and the forests are more tropical. At low temperatures, the grasslands are mainly C_3 species and the forests are hardwoods and conifers. (Adapted from Whittaker, 1972.)

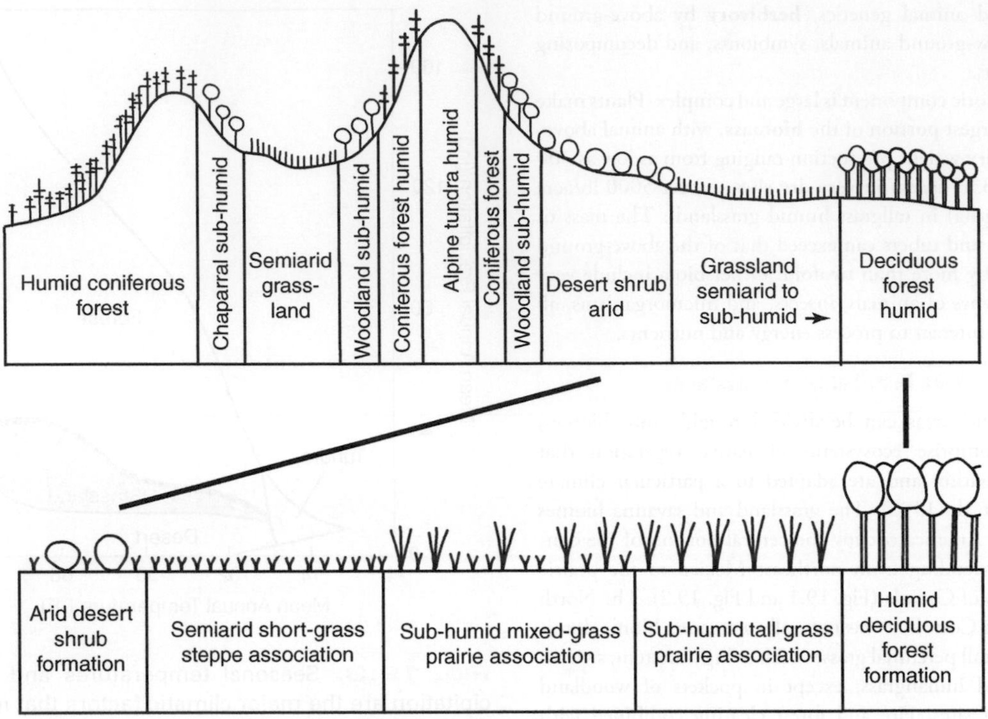

FIG. 19.4. A diagrammatic cross section of the USA showing where grassland ecosystems develop. Major grasslands form in rain shadows on leeward sides of major mountain ranges, such as the intermontane region in the West and east of the Rocky Mountains. Such regions are too warm for tundra, too wet for deserts, and too dry for forest development. Note the change in grassland types as rainfall increases going east, until the climate becomes favorable for forests. (Adapted from Harlan, 1956.)

Woodland is associated with sub-humid to humid zones, whereas grassland is associated with arid, semiarid, and sub-humid zones. The intermontane Great Basin, designated as semiarid grassland or **chaparral**, receives an average annual rainfall of 4–12 in. (100–305 mm); it is dominated by shrub-bunchgrass vegetation, such as sagebrush.

How Can Evapotranspiration Be Optimized?

Evapotranspiration (ET) is the sum of evaporation from the soil and transpiration from the plants. Thus it represents the total water loss from the soil profile to the atmosphere. The soil profile is recharged by rainfall and snow melt. This relationship can be expressed as follows:

$$ET = P - (R + \Delta SW)$$

where P is precipitation, R is run-off, and ΔSW is the change in soil water content. In this case, the soil serves as a water reservoir to buffer daily or seasonal changes in ET or P. Daily ET depends largely on solar radiation, with lesser effects from temperature, relative humidity, and wind. Potential ET reflects the suction power of the atmosphere to extract water from the grassland. Low rainfall relative to ET causes soil water deficit.

In general, plants are better water managers than soil. They help to reduce run-off, they shade the soil from radiation, thereby lowering evaporation, and they close their stomata at night to reduce transpiration by more than 90%. In contrast, water evaporates from the soil both day and night.

The goal is to manage the forage or grassland so that as much water as possible is captured in the soil and then passes through the plant, rather than running off or evaporating from the soil surface. Plants differ in rooting depth (see Chapters 2 and 3), and rooting is affected by defoliation (see Chapter 2). In addition, under conditions that are more arid, root growth is favored over top growth, which increases the depth of the reservoir of available soil water per unit of top growth.

The Great Plains ecoregion extends from just east of the Rocky Mountains to the deciduous forests of the eastern states, and it contains the widest diversity of grassland types on earth (expanded portion of Fig. 19.4). Sparse, short grasses that grow to a height of 6–12 in. (15–30 cm), such as blue grama, buffalograss, and western wheatgrass, are dominant in the semiarid short-grass steppe. The west-to-east gradient of increasing precipitation is associated with increasing canopy height and grassland productivity.

The tallgrass prairie of the USA developed in the eastern region of the grassland and savanna biomes. Here the dominant species, including big bluestem, switchgrass, and indiangrass, can grow to a height of 6 ft (183 cm) and attain annual top-growth yields of 5000 lb/acre. The central grassland biome includes the mixed-grass prairies, dominated by both tallgrass and shortgrass species together with wheatgrasses and needlegrasses.

Water availability increases from west to east across the grassland biome, whereas an increasing temperature gradient extends from north to south. Grasses are classified as C_3 or C_4 species (see Chapter 4). The C_3 species have temperature optima of 65–80°F (18–27°C), grow better in cool environments, are frost tolerant, and display optimum growth during spring and early autumn. The C_4 species have temperature optima of 80–100°F, flower later, and reach maximum productivity during summer. Although most of the North American grassland biome is dominated by C_4 grasses, C_3 species are more important at higher latitudes and elevations, due to the cooler climate (Fig. 19.5). In terms of cooling, an increase in elevation of 1000 ft (305 m) is similar to an increase in latitude of 3°.

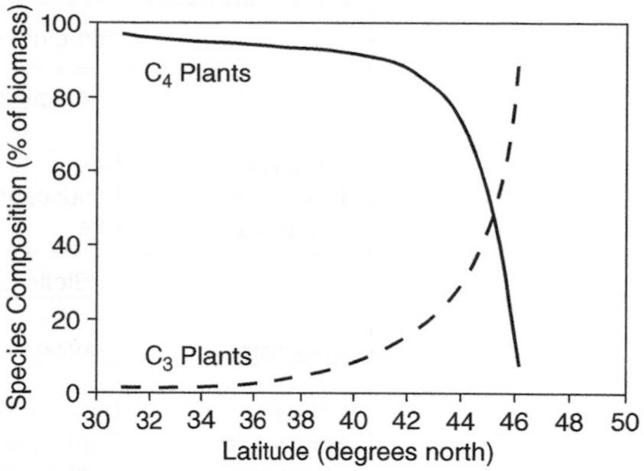

Fig. 19.5. Cool-season species (C_3) are better adapted to cool climates than are warm-season species (C_4). As latitude increases, average temperature decreases and C_3 species become more dominant in the grassland ecosystem. The C_4 species are more sensitive to low night-time temperatures, and have higher temperature optima for photosynthesis. Data are from biome sites in the western USA; the crossover point was in South Dakota. (Adapted from Sims, 1988.)

Energy Flow in Grassland Ecosystems

Ecologists describe ecosystems in terms of structure and function. *Structure* refers to the range of chemical and physical conditions (abiotic components) that limit the activity of living organisms (the biotic component) (Fig. 19.6), and to the hierarchy and mass of organisms present. *Function* refers to the roles that these organisms play in the transfer of energy and nutrients in the ecosystem. It also refers to the interactions between abiotic and biotic components in their responses to disturbances (e.g., grazing, fire), efficiencies of resource conversion (e.g., photosynthetic pathways), species interactions (e.g., symbiosis, competition), and population shifts (e.g., weed invasion).

Organisms are either **autotrophs** (self-nourishing) or **heterotrophs** (other-nourishing). Through photosynthesis, autotrophs (green plants or primary producers) convert solar energy into chemical energy, part of which is consumed by and transferred to herbivores, or secondary producers (Fig. 19.7). A chain of consumers, including **carnivores**, **insectivores**, and **omnivores** (animals that consume both plants and other animals), provides additional steps or **trophic levels** along a food chain, with losses of energy occurring at each transfer step.

Energy transfer from solar radiation through primary production to secondary production is inherently inefficient. Often less than 0.01% of the available radiation energy is captured in the desired animal product in arid environments, due to ineffective energy capture (e.g., leaf loss due to drought or incomplete grazing of forage) and biochemical barriers to energy conversion (e.g., respiratory losses of energy during tissue synthesis and turnover in plants or indigestibility of lignin in herbivores).

Some primary production avoids animal consumption and passes directly through the decomposer food chain (Fig. 19.7). Plant matter represents the largest mass of a grassland ecosystem. The second largest group of organisms in terms of mass is not, as one might expect, the mammals and birds, but the billions of decomposers. The combined weight of animal decomposers (mainly insects, worms, and nematodes) may be three to four times that of grazing herbivores at average stocking rates. Insects and other invertebrates, fungi, and bacteria ultimately break down plant and animal organic matter to carbon dioxide (CO_2) and inorganic forms of nitrogen (N), phosphorus (P), sulfur (S), and other minerals. This decomposition leads to energy release and nutrient recycling.

Grasses alter the partitioning of photosynthate between above-ground and below-ground plant parts in response to various stresses. Grazing or low light availability causes reduced root growth relative to top growth (see Chapter 2).

Structure of Ecosystems

Abiotic Components

Radiation	Soil	Wind speed
Precipitation	Landscape	Carbon dioxide
Temperature	Fire	Relative humidity

Biotic Components

Organism	Function	
Plants	Producers	Autotrophs
Herbivores	Primary consumers	Heterotrophs
Carnivores	Secondary consumers	Heterotrophs
Decomposers	Primary, secondary, and tertiary consumers	Heterotrophs

Fig. 19.6. General description of ecosystem structure. The abiotic components describe the chemical and physical environments that the biotic components inhabit. The biotic components describe the food chain and decomposers that release minerals, organic matter, and CO_2. (Adapted from Briske and Heitschmidt, 1999.)

Conversely, drought increases root growth relative to top growth (see Chapter 5). In arid areas, the plant mass that is underground in the form of roots and rhizomes can greatly exceed annual productivity of the above-ground biomass, but most of that below-ground mass was accumulated during previous years and turns over slowly unless the soil is tilled or disturbed.

Most soil organic matter is derived from below-ground biomass. The fertile soils of the world's richest agricultural regions, such as the Midwestern USA, Ukraine, and the pampas of South America, were all produced by long-term grasslands (Fig. 19.1). Grasslands are also effective in rebuilding soil organic matter after it has been depleted by crop production or erosion. Three factors contribute to this process in grassland regions, namely high rates of below-ground production, slow decomposition of grass roots and litter, and low to moderate rainfall (which limits leaching of nutrients and minerals).

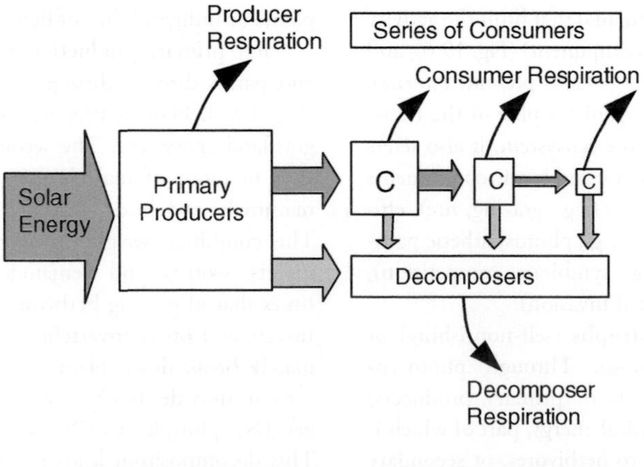

Fig. 19.7. Schematic diagram of energy flow through a grassland ecosystem. Solar energy is captured by a primary producer and is transferred (gray arrows) through one or more consumer (C) feeding levels. Each consumer respires, giving rise to CO_2 losses (black arrows). Some energy captured by the primary producer is used directly by decomposers (e.g., root turnover). (Adapted from the following: Briske and Heitschmidt, 1999; Whittaker, 1972.)

Nutrient Cycling in Grassland Ecosystems

Rangeland productivity is limited mainly by insufficient precipitation in relation to potential evapotranspiration. In higher-rainfall areas, pastureland productivity is limited mainly by N supply, because of the high demand by plants and its low natural availability. Thus pastures are frequently fertilized with N and other nutrients to enhance the carrying capacity of the land and obtain a larger economic return on livestock production. In range and pasture, the cycling of plant nutrients from the soil to the plant and animal, and then back to the soil is critical for sustaining plant growth to support livestock at the lowest cost.

Cycling of N in grasslands helps to regulate primary productivity and animal productivity because N dynamics are closely coupled to the flow of energy in the form of C. Primary producers capture solar radiation in C bonds of sugars via photosynthesis. Plants take N from the soil as ammonium (NH_4^+) and nitrate (NO_3^-), or as dinitrogen (N_2) from the air (Fig. 19.8), and combine it with C from sugars to produce proteins and nucleic acids required for plant growth. Grazing animals consume N compounds and synthesize animal proteins for meat, milk, and wool, and then excrete 70–90% of the consumed N as waste, mainly as urea in urine. The C–N bonds formed in the plants and animals must be broken during decomposition of urine in the soil, such that urea and waste proteins are converted back to NH_4^+ and NO_3^- for plant use. However, these processes also expose the N to gaseous loss of ammonia and leaching of NO_3^-.

Grasslands are characterized by relatively closed N cycles. These ecosystems have low inputs and low losses of N, which allow large amounts of N to be stored in soil organic matter, plant biomass, and decomposers (Fig. 19.8). The plants are usually limited in terms of N, and the roots scavenge the meager supplies of released N, so little soil N is lost. In contrast, eastern grasslands managed for high output of animal products, such as improved pastures or hayfields, are characterized by more open N cycles with higher annual flow of N through the system. High N inputs are matched by relatively high outputs of hay

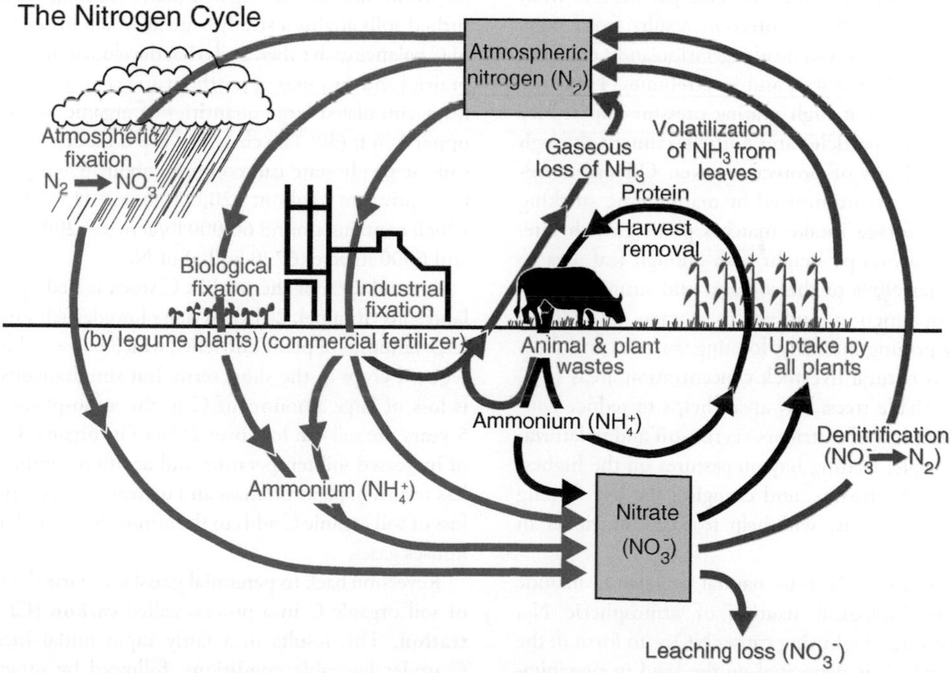

Fig. 19.8. Nitrogen gas (N_2) makes up 78% of the earth's atmosphere. Energy from lightning, fossil fuel use, and photosynthesis for legumes is used to convert N_2 to useable forms, ammonium (NH_4^+) and mainly nitrate (NO_3^-), for plant use (green lines). The N compounds in plant residues and animal excrement (black lines) are broken down in the soil to NH_4^+, with some loss of NH_3, and then to NO_3^-, which is water soluble. Most soil NO_3^- is used by actively growing plants, while high precipitation can leach some below the root zone or carry it in run-off to a water body, at which point it is lost from the N cycle (red lines). Some is converted to N_2 and returned to the atmosphere. The agricultural goal is to use as much N as possible for plant growth, and to use legumes to reduce the need for costly industrial production. (Adapted from Ball et al., 1996. Reproduced with permission of the International Plant Nutrition Institute (IPNI).)

or animal product, but with greater potential for N losses (Ball and Ryden, 1984). The challenge is to manage grasslands for efficient conversion of N inputs into economic products while minimizing losses to the environment.

With low **grazing pressure**, most plant C–N compounds are not consumed, and return directly to the soil as plant debris, where they decompose slowly to release N for plant uptake or incorporation into soil organic matter. With high grazing pressure, more of the plant C–N is consumed and passed through the animal's digestive tract, resulting in breakage of some of the C–N bonds. Since 60–80% of total N excretion is via urine, excreted N is quickly converted to inorganic forms in the soil. Furthermore, grazing livestock tend to concentrate their excreta within small areas (e.g., shade, water sites), which results in the accumulation of inorganic N to levels that exceed the plants' ability for efficient uptake. Substantial leaching of NO_3^- or gaseous loss of N may ensue.

The low-disturbance condition of perennial pastures, relative to cultivated annual cropland, functions nearly as effectively as in natural grasslands with regard to mitigating soil erosion and filtering nutrients and particulates from entering waterbodies. Soil protection results from yearlong, dense cover of live and dead vegetation and a complex community of soil microbes and invertebrates. However, continuous stocking at high grazing pressure can reduce the stability of soil particles and nutrient content through compaction and loss of protective cover. Off-site movement of nutrients is minimized by maintaining stocking rates at which forage intake matches plant growth rate, while keeping a canopy height with enough leaf area to cover more than 90% of the soil area and sustain photosynthesis, root function, and energy storage. In addition to governing grazing intensity, locating waterers and feeders away from natural livestock concentration areas (e.g., streambanks, shade trees, low areas) helps to reduce concentrations and loss of nutrients via run-off and volatilization. For example, feeding hay on pastures on the highest points away from streams, and changing the hay-feeding site on a regular basis, will help to keep nutrients in place.

Primary inputs of N into natural grasslands include lightning and biological fixation of atmospheric N_2, mainly by legumes. Lightning causes NO_3– to form in the atmosphere, which is deposited on the land in precipitation, but usually at no more than 5 lb/acre (5.6 kg/ha) annually. Therefore natural N inputs in unfertilized grasslands tend to be small compared with release (**mineralization**) of NH_4^+ and NO_3^- from plant litter and from soil organic matter by microbial decomposers. The total annual supply of available N from mineralization, fixation, and atmospheric deposition ranges from 25 lb/acre (28 kg/ha) in semiarid rangeland with few legumes to 200 lb/acre (224 kg/ha) in humid-area pastures with large amounts of legumes.

In soils that are well aerated and near neutral in pH, excess NH_4^+ is converted by bacteria to NO_3^-, which is water soluble and can leach below the rooting zone (Fig. 19.8). Although NO_3^- leaching is generally low in unfertilized grasslands, losses can be significant under high **stocking density** or when N inputs from manure, fertilizer, or pollutants exceed the uptake capacity of plants. Under these conditions, the volatilization of NH_3 and other nitrogenous gases (e.g., nitrous oxide, N_2O) also leads to significant losses of N from the ecosystem.

Soil Carbon and Climate Change

Soil organic matter levels in grassland soils are of interest for two key reasons. First, the organic material stabilizes soil aggregates, which resist erosion and maintain macropores, harbor diverse microbiota essential for nutrient transformations, and retain water and nutrients. Second, the enormous amount of C in the organic matter offers some buffer against the atmospheric build-up of CO_2, the most common gas linked to climate warming via its greenhouse effect. Organic matter in long-term undisturbed soils attains a steady state at which losses and gains of C balance each other. Soils that developed under tallgrass prairies and savannas of northern and central North America accumulated large quantities of organic matter in the upper 1–4 ft (30–120 cm). The top 6 in. (15 cm) of these soils at steady state can comprise about 6% organic matter, equivalent to about 120,000 lb/acre (134,000 kg/ha), which contains around 60,000 lb/acre (67,200 kg/ha) of C and 6000 lb/acre (6720 kg/ha) of N.

Around 20% of the world's C stock is tied up in grassland soils (Rumpel, 2011). When plowed, such soils release large amounts of plant nutrients, which boost yields of subsequent crops in the short term, but simultaneously there is loss of large amounts of C to the atmosphere. Within 5 years the soil can lose over 25% of its organic C because of increased soil temperature and aeration combined with less return of plant biomass and increased soil erosion. The loss of soil organic C adds to the atmospheric pool of greenhouses gases.

Reversion back to perennial grassland starts the recovery of soil organic C in a process called **carbon (C) sequestration**. This results in a fairly rapid initial increase in C under favorable conditions, followed by progressively slower sequestration to reach a new steady state within 50–200 years. Annual rates of C sequestration vary widely, from 100 lb/acre (112 kg/ha) to over 1500 lb/acre (1680 kg/ha) in the first 5 years of reversion to grassland, depending on climate, soil fertility, and grass types (Schnabel et al., 2001). Sequestration of C into the long-term organic matter has been touted as a means of countering the build-up of atmospheric CO_2 from other sources (Follett, 2001). A mere 1% increase in the C stocks of the top 4 in. (10 cm) of US grassland soil would counteract the entire annual emission of CO_2 from US cropland agriculture. Old, well-managed grazing lands are considered to be near or at their potential steady-state C level, and therefore they offer little potential for sequestering more C. A greater opportunity for C sequestration exists in the long-term Conservation Reserve Program, by which the widespread conversion of erosive cropland with low organic matter to perennial grass and woodland sequesters around 18 million tons of C in US soils annually (National Resources Conservation Service, 1995).

Climate change will probably affect the composition and function of managed pastures and native rangelands. The accumulation of greenhouse gas is predicted to push average global temperatures up by 2.7–3.6°F (1.5–2.0°C) by 2060, and to extend growing-season length by about 10 days (Hatfield, 2011). Furthermore, precipitation events will probably become more severe and sporadic, leading to more flooding and extended drought periods. Such changes could increase survival and competitiveness of warm-season, C_4 grasses at higher latitudes and altitudes, slow down C build-up in restored grasslands, increase invasion by stubborn weeds, reduce persistence of N_2-fixing legumes, and intensify heat stress on grazing livestock. Possible benefits are longer grazing seasons (and thus less reliance on stored feeds) and higher photosynthetic rates of C_3 species (due to higher CO_2 levels in the atmosphere). A high degree of climate variability will require the avoidance of excessive grazing pressure in order to maintain grassland resilience by sustaining rooting depth and vegetative cover.

Grasslands and Wildlife

Cows may not like to graze the stiff stems of grasses, but stems contribute to desirable wildlife habitat. Stems of tall grasses serve as perches for small birds, and stems that break over well above ground level form canopies for winter cover and protected runways for rodents and other wildlife. Grasslands have a great diversity of plant species, including a range of seed producers that provide food directly to birds and small animals while supporting insect life—all vital parts of the wildlife food chain.

Most forages decrease in quality when the plants flower and begin to produce seed, but some farmers and ranchers consider wildlife habitat in their management systems. For example, delaying hay harvest in the north allows ground-nesting birds such as pheasants to hatch their eggs and protect their young. Increasing the proportion of species such as annual lespedeza or Illinois bundleflower is associated with quail success, and establishing food plots attracts deer. Charging a fee to hunters or bird watchers can offset the loss in livestock performance when the pasture is managed for wildlife. The public is becoming more cognizant of the multiple uses of grasslands, and now demands green spaces to support wildlife habitat and provide aesthetic surroundings.

Herbivory in Grassland Ecosystems

Grasslands are often subject to high levels of herbivory. In the range of 20–60% of **net primary production** is consumed by vertebrate herbivores (livestock and wildlife) and invertebrate herbivores (mainly insects). In contrast, 50–75% of above-ground net primary production in intensively managed temperate pastures is usually consumed by **grazers.** Invertebrate herbivores may consume about 10% of the net primary production in natural grasslands, whereas soil-borne herbivores often consume 25% of below-ground plant material, with nematodes alone consuming 10–15% (Detling, 1988). However, these invertebrates contribute to soil quality and subsequent herbage production.

Grazing can diminish plant growth and vigor by reducing leaf area, removing the active **shoot apex,** depleting reserves of mineral nutrients and energy, and shifting those reserves from the root to the shoot to replace lost photosynthetic tissues. Conversely, grazing can benefit plants by increasing light availability to lower parts of the plant canopy, removing older leaves to expose younger leaves with higher photosynthetic rates, and stimulating tiller and rhizome development by activating axillary meristems in crowns.

Grasses are particularly capable of enduring frequent defoliation because they have **intercalary meristems** near soil level for leaf and stem growth (see Chapter 2), they often have dense tillering from a single plant, and some can survive drought by enduring severe shoot desiccation (Volaire et al., 2013). Conversely, the elevated shoot apices of legumes, forbs, and shrubs are removed when they are grazed (see Chapter 3); however, many species have defenses against herbivory in the form of thorns and bitter-tasting chemicals.

Herbivory also affects the spatial pattern of nutrient cycling (Fig. 19.8). Grazers gather forage N (largely plant protein) that is fairly uniformly distributed across the grazing area, and then concentrate it into small areas, especially where animals congregate, as urine and feces. "Application rates" of N in urine patches can exceed 300 lb/acre (336 kg/ha), far in excess of the potential for plant uptake in a short time. The resulting high levels of labile N compounds in the soil solution are readily volatilized as NH_3 or leached as NO_3^-. Thus grazing alters N distribution within the grazing area, hastens N cycling, and increases the vulnerability of the grassland to N losses. Likewise, other mineral nutrients, such as concentrated P, K, and S in dung patches, become unevenly redistributed as a result of livestock herbivory. A **grazing system** that involves short-term rotation among numerous paddocks results in a more even distribution of excreted nutrients, and presumably more efficient nutrient recycling to the forage (see Chapter 18).

Grazing often alters the species composition of the plant community. Plant species with either low palatability or good tolerance of high grazing pressure can become dominant over more palatable or less grazing-tolerant species. This situation can easily lead to invasion by highly competitive, unpalatable species that are difficult and expensive to eradicate, and which threaten the long-term sustainability of the grassland. High grazing pressure also shifts the competitive balance within desirable mixtures to favor short or prostrate species such as white clover, kentucky bluegrass, buffalograss, and blue grama over taller species such as orchardgrass, big bluestem, and alfalfa. Low grazing pressure, rotational stocking, or leaving some areas ungrazed allows taller species to intercept light at the expense of shorter species, and improves sustainability by preserving forage leaf area, energy reserves, and resilience of the desired mixture.

Plant Responses to Managed Herbivory

Persistent grassland plants evolved under the selection pressure of herbivores to either tolerate or avoid excessive herbivory. Many grasses tolerate grazing by maintaining numerous basal meristems that efficiently replace grazed leaf tissue. Other plants avoid grazing by deploying either physical deterrents, such as spines, or chemical deterrents, such as alkaloids and tannins. Although chemical defenses against herbivory are less common in grasses than in other plant types, *Neotyphodium* fungal endophytes produce alkaloids in tall fescue and perennial ryegrass leaves that cause animal toxicosis and lead to reduced forage intake (see Chapter 16). However, the presence of endophytes in these grasses also inhibits insect and nematode herbivory, thus conferring another strong competitive advantage on infected plants.

The grassland manager decides what constitutes a sustainable stocking rate for a particular environment and plant community (to maximize the conversion of solar energy and precipitation to marketable animal product while maintaining a vigorous plant community of desirable species). Although grazing generally benefits grassland productivity, overstocking causes plants to be grazed too short and/or too frequently for maintenance of leaf area, meristems, and energy reserves, resulting in decreased productivity. Continual overstocking and declining productivity lead to land degradation and **desertification** in low-rainfall grasslands.

The rule of thumb in semiarid rangeland areas is to "graze half and leave half" to control stocking so that half of the above-ground plant mass is left ungrazed to sustain leaf area and root mass (see Chapter 2). In higher-rainfall areas, high stocking rates on long-term grasslands favor the dominance of low-growing, introduced grasses, such as kentucky bluegrass and bermudagrass. These grasses can tolerate removal of more than half their forage mass by cattle and still remain vigorous. Sheep graze closer to ground level.

Fire in Grassland Ecosystems

Native Americans learned that fire was critical for maintaining the grasslands and savannas of central North America long before European settlement. Fire also maintained production of the high velds of South Africa, and the pampas of South America. If fire, mowing, or grazing is absent, savannah or forest can displace grasslands within just a few decades. Timely fires restrict invasion of woody and weedy species.

In water-limited shortgrass communities, forage availability is low, but forage quality tends to be good, even in senesced plant material. Consequently, fire in arid grasslands is detrimental to livestock production because it removes valuable forage. In contrast, in sub-humid to humid grasslands, there is a buildup of C_4 grass biomass that contains low concentrations of N and high concentrations of high-lignin fiber, both of which reduce the rate of plant litter decomposition. Fire improves forage quality and animal performance by removing the low-quality dead biomass and recycling the mineral nutrients to support regrowth of nutritious young forage (Fig. 19.9). Therefore fire is a practical, low-input tool for reducing undesirable vegetation and improving forage quality.

Managing Grasslands for Sustainability

The science of grassland ecology underpins the approaches used to preserve and improve grasslands. Historically this was aimed at increasing economic returns from grazing cattle and sheep, and research on management emphasized the long-term profitability of the farm or ranch enterprise across wide fluctuations in markets and weather conditions. Now, sustainability demands consideration of environmental quality from field to watershed level, and the social viability of rural communities. Non-agricultural goals include maintaining high species diversity, preserving near-natural conditions for wildlife and pollinator habitat, improving recreational use, and enhancing landscape beautification.

Agricultural improvement strategies involve the introduction of plant species or cultivars with higher nutritional quality, higher yield, and longer stand life, and the use of animal species and breeds with more efficient feed conversion. To alleviate limitations to plant and animal growth, inputs such as fertilizers to increase yield and pesticides to control undesirable competitors, predators, and parasites can be used. In addition, grassland improvement entails controlling the intensity and timing of grazing, mitigating soil erosion, and improving nutrient cycling from animal excreta back into the forage.

Increased management inputs can have negative environmental impacts. Introduced ("improved") species show a sharp yield increase up to medium levels of inputs (Fig. 19.10), but when the yield response levels off, higher

Fig. 19.9. Prescribed burning of upright warm-season grasses every 2 or 3 years is a practical, low-input tool for reducing undesirable and dead vegetation and enhancing the vigor of the stand. Dry growth from the previous fall is burned in spring when leaf growth of desired grasses has begun, but the shoot apex remains protected at or below soil level. The back-burn fire (against a mild wind) moves slowly and kills or retards small trees, shrubs, and broad-leaved weeds which have shoot apices above soil level. The ashes contain mineral nutrients and absorb radiation to warm the soil and stimulate early growth and competitiveness, thereby increasing yield and forage quality. Burning requires safety equipment and training, and must comply with local regulations on fire prevention and air quality. (Photo courtesy of the Natural Resource Conservation Service.)

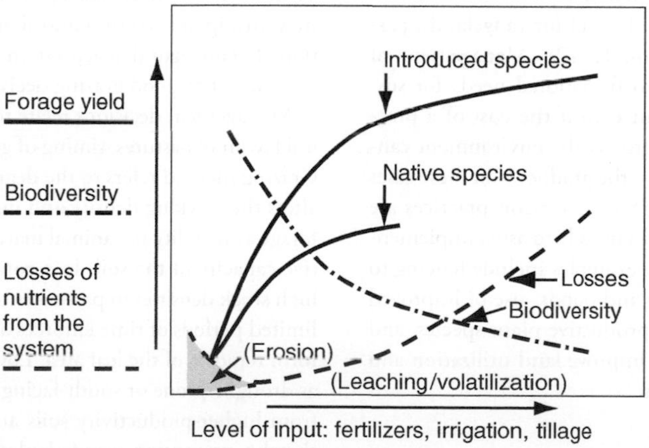

Fig. 19.10. Schematic diagram to show how increasing the level of management inputs increases forage yield and decreases biodiversity as some species are favored. Nutrient losses that occur from wind or water erosion in denuded grasslands can be curtailed by modest inputs and improved grazing management. Nutrients are used most efficiently at low levels. With further inputs, biodiversity decreases, yield increase per unit input decreases (especially for native species), and nutrient losses increase due to leaching and volatilization.

applications of nutrients such as N and P "leak" off-site and can contaminate groundwater and surface waters. Another nutrient leakage route is volatilization of NH_3 and the greenhouse gas N_2O. Native species generally show less yield response to inputs. High inputs of nutrients, water, and tillage reduce the forage species biodiversity of the grassland. Rangelands receive very few inputs, due to the large land area and low opportunity for economic return in the absence of adequate water. Such grasslands are improved mainly by controlling stocking rates, deferring grazing to allow recovery of plant vigor, and prescribing fire in critical sites.

Managing for sustainability is a balancing act between exploitation for economic efficiency and conservation of complex plant, soil, and animal ecosystems to prevent an off-site decline in environmental quality, all in a socially acceptable way. The USDA Natural Resources Conservation Service (NRCS) has formulated voluntary practices, including the very popular Conservation Reserve Program, which is designed to exploit the benefits of grasslands to conserve soil, protect water quality, and provide habitat for pollinators and migrating birds. These programs, which are aimed at achieving a balance between economic and environmental sustainability, include stabilizing soil as a storage medium for nutrients, water, and beneficial microbes, managing grazing to maintain vegetative ground cover, grassland renovation, and leveraging the growth habits of diverse plant communities to protect sensitive landscape features and wildlife.

Detailed analyses of desired conservation outcomes from many of the individual practices have been reviewed for rangelands by Briske (2011), and for rangelands, pastures, and hayland by Nelson (2012). Management of range and grasslands needs to fit national needs for sustainability of the environment even if the cost of a practice that is implemented to protect the environment cannot be recovered in the value of the product—in these cases mainly sale of animals. NRCS conservation practices are voluntary and are subsidized financially to assist implementation of the programs. Some examples include fencing to keep livestock out of streams and ponds, use of improved grazing systems to maintain productive plant species, and use of rotational stocking to improve land utilization and reduce mineral redistribution.

Judicious Use of Fertilizers

Chemical or organic fertilizers, such as animal wastes, are frequently applied to long-term pastures to remove a major limitation to plant growth, where precipitation is adequate to favor an economic response. Nitrogen is the nutrient most frequently applied to grass-dominated pastures, because of the high N requirement of grasses for growth. Potassium (K) and P are more likely to be limiting in legume-dominated systems. Optimization of soil fertility and pH enhances plant growth and allows an increase in animal-carrying capacity, but it often has little or no effect on forage quality. Conversely, fertilization with P and K may improve individual animal performance by increasing the legume composition of a legume–grass mixture (see Chapter 12). Adding magnesium (Mg) fertilizer may reduce the risk of grass tetany (see Chapter 16).

Grasses and legumes exploit different niches with regard to N acquisition. Grasses depend on soil NO_3^- and NH_4^+ for protein synthesis, which reduces the steep decrease in crude protein concentration as leaves age and the plant begin to flower. Legumes can also take up NO_3^- and NH_4^+ from the soil, but when these forms are scarce they can assimilate atmospheric N via symbiosis with *Rhizobium* and *Bradyrhizobium* species (see Chapters 2 and 4, and Fig. 19.8) and maintain high protein levels. In turn, some of the fixed N is released to associated grasses. Maintaining at least 30% legumes in pastures reduces fertilizer costs and enhances the flow of solar energy and nutrients to a marketable animal product.

Grazing Management for Sustainability

The objectives of grazing management usually include optimization of both production per animal and animal production per unit land area. Other objectives include managing animal breeding, controlling weeds, balancing the animals' nutrition, forcing hoof action to lightly disturb the canopy and soil, and allowing recuperation of the ecosystem from severe drought. Grazing management is one of the few economical inputs available that allows dramatic manipulation of the grassland ecosystem to achieve a balance between livestock production and environmental stewardship. It is via the control of leaf removal and restitution of plant vigor that ecosystem stability and productivity are realized by good grazing decisions.

Management decisions relate to grazing methods, physical layout of pastures, timing of grazing, and stocking rate. Grazing method refers to the degree to which the manager alters the stocking density and duration of grazing so that forage availability per animal matches the localized productive capacity of the soil. In a **rotational stocking** system, high stock densities in pasture subdivisions (**paddocks**) for limited periods of time enforces a more complete and uniform removal of the leaf area. Low-productivity soils, such as drought-prone or south-facing slopes, can be fenced off from higher-productivity soils and grazed less frequently, thereby preventing stand depletion during the summer months. Timely moving of the livestock to fresh paddocks prevents chronic overgrazing and allows plants to rapidly recover leaf area and resume photosynthesis. The results are a high proportion of plant energy directed through the animal, low wastage of forage, and maintenance of stand vigor on marginal soil sites.

The intensity of rotation ranges widely, from a simple two-paddock method at fairly low stocking density to a mob-grazing or strip-grazing method using ultra-high density for one day or less per paddock. **Continuous stocking**, involving no paddocks and allowing free movement of livestock at low stocking density, can be controlled to a limited degree by carefully monitoring grazing pressure, stand condition, and placement of water, feed, and shade. Rotational stocking permits site-specific grazing control, and can minimize stresses due to severe weather events. It can also be used for recovery from situations resulting from mismanagement.

New Plantings or Renovation for Diversity

Biodiversity in grasslands was high before Europeans settled in North America and introduced domestic livestock. The boosting of livestock production by clearing forests, cropping, fertilizing, controlling weeds, introducing exotic species, and increasing grazing pressure resulted in depleted soils and reduced plant diversity in managed grasslands and pastures compared with the earlier native grasslands. For example, over 500 plant species are found in the native Konza Prairie of central Kansas that depend on judicious management that was unknown by Europeans. With introduced species, primarily from Europe, the long-term pastures in more humid areas of the eastern USA typically contain around 10–20 species.

Grasslands can be improved by seeding new species or cultivars with enhanced yield, forage quality, grazing tolerance, resistance to pests, tolerance of climatic extremes, and a longer growing season. New plantings are also useful for non-economic objectives, such as increasing soil organic matter and fertility, reducing soil and streambank erosion, restoring native plant and animal communities, promoting pollinator insects and birds, and beautifying the landscape. New plantings typically involve introduced species for humid environments, to exploit their selected traits for quick establishment and high productivity. Native species are usually preferred in semiarid rangelands, because of their proven adaptation to the highly variable temperature and precipitation regimes.

Seedbed preparation and replanting on water-limited rangeland and erosive pasture sites carry a substantial risk of failure. Therefore minimum tillage or no-till methods are preferred for establishment (see Chapter 11), and successful results are not usually realized before 2 or 3 years. Replanting must be viewed as merely one component of a package of improvement practices, such as brush control and controlled stocking. Rather than sowing the same species in all terrains, appropriate species should be matched to specific soil and microclimatic conditions at each landscape position (see Chapters 6 to 10).

For new plantings there are choices with regard to the use of native or introduced (exotic) plant species and the diversity of species. Humid pastures are usually seeded with introduced species, whereas tall-growing C_4 native grasses such as switchgrass and big bluestem also have useful roles. These grasses are indigenous to humid zones, and therefore they are adapted. Moreover, native grasses generally have a complementary growth period, require less N fertilizer than introduced grasses such as bermudagrass and tall fescue, exhibit good drought tolerance, and provide good nesting habitat due to their upright, bunchy growth. The use of native grasses is encouraged for NRCS-sponsored Conservation Reserve Program grassland plantings, which can be used for forage or biofuels after the contract period.

The issue of plant diversity hinges on the ecological theory that species diversity promotes stability of the plant community in the face of disturbances such as drought, and can favor greater biomass productivity and a reduction in nutrient loss (Sanderson et al., 2004). **Species richness** refers to the number of different species that are present in a grassland. **Functional richness** refers to the number of different groups of species as defined by their growth habit. For example, warm- and cool-season grasses and warm- and cool-season legumes constitute four distinct functional groups. Having two species from each functional group in the same pasture (with a species richness of eight) is potentially more beneficial than having the same richness of eight species from only one functional group. In the former case, more resource niches would be exploited with minimal overlap among species, whereas in the latter case, several species would overlap in their use of resources and become competitive. The benefits of high species diversity generally hold in low-resource environments, whereas high diversity is less important in humid, high-fertility pasture systems that are well managed.

Integration of Approaches to Manage Grassland Sustainability

Grassland and livestock production systems are ecologically complex and challenging because they contend with diverse land features, plant qualities and growth habits, and types of herbivores. The ecological benefits of harnessing perennial vegetation to restore and stabilize soil and water are difficult to demonstrate economically. Overcoming these challenges and their complexity entails broader ecological roles of grasslands extending beyond a given field to include watershed protection, economic vitality of rural communities through crop and livestock diversification, and value-added goods to society. For example, the critical role of certain insects and birds in pollinating plants is important in food production, and maintaining native communities that support wildlife has recently received greater attention as a result of government incentive programs. Forage legumes such as alfalfa and clovers and native forbs such as sunflowers attract bees for the nectar, and provide habitat for pollinators to aid seed production by neighboring plants. Restoring grasslands to mimic predevelopment vegetative diversity attracts birds and mammals, thereby diversifying the recreational value of grasslands.

Today's concept of grassland sustainability calls for a holistic set of ethics that fosters soil and water protection, C sequestration, plant communities with more native species allowing only minimal leakage of water-soluble nutrients, habitat and food for wildlife and other herbivores, and a pastoral aesthetic in the landscape. Strategies for combining this ethic with profitable entrepreneurship are now targeted to fit consumers who are demanding foods that are produced using more natural processes of nutrient and energy flow, and which are perceived to be less stressful to animals. For example "natural milk" and "grass-fed beef" are produced on all-forage or high-forage diets, mainly by grazing, to exploit the natural fermentation function of the rumen and to produce a more healthy fatty acid profile during rumen digestion. These foods are typically produced locally, with few or no synthetic or industrial inputs such as fertilizers, pesticides, or growth stimulants, and by feeding little or no grain in order both to avoid competitive demand for grains as human food and to reduce the amount of greenhouse gas emissions from grain production, transport, storage, and processing.

The Demand for Organic Products

The most highly refined and regulated version of the natural-food production chain is **organic agriculture**. Products that earn the official "organic" label in the USA must be certified as such by the USDA according to very high standards that meet consumer expectations with regard to purity, production methods, genetic materials, and livestock living conditions, among others. Grassland-based livestock production lends itself quite well to the unofficial "natural" label or the official "organic" label, because the ecosystems are quasi-natural already by virtue of their perennial nature, inclusion of legumes as the major N source, and effective internal nutrient cycling.

Challenges for natural or organic foods include the lower rate of production compared with the high nutrient density of feed grains and protein supplements, and the limited options for controlling parasites and other pests without synthetic chemicals. Overall, a high level of management expertise is needed to balance the system, but the market price of the foods produced is markedly higher, which balances the effort.

Another enterprising opportunity that complements organic and natural meat and milk production is agritourism, or farm visits by urban dwellers to relish the personal experience of seeing how their food is produced and then buying directly from the farmer. The rapid rise in farmers' markets in urban areas has also been linked to consumers forming personal associations with specific farmers for their products. This activity helps non-farm consumers to understand the wholesomeness and beauty of grassland agriculture and the care provided by the landowner to meet their demands for freshness and quality.

> **Riparian Strips Protect Water Quality**
>
> Riparian strips are narrow bands of grassland or shrubs that occur naturally or are planted along a streambank to intercept water run-off from crop fields and so reduce the amount of sediment and chemical residues that enters streams. Planting forage grasses and shrubs along the streambank can abate this problem. The plant roots hold the streambank soil together and reduce the erosion of soil particles into the water. The stems and leaves slow the flow of surface water from fields, holding back solid particles like a sieve, and allow settling of the sediments.
>
> As water flows across fields, it carries some phosphorus that is loosely bound to organic materials, such as manure particles, and some phosphorus that is strongly bound to clay particles. Soils with slow water infiltration and that are easily eroded are especially prone to losses of nutrients with water run-off. Lateral flow of infiltrated water is intercepted below the soil surface by plant roots that can scavenge nutrients before they enter the stream.
>
> Permanent pastures on and above streambanks protect the water, especially if the livestock are excluded from the stream. Livestock can cause soil compaction and loss of vegetative cover if the pasture is overgrazed, which aggravates run-off and sediment entering the stream. In addition, unless they are excluded from the stream, livestock excrete waste directly into the water, bypassing the cleansing function of the riparian strip.
>
> How would the slope of nearby land affect the required width of the riparian strip? How would the soil type affect this width?

Summary

Grasslands represent the major land form used for grazing livestock production. The grassland and savanna biomes of North America extend from northeast Mexico to the prairie provinces of Canada. These grasslands consist mainly of grasses, but usually also include legumes, composites, sedges, shrubs, and sparse trees. Rangelands are extensively grazed grassland areas in drier environments that are dominated by native species. In the region of 20–60% of the net primary production of grasslands is consumed by herbivores.

Cycling of N in grasslands helps to regulate primary productivity and animal productivity. The relatively undisturbed condition of perennial pastures is nearly as effective as natural grasslands in mitigating soil erosion and filtering nutrients and particulates, thereby preventing them from entering waterbodies. Soil organic matter stabilizes soil aggregates, harbors diverse microbiota essential for nutrient transformations, and retains water and nutrients.

Questions

1. Outline the major abiotic and biotic factors that determine where grasslands form. Give particular emphasis to major climatic factors.
2. List the dominant grass species in the short-, mid-, and tallgrass prairies.
3. How do naturalized grasslands differ from native grasslands?
4. What is the consequence of prolonged overgrazing? Will a grassland necessarily return to a productive status once grazers have been removed? Describe how grazing can benefit grasslands.
5. Propose a simplified N-flow diagram for a grassland in your region. Take into account the species and conditions that occur in your area.
6. Describe how fire is beneficial to grassland development.
7. Discuss improvement techniques for (a) grasslands with declining populations of desirable species, (b) grasslands that have lost certain key species, (c) grasslands suffering from lack of prolonged production or from extensive erosion, and (d) denuded grasslands.
8. Explain the value of species diversity (biodiversity) in grasslands.
9. List some regions where extensive management of naturalized grasslands may be more economic than increasing inputs and intensifying management.

References

Bailey, RG. 1995. Description of the Ecoregions of the United States. Available at: www.fs.fed.us/land/ecosysmgmt/ (accessed August 29, 2016).

Ball, DM, CS Hoveland, and GD Lacefield. 1996. Southern Forages, 2nd ed. Potash & Phosphate Institute and Foundation for Agronomic Research, Norcross, GA.

Ball, PR, and JC Ryden. 1984. Nitrogen relationships in intensively managed temperate grasslands. Plant Soil 76:23–33.

Briske, DD. (ed.) 2011. Conservation Benefits of Rangeland Practices: Assessment, Recommendations, and Knowledge Gaps. Allen Press, Lawrence, KS.

Briske, DD, and RL Heitschmidt. 1999. An ecological perspective. In RK Heitschmidt and JW Stuth (eds.), Grazing Management: An Ecological Perspective, pp. 11–26. Timber Press, Portland, OR.

Detling, JK. 1988. Grasslands and savannas: regulation of energy flow and nutrient cycling by herbivores. In LR Pomeroy and JJ Alberts (eds.), Concepts of Ecosystem Ecology: A Comparative View, pp. 131–154. Springer-Verlag, New York.

Follett, RF. 2001. Organic carbon pools in grazing land soils. In RF Follett, JM Kimble, and R Lal (eds.), The Potential of U.S. Grazing Lands to Sequester Carbon and Mitigate the Greenhouse Effect, pp. 65–86. CRC Press, Boca Raton, FL.

Harlan, JR. 1956. Theory and Dynamics of Grassland Agriculture. Van Nostrand Co., Princeton, NJ.

Hatfield, JL. 2011. Climate impacts on agriculture in the United States: the value of past observations. In D Hillel and C Rosenzweig (eds.), Handbook of Climate Change and Agroecosystems: Impacts, Adaptation, and Mitigation, pp. 239–254. Imperial College Press, London, UK.

Kephart, KD, CP West, and DA Wedin. 1995. Grassland ecosystems and their improvement. In RF Barnes, DA Miller, and CJ Nelson (eds.), Forages: An Introduction to Grassland Agriculture, 5th ed., pp. 141–153. Iowa State University Press, Ames, IA.

Natural Resources Conservation Service. 1995. Grazing Lands. RCA Issue Brief #6 November 1995. Available at: www.nrcs.usda.gov/wps/portal/nrcs/detail/national/technical/nra/rca/?cid=nrcs143_014209 (accessed September 7, 2016).

Nelson, CJ (ed.). 2012. Conservation Outcomes from Pastureland and Hayland Practices: Assessment, Recommendations, and Knowledge Gaps. Allen Press, Lawrence, KS.

Raven, PH, RF Evert, and SE Eichhorn. 1992. Biology of Plants. Worth Publishers, New York.

Rumpel, C. 2011. Carbon storage and organic matter. In G Lemaire, J Hodgson, and A Chabbi (eds.), Grassland Productivity and Ecosystem Services, pp. 65–72. CABI Publishing, Wallingford, UK.

Sanderson, MA, RH Skinner, DJ Barker, GR Edwards, BF Tracy, and DA Wedin. 2004. Plant species diversity and management of temperate forage and grazing land ecosystems. Crop Sci. 44:1132–1144.

Schnabel, RR, AJ Franzluebbers, WL Stout, MA Sanderson, and JA Stuedemann. 2001. The effects of pasture management practices. In RF Follett, JM Kimble, and R Lal (eds.), The Potential of U.S. Grazing Lands to Sequester Carbon and Mitigate the Greenhouse Effect, pp. 291–322. CRC Press, Boca Raton, FL.

Sims, PL. 1988. Grasslands. In MG Barbour and WD Billings (eds.), North American Terrestrial Vegetation, pp. 266–286. Cambridge University Press, New York.

Volaire, F, K Barkaoui, and M Norton. 2013. Designing resilient and sustainable grasslands for a drier future: adaptive strategies, functional traits and biotic interactions. Eur. J. Agron. 52:81–89.

Whittaker RH. 1972. Evolution and measurement of species diversity. Taxon 21:213–251.

Appendix: Common and Botanical Names of Forages

The accompanying list gives the common and scientific names of the grasses, legumes, and other plants commonly encountered in the study of forages. Botanical names are important because many forage crop species have different common names in different parts of the USA or in different countries around the world. Common names are listed alphabetically by the name most commonly used (e.g., alfalfa), followed by the less frequently used names. The initial letter of the first part of the scientific name, the genus, is always capitalized (e.g., *Medicago*). The second part, the species epithet, is written entirely in lower case (e.g., *sativa*). The genus corresponds roughly to a last name and the species epithet is comparable to a first name. Thus, for example, *Zea mays* would correspond to Brown, John. The scientific name is followed by an abbreviation or name to identify the person credited with naming the species. Thus *Zea mays* L. indicates that this species was named by Linnaeus. However, *Cynodon dactylon* (L.) Pers. was first described as *Panicum dactylon* by Linnaeus, but was later transferred to another genus, *Cynodon*, by Persoon.

The following abbreviations and symbols are used: syn. = synonym, in which case there are two accepted scientific names; × before the genus name indicates a hybrid between two genera; × before the species epithet or subspecies name refers to a hybrid between two species or two subspecies.

Scientific names continue to change as new technologies for classification are developed. For example, the taxonomy and nomenclature of the perennial grasses of the tribe Triticeae were revised by Barkworth and Dewey (1985). The scientific name for tall fescue, *Festuca arundinacea* Schreb., has two synonyms, *Lolium arundinaceum* (Schreb.) S.J. Darbyshire and *Schedonorus arundinaceus* (Schreb.) Dumort., which are gaining acceptance. The possibility of moving tall fescue to the genus *Lolium* is under consideration because it readily hybridizes with perennial ryegrass, and the two species are infected by endophytic fungi that are genetically similar. The Crop Science Society of America, the Weed Science Society of America, and the USDA Germplasm Resources Information Network (GRIN) database regularly update listings.

Common Names	Scientific Names
aeschynomene (*see* jointvetch)	*Aeschynomene americana* L.
alfalfa	*Medicago* spp.
alfalfa	*M. sativa* L.
variegated	*M. sativa* × *varia* (Martyn) Arcang. (syn. *M. media* Pers.)
yellow	*M. sativa* subsp. *falcata* (L.) Arcang.
alyceclover	*Alysicarpus vaginalis* (L.) DC.
American jointvetch	*Aeschynomene americana* L.
artichoke	*Cynara cardunculus* L.

(*continued*)

Forages: An Introduction to Grassland Agriculture, Seventh Edition. Edited by Michael Collins, C. Jerry Nelson, Kenneth J. Moore and Robert F Barnes.

Common Names	Scientific Names
aster, heath	*Aster ericoides* L.
bahiagrass	*Paspalum notatum* Flüggé
barley	*Hordeum* spp.
common	*H. vulgare* L.
foxtail	*H. jubatum* L.
meadow	*H. brachyantherum* Nevski
beggarweed (tickclover)	*Desmodium* spp.
beggarweed, Florida	*D. tortuosum* (Sw.) DC.
creeping (kaimi clover)	*D. incanum* (G. Mey.) DC.
bentgrass, creeping	*Agrostis stolonifera* L. var. *palustris*
bermudagrass	*Cynodon dactylon* (L.) Pers.
black sampson	*Echinacea angustifolia* DC.
bladygrass (cogongrass)	*Imperata cylindrica* (L.) P. Beauv.
bluegrass	*Poa* spp.
annual	*P. annua* L.
Canada	*P. compressa* L.
Kentucky	*P. pratensis* L.
Texas	*P. arachnifera* Torr.
bluejoint (bluejoint reedgrass)	*Calamagrostis canadensis* (Michx.) P. Beauv.
bluestem	
big (turkey foot)	*Andropogon gerardii* Vitman
Caucasian	*Bothriochloa bladhii* (Retz.) S.T. Blake
little	*Schizachyrium scoparium* (Michx.) Nash
sand	*Andropogon hallii* Hack
silver	*Bothriochloa saccharoides* (Sw.) Rydb.
bristlegrass, plains	*Setaria leucopila* (Scribn. & Merr.) K. Schum.
bromegrass	*Bromus* spp.
cheatgrass	*B. tectorum* L.
chess	*B. secalinus* L.
downy (cheatgrass)	*B. tectorum* L.
field	*B. arvensis* L.
Japanese	*B. japonicus* Houtt.
Matua (prairie grass)	*B. catharticus* Vahl var. *catharticus*
meadow	*B. riparius* Rehmann
mountain	*B. carinatus* Hook. & Arn. var. *marginatus* (Steud.) Barkworth & Anderton
ripgutgrass	*B. diandrus* Roth var. *rigidus* (Roth) Sales
smooth	*B. inermis* Leyss.
soft chess	*B. hordeaceus* L.
broomsedge	*Andropogon virginicus* L.
brownseed paspalum	*Paspalum plicatulum* Michx.
buckbrush (coralberry)	*Symphoricarpos orbiculatus* Moench.
buffalo clover (alyceclover)	*Alysicarpus vaginalis* (L.) DC.
buffalograss	*Buchloe dactyloides* (Nutt.) Columbus
buffelgrass	*Pennisetum ciliare* (L.) Link
bundleflower, Illinois	*Desmanthus illinoensis* (Michx.) MacMill. ex B. L. Rob. & Fernald
burclover	*Medicago* spp.
little	*M. minima* (L.) Bartal.
spotted	*M. arabica* (L.) Huds.
butterfly pea	*Clitoria ternatea* L.
buttonclover	*Medicago orbicularis* (L.) Bartal.
cabbage	*Brassica oleracea* L. var. *capitata* L.
Caley pea (roughpea)	*Lathyrus hirsutus* L.

Common Names	Scientific Names
canarygrass	*Phalaris* spp.
annual	*P. canariensis* L.
reed	*P. arundinacea* L.
canola (rape)	*Brassica napus* L. var. *napus*
carpetgrass	*Axonopus* spp.
common	*A. fissifolius* (Raddi) Kuhlm.
tropical	*A. compressus* (Sw.) Beauv.
carpon desmodium (*see* desmodium)	
catclaw acacia	*Acacia greggii* A. Gray
cedar, eastern red	*Juniperus virginiana* L.
centipedegrass	*Eremochloa ophiuroides* (Munro) Hack.
cheat, downy (cheatgrass)	*Bromus tectorum* L.
chess	*Bromus secalinus* L.
hairy	*B. commutatus* Schrad.
Japanese	*B. japonicus* Houtt.
ripgutgrass	*Bromus diandrus* Roth var. *rigidus* (Roth) Sales
soft	*B. hordeaceus* L.
chickpea (garbanzo)	*Cicer arietinum* L.
chickweed, common	*Stellaria media* (L.) Vill.
chicory (succory)	*Cichorium intybus* L.
Chinese cabbage	*Brassica rapa* L. subsp. *pekinensis* (Lour) Hanelt
cicer milkvetch	*Astragalus cicer* L.
clover	*Trifolium* spp.
alsike	*T. hybridum* L.
arrowleaf	*T. vesiculosum* Savi
ball	*T. nigrescens* Viv.
berseem (Egyptian)	*T. alexandrinum* L.
bigflower (mikes)	*T. michelianum* Savi
buffalo	*T. reflexum* L.
crimson	*T. incarnatum* L.
Egyptian (berseem)	*T. alexandrinum* L.
hop	*T. aurem* Pollich
Hungarian	*T. pannonicum* Jacq.
kura	*T. ambiguum* M. Bieb.
ladino	*T. repens* L.
large hop	*T. campestre* Schreb.
Persian	*T. resupinatum* L.
red	*T. pratense* L.
rose	*T. hirtum* All.
small hop	*T. dubium* Sibth.
strawberry	*T. fragiferum* L.
striate	*T. striatum* L.
subterranean	*T. subterraneum* L.
white (white dutch, common)	*T. repens* L.
zigzag	*T. medium* L.
cocksfoot (orchardgrass)	*Dactylis glomerata* L.
cogongrass (bladygrass)	*Imperata cylindrica* (L.) P. Beauv.
cordgrass	*Spartina* spp.
prairie	*S. pectinata* Bosc ex Link
spike (California)	*S. foliosa* Trin.

(*continued*)

Common Names	Scientific Names
corn (maize)	*Zea mays* L.
cottontop	*Digitaria californica* (Benth.) Henr.
couchgrass (quackgrass)	*Elytrigia repens* (L.) Gould
cowpea	*Vigna unguiculata* (L.) Walp.
crabgrass	*Digitaria* spp.
hairy	*D. sanguinalis* (L.) Scop.
southern	*D. ciliaris* (Retz.) Koel.
creosotebush	*Larrea tridentata* (DC.) Coville
crownvetch	*Securigera varia* (L.) Lassen
curly mesquite	*Hilaria belangeri* (Steud.) Nash
dallisgrass	*Paspalum dilatatum* Poir.
dandelion	*Taraxacum officinale* F.H. Wigg.
darnel	*Lolium temulentum* L.
deertongue	*Dichanthelium* clandestinum (L.) Gould
deervetch, big	*Lotus crassifolius* (Benth.) Greene
desmodium	*Desmodium* spp.
carpon	*D. heterocarpon* (L.) DC.
greenleaf	*D. intortum* (Mill) Urb.
hetero	*D. heterophyllum* (Willd.) DC.
silverleaf	*D. uncinatum* (Jacq.) DC.
digitgrass (fingergrass)	*Digiteria eriantha* Steud.
dock, curly	*Rumex crispus* L.
dodder	*Cuscuta* spp.
dogfennel	*Eupatorium capillifolium* (Lam.) Small
dotted gayfeather	*Liatris punctata* Hook.
dropseed	*Sporobolus* spp.
giant	*S. giganteus* Nash
meadow	*S. compositus* (Poir.) Merr.
prairie	*S. heterolepis* (A. Gray) A. Gray
sand	*S. cryptandrus* (Torr.) A. Gray
spike	*S. contractus* Hitchc.
tall	*S. compositus* (Poir.) Merr.
drunken horse grass	*Achnatherum inebrians* [Hance] Keng
eastern gamagrass	*Tripsacum dactyloides* (L.) L.
elephantgrass (napiergrass)	*Pennisetum purpureum* Schumach.
falcon-pea (flatpod peavine)	*Lathyrus cicera* L.
fescue	*Festuca* spp.
Arizona	*F. arizonica* Vasey
Chewing's	*F. rubra* L. Nyman
Idaho	*F. idahoensis* Elmer
meadow	*F. pratensis* Huds.
red	*F. rubra* L.
tall	*F. arundinacea* Schreb.
sixweeks	*Vulpia octoflora* (Walt.) Rydb.
flaccidgrass	*Pennisetum flaccidum* Griseb.
fodder beet	*Beta vulgaris* L.
flat pea (Wagner pea)	*Lathyrus sylvestris* L.
foxtail	
creeping	*Alopecurus arundinaceus* Poir.
giant	*Setaria faberi* R.A.W. Herrm.
meadow	*Alopecurus pratensis* L.
goldenrod, Missouri	*Solidago missouriensis* Nutt.

Common Names	Scientific Names
grama	*Bouteloua* spp.
black	*B. eriopoda* (Torr.) Torr.
blue	*B. gracilis* (Willd. ex Kunth) Lag. ex Griffiths
hairy	*B. hirsuta* Lag.
sideoats	*B. curtipendula* (Michx.) Torr.
slender	*B. repens* (Kunth) Scribn. & Merr.
grass pea	*Lathyrus sativus* L.
groundsel, common	*Senecio vulgaris* L.
guineagrass	*Megathyrsus maximus* (Jacq.) B. K. Simon & S. W. L. Jacobs
halogeton	*Halogeton glomeratus* (Bieb.) C.A. Mey.
hardinggrass	*Phalaris aquatica* L.
hemlock, poison	*Conium maculatum* L.
hemlock, water	*Cicuta maculata* L.
herdgrass (timothy)	*Phleum pratense* L.
horsenettle	*Solanum carolinense* L.
hound's tongue	*Cynoglossum officinale* L.
indiangrass	*Sorghastrum nutans* (L.) Nash
indigo, hairy	*Indigofera hirsuta* L.
ironweed, western	*Vernonia baldwinii* Torr.
Japanese millet	*Echinochloa frumentacea* Link
Jerusalem artichoke	*Helianthus tuberosus* L.
jimsonweed	*Datura stramonium* L.
johnsongrass	*Sorghum halepense* (L.) Pers.
jointvetch	*Aeschynomene* spp.
American	*A. americana* L.
Australian	*A. falcata* (Poir.) DC.
junegrass (prairie junegrass)	*Koeleria macrantha* (Ledeb.) Schult.
kaimi clover (creeping beggarweed)	*Desmodium incanum* (G. Mey.) DC.
kale	*Brassica oleracea* L.
kikuyugrass	*Pennisetum clandestinum* Hochst. ex Chiov.
kleingrass	*Panicum coloratum* L.
knotgrass	*Paspalum distichum* L.
kochia	*Kochia scoparia* (L.) Schrad.
kohlrabi	*Brassica oleracea var. gongylodes* L.
kudzu	*Pueraria lobata* (Willd.) Ohwi
tropical (puero)	*P. phaseoloides* (Roxb.) Benth.
lablab bean	*Dolichos lablab* (L.) Sweet
lambsquarters, common	*Chenopodium album* L.
lawngrass, Japanese	*Zoysia japonica* Steud.
leadplant	*Amorpha canescens* Pursh
leafy spurge	*Euphorbia esula* L.
lespedeza	
annual; Korean	*Kummerowia stipulacea* (Maxim.) Makino
annual; striate (common)	*Kummerowia striata* (Thunb.) Schindler
bicolored	*Lespedeza bicolor* Turcz.
sericea	*Lespedeza cuneata* (Dum. Cours.) G. Don
leucaena	*Leucaena leucocephala* (Lam.) de Wit
limpograss	*Hemarthria altissima* (Poir.) Stapf & Hubb.
lovegrass	*Eragrostis* spp.
boer	*E. curvula* (Schrad.) Nees
Lehmann	*E. lehmanniana* Nees

(*continued*)

Common Names	Scientific Names
sand	*E. trichodes* (Nutt.) Alph. Wood
weeping	*E. curvula* (Schrad.) Nees
Wilman	*E. superba* Peyr.
lucerne (alfalfa)	*Medicago sativa* L.
lupine	*Lupinus* spp.
blue	*L. angustifolius* Eastw.
Texas (bluebonnet)	*L. subcarnosus* Hook.
white	*L. albus* L.
yellow	*L. luteus* L.
maize; corn	*Zea mays* L.
medic	*Medicago* spp.
barrel	*M. truncatula* Gaertn.
black	*M. lupulina* L.
burr	*M. polymorpha* L.
snail	*M. scutellata* (L.) Mill.
mesquite	*Prosopis* spp.
algarrobo	*P. pallida* (Humb. & Bonpl. ex Willd.) Kunth
honey	*P. glandulosa* Torr.
velvet	*P. velutina* Woot.
milkvetch	
cicer	*Astragalus cicer* L.
ruby	*Oxytropis riparia* Litv.
sicklepod (sickle)	*Astragalus falcatus* Lam.
milkweed, whorled	*Asclepias verticillata* L.
millet	
broom or broomcorn	*Panicum miliaceum* L.
browntop	*Urochloa ramosa* (L.) Nguyen
foxtail (Italian)	*Setaria italica* (L.) P. Beauv.
Japanese	*Echinochloa frumentacea* Link
pearl millet	*Pennisetum glaucum* [L.] R. Br.
proso (hog)	*Panicum miliaceum* L.
miscanthus	
giant	*Miscanthus × giganteus*
Japanese	*Miscanthus sinensis* Andersson
muhly	*Muhlenbergia* spp.
bush	*M. porteri* Scribn. ex Beal
mountain	*M. montana* (Nutt.) Hitch.
sandhill	*M. pungens* Thurb.
spike	*M. wrightii* Vasey ex J. M. Coult.
napiergrass (elephantgrass)	*Pennisetum purpureum* Schumach.
needle-and-thread	*Hesperostipa comata* (Trin. & Rupr.) Barkworth
needlegrass	
California	*Achnatherum occidentale* (Thurb.) Barkworth subsp. *californicum* (Merr. & Burtt Davy) Barkworth
green	*Nassella viridula* (Trin.) Barkworth
needlegrass (needle-and-thread)	*Hesperostipa comata* (Trin. & Rupr.) Barkworth
Texas wintergrass	*Nassella leucotricha* (Trin. & Rupr.) R.W. Pohl
nimblewill	*Muhlenbergia schreberi* J.F. Gmel.
nutsedge, yellow	*Cyperus esculentus* L.
oat	*Avena* spp.
cultivated	*A. sativa* L.
slender	*A. barbata* Pott ex Link
wild	*A. fatua* L.

Common Names	Scientific Names
oatgrass, tall	*Arrhenatherum elatius* (L.) P. Beauv. ex J. S. & C. Presl.
old world bluestems	*Bothriochloa* spp.
orchardgrass (cocksfoot)	*Dactylis glomerata* L.
palisade grass	*Urochloa brizantha* (Hochst. ex A. Rich.) R. D. Webster
pampas grass	*Cortaderia selloana* (Schult. & Schult. f.) Asch. & Graebn.
Pangola-grass	*Digitaria eriantha* Steud.
panicgrass	*Panicum* spp.
blue	*P. antidotale* Retz.
green	*Megathyrsus maximus* (Jacq.) B. K. Simon & S. W. L. Jacobs
paragrass	*Urochloa mutica* (Forssk.) T. Q. Nguyen
partridge pea	*Chamaecrista fasciculata* (Michx.) Greene
paspalum	*Paspalum* spp.
brownseed	*P. plicatulum* Michx.
field	*P. laeve* Michx.
knotgrass	*P. distichum* L.
ribbed	*P. malacophyllum* Trin.
sand	*P. setaceum Michx. var. stramineum* (Nash) D. J. Banks
pea	
field	*Pisum sativum* L.
singletary (rough)	*Lathyrus hirsutus* L.
peanut	*Arachis hypogaea* L.
perennial	*A. glabrata* Benth.
pinto	*A. pintoi* Krapov. & W.C. Greg.
rhizoma	*A. glabrata* Benth.
pearlmillet	*Pennisetum glaucum* (L.) R. Br.
peavine, flatpod	*Lathyrus cicera* L.
pennycress	*Thlaspi arvense* L.
penstemon, shell-leaf	*Penstemon grandiflorus* Nutt.
persimmon	*Diospyros virginiana* L.
phasey bean	*Macroptilium lathyroides* (L.) Urb.
pigeonpea	*Cajanus cajan* (L.) Millsp.
pigweed	*Amaranthus* L.
prostrate	*A. blitoides* S. Watson
redroot	*A. retroflexus* L.
pinegrass	*Calamagrostis rubescens* Buckley
pinto peanut	*Arachis pintoi* Krapov. & W.C. Greg.
plantain, buckhorn	*Plantago lanceolata* L.
porcupinegrass	*Hesperostipa spartea* (Trin.) Barkworth
povertygrass	*Danthonia spicata* (L.) Beauv. ex Roem. & Schult.
prairie clover, purple	*Dalea purpurea* Vent.
prairie grass	*Bromus catharticus* Vahl. var. *catharticus*
purpletop	*Tridens flavus* (L.) Hitchc. var. *flavus*
quackgrass (couchgrass)	*Elytrigia repens* (L.) Gould
quakinggrass	*Briza media* L.
big	*B. maxima* L.
little	*B. minor* L.
ragweed	*Ambrosia* spp.
common	*A. artemisiifolia* L.
lanceleaf	*A. bidentata* Michx.
western	*A. psilostachya* DC.
rape (canola)	*Brassica napus* L. var. *napus*

(*continued*)

Common Names	Scientific Names
rattailgrass (smutgrass)	*Sporobolus indicus* (L.) R. Br. var. *indicus*
redtop	*Agrostis gigantea* Roth
reed canarygrass	*Phalaris arundinacea* L.
rescuegrass	*Bromus catharticus* Vahl.
rhizoma peanut	*Arachis glabrata* Benth.
rhodesgrass	*Chloris gayana* Kunth
Ricegrass	
Indian	*Achnatherum hymenoides* (Roem. & Schult.) Barkworth
mandan	X *Achnella caduca* (Beal) Barkworth
ripgutgrass	*Bromus diandrus* Roth var. *rigidus* (Roth) Sales
rivergrass, common	*Scolochloa festucacea* (Willd.) Link
rose, multiflora	*Rosa multiflora* Thunb.
roughpea (caleypea; singletarypea)	*Lathyrus hirsutus* L.
rushes	*Juncus* spp.; *Eleocharis* spp.
rushgrass, longleaf (tall dropseed)	*Sporobolus compositus* (Poir.) Merr.
ruzigrass	*Urochloa ruziziensis* (R. Germ. & C. M. Evrard) Crins
rye	*Secale cereale* L.
ryegrass	*Lolium* spp.
annual (Italian)	*L. multiflorum* Lam.
Italian (annual)	*L. multiflorum* Lam.
perennial	*L. perenne* L.
sacaton	*Sporobolus* spp.
alkali	*S. airoides* (Torr.) Torr.
big	*S. wrightii* Munro ex. Scribn.
sagebrush	*Artemisia* spp.
big	*A. tridentata* Nutt.
black	*A. nova* A. Nelson
sand	*A. filifolia* Torr.
threetip	*A. tripartita* Rydb.
sainfoin	*Onobrychis* spp.
common	*O. viciifolia* Scop.
Russian	*O. transcaucasica* Grossh.
Siberian	*O. arenaria* (Kit.) DC.
St. Augustine grass	*Stenotaphrum secundatum* (Walter) Kuntze
saltbush	*Atriplex* spp.
saltgrass, inland	*Distichlis spicata* (L.) Greene var. *stricta* (Torr.) Scribn.
sandreed	*Calamovilfa* spp.
big	*C. gigantea* (Nutt.) Scribn. & Merr.
prairie	*C. longifolia* (Hook.) Scribn.
scrobicgrass, creeping paspalum	*Paspalum scrobiculatum* L.
sedge, threadleaf	*Carex filifolia* Nutt.
sesbania	*Sesbania sesban* (L.) Merr.
hemp sesbania	*S. exaltata* (Raf.) Rybd.
setaria (golden timothy)	*Setaria sphacelata* (Schum.) Stapf & C.E. Hubb.
shadscale (shadscale saltbush)	*Atriplex confertifolia* (Torr. & Frem.) S. Wats.
shepherd's purse	*Capsella bursa-pastoris* (L.) Medik.
signalgrass	*Urochloa decumbens* (Stapf) R.D. Webster
creeping (koroniviagrass)	*U. humidicola* (Rendle) Morrone & Zuloaga
siratro	*Macroptilium atropurpureum* (DC.) Urb.
sleepygrass	*Achnatherum robustum* (Vasey) Barkworth
sloughgrass, American	*Beckmannia syzigachne* (Steud.) Fernald
smartweed, Pennsylvania	*Persicaria pensylvanica* (L.) M. Gómez
sneezeweed, bitter	*Helenium amarum* (Raf.) H. Rock

Common Names	Scientific Names
snow-on-the-mountain	*Euphorbia marginata* Pursh
sorghum	*Sorghum bicolor* (L.) Moench
sorghum × sudangrass hybrid	*Sorghum bicolor* (L.) Moench
sourclover	*Melilotus indicus* (L.) All.
soybean	*Glycine max* (L.) Merr.
soybean, perennial (glycine)	*Neonotonia wightii* (Wight & Arn.) J.A. Lackey
speargrass	*Heteropogon contortus* (L.) Beauv. ex Roem. & Schult.
sprangletop, green	*Disakisperma dubia* (Kunth) P. M. Peterson & N. Snow
squirreltail	*Elymus elymoides* (Raf.) Swezey
stargrass	*Cynodon nlemfuensis* Vanderyst
giant	*C. aethiopicus* Clayton & J.R. Harlan
Naivasha	*C. plectostachyus* (K. Schum.) Pilg.
stylo	*Stylosanthes guianensis* (Aubl.) Sw.
Caribbean	*S. hamata* (L.) Taub.
shrubby	*S. scabra* Vogel
Townsville	*S. humilis* Kunth
sudangrass	*Sorghum bicolor* (L.) Moench
sugarbeet	*Beta vulgaris* L.
sugarcane	*Saccharum officinarum* L.
sulla (sulla sweetvetch)	*Hedysarum coronarium* L.
sumac	*Rhus* spp.
sunflower	*Helianthus annuus* L.
plains	*H. petiolaris* Nutt.
stiff	*H. pauciflorus* Nutt.
swede	*Brassica napus subsp. Rapifera* Metzg.
sweetclover	*Melilotus* spp.
annual yellow	*M. indicus* (L.) All.
white	*M. albus* Medik.
yellow	*M. officinalis* (L.) Lam.
switchgrass	*Panicum virgatum* L.
tanglehead	*Heteropogon contortus* (L.) P. Beauv. ex Roem. & Schult.
tansymustard	*Descurainia pinnata* (Walter) Britton
tarweed	*Amsinckia intermedia* Fisch. & C. A. Mey.
teosinte	*Zea mays* subsp. *mexicana* (Schrad.) H.H. Iltis
thistle	
Canada	*Cirsium arvense* (L.) Scop.
musk	*Carduus nutans* L.
three-awn	*Aristida* L.
pineland	*A. stricta* Michx.
purple	*A. purpurea* Nutt.
red	*Aristida purpurea* Nutt. var. *longiseta* (Steud.) Vasey
tickclover, tall	*Desmodium tortuosum* (Sw.) DC.
ticklegrass (winter bentgrass)	*Agrostis hyemalis* (Walter) Britton et al.
timothy	*Phleum pratense* L.
alpine	*P. alpinum* L.
turf	*P. bertolonii* DC.
torpedograss	*Panicum repens* L.
trefoil	*Lotus* spp.
big	*L. uliginosus* Schkuhr.
birdsfoot	*L. corniculatus* L.
narrowleaf birdsfoot	*Lotus tenuis* Waldst. & Kit. ex Willd.

(*continued*)

Common Names	Scientific Names
triticale	*X Triticosecale* spp.
tunisgrass	*Sorghum bicolor* (L.) Moench subsp. *verticilliflorum* (Steud.) de Wet ex Wiersema & J. Dahlb.
turkeyfoot (big bluestem)	*Andropogon gerardii* Vitman
turnip	*Brassica rapa* L. subsp. *rapa*
vaseygrass	*Paspalum urvillei* Steud.
veldtgrass	*Ehrharta calycina* Sm.
velvetbean, Florida	*Mucuna pruriens* (L.) DC.
velvetgrass	*Holcus* spp.
common	*H. lanatus* L.
German	*H. mollis* L.
velvetleaf	*Abutilon theophrasti* Medik.
vernalgrass, sweet	*Anthoxanthum odoratum* L.
vervain, hoary	*Verbena stricta* Vent.
vetch	*Vicia* spp.
bird (cow)	*V. cracca* L.
bitter	*V. ervilia* (L.) Willd.
common	*V. sativa* L.
cordateleaf, common	*V. sativa* subsp. *cordata* (Wulfen ex Hoppe) Batt.
grandiflora (bigflower)	*V. grandiflora* Scop.
hairy	*V. villosa* Roth
Hungarian	*V. pannonica* Crantz
narrowleaf	*V. sativa L.* subsp. *nigra (L.)* Ehrh
purple	*V. benghalensis* L.
single-flowered	*V. articulata* Hornem.
winter	*V. villosa* Roth
vine mesquitegrass	*Hopia obtusa* (Kunth) Zuloaga & Morrone
wheat	*Triticum aestivum* L.
durum	*T. turgidum* L.
goatgrass	*Aegilops tauschii* Coss.
wheat-grass	
arctic	*Elymus macrourus* (Turcz. ex Steud.) Tzvelev
bearded	*Elymus caninus* (L.) L.
bluebunch	*Pseudoroegneria spicata* (Pursh) Á. Löve
crested	*Agropyron cristatum* (L.) Gaertn.
intermediate	*Thinopyrum intermedium* (Host) *Barkworth & D.R. Dewey* subsp. *intermedium*
pubescent	*Thinopyrum intermedium* (Host) Barkworth & D.R. Dewey subsp. *barbulatum* (Schur) Barkworth & D. R. Dewey
Siberian	*Agropyron fragile* (Roth) P. Candargy
slender	*Elymus trachycaulus* (Link) Gould ex Shinners
standard crested	*Agropyron desertorum* (Fisch. ex Link) Schult.
tall	*Thinopyrum ponticum* (Podp.) Barkworth & D.R. Dewey
thickspike	*Elymus lanceolatus* (Scribn. & J.G. Smith) Gould
western	*Pascopyrum smithii* (Rydb.) Barkworth & D. R. Dewey
white cockle	*Silene latifolia* Poir.
wildbean, trailing	*Strophostyles helvola* (L.) Elliott
wildrice, cultivated	*Zizania palustris* L.
wildrye	
altai	*Leymus angustus* (Trin.) Pilg.
beardless	*Leymus triticoides* (Buckley) Pilg.
blue	*Elymus glaucus* Buckley
Canada	*Elymus canadensis* L.

Common Names	Scientific Names
dune	*Leymus mollis* (Trin.) Pilg.
giant	*Leymus condensatus* (J. Presl) Á. Löve
great basin	*Leymus cinereus* (Scribn. & Merr.) Á. Löve
Russian	*Psathyrostachys juncea* (Fisch.) Nevski
Siberian	*Elymus sibiricus* L.
willow	*Salix* spp.
winterfat	*Krascheninnikovia lanata* (Pursh) A. Meeuse & A. Smit
yellow rocket	*Barbarea vulgaris* W. T. Aiton
yorkshire fog	*Holcus lanatus* L.
zoysia	*Zoysia japonica* Steud.

References

Bailey LH, and Hortorium Staff. 1976. Hortus Third: A Concise Dictionary of Plants Cultivated in the United States and Canada. Macmillan, New York.

Barkworth, ME, and DR Dewey. 1985. Genomically based genera of the perennial Triticeae of North America: Identification and membership. Amer. J. Bot. 72:767–776.

Bayer, AG, Agrichemicals Division. 1986. Important Crops of the World and Their Weeds. Bayer AG, Leverkusen, Germany.

Cronquist, A. 1981. An Integrated System of Classification of Flowering Plants. Columbia University Press, New York.

Encke, F, G Buchheim, and S Seybold. 1984. Zander-Handworterbuch der Pflanzennamen, 13th ed. Eugen Ulmer, Stuttgart, Germany.

Hitchcock, AS (revised by A Chase). 1951. Manual of the Grasses of the United States. U.S. Department of Agriculture Miscellaneous Publication 200. US Government Printing Office, Washington, DC.

Patterson, DT (chairman). 1984. Composite list of weeds. Weed Sci. 32 (Suppl. 2):1–137.

Terrell, EE, SR Hill, JH Wiersema, and WE Rice. 1986. A Checklist of Names of 3000 Vascular Plants of Economic Importance. US Department of Agriculture, Washington, DC.

Voss, EG, et al. (eds.). 1983. International Code of Botanical Nomenclature. Regnum Vegetabile 111. Bohn, Scheltema, and Holkema, Utrecht, Netherlands.

Weed Science Society of America. 1989. Composite List of Weeds. Weed Science Society of America, Champaign, IL.

Websites

Germplasm Resources Information Network (GRIN) Taxonomy for Plants. (www.ars-grin.gov/cgi-bin/npgs/html/tax_search.pl). National Genetic Resources Program, USDA-ARS, Beltsville, MD.

PLANTS Database, Version 3.1 (http://plants.usda.gov/java/). National Plant Data Team, USDA-NRCS, Greensboro, NC.

Glossary

abiotic—The non-living components of the environment, such as water, solar radiation, oxygen, organic compounds, and soil nutrients.

abomasum—The fourth compartment of the ruminant stomach, comprising the true stomach, where digestive processes similar to those found in the non-ruminant stomach occur.

absorption—The movement of nutrients from the digestive tract into the bloodstream for transport to tissues and organs.

achene—A dry, one-seeded indehiscent fruit.

acid detergent fiber (ADF)—The insoluble residue that remains following extraction of herbage with acid detergent (van Soest); it consists of the cell wall constituents minus hemicellulose.

acid detergent lignin (ADL)—The insoluble residue that remains after a forage sample has been extracted with a dilute acid detergent followed by treatment with a strong acid (72% H_2SO_4).

action threshold—The number of pests present or the amount of damage that must have occurred before action to control the pests is required.

***ad libitum* intake**—Consumption by an animal of a feed or forage offered in excess of what the animal can consume.

adventitious buds—Buds that arise occasionally from the stem at non-nodal positions (i.e., not from an axillary bud), or more frequently from root tissue. Random cells in the endodermis or pericycle of the root are stimulated to divide and differentiate into a normal bud that pushes outward and upward by forming nodes and internodes to extend above the soil.

adventitious roots—Roots that emerge from nodes at the base of vertical tillers and the nodes of rhizomes and stolons. They become the dominant root system for established grasses.

aerobic—Living in the presence of free oxygen; the opposite of anaerobic.

aerobic respiration—Energy-producing biochemical reactions in cells that utilize O_2 to oxidize carbohydrates and lipids to produce ATP, carbon dioxide, and water.

agroforestry—A land-use system in which tree crops are combined with agricultural crops including forage crops, and/or animal production.

allelomimicry—Mimicking of another (e.g., a foal mimicking its dam's grazing behavior).

anaerobic—Living in the absence of free oxygen; the opposite of aerobic.

anaerobic respiration—Respiration in the absence of oxygen, which is less efficient than aerobic respiration in terms of ATP production.

anemia—A condition of animals that is characterized by a lack of hemoglobin or a deficiency of red blood cells, which limits oxygen supply to body tissues.

animal performance—Production per animal (weight change or animal products) per unit of time.

animal unit—One mature non-lactating cow weighing 1100 lb (500 kg) and fed at maintenance level or the equivalent, expressed as $(\text{weight})^{0.75}$, for comparison with other kinds or classes of animals.

annual—A plant that completes its life cycle within less than 1 year. Summer annuals germinate in the spring, produce seed in summer or fall, and then die. Winter annuals germinate in the fall, overwinter, and grow and produce seed the following spring or summer.

antagonistic—Relating to a biological association between two species in which one organism benefits from the relationship and the other organism is negatively affected.

anthesis—The period when a flower (e.g., the lemma and palea in grasses) is open, the anthers and stigma are

Forages: An Introduction to Grassland Agriculture, Seventh Edition. Edited by Michael Collins, C. Jerry Nelson, Kenneth J. Moore and Robert F Barnes.

mature, and pollen is shed. In self-pollinated plants, this occurs before flower opening.

anti-quality factors—Chemical compounds that have negative effects on forage intake or that have adverse effects on animals consuming the forage.

apomictic—Reproducing by apomixis.

apomixis—Asexual production of seeds in the normal area of sexual production (floret and spikelet), which does not involve the fusion of male and female gametes.

asphyxiation—A state of unconsciousness caused by a lack of oxygen.

ATP (adenosine triphosphate)—A compound that contains chemical energy which can be used within cells. ATP is synthesized during respiration of carbohydrates and lipids.

auricle—An ear-like projection at the base of the grass leaf blade.

autotoxicity—A specific type of allelopathy in which the presence of adult plants interferes with the germination and development of seedlings of the same species.

autotroph—A plant that is able to synthesize its own organic food supply, especially by photosynthesis.

awn—A bristle-like structure originating from the lemma or glume.

axillary bud—The meristematic apex located at the junction of the leaf and stem. It gives rise to tillers in grasses, and to branches and flowers in dicots.

bacteroid—A nitrogen-fixing organelle derived from rhizobia bacteria residing in the root nodules of host legume plants.

biennial—A plant that completes its life cycle in 2 years. A true biennial grows vegetatively during the first growing season, and reproduces and dies during the second growing season.

biofuel—A fuel that is produced from biological material of recent origin.

biomass—The weight of living organisms (plants and animals) in an ecosystem at a given point in time, expressed as either fresh or dry weight.

biome—An ecological region that is often defined according to its predominant vegetation, such as grassland, temperate deciduous forest, or desert.

biorefinery—A facility for refining biofeedstocks into fuels and other industrial products.

biotic—Relating to living components of the environment, such as higher plants, algae, microorganisms, nematodes, worms, insects, birds, and mammals.

bite mass—The amount of forage contained in a single bite by a grazing animal.

blade—The flat, expanded part of a leaf above the sheath or petiole; the major photosynthetic organ.

bloat—Excessive accumulation of gases in the rumen of animals because normal escape through the esophagus is impaired, causing distension of the rumen.

bloom—The developmental stage of maximum flowering within an individual plant or plant community; it is often specified by a fraction representing the number of reproductive stems in flower, or by designations of early, mid, or full bloom.

body weight—(also called live weight) The gross weight of an animal before it has been shrunk or slaughtered.

bolus—A wad of herbage that accumulates in the mouth of a ruminant from a number of bites in preparation for swallowing.

boot stage—The stage of maturity of a grass tiller when the inflorescence is in the sheath of the flag leaf (the last leaf produced by the tiller).

bound water—Water that is incapable of forming ice crystals because it is held tightly to cellular constituents.

bovine—Relating to cattle.

brace root—A root that originates from a node above the ground but that penetrates the soil.

bract—A leaf-like structure that subtends a flower or an inflorescence.

browse—(1) *noun* Leaf and twig growth of shrubs, woody vines, trees, cacti, and other non-herbaceous vegetation available for animal consumption. (2) *verb* To consume woody vegetation.

buffer stocking—The practice of using temporary fencing to adjust the pasture area available to animals.

buffering capacity—The relative ability of a solution or material to resist change in pH.

bundle sheath—The one layer (in C_4 plants) or two layers (in C_3 plants) of cells that surround a vascular bundle in a leaf. In C_4 plants, the bundle sheath contains chloroplasts in which CO_2 is reduced to carbohydrate.

bypass protein—Dietary protein that passes from the rumen to the abomasum without being degraded by rumen microorganisms.

carbon (C) sequestration—The net removal of CO_2 from the atmosphere into long-lived pools of carbon in terrestrial ecosystems. The pools can be living, above-ground biomass (e.g., trees), products with a long, useful life created from biomass (e.g., lumber), living biomass in soils (e.g., roots, microorganisms), or recalcitrant organic and inorganic carbon in soils and deeper subsurface environments.

C_3 plant—A plant in which ribulose-1,5-bisphosphate carboxylase is the primary CO_2-capturing enzyme, with the first product being a 3-carbon acid. C_3 plants exhibit photorespiration.

C_4 plant—A plant in which phosphoenolpyruvate (PEP) carboxylase is the primary CO_2-capturing enzyme, with the first product being a 4-carbon acid. C_4 plants do not exhibit photorespiration.

canopy—The aerial portion of plants in their natural growth position. It is usually expressed as the percentage of ground occupied, or as leaf area index.

canopy morphology See **canopy structure**.

canopy structure—The spatial (three-dimensional) physical arrangement of leaves and stems of different species that make up a pasture sward.

capsule—A dry, dehiscent fruit containing two or more seeds.

carbohydrate—A compound composed of carbon, hydrogen, and oxygen in the ratio of CH_2O (e.g., sugar, starch, cellulose).

carnivore—An organism that consumes animal tissues as its main diet.

carrying capacity—(also called grazing capacity) The maximum stocking rate (i.e., number of animals per acre) that will achieve a target level of animal performance using a specified stocking method, and that can be applied over a defined period of time without deterioration of the ecosystem. Carrying capacity does not remain static from season to season or from year to year, and may be defined over fractional parts of years. *Average carrying capacity* is a long-term carrying capacity averaged over years. *Annual carrying capacity* refers to a specific year.

caryopsis—The single-seeded fruit or bare seed of a grass, borne between the lemma and the palea.

cataphyll—The reduced, often scaly leaf structure located at each node on a rhizome.

cation-exchange capacity (CEC)—The weak electrostatic charge of soil particles resulting from loss of H^+ ions, which attracts soil cations, holding them in a plant-available form.

caudex—An underground stem base of a herbaceous plant that is usually woody and from which new branches can arise.

cecum—The intestinal pouch located at the junction of the large and small intestines of non-ruminants, which functions somewhat like a rumen. Usually it is much larger in the herbivorous horse than in non-herbivorous monogastric animals.

cell wall—The rigid wall of plant cells, which is composed primarily of cellulose, hemicellulose, pectin, and lignin.

cell wall constituents—The compounds of which plant cell walls are composed, including cellulose, hemicellulose, lignin, pectin, and minerals (ash).

cellulose—A carbohydrate formed from glucose that is linked by beta-1,4 bonds, which is a major constituent of plant cell walls. It is a colorless solid that is insoluble in water.

chaparral—An area of grassland in a semiarid region that is characterized by a mixture of woody shrubs, scrub trees, and low-growing herbaceous species, mainly grasses.

chilling injury—Damage to plants at low temperatures in the absence of freezing. It commonly occurs in plants of tropical or subtropical origin at temperatures below 50°F. The cause of chilling injury is usually a change in viscosity of the lipids in membranes.

chloroplast—The cellular organelle in which photosynthesis occurs.

climate—A characteristic condition or pattern of the various elements of weather for a given geographic area or region of the earth.

clostridia—Gram-positive, spore-forming, anaerobic bacteria of the genus *Clostridium* that typically carry out butyric fermentation in silage, which has detrimental effects on silage quality.

cold resistance—The ability of plants to resist cold temperature stress (temperatures below 32°F). It is often manifested as cold avoidance or cold tolerance.

cold tolerance—The ability of plants to tolerate stresses when exposed to temperatures below 32°F.

coleoptilar node—The node above the seed on a developing grass seedling where the coleoptile is attached. It forms a large number of adventitious roots.

coleoptile—The specialized leaf consisting of a modified sheath that is attached to the coleoptilar node. It elongates through the seed coat of grasses and protects the shoot and its tender leaves as they are pushed through the soil to emerge above ground.

coleorhiza—The protective sheath around the radicle within the seed. The radicle grows through the coleorhiza during germination, after which the coleorhiza seals around the root, reducing pathogen entry to the seed.

collar—The area at the junction of the grass leaf blade and sheath.

colon—The portion of the large intestine between the cecum and the rectum.

companion crop—(also called nurse crop) A fast-growing annual crop, often a cereal crop such as oat, that is planted in spring with a small-seeded forage crop to help to control weeds and soil erosion while the forage seedlings germinate and develop. The companion crop is either harvested early for forage or left to grow to maturity.

compensatory gain—The period of increased growth rate that occurs when nutrient intake returns to normal following a period of restricted intake. The increase in growth rate recovers some, but not all, of the lost gain.

competition—The mutually adverse effects on plants that are utilizing a resource which is in short supply.

composite—A very large family of dicotyledonous plants that are characterized by a head-type inflorescence with many flowers borne on a large receptacle (e.g., dandelion, sunflower, aster).

conditioning—Mechanical or chemical treatments designed to reduce restrictions to moisture loss to the atmosphere from cut plants.

continuous stocking—A method of grazing livestock on a given unit of land in which animals have unrestricted and uninterrupted access throughout the time period when grazing is allowed.

cool-season—Relating to plant species that grow best during cool, moist periods of the year. They commonly have temperature optima in the range of 59–77°F and exhibit C_3 photosynthesis.

cool-season grass—Any grass species that grows best during cool, moist periods of the year. They commonly have temperature optima in the range of 59–77°F and exhibit C_3 photosynthesis.

coprophagy—Eating of dung.

corm—A solid, swollen underground stem base that functions as a storage tissue.

corrosiveness—The ability of a substance to cause gradual breakdown or destruction by chemical means.

cotyledon—The specialized seed leaf in the embryo that serves as a storage organ in dicots, and in nutrient absorption in monocots such as grasses. There are two cotyledons in dicots and only one in monocots. See also **scutellum**.

coumarin—An anti-quality component of sweetclover; it is a white crystalline compound with a vanilla-like odor that gives sweetclover its characteristic scent. It is now used as a rodent poison and in medicine.

cover crop—A crop, usually a grass or legume, that is grown in waterways or between annual crops in summer or winter primarily to manage soil erosion, suppress weeds, improve soil fertility and quality, and control diseases and insect pests.

creep stocking—The practice of allowing juvenile animals to graze areas that their dams cannot access at the same time.

critical leaf area index—The leaf area index at which 95% of light is intercepted by the canopy. See also **leaf area index**.

crop residue—The portion of plants that remains after seed harvest. The term is mainly used with reference to grain crops such as corn stover or small-grain straw and stubble.

cropland—Land that is used for the production of cultivated crops.

cropland pasture—Cropland on which grazing occurs but is generally of limited duration.

crown—The persistent base of a herbaceous perennial plant, located between the soil surface and mowing/grazing height. It consists of stem bases, including tiller and rhizome buds.

crude fiber—The coarse, fibrous constituents of plants, such as cellulose, that are partially digestible and have a relatively low nutritional value. In chemical analysis, it is the residue obtained after boiling plant material with dilute acid and then with dilute alkali.

crude protein—An estimate of the protein concentration in a feed or forage that is based on total nitrogen (N) concentration. It is generally calculated as total N × 6.25, because plant proteins are assumed to contain around 16% N.

culm—The elongated, jointed stem of grasses, generally terminating in a seed head.

cultivar—(1) A variety, strain, or race that has originated and persisted under cultivation or that was specifically developed for the purpose of cultivation. (2) For cultivated plants, the equivalent of botanical variety.

cuticle—A waxy layer secreted by epidermal cells on the outer surface of plants.

cutin—A waxy, somewhat waterproof material that provides an outer covering on plants.

cyathium—An inflorescence in which a single female flower and several male flowers are enclosed within a cluster of modified leaves.

cyme—A branched, relatively flat-topped inflorescence in which the central flower opens first.

cytosol—The liquid matrix of the cell protoplasm in which organelles, proteins, salts, and sugars are suspended.

dam—The female parent of an animal.

damping-off—A seedling disease characterized by necrosis of the stem or hypocotyl tissue at the soil surface, often occurring in poorly drained soils. It is caused by several different fungal organisms.

day-neutral plant—A plant that flowers independently of photoperiod. Day-neutral plants often flower at a particular developmental stage.

dehiscence—The natural release of seed as a result of the opening or splitting of a seedpod or other seed-containing structure, generally as a result of drying.

denitrification—The reduction of NO_3^- to N_2O, NO_2, and N_2, which are lost to the atmosphere.

dental pad—Cornified gum tissues of the upper jaw that substitute for upper incisors in ruminants.

desertification—The ecological process by which an ecosystem becomes desert-like and degrades to a more xeric, less productive condition.

desiccant—A drying agent.

determinate—Relating to a plant growth habit in which vegetative growth is of fixed duration. When a stem terminates in a floral bud, its growth is determinate.

diecious—Having male (staminate) and female (pistillate) flowers on separate plants.

diet learning—Learning processes in animals that lead to the selection of familiar feedstuffs and refusal of toxic ones for ingestion.

diet quality—The forage quality of the dry matter consumed by an animal, generally in a grazing context, as distinct from the quality of the total available forage.

diet selection—Expression of diet learning during feeding.

digesta—The contents of the digestive tract of animals, consisting mainly of non-digested and partially digested food.

digestibility—The proportion of dry matter or constituent digested within the digestive tract of an animal.

digestible—Capable of being digested by the gastrointestinal system of an animal.

digestible energy (DE)—Feed-intake gross energy minus fecal energy, expressed as calories per unit feed dry matter consumed.

digestion—The conversion of complex, generally insoluble foods to simple substances that are soluble in water.

digestion rate—The speed at which digestion processes proceed.

distal esophageal sphincter—A constriction in the esophagus separating the esophagus and reticulum that regulates reflux.

diurnal—Relating to the change in a process during a 24-hour period, mainly due to the differences between day and night.

dormancy—A period of arrested growth and development due to physical or physiological factors.

drought avoidance—The ability of a plant to avoid exposure to water-deficit stress. It often involves increased rooting or reduced transpiration.

drought resistance—The ability of a plant to resist injury caused by drought stress. It often involves drought avoidance or drought tolerance.

drought tolerance—The ability of a plant to tolerate water-deficit stress when it occurs.

drying rate—The speed with which moisture moves from cut plants to the atmosphere.

duodenum—The first part of the small intestine.

economic injury level (EIL)—The number of pests at which the cost of controlling them becomes less than the loss of value caused by the plant damage that they cause.

economic threshold—The number of pests at which action to control them is required to prevent economic loss.

ecosystem—A living community and all the factors in its non-living environment. An ecosystem can be natural or managed for a purpose, such as forest, grassland, or agricultural crops.

ecosystem services—The human benefits derived from the functioning of an ecosystem. They include production (e.g., of food and water), regulation (e.g., control of pests), supporting (e.g., nutrient cycles), and cultural (e.g., recreational) benefits.

effluent—The liquid, which contains some nutrients and other solids, that is lost from silage.

embryo—A young plant that exists in an arrested state of development within a seed.

embryo axis—The growing parts of the seed, including the radicle, coleorhiza, coleoptilar node, and epicotyl.

endosperm—In seeds plants, the nutritive tissue formed within the embryo sac by the union of a male nucleus with two polar nuclei of the female.

enzyme—A protein that catalyzes one or more biochemical reactions within a living cell.

epicotyl—The region of the embryonic axis that is located between the cotyledonary node and the shoot apical meristem.

epidermis—The tissue composed of parenchyma cells that makes up the outer covering of plants.

epigeal germination—A type of germination in which the cotyledons are raised above the ground by elongation of the hypocotyl.

equid—Any animal belonging to the family Equidae. The term is often used to refer to a horse.

equilibrium moisture—A hay-curing term that describes the moisture concentration at which no further exchange of water occurs between the drying crop and the atmosphere.

eructation—The belching of fermentation gases in the rumen through the esophagus and mouth.

escape protein See **bypass protein.**

esophagus—The muscular tube that conveys food from the mouth of an animal to its digestive system.

ether extract—Fats, waxes, oils, and similar plant components extracted with warm ether in chemical analysis.

evapotranspiration—The water loss from a soil to the aerial environment due to the combination of direct evaporation from the soil and transpirational loss through the plants that are rooted in that soil.

extensive grazing management—Grazing management that utilizes relatively large land areas per animal, and a relatively low level of labor, resources, and capital.

extravaginal tiller—A lateral tiller that penetrates the leaf sheaths of the parent tiller.

Fabaceae—A plant family, commonly referred to as the legumes, that includes peas, beans, and many important forage species that form symbiotic associations with nitrogen-fixing bacteria.

feeding deterrents—Chemical and physical barriers to ingestion.

fermentation—Anaerobic chemical conversion of sugars to carbon dioxide and alcohol, induced by the activity of enzyme systems of microorganisms (e.g., yeast).

fertilizer—Mineral or organic source of nutrients applied to soil to increase its fertility.

festucoid—Relating to a type of grass seedling development in which there is a long coleoptile and generally little or no subcoleoptile internode elongation. It is characteristic of grasses belonging to the subfamily Festucoideae.

fiber—A nutritional entity that is relatively resistant to digestion and is slowly and only partially degraded by herbivores. Forage fiber is composed of structural polysaccharides, cell wall proteins, and lignin.

fibrous—Containing high concentrations of fiber.

first–last stocking—A method of utilizing two or more groups of animals, usually with different nutritional requirements, to graze sequentially on the same land area. First grazers can be more selective and consume a higher-quality diet.

flag leaf—The topmost or last leaf on a reproductive grass tiller (culm).

flagging—The bending of the distal part of a grass leaf blade, causing it to have a flag-like appearance.

flooding tolerance—The ability of a plant to tolerate injury associated with anoxia and other stresses caused by lack of soil air and by water inundation.

floret—The grass flower, consisting of the lemma and palea with included stamens and pistil (later including the caryopsis).

flowering—The physiological stage of a grass plant at which anthesis (blooming) occurs, or the stage at which flowers are visible in non-grass plants.

follicle—A dry dehiscent fruit that splits along one side.

forage—(1) *noun* The edible parts of plants, other than separated grain, that can provide feed for grazing animals or that can be harvested for feeding, including browse, herbage, and mast. Generally, the term refers to more digestible material (e.g., pasturage, hay, silage, green chop), in contrast to less digestible plant material, which is known as roughage. (2) *verb* (Of animals) to search for or consume forage. See also **browse, graze.**

forage allowance—The relationship between the weight of forage dry matter per unit area and the number of animal units or forage intake units at any one point in time; a forage-to-animal relationship. Forage allowance is the inverse of grazing pressure.

forage mass—The total dry weight of forage per unit area of land, usually above ground level and at a defined reference level.

forage nutritive value See **nutritive value.**

forage quality—The characteristics that make forage valuable as a source of nutrients to animals; the combination of chemical and biological characteristics of forage that determines its potential to produce meat, milk, wool, or physical work. Some authors consider forage quality to be synonymous with feeding value and nutritive value.

forage yield—The aggregate of products resulting from the growth of a crop. Forage yield is the total mass of forage produced per unit of land area over a period of time.

forb—Any herbaceous broadleaf plant. When the term is used to describe cultivated forages, legumes are often considered to be a separate class; in this context, "forb" refers to non-leguminous herbaceous broadleaf plants.

forest range—A forest ecosystem that produces, at least periodically, an understory of natural herbaceous or shrubby plants that can be grazed or browsed.

forestland—Land on which the vegetation is dominated by forest or, if trees are lacking, the land bears evidence of former forest and has not been converted to other vegetation.

free water—Tissue water that is not associated with cellular constituents and which can freeze when temperatures fall below 32°F.

frontal stocking—A stocking method in which forage within a land area is allocated by periodically moving a sliding fence such that livestock can advance to gain access to new increments of ungrazed forage.

fructan—A carbohydrate polymer of fructose that includes one glucose molecule. Synthesized from sucrose, this polymer is water soluble, is stored in vacuoles, and mainly functions as a storage carbohydrate.

functional richness—The diversity of functional traits conferred by species inhabiting an ecosystem.

gasification—Thermochemical decomposition of biological feedstocks under limited oxygen that produces combustible gases.

gastric lysozymes—Enzymes secreted by the stomach that degrade bacterial cell walls.

genera—Plural of **genus.**

genetic engineering—The transfer of selected genes between species and genera using a vector, without the involvement of gametes.

genotype—The genetic make-up of an individual or group.

genus—A taxonomic category that designates a closely related and definable group of plants, including one or more species. The name of the genus becomes the first word of the binomial employed in scientific literature.

germination—Resumption of active growth of a seed, which results in rupture of the seed coat and emergence of the radicle.

gestation—In mammals, the time interval between conception and birth, during which the offspring develops.

glandular stomach—A stomach that secretes enzymes and acids from its glands (e.g., the abomasum of ruminants).

glumes—The pair of bracts subtending the spikelet.

glyphosate—The active ingredient in a broad-spectrum herbicide that kills most non-woody plants when sprayed on leaves and stems. It was first marketed by Monsanto as Roundup®.

grana—Stacks of thylakoid membranes in the stroma of chloroplasts.

grass—Any member of the plant family Poaceae.

grass tetany—(also called hypomagnesemic tetany) A livestock disorder caused by low blood levels of magnesium.

grassland—(1) Land on which the vegetation is dominated by grasses. (2) Any plant community in which the dominant vegetation is composed of grasses and/or legumes.

grassland agriculture—A farming system that emphasizes the importance of grasses, legumes, and other forages in livestock and land management.

graze—(of animals) To consume forage *in situ.* See also **browse, forage.**

grazed horizon—The layer of a sward that is removed in a grazing event. It is the depth from the top of the plant canopy that grazing animals can consume with one biting motion.

grazer—An animal that grazes standing herbage.

grazing cycle—The time that elapses between the beginning of one grazing period and the beginning of the next grazing period in the same paddock where the forage is regularly grazed and rested.

grazing management—The manipulation of animal grazing in pursuit of a defined management objective.

grazing management unit—The grazing land used to support a group of grazing animals for a grazing season. It may be a single area or it may have a number of subdivisions.

grazing pressure—The relationship between the number of animal units or forage intake units and the weight of forage dry matter per unit area at a given time; an animal-to-forage relationship. Grazing pressure is the inverse of forage allowance.

grazing strategy—The management of a grazing system in order to achieve specific production and other goals.

grazing system—A defined, integrated combination of animal, plant, soil, and other environmental components and the grazing method(s) by which the system is managed in order to achieve specific results or goals.

green manure—Plantings of annual crops, especially legumes, that are grown for the purpose of producing biomass that can be tilled into the soil to increase soil organic matter and fertility status prior to planting the desired economic crop or forage.

greenhouse gas—A gas that absorbs heat radiated from the earth and thus contributes to the warming of the atmosphere.

growing degree day—Accumulation of daily heat units above a baseline at which growth is near zero. The baseline is usually 32°F for cool-season grasses, 41°F for alfalfa, and 50°F for corn. Daily units are calculated by summing the average of the maximum and minimum daily temperatures minus the baseline over time. For example, $(70°F + 46°F)/2 = 58°F - 41°F = 17$ growing degrees for that day for alfalfa.

growth respiration—The portion of aerobic respiration that is used for growth processes such as the synthesis of cell walls.

gut fill—The volume of a forage that equals the fill capacity of an animal's digestive system.

hard seed—Seeds that remain dormant under conditions suitable for germination. This is often associated with a seed coat that is impervious to moisture.

hay—Forage preserved by field drying to moisture levels low enough to prevent microbial activity that would lead to spoilage.

hay preservative See **preservative.**

haylage—The product that results from ensiling forage with about 45% moisture in the absence of oxygen.

head—A type of inflorescence in which individual flowers are sessile at their points of attachment to an enlarged receptacle.

hemicellulose—A polysaccharide that is associated with cellulose and lignin in the cell walls of green plants. It differs from cellulose in that it is soluble in alkali, and on acid hydrolysis gives rise to uronic acid, xylose, galactose, and other carbohydrates, as well as glucose.

herbaceous—Having little if any woody tissue. A herbaceous plant usually dies back to the ground each year, in contrast to a woody plant.

herbage—The biomass of herbaceous plants. The term generally refers to edible plant parts other than grain.

herbicide—A chemical that is used to control weeds in croplands and grasslands.

herbivore—A mammal, insect, or other higher organism that subsists primarily on plants or plant products.

herbivory—The consumption of a diet that consists mainly or solely of plants.

heterotroph—An organism that is not capable of synthesizing its own organic food supply, and that is dependent on another organism or its products for its energy source.

high-moisture silage—Silage prepared from plant material without wilting or otherwise drying it before ensiling.

hindgut fiber digester.—A monogastric animal which has a modified hindgut that can accommodate microbial digestion of fiber after it has passed through the foregut. Common examples include horses and rabbits.

humic acid—Any of the organic acids present within humus.

humus—The organic fraction of the soil that is resistant to further decomposition.

hypocotyl—The region of the embryonic axis that is located between the cotyledonary node and the radicle.

hypogeal germination—A type of germination in which the cotyledons remain below the ground while the epicotyl grows and emerges above the ground.

hypomagnesemic tetany See **grass tetany.**

hypoxia—Conditions of oxygen deficiency in an organism or environment.

hypsodont teeth—Teeth with relatively large crowns and short roots that are characteristic of herbivores.

ice sheet—A relatively thin ice layer covering the soil surface and the plants present at a particular location and restricting atmospheric gas exchange, which leads to hypoxia.

immobilization—Binding of soil N by carbon, typically when the C:N ratio exceeds 10: 1.

***in-vitro* dry matter disappearance (IVDMD)**—A gravimetric measurement of the amount of dry matter lost upon filtration following the incubation of forage in test tubes with rumen microflora. It is usually expressed as a percentage (i.e., as the dry sample weight minus the residue weight divided by the dry sample weight).

indeterminate—Relating to a plant growth habit characterized by continuation of apical vegetative growth during the differentiation of lateral apices into inflorescences.

indigenous—Originating or produced naturally in a particular land, region, or environment. See also **native forage.**

induction—The change in status of a shoot apex that gives it the potential to flower. The response is stimulated by exposure to a prolonged cold period. See **vernalization.**

inflorescence—The mode of arrangement of flowers or spikelets on a plant.

infrared—Relating to radiation wavelengths that are longer than 1000 nm.

inoculation—The act of introducing or placing specific microorganisms or bacteria on a plant part, especially the placing of rhizobia bacteria in or on a legume seed.

insectivore—An organism, generally a bird or mammal (e.g., shrew, mole, hedgehog), that consumes mainly insects.

intake—The quantity of forage consumed by an animal during a specified period; it is usually expressed as lb/day or kg/day.

integrated pest management—An approach to pest management that is based on biological knowledge of the pest and host, observations of conditions in the field, and economic assessment of alternative controls in order to select among several control procedures, including biological, cultural, genetic, and pesticide control methods.

intensive grazing management—Grazing management that attempts to increase production or utilization per unit area or production per animal through a relative increase in stocking rates, forage utilization, labor, resources, or capital.

intercalary meristem—Any zone of cell division and cell elongation in grass shoots that is not part of the shoot apex. It is mainly responsible for growth of the leaf blade, leaf sheath, and culm internodes.

intravaginal tiller—An upright tiller that does not penetrate the leaf sheaths of the parent tiller. It emerges at the collar of the subtending leaf.

killing frost—A temperature low enough to affect the shoot apex sufficiently to stop growth, but that does not kill all of the leaves. It is generally considered to be about 24°F for upright legumes that have apices near the top of the canopy.

lactation—The secretion of milk by female mammals after giving birth.

laminitis—A disease of hooved animals, characterized by inflammation of the epidermal laminae that connect the hoof wall to the hoof bones.

latitude—The angular distance north or south from the earth's equator measured in degrees.

leaf area index (LAI)—The ratio of the leaf surface area of plants to the land area on which the plants are growing. It is a measure of the relative density of leaves within a canopy.

leaf shattering—Physical detachment and loss of portions of a hay or silage crop due to either excessive drying or mechanical treatments such as raking or tedding. Legumes are generally more susceptible to shattering loss than are grasses.

leghemoglobin—An O_2 carrier in legume root nodules that is used to capture and supply O_2 for respiration while maintaining a low O_2 concentration within nodule cells.

legume—A member of the plant family Fabaceae.

lemma—The outer or lower covering of the grass floret. It is larger and heavier than the palea.

lethal dose (LD_{50})—A measure of toxicity expressed as the dosage required to kill 50% of the affected population.

ley—The forage component of a crop rotation that includes cultivated grain crops.

lignified—Relating to plant cells that have undergone secondary thickening.

lignify—To make woody as a result of the thickening, hardening, and strengthening of plant cells by the deposition of lignin on and within the cell walls.

lignin—A complex polyphenolic compound that adds strength and rigidity to plant cell walls. Extremely resistant to digestion, it also lowers the digestibility of cell wall carbohydrates.

ligule—The membrane-like projection on the inner side of the leaf sheath arising at the collar.

lime—Pulverized limestone that provides calcium carbonate ($CaCO_3$) when applied to the soil to reduce acidity.

lipid—An organic compound that contains long-chain aliphatic hydrocarbons and their derivatives, such as fatty acids, alcohols, amines, amino alcohols, and aldehydes. Lipids include waxes, fats, and derived compounds.

lodging—The falling down of a crop due to either stalk breakage or uprooting.

lodicule—A small sac in the base of the grass flower that helps to force open the lemma and palea at anthesis.

long-day plant—A plant that flowers under long photoperiods (i.e., short nights).

low-moisture silage—Silage prepared from relatively dry plant material, usually containing less than 50% moisture.

low-molecular-weight phenolic compounds—Secondary metabolites that contain one or more aromatic rings and that have a molecular weight of less than 900 Da.

lumen—The inner space of a tubular structure, such as the esophagus.

macroclimate—The climate that occurs on a large geographic scale that is independent of local topography and vegetation.

macronutrient—A nutrient that is required in relatively large amounts (> 1000 ppm) in plants. Plant macronutrients include N, P, K, Ca, Mg, and S.

Maillard reaction—The reaction between reducing sugars and free amino groups in proteins to form a complex that undergoes a series of reactions to produce brown polymers. Higher temperatures and basic pH favor the reaction. The process reduces the digestibility of the reactants.

maintenance phosphorus—An annual application of typically smaller rates of P with the aim of maintaining optimum soil P status.

maintenance respiration—The portion of aerobic respiration that is used to support ongoing functions of non-growing tissues.

managed agroecosystem—An ecosystem of plants and animals coexisting in nature that is being managed by human activity to provide economic and environmental benefits.

mast—The fruits and seeds of shrubs, trees, and other non-herbaceous vegetation available for animal consumption.

mastication—Initial chewing prior to swallowing; in ruminants, it also refers to chewing the cud after regurgitation of a bolus.

meadow—A grassland site where forage productivity is influenced by landscape position, hydrology, or intended use (e.g., mountain meadow, wet meadow, hay meadow, native meadow).

meristem—A localized group of small, undifferentiated, rapidly dividing cells from which plant tissue systems (e.g., root, shoot, leaf, inflorescence) are derived.

meristematic—Containing small, undifferentiated, rapidly dividing cells from which plant tissue systems arise.

mesocotyl—An alternative term for the subcoleoptilar internode, located between the scutellar node and the coleoptilar node.

mesophyll—The leaf tissue composed of thin-walled cells that contain chloroplasts. It is located between the upper and lower epidermis of the leaf.

metabolic body weight—The basal metabolic rate, or energy expenditure per unit body weight per unit time (expressed in kcal heat/kg body weight/day). It varies as a function of a fractional power of body weight, usually determined to be body weight raised to the 0.75 power.

metabolizable energy (ME)—Digestible energy (DE) minus the energy lost as methane from the rumen and the energy lost in urine by ruminant animals.

methane—A gas (CH_4) that is produced naturally by respiration under anaerobic conditions (e.g., in the rumen of ruminant animals, or in wetlands).

methanogenesis—The anaerobic generation of methane.

microclimate—The climate that occurs on a small geographic scale and that is influenced by local topography and vegetation.

microfibril—An aggregation of cellulose molecules.

micronutrient—A nutrient that is required in relatively small amounts (< 100 ppm) in plants. Plant micronutrients include B, Cl, Cu, Fe, Mn, Mo, Ni, Co, and Zn.

mimosine—A non-protein amino acid in leucaena that is metabolized in the rumen to 3-hydroxy-4(1H)-pyridone (3,4-DHP), which is a potent goitrogen and reduces feed intake.

mineralization—The release of nutrients (especially N) from either organic or inorganic sources in plant-available forms by soil microorganisms.

mitochondria—The organelles in all living cells where aerobic respiration produces ATP from carbohydrates and lipids.

mixed stocking—Grazing by two or more species of grazing animals on the same land unit, not necessarily at the same time but within the same grazing season.

mob stocking—In the management of a grazing unit, grazing by a relatively large number of animals at a high stocking density for a short time.

monecious—Having female (pistillate) and male (staminate) organs in separate flowers but on the same plant.

monensin—A broad-spectrum antibiotic that is used as a feed additive in cattle diets.

monoculture—The practice of producing a single crop species from an area of land within a given period of time. Monoculture is often used to simplify management by optimization of practices for a specific crop.

monogastric—Relating to a simple-stomached animal that lacks significant structural carbohydrate (fiber) digestion capability.

morphology—The features comprising the form and structure of an organism or any of its parts.

mucopolysaccharide—A high-molecular-weight protein–polysaccharide complex composed primarily of long chains of sugar molecules, and containing hexosamine.

mutualistic—Relating to a biological association between two species in which both organisms benefit from the relationship.

muzzle—The mouth or snout of an animal.

mycorrhiza—An association between a variety of fungal species and plant roots in which plants provide a protected environment and fungal hyphae grow into the soil, effectively extending the zone of soil contact of the root system and facilitating uptake of nutrients (especially P).

mycotoxin—A toxin or toxic substance produced by a fungus.

NADH (nicotinamide adenine dinucleotide)—An electron acceptor that functions in respiration to reduce other compounds.

NADPH (nicotinamide adenine dinucleotide phosphate)—An electron acceptor that functions in photosynthesis to reduce other compounds.
native forage—Any forage species that is indigenous to an environment or area (i.e., not introduced from another environment or area).
naturalized—relating to pasture, grassland, or rangeland consisting of plants that have been introduced or which have naturally migrated from other regions, and that have become adapted, self-sustaining populations in a given region. Examples include pastures made up of white clover and kentucky bluegrass or bermudagrass, and rangeland composed of species such as weeping lovegrass, old world bluestems, crested wheatgrass, and cheat.
near-infrared reflectance spectroscopy (NIRS)—A method of forage quality analysis based on spectrophotometry at wavelengths in the near-infrared region.
net energy (NE)—Metabolizable energy minus the energy lost in the heat increment.
net energy of lactation (NEL)—Feed energy available for maintenance and milk production after digestive and metabolic losses have been deducted.
net primary production—The net assimilation of atmospheric CO_2 into organic matter by plants after losses due to respiration. It is measured in units of mass per area per unit time.
neurohormone—Any chemical compound associated with the functions of neural systems.
neutral detergent fiber (NDF)—The portion of a forage that is insoluble in neutral detergent; synonymous with cell wall constituents.
nitrate—NO_3^-, a common form of nitrogen in the soil, that is readily taken up by plants.
nitrification—The oxidation of NH_4^+ to NO_3^- by the free-living soil bacteria *Nitrosomonas* and *Nitrobacter.*
nitrogen fixation—The process by which atmospheric nitrogen is made available to plants by rhizobia that reduce N_2 to two molecules of NH_3.
nitrogen-free extract (NFE)—The highly digestible portion of a plant, consisting mostly of carbohydrates, that remains after the protein, ash, crude fiber, ether extracts, and moisture content have been determined.
nitrogenase—An enzyme that catalyzes the reduction of atmospheric nitrogen (N_2) to ammonia (NH_3).
non-protein nitrogen (NPN)—The soluble fraction of nitrogen in the plant that includes inorganic forms of nitrogen, and nitrogen contained in low-molecular-weight compounds such as amino acids.
non-structural carbohydrates—Sugars, starches, fructans, and other carbohydrates available for metabolic activity within plants. They are rapidly and completely digested by herbivores.
nurse crop—A fast-emerging crop that is planted with a slower-emerging forage crop to provide protection from soil and wind erosion, frost, and weed competition.
nutrient, animal—Any food constituent that is required to support animal life (e.g., vitamins, amino acids, minerals).
nutrient, plant—Any element, or compound containing that element, that is required to support plant life (e.g., nitrogen, but also NO_3^- and NH_4^+, which are compounds containing N).
nutrient balance—The net difference between the cumulative inputs of plant nutrients to a system (e.g., field, farm, watershed) and removals from that system. Examples of removal include harvested crops, animal products, and losses to the environment.
nutritive value—The chemical composition, digestibility, and nature of digested products of a forage.
occlusal—Related to the grinding surfaces of teeth.
omasum—The third chamber of the ruminant stomach, where the contents are mixed until they are in a more or less homogeneous state.
omnivore—An organism that eats both plants and animals.
oral cavity—Mouth cavity.
organic agriculture—Farming systems that emphasize ecosystem management over external inputs in crop production.
organic matter—The soil fraction (usually 3–6%) that consists of plant or animal tissue in various stages of decomposition.
organic reserves—Sugars, polysaccharides, amino acids, and storage proteins that are accumulated in vegetative tissues and that can be translocated to and used by other plant organs.
osmolarity—The molar concentration of dissolved substances.
osmotica—Solutes that increase the osmotic activity of plant cells in order to reduce water loss in plants exposed to drought or freezing stress, and lower the freezing point during cold stress.
ovary—The organ that produces the female gametes and contains one or more ovules (one ovule in grasses). It develops into a single-seeded grass fruit (caryopsis) or a multi-seeded legume (pod).
overgrazing—The long-term overutilization of a forage resource. Overgrazing occurs when the carrying capacity of the forage resource is repeatedly exceeded.
overseeding—The process of scattering seed on the non-tilled soil surface, usually in late winter, to add a new species or thicken the plant density in a pasture or hayfield. Late winter seeding uses alternate freezing and thawing to roughen the soil surface, and rainfall to move soil over the seed, allowing the seed to germinate.
ovine—Relating to sheep.
ovule—The structure in the plant ovary that contains the egg cell and develops into a seed when mature.
paddock—A grazing area that is a subdivision of a grazing management unit and which is enclosed and separated from other grazing areas by a fence or barrier.

palatability—An animal preference that is based on plant characteristics eliciting a choice between two or more forages or parts of the same forage. It is conditioned by the animal and environmental factors that stimulate a selective intake response.

palatable—Acceptable for consumption by an animal.

palea—The inner or upper covering of the grass floret. It is more membranous than the lemma.

panicle—A grass inflorescence in which the spikelets are attached to pedicels on a subdivided or branched axis; a branched inflorescence.

panicoid—Relating to a type of grass seedling development in which a short coleoptile and considerable elongation of the subcoleoptilar internode pushes the coleoptilar node, coleoptile, and enclosed shoot through the soil. It is characteristic of grasses belonging to the subfamily Panicoideae.

parenchyma—In higher plants, a tissue consisting of thin-walled living cells that are active in photosynthesis and/or storage. It generally has very high digestibility.

parent material—Source material that is weathered and otherwise modified (e.g., by organic matter) to form soil.

pasturage—Vegetation that is grazed by animals, including grasses, legumes, forbs, and shrubs. (Not a recommended term. See **forage.**)

pasture—A type of grazing management unit that is enclosed and separated from other grazing areas by fencing or other barriers, and dedicated to the production of forage for harvest primarily by grazing.

peduncle—The stalk of an inflorescence, or of a single flower that is not part of an inflorescence.

PEP carboxylase—The enzyme that captures CO_2 for photosynthesis in C_4 plants.

pepsin—an enzyme that is secreted in the stomach of an animal, and that breaks down proteins to polypeptides.

perennation—The self-perpetuation of a plant by reseeding or vegetative reproduction.

perennial—A plant that lives for more than 1 year by means of persisting organs, such as rhizomes, stolons, or crowns that contain tiller buds.

permanent pasture—Pasture that is maintained indefinitely for the purpose of grazing. It usually consists of naturalized perennial or self-seeding annual species, with modest management on lower-productivity sites that are not part of the crop rotation system.

persistence, plant—The ability of perennial plants to remain alive and productive over long periods of time.

persistence, stand—A trait that allows a forage stand to be productive for several years. Stand persistence can occur due to plant persistence (e.g., in the case of alfalfa or red clover) or natural reseeding (e.g. in the case of annual lespedeza or birdsfoot trefoil).

petiole—The stalk of a leaf that attaches the leaf blade to the stem.

petiolule—The stalk that attaches a leaflet to the petiole of a compound leaf.

phloem—In vascular plants, a conducting tissue that is mainly concerned with the transport of sugars and other organic food materials within the plant. When fully developed, the phloem consists of sieve tubes and parenchyma, generally with companion cells.

3-phosphoglycerate (3PGA)—A three-carbon molecule involved in the photosynthetic pathway.

photon—A particle of light.

photoperiod—The duration of a plant's daily exposure to light.

photoperiodism—The regulation of plant responses by the duration of daylight.

photorespiration—The metabolic process resulting in the loss of CO_2 that occurs in C_3 plants when rubisco reacts with O_2 instead of CO_2.

photosensitization—In animals, a condition in which the skin is very sensitive to ultraviolet light due to the presence in some plants of photodynamic agents that react with light to cause symptoms such as severe sunburn and skin sloughing.

photosynthate—A chemical product derived from photosynthesis.

photosynthesis—The production of carbohydrates that occurs in chloroplasts when energy from light is used to reduce CO_2. Photosynthesis is the foundation of the food chain for all life.

phototropism—A directional growth response of plants to light.

physiology—The processes in a plant or animal that involve metabolism, the dynamics of growth and development, and interactions with the environment.

phytochrome—A protein pigment that absorbs red and far-red light. It is the major pigment involved in monitoring the length of the photoperiod or light penetration through a canopy.

phytomer—The unit of structure of a grass tiller. It consists of a leaf blade and sheath, the internode, node, and associated axillary bud below the point of sheath attachment.

pistil—The female reproductive structure of a flower, consisting of the stigma, style, and ovary.

pistillate—Relating to plants, inflorescences, or flowers that have pistils only (female).

plant heaving—The upward movement of plants caused by the alternate freezing and thawing of wet soils.

plasmodesmata—Connections between living cells, via the cell walls, that allow substances such as sugars, proteins, and minerals to move between cells.

plumule—The embryonic shoot of a germinating seed that develops into the epicotyl (shoot).

pollination—The process of transferring pollen to the stigma of either the same plant (self-pollination) or a different plant (cross-pollination).

prairie—Nearly level or rolling grassland that was originally treeless; it is usually characterized by fertile soil.

preservative—An additive that is used to protect against decay, discoloration, or spoilage.
primordium (pl. primordia)—A plant organ or structure at the earliest stage at which its differentiation can be detected.
progenitor—An ancestor or parent of a plant or animal that is considered to be the source of a genetic trait.
protease—An enzyme that causes the degradation or hydrolysis of proteins to simpler substances.
proteolysis—The hydrolysis of peptide bonds in proteins or peptides.
proximate analysis—An analytical system for feedstuffs that includes the determination of ash, crude fiber, crude protein, ether extract, moisture (dry matter), and nitrogen-free extract.
pseudostem—The sheaths of successive leaves of a vegetative grass tiller that form the lower portion of the canopy.
psychogenic—Originating in the mind rather than having a physical cause.
pubescent—Covered with fine, soft, short hairs or trichomes.
pure live seed—The percentage of pure germinating seed, which is calculated by multiplying the seed purity by the seed germination: pure seed percentage × (germination percentage)/100.
pyrolysis—A thermochemical conversion process that involves combustion of biomass at high temperatures and pressures in the absence of oxygen. It produces bio-oil, biochar, and some combustible gases.
raceme—An unbranched inflorescence in which the spikelets are attached to the rachis by pedicels.
rachis—The central axis of an inflorescence.
radicle—In seed plants, the embryonic root that emerges first through the seed coat during germination (primary root). It develops as the main taproot of legumes, forbs, and other species, but is short-lived in grasses and is replaced with an adventitious root system.
range—Land supporting indigenous vegetation that is grazed or that has the potential to be grazed, and is managed as a natural ecosystem. It includes grazable forestland and rangeland.
rangeland—Land on which the indigenous vegetation is predominantly composed of grasses, grass-like plants, forbs, and shrubs. It includes lands revegetated naturally or artificially when routine management is primarily grazing. It is not a use of land, but a type of land.
rate of digestion—The relative rate at which a forage is digested.
regurgitation—The return of partially digested forage or feed from the rumen to the mouth of a ruminant animal.
relative humidity—Atmospheric water content expressed as a percentage of the water content at complete saturation.
reseeding annual—A forage that completes its life cycle in one growing season and produces seed from which it may re-establish the following growing season.
resistance—(1) The ability of a plant or crop to grow and produce even though it is heavily inoculated or actually infected or infested with a pest. (2) The ability of a plant to survive a period of stress, such as drought, cold, or heat.
respiration—The process by which tissues and organisms gain usable energy, generally associated with exchange of gases and oxidation of sugars to release energy to enable the plant to grow and reproduce. See also **aerobic respiration** and **anaerobic respiration.**
rest—To leave an area of grazing land ungrazed or unharvested for a specific time, such as a year, a growing season, or a specified period required within a particular management practice.
rest period—The length of time for which a specific land area is allowed to rest.
reticular groove—In suckling ruminants, a channel formed by a groove located along the upper part of the foregut that conveys milk directly to the abomasum.
reticulorumen—The first chamber of the ruminant digestive tract, composed of the rumen and the reticulum.
reverse peristalsis—Waves of contractions of the esophagus wall that carry boluses of ingesta from the rumen to the mouth.
rhizobia—Bacteria belonging to the genus *Rhizobium* that form nodules on legume roots and symbiotically fix N_2 from the air into forms that can be utilized by the plant host.
rhizome—A below-ground horizontal stem with scale-like leaves (cataphylls) and axillary buds at the nodes from which new tillers or rhizomes can arise.
riparian—Relating to an area of land adjacent to a natural waterway.
rotational pasture—A pasture area that is managed relatively intensively for a few years as it is part of the crop rotation on good soil sites and is used for animal production and benefits to subsequent crops.
rotational stocking—A grazing stocking method that utilizes recurring periods of grazing and rest among two or more paddocks in a grazing management unit throughout the period when grazing is allowed.
rubisco (ribulose-1,5-bisphosphate carboxylase/oxygenase)—The enzyme that captures CO_2 for photosynthesis in C_3 plants, but which may also react with O_2 when the CO_2 concentration is low. It is very abundant in plants, representing about 40% of the soluble protein in the leaves of C_3 plants.
rumen—The first and largest compartment of the stomach of a ruminant or cud-chewing animal. It is the site of microbial fermentation.
rumen cellulolytic bacteria—Bacteria inhabiting the rumen which secrete enzymes that break down cellulose.

rumen degradable protein—The portion of dietary protein that can be degraded in the rumen by microorganisms that utilize the protein breakdown products to manufacture microbial cell proteins.

rumen detoxification—Modification or degradation of ingested toxins in the rumen.

rumen motility—Movements of digesta promoted by contractions of the rumen wall.

ruminant—A suborder of mammals that have a complex multi-chambered stomach. Ruminants primarily use forages as feedstuffs.

rumination—In ruminants, regurgitation and remastication of food in preparation for true digestion.

savanna—Grassland with scattered trees or shrubs. Savanna is often a transitional type of vegetation between true grassland and forestland, and is accompanied by a climate with alternating wet and dry seasons.

scarification—The process of scratching or abrading the seed coat of seed of certain species to allow uptake of water and gases as an aid to seed germination.

scarified—relating to seed that has been subjected to scarification.

sclerenchyma—In plants, a strengthening tissue composed of cells with heavily lignified cell walls. Its functions are to support and protect the softer tissues of the plant.

scutellar node—In developing grass seedlings, the node of the embryo axis where the scutellum is attached. It is designated as the first node.

scutellum—The single cotyledon in a monocot.

secondary metabolite—Any chemical substance produced by an organism that is not involved in the fundamental metabolic pathways that sustain life. In plants, secondary metabolites are stored in the vacuole and often serve as a defense mechanism against other organisms.

sedge—A grass-like plant, generally with a three-sided stem, that is a member of the family Cyperaceae.

seed—*noun* A ripened (mature) ovule consisting of an embryo, a seed coat, and a supply of food, which in some species is stored in the endosperm. *verb* To sow (e.g., to broadcast or drill small-seeded grasses and legumes or other crops).

seedbed—The upper portion of the soil into which seeds are placed for germination and growth.

selective grazing—The expression of diet learning by grazing herbivores.

seminal roots—The roots of a grass seedling that emerge from the cotyledonary node shortly after germination, but live for only a few weeks. There are generally three to four seminal roots per seedling.

senescence—The natural process of aging during which plant tissues exhibit changes in physiological activity in order to recover non-structural proteins, carbohydrates, nucleic acids, and mineral nutrients from plant organs prior to death.

sequence stocking—The grazing of two or more land units in succession that differ in forage species composition. Sequence stocking exploits the differences among forage species and species combinations grown in separate areas for management purposes to extend the grazing season or enhance forage quality.

sessile—Directly attached to a central axis without a stalk.

sheath—In grasses, the tubular basal portion of the leaf that encloses the stem on reproductive tillers.

shoot—A stem and the leaves associated with it (this may also include flowers and reproductive structures) that arise from the seed or an axillary bud. The term "shoot" is often used for dicots, whereas the term "tiller" is generally used for grasses.

shoot apex—The meristematic area at the end of a stem that initiates leaf primordia, nodes, internode initials, and axillary buds. It differentiates into an inflorescence in grasses and other determinate plants.

short-day plant—A plant that flowers under short photoperiods (i.e., long nights).

shrub—Any low-growing, woody plant that produces multiple stems.

silage—Forage that is preserved at low pH in a succulent condition due to production of organic acids by partial anaerobic fermentation of sugars in the forage.

silvopasture—A combination of trees, improved pasture plants, and grazing livestock in a carefully defined agroforestry practice that integrates intensive animal husbandry, silviculture, and forage management.

sink—An area of metabolic activity or storage to which organic materials and nutrients are translocated.

sod seeding—Mechanical placement of seed, usually of legumes or small grains, directly into a grass sod.

sod-bound—The condition in which the upper soil profile is filled with live and dead roots, making it impermeable to water and low in productivity due to lack of available nitrogen.

soil fertility—The ability of a soil to provide the nutrients that are essential for plant growth.

soil health—The capacity of a soil to function within natural or managed ecosystem boundaries to sustain plant and animal productivity, maintain or enhance water and air quality, and support human health and habitation.

soil organic matter—Carbon-based material (containing 45–55% carbon) in the soil that has resulted directly or indirectly from the decay and breakdown of plant, animal, or microbial tissue.

soil pH—The relative acidity or alkalinity of the soil solution.

soil solution—The layer of water that covers the soil particles and in which nutrients are dissolved prior to uptake by plants.

solar radiation intensity—The rate at which solar energy is received by a given surface area per unit time.

spatial perception—The ability to sense spatial relationships such as size, shape, and movement of objects in the environment.

species—A taxonomic category that ranks immediately below a genus and includes closely related and morphologically similar individuals that can interbreed.

species epithet—The second word of the binomial name used to indicate a plant species.

species richness—The number of species present in an ecological community. The term is often used to refer to the number of plant species in a plant community.

spike—An inflorescence in which the spikelets are attached directly to the rachis (i.e., they are sessile).

spikelet—In a grass inflorescence, the unit consisting of (generally) two subtending glumes and one or more florets.

spot grazing—Behavior in which grazing animals, especially horses, preferentially graze some areas within a pasture but avoid others because of differences in forage species, effects of previous defoliation, fouling by dung and urine, or other reasons.

stamen—In flowering plants, the male reproductive structure, consisting of an anther borne on a filament.

staminate—Relating to plants, inflorescences, or flowers that have stamens only (male).

starch—An insoluble but readily digested storage carbohydrate composed of hundreds of linked glucose units; it includes amylose and amylopectin.

steppe—vast semiarid grass-covered plain, usually lightly wooded; semiarid grassland characterized by short grasses occurring in scattered bunches with other herbaceous vegetation, and occasional woody species.

stigma—The top part of the pistil, where pollen is deposited or captured.

stocker—Young cattle post weaning that are generally being grown on forage diets to increase their size before going to concentrate feed in feedlots.

stocking density—The relationship between the number of animals and the area of land at any point in time, or the grazing management unit utilized over a specified time period.

stocking method—A defined procedure or technique of grazing management designed to achieve one or more specific objectives. One or more stocking methods can be utilized within a grazing system.

stocking rate—The relationship between the number of animals and the grazing management unit utilized over a specified time period.

stockpile—To allow forage to accumulate for grazing at a later date. Forage is often stockpiled for autumn and winter grazing after or during dormancy or semi-dormancy, but stockpiling may occur at any time of the year as part of a management plan. Stockpiling can be described in terms of deferment and forage accumulation.

stockpiled forage—Forage that has been allowed to accumulate for use at a later date.

stolon—Above-ground lateral stems with nodes at which buds can form with the potential to develop into new plants; a form of vegetative reproduction.

stratification—The process of exposing imbibed seeds to cool temperature conditions in order to break seed dormancy.

strip stocking—A stocking method in which animals are confined to a small area of a pasture for a relatively short period of time. It is usually accomplished by means of a temporary fence that is moved in increments across the pasture as the available forage is consumed.

stroma—The aqueous inner matrix of chloroplasts where reduction of CO_2 to carbohydrate occurs. The stroma surrounds the thylakoid membranes.

structural carbohydrates—The carbohydrates that are present in plant cell walls (e.g., hemicellulose, cellulose).

stubble—The residual material, usually consisting of stems, that remains after crop harvest.

style—The stalk of the pistil, which connects the stigma and the ovary.

supplement—Any nutritional additive (e.g., salt, protein, phosphorus) that is intended to improve nutritional balance and remedy dietary deficiencies.

supplementation—Provision of a nutrient supplement. See **supplement.**

sward—A population of herbaceous plants characterized by a relatively short growth habit and relatively continuous ground cover, including both above-ground and below-ground parts.

sward surface height—The height above the soil surface of the top of the plant canopy.

swath—A layer of forage material left by mowing machines or self-propelled windrowers. Swaths are wider than windrows, and have not been subjected to raking.

symbiotic—Relating to a mutually beneficial relationship between two organisms, such as that between legumes and rhizobia bacteria.

tannin—Any of a broad class of soluble polyphenols that occur naturally in many forage plants, and that share the common property of condensing with protein to form a leather-like substance that is insoluble and of low digestibility.

taproot—In plants, a main or single root that has few branches and grows mainly in length to take up water and nutrients from deeper in the soil.

terminal meristem—The meristematic area that is located at the end of a stem and that initiates leaf primordia, nodes, internode initials, and axillary buds. It differentiates into an inflorescence in grasses and other determinate plants.

thermoneutral zone—The temperature range within which an animal is able to maintain its core temperature without expenditure of energy. In cattle, it is in the range of 60–75°F (16–24°C).
thermophilic—Relating to organisms that grow well at high temperatures.
thylakoid membrane—The membrane that is the location of the photosynthetic chlorophyll and carotenoid pigments in the chloroplast. Thylakoid membranes are the site of the capture and conversion of solar energy to the chemical energy of ATP and NADPH.
tiller—The new leaves and stem that originate from an axillary bud, especially on grasses, where tillers are further classified as vegetative (having unelongated internodes) or reproductive (having a culm).
total digestible nutrients (TDN)—The sum total of the digestibility of the organic components of plant material and/or seed (e.g., crude protein + nitrogen-free extract + crude fiber + fat).
toxic—Causing injury, impairment, or death.
toxicants—Substances that are poisonous to living organisms. It is the preferred term for toxins.
trace element See **micronutrient.**
translocation—The movement of organic nutrients from regions of synthesis or deposition to sites of utilization (e.g., from leaves to seeds, or from storage organs to regrowing shoots or tillers).
transpiration—The movement of water from the soil through plants and ultimately (primarily via the stomata) to the atmosphere.
trichome—A filamentous outgrowth; an epidermal hair structure on a plant.
trophic level—A category of individual organisms that are defined by their position in the food chain.
tropical—Relating to or having characteristics of the tropics; having a frost-free climate with temperatures high enough to support year-round plant growth.
tuber—An undergound stem that is usually short and fleshy with scale-like leaves that bear axillary buds.
turgor—The force (typically positive) that cellular water exerts on the cell wall to drive cell growth and assist in preventing mature tissue and organs from wilting.
ultraviolet—Radiation wavelengths shorter than 400 nm.
umbel—A type of inflorescence in which individual flowers are attached to the tip of a peduncle by pedicels of equal length.
utricle—A small, one-seeded indehiscent fruit with a thin membranous wall.
vacuole—A large organelle (representing up to 90% of cell volume) surrounded by a single membrane and containing water and dissolved salts, pigments, and other organic compounds. Water uptake into the vacuole drives cell expansion by generating turgor pressure.
vascular—Relating to conducting tissue containing vessels or ducts.
vascular bundle—An elongated strand containing xylem and phloem, the conducting tissues of plants that transport water and food.
vegetative—Relating to non-reproductive plant parts (e.g., leaf, stem) or stages of plant development, in contrast to reproductive plant parts (e.g., flower, seed) or stages of plant development.
vegetative propagation—Asexual propagation of a plant from growing points located on vegetative organs.
vernalization—A cold treatment required by shoot apices of certain plant species in order for them to flower.
volatility—The rate of evaporation of a substance under ambient environmental conditions.
volatilization—The process whereby some applied fertilizers become converted to free ammonia (NH_3) gas, which is lost to the atmosphere.
voluntary intake—*Ad-libitum* (voluntary) intake of a feed or forage achieved when an animal is offered an excess of a single feed or forage.
warm-season—Relating to plant species that grow best during warm periods of the year. They commonly have temperature optima in the range of 86–95°F, and exhibit C_4 photosynthesis.
water potential—A relative measure of plant water status. It ranges from values just below zero for well-watered plants and soils to more negative values as water-deficit stress increases.
weather—The state of the atmosphere, mainly with regard to its effect upon life and human activities. Weather consists of the short-term (from minutes to months) variations in the atmosphere, whereas climate is the long-term average of weather events.
wilted silage—Silage prepared from plant material at intermediate moisture levels, usually in the range of 50–70%.
windrow—The narrow band of forage material that remains after raking a swath or field of forage in preparation for baling or chopping.
winter hardiness—The ability of a plant to survive winter.
winter survival—The ability of a plant to remain alive over winter. It is broader in scope than winter hardiness, which refers to the ability to survive cold. Winter survival also includes plant death due to lack of ground cover, plant heaving, and disease or insect problems.
xeric—Extremely dry, as opposed to hydric (normally very wet) or mesic (neither very wet nor very dry).
xylem—The portion of vascular tissue that has thick, lignified cell walls and is specialized for the conduction of water and minerals.

References

Allen VG, C Batello, EJ Berretta, J Hodgson, M Kothmann, X Li, J McIvor, J Milne, C Morris, A Peeters, and M Sanderson. 2011. An international terminology for grazing lands and grazing animals. Grass Forage Sci. 66:2–28.

Barnes, RF, and JB Beard. 1992. Glossary of Crop Science Terms. Crop Science Society of America, Madison, WI (www.crops.org/publications/crops-glossary/).

Barnes, RF, CJ Nelson, M Collins, and KJ Moore (eds.). 2004. Forages. Volume I: An Introduction to Grassland Agriculture, 6th ed. Iowa State University Press, Ames, IA.

Certified Crop Adviser Program. 2013. Glossary. International Certified Crop Adviser Program, Madison, WI (www.certifiedcropadviser.org/files/certifications/icca-glossary.pdf).

Glossary of Soil Science Terms Committee. 2008. Glossary of Soil Science Terms. Soil Science Society of America, Madison, WI (www.crops.org/publications/soils-glossary).

Index

Note: Page numbers followed by "t" indicate tables.

Forages: An Introduction to Grassland Agriculture, Seventh Edition. Edited by Michael Collins, C. Jerry Nelson, Kenneth J. Moore and Robert F Barnes.

Printed and bound by CPI Group (UK) Ltd, Croydon, CR0 4YY
27/10/2024
14580275-0001